Heavy Equipment Operations

Level Two

Trainee Guide
Third Edition

PEARSON

Boston Columbus Indianapolis New York San Francisco Upper Saddle River
Amsterdam Cape Town Dubai London Madrid Milan Munich Paris Montreal Toronto
Delhi Mexico City São Paulo Sydney Hong Kong Seoul Singapore Taipei Tokyo

NCCER
President: Don Whyte
Director of Product Development: Daniele Stacey
Heavy Equipment Operations Project Manager: Patty Bird
Senior Manager: Tim Davis
Quality Assurance Coordinator: Debie Ness
Desktop Publishing Coordinator: James McKay
Permissions Specialist: Amanda Werts
Production Specialist: Megan Casey
Editor: Chris Wilson

Writing and development services provided by Topaz Publications, Liverpool, NY
Lead Writer/Project Manager: Tom Burke
Desktop Publisher: Joanne Hart
Art Director: Alison Richmond
Permissions Editors: Toni Burke, Andrea LaBarge
Writers: Thomas Burke, Roy Parker, Troy Staton

Pearson Education, Inc.
Editorial Director: Vernon R. Anthony
Executive Editor: Alli Gentile
Editorial Assistant: Douglas Greive
Program Manager: Alexandrina B. Wolf
Operations Supervisor: Deidra M. Skahill
Art Director: Jayne Conte
Director of Marketing: David Gesell
Executive Marketing Manager: Derril Trakalo
Marketing Manager: Brian Hoehl
Marketing Coordinator: Crystal Gonzalez
Composition: NCCER
Printer/Binder: Courier/Kendallville, Inc.
Cover Printer: Lehigh-Phoenix Color/Hagerstown
Text Fonts: Palatino and Univers

Credits and acknowledgments for content borrowed from other sources and reproduced, with permission, in this textbook appear at the end of each module.

---

10 9 8 7 6 5 4 3 2

PEARSON

Perfect bound ISBN-13: 978-0-13-340251-3
ISBN-10: 0-13-340251-7

Preface

To the Trainee

Welcome to your second year of training in heavy equipment operations. If you are training under an NCCER Accredited Training Program Sponsor, you have successfully completed Heavy Equipment Operations Level One, and are well on your way to more advanced training.

Heavy equipment operators work on a wide variety of projects, including building construction, and on roads, bridges, mining, and timber operations, just to name a few. New construction and infra-structure projects continue to increase the demand for qualified operators. The skills qualified operators provide are vital for clearing sites, moving materials, or any earthmoving operations.

New with *HEO Level Two*

NCCER is proud to release the newest edition of *Heavy Equipment Operations Level Two* in full color with updates to the curriculum that will engage you and give you the best training possible. In this edition, you will find that the layout has changed to better align with the learning objectives. There are also new end-of-section review questions to compliment the module review. *Introduction to Earthmoving*, which was in Level Two, has been moved to Level One. *Soils*, a module formally found in Level Three, has been moved to Level Two. Also new is the *Skid Steers* module that focuses on the safety, operation, and preventive maintenance of skid steer loaders their attachments. The modules *Grades Part Two* and *Advanced Operational Techniques* have been combined to create a new *Site Work* module. The Level Two dump trucks module has been revised to focus on on-road dump trucks with off-road dump trucks moving to Level Three.

We invite you to visit the NCCER website at **www.nccer.org** for information on the latest product releases and training, as well as online versions of the *Cornerstone* newsletter and Pearson's NCCER product catalog.

Your feedback is welcome. You may email your comments to **curriculum@nccer.org** or send general comments and inquiries to **info@nccer.org**.

NCCER Standardized Curricula

NCCER is a not-for-profit 501(c)(3) education foundation established in 1996 by the world's largest and most progressive construction companies and national construction associations. It was founded to address the severe workforce shortage facing the industry and to develop a standardized training process and curricula. Today, NCCER is supported by hundreds of leading construction and maintenance companies, manufacturers, and national associations. The NCCER Standardized Curricula was developed by NCCER in partnership with Pearson, the world's largest educational publisher.

Some features of the NCCER Standardized Curricula are as follows:

- An industry-proven record of success
- Curricula developed by the industry for the industry
- National standardization providing portability of learned job skills and educational credits
- Compliance with the Office of Apprenticeship requirements for related classroom training *(CFR 29:29)*
- Well-illustrated, up-to-date, and practical information

NCCER also maintains a National Registry that provides transcripts, certificates, and wallet cards to individuals who have successfully completed modules of NCCER's Curricula. *Training programs must be delivered by an NCCER Accredited Training Sponsor in order to receive these credentials.*

Special Features

In an effort to provide a comprehensive, user-friendly training resource, we have incorporated many different features for your use. Whether you are a visual or hands-on learner, this book will provide you with the proper tools to get started in heavy equipment operations.

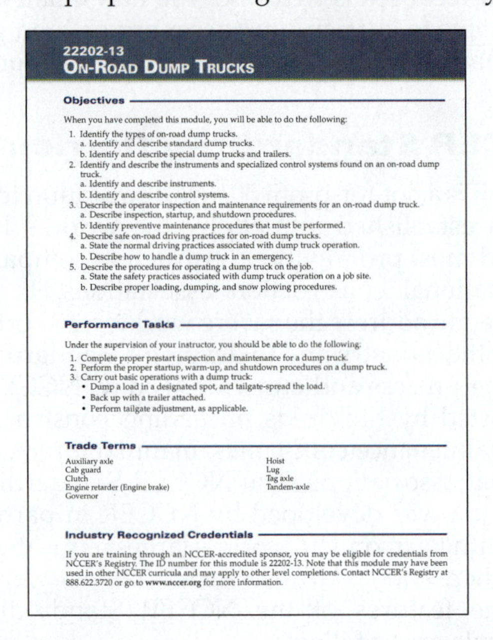

22202-13
ON-ROAD DUMP TRUCKS

Objectives

When you have completed this module, you will be able to do the following:

1. Identify the types of on-road dump trucks.
   a. Identify and describe standard dump trucks.
   b. Identify and describe special dump trucks and trailers.
2. Identify and describe the instruments and specialized control systems found on an on-road dump truck.
   a. Identify and describe instruments.
   b. Identify and describe control systems.
3. Describe the operator inspection and maintenance requirements for an on-road dump truck.
   a. Describe inspection, startup, and shutdown procedures.
   b. Identify preventive maintenance procedures that must be performed.
4. Describe safe on-road driving practices for on-road dump trucks.
   a. State the normal driving practices associated with dump truck operation.
   b. Describe how to handle a dump truck in an emergency.
5. Describe the procedures for operating a dump truck on the job.
   a. State the safety practices associated with dump truck operation on a job site.
   b. Describe proper loading, dumping, and snow plowing procedures.

Performance Tasks

Under the supervision of your instructor, you should be able to do the following:

1. Complete proper prestart inspection and maintenance for a dump truck.
2. Perform the proper startup, warm-up, and shutdown procedures on a dump truck.
3. Carry out basic operations with a dump truck:
   - Dump a load in a designated spot, and tailgate-spread the load.
   - Back up with a trailer attached.
   - Perform tailgate adjustment, as applicable.

Trade Terms

Auxiliary axle
Cab guard
Clutch
Engine retarder (Engine brake)
Governor
Hoist
Lug
Tag axle
Tandem-axle

Industry Recognized Credentials

If you are training through an NCCER-accredited sponsor, you may be eligible for credentials from NCCER's Registry. The ID number for this module is 22202-13. Note that this module may have been used in other NCCER curricula and may apply to other level completions. Contact NCCER's Registry at 888.622.3720 or go to **www.nccer.org** for more information.

Introduction

This page is found at the beginning of each module and lists the Objectives, Performance Tasks, Trade Terms, and Required Trainee Materials for that module. The Objectives list the skills and knowledge you will need in order to complete the module successfully. The Performance Tasks give you an opportunity to apply your knowledge to the real-world duties that heavy equipment operators perform. The list of Trade Terms identifies important terms you will need to know by the end of the module. Required Trainee Materials list the materials and supplies needed for the module.

Special Features

Features provide a head start for those entering heavy equipment operations by presenting technical tips and professional practices from operators in a variety of disciplines. These features often include real-life scenarios similar to those you might encounter on the job site.

Weigh Stations

Anyone who travels on the Interstate Highway System has seen the weigh stations at intervals along the highway. Commercial vehicles using the highways are required to pass through these stations. The more modern weigh stations can record the weight while the vehicle is moving through the station. Others require the vehicle to stop on a scale. The primary purpose of these

Color Illustrations and Photographs

Full-color illustrations and photographs are used throughout each module to provide vivid detail. These figures highlight important concepts from the text and provide clarity for complex instructions. Each figure reference is denoted in the text in *italic type* for easy reference.

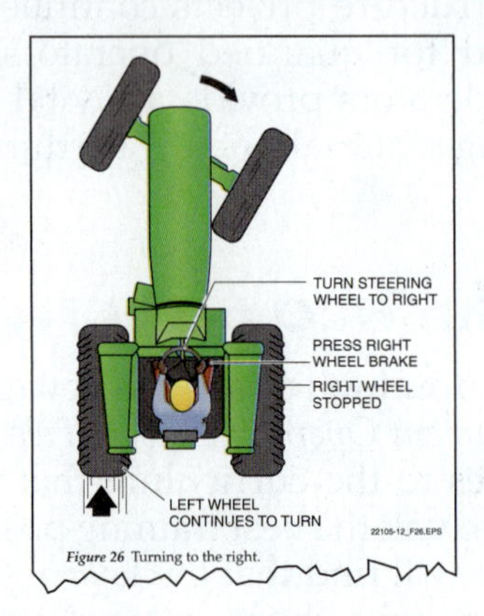

Figure 26 Turning to the right.

Notes, Cautions, and Warnings

Safety features are set off from the main text in highlighted boxes and are organized into three categories based on the potential danger of the issue being addressed. Notes simply provide additional information on the topic area. Cautions alert you of a danger that does not present potential injury but may cause damage to equipment. Warnings stress a potentially dangerous situation that may cause injury to you or a co-worker.

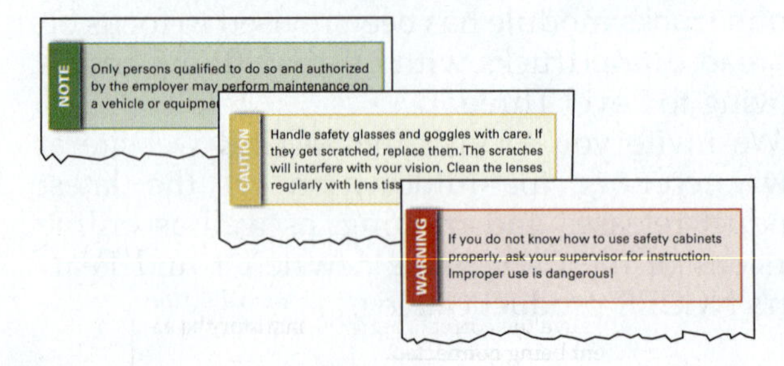

Going Green

Going Green looks at ways to preserve the environment, save energy, and make good choices regarding the health of the planet. Through the introduction of new construction practices and products, you will see how the "greening of America" has already taken root.

GOING GREEN
Hybrid Excavator
As part of an ongoing effort to reduce emissions and improve fuel efficiency, heavy equipment manufacturers are looking for new ways to power their equipment. This Komatsu hybrid excavator uses an electric motor to drive the upper structure. The energy created when the structure brakes and comes to a stop is converted to electricity, which is stored in an electronic device known as an ultra-capacitor. The stored energy is then released and used to drive the swing and to assist the vehicle's engine. Fuel savings can be 20 to 40 percent and the carbon footprint is reduced by a like amount.
KOMATSU
Hybrid

Did You Know?

The Did You Know? features offer hints, tips, and other helpful bits of information from the trade.

Did You Know?

Each year in the United States nearly 100 workers are killed and another 20,000 are seriously injured in forklift-related incidents. The most frequent type of accident is a forklift striking a pedestrian. This accounts for 25 percent of all forklift accidents. However, these figures include all forklifts. Rough-terrain models have a lower incidence of accidents, especially in striking pedestrians, since they are generally used outdoors where visibility is better. The tight confines and blind corners of a warehouse significantly increase the accident rate.

Step-by-Step Instructions

Step-by-step instructions are used throughout to guide you through technical procedures and tasks from start to finish. These steps show you not only how to perform a task but also how to do it safely and efficiently.

5.6.0 Attaching a PTO

Use the following guidelines to attach a PTO. Be sure to have the correct rpm PTO shaft for the attachment being connected.

Step 1 Set the wheel brakes.

Step 2 Disengage the power to the PTO shaft.

Step 3 Shut off the engine.

WARNING! Always shut off the engine and disengage the PTO before attempting to connect or disconnect attachments.

Trade Terms

Each module presents a list of Trade Terms that are discussed within the text and defined in the Glossary at the end of the module. These terms are denoted in the text with **bold, blue type** upon their first occurrence. To make searches for key information easier, a comprehensive Glossary of Trade Terms from all modules is located at the back of this book.

Before operating a bulldozer, the operator must understand how to operate the blade and its controls. The blade position can be changed in lift, angle, tilt, and **pitch**. Changing the position of the blade allows the bulldozer to perform different grading operations. Refer to the O&M manual for each dozer for the location and operation of the blade controls.

The lift control lowers or raises the blade. Lowering the blade allows the operator to change the amount of bite or depth to which the blade will dig into the material. Raising the blade permits the operator to travel, shape slopes, or create stockpiles. The lift lever can also be set to **float**. The blade adjusts freely to the contour of the

Review Questions

Review Questions are provided to reinforce the knowledge you have gained. This makes them a useful tool for measuring what you have learned.

Review Questions

1. Loaders are grouped into how many main categories?
   a. One
   b. Two
   c. Three
   d. Four

2. What plays a major role in the breakout force and the tipping load abilities of a loader?
   a. The type and size of the engine
   b. The type and size of the loader's frame
   c. The type of lift arms and hydraulic components
   d. The size of the bucket and the arms that lift and tilt it

3. Bucket controls (levers or joystick) are typically located _____.
   a. on the left armrest
   b. on the right armrest
   c. in front of or slightly to the left of the left armrest
   d. in front of or slightly to the right of the right armrest

4. What is the term used for a control position that allows the hydraulic fluid to the lift arm hydraulic cylinders to flow in and out both ends of the cylinders so that the bucket can follow the contour of the ground as the loader moves forward or backward?
   a. Tilt
   b. Hold
   c. Float
   d. Skim

5. What control term used with a loader's arm controls is considered to be a neutral position?
   a. Hold
   b. Float
   c. Stay
   d. Neutral

6. If a loader's diesel engine runs out of fuel, refuel the loader and _____.
   a. restart the engine
   b. clean the injectors
   c. allow the engine to completely cool before restarting the engine
   d. bleed the air out of the fuel lines and injectors before restarting the engine

7. A flashing indicator light means _____.
   a. that the system associated with the indicator needs attention
   b. to recheck the machine system associated with the indicator
   c. to check all machine temperatures and levels
   d. to stop all operations immediately

8. The width of a loader's bucket is normally the same as the _____.
   a. loader's back wheels
   b. loader's front wheels
   c. width of the loader's chassis or frame
   d. length of the loader's chassis or frame

9. A broom attachment can be angled horizontally up to how many degrees on either side?
   a. 10 degrees
   b. 15 degrees
   c. 20 degrees
   d. 30 degrees

10. A grouser is part of the _____.
    a. track
    b. idler
    c. sprocket
    d. ROPS

11. On a track loader, what component(s) keep tension on the tracks to keep them from jumping off?
    a. The idlers
    b. The sprockets
    c. The grousers
    d. The tensioning springs on the axles

NCCER Standardized Curricula

NCCER's training programs comprise more than 80 construction, maintenance, pipeline, and utility areas and include skills assessments, safety training, and management education.

Boilermaking
Cabinetmaking
Carpentry
Concrete Finishing
Construction Craft Laborer
Construction Technology
Core Curriculum:
  Introductory Craft Skills
Drywall
Electrical
Electronic Systems Technician
Heating, Ventilating, and
  Air Conditioning
Heavy Equipment Operations
Highway/Heavy Construction
Hydroblasting
Industrial Coating and Lining
  Application Specialist
Industrial Maintenance
  Electrical and Instrumentation
  Technician
Industrial Maintenance
  Mechanic
Instrumentation
Insulating
Ironworking
Masonry
Millwright
Mobile Crane Operations
Painting
Painting, Industrial
Pipefitting
Pipelayer
Plumbing
Reinforcing Ironwork
Rigging
Scaffolding
Sheet Metal
Signal Person
Site Layout
Sprinkler Fitting
Tower Crane Operator
Welding

Maritime

Maritime Industry Fundamentals

Green/Sustainable Construction

Building Auditor
Fundamentals of Weatherization
Introduction to Weatherization
Sustainable Construction
  Supervisor
Weatherization Crew Chief
Weatherization Technician
Your Role in the Green
  Environment

Energy

Alternative Energy
Introduction to the Power
  Industry
Introduction to Solar
  Photovoltaics
Introduction to Wind Energy
Power Industry Fundamentals
Power Generation Maintenance
  Electrician
Power Generation I&C
  Maintenance Technician
Power Generation Maintenance
  Mechanic
Power Line Worker
Power Line Worker: Distribution
Power Line Worker: Substation
Power Line Worker:
  Transmission
Solar Photovoltaic Systems
  Installer
Wind Turbine Maintenance
  Technician

Pipeline

Control Center Operations,
  Liquid
Corrosion Control
Electrical and Instrumentation
Field Operations, Liquid
Field Operations, Gas
Maintenance
Mechanical

Safety

Field Safety
Safety Orientation
Safety Technology

Management

Fundamentals of Crew
  Leadership
Project Management
Project Supervision

Supplemental Titles

Applied Construction Math
Careers in Construction
Tools for Success

Spanish Translations

Basic Rigging
  (Principios Básicos de
  Maniobras)
Carpentry Fundamentals
  (Introducción a la
  Carpintería, Nivel Uno)
Carpentry Forms
  (Formas para Carpintería,
  Nivel Trés)
Concrete Finishing, Level One
  (Acabado de Concreto,
  Nivel Uno)
Core Curriculum:
  Introductory Craft Skills
  (Currículo Básico:
  Habilidades Introductorias del
  Oficio)
Drywall, Level One
  (Paneles de Yeso, Nivel Uno)
Electrical, Level One
  (Electricidad, Nivel Uno)
Field Safety
  (Seguridad de Campo)
Insulating, Level One
  (Aislamiento, Nivel Uno)
Ironworking, Level One
  (Herrería, Nivel Uno)
Masonry, Level One
  (Albañilería, Nivel Uno)
Pipefitting, Level One
  (Instalación de Tubería
  Industrial, Nivel Uno)
Reinforcing Ironwork, Level One
  (Herreria de Refuerzo,
  Nivel Uno)
Safety Orientation
  (Orientación de Seguridad)
Scaffolding
  (Andamios)
Sprinkler Fitting, Level One
  (Instalación de Rociadores,
  Nivel Uno)

Acknowledgments

This curriculum was revised as a result of the farsightedness and leadership of the following sponsors:

Bridgerland Applied Technical College
Carolina Bridge Company, Inc.
Caterpillar, Inc.
John Deere
Phillipps and Jordan Inc.
Skyview Construction and Engineering, Inc.
Southland Safety, LLC

This curriculum would not exist were it not for the dedication and unselfish energy of those volunteers who served on the Authoring Team. A sincere thanks is extended to the following:

Roger Arnett
Jonathan Goodney
Paul James
Mark Jones
Dan Nickel
Larry Proemsey
Joseph Watts

NCCER Partners

American Fire Sprinkler Association
Associated Builders and Contractors, Inc.
Associated General Contractors of America
Association for Career and Technical Education
Association for Skilled and Technical Sciences
Carolinas AGC, Inc.
Carolinas Electrical Contractors Association
Center for the Improvement of Construction Management and Processes
Construction Industry Institute
Construction Users Roundtable
Construction Workforce Development Center
Design Build Institute of America
GSSC – Gulf States Shipbuilders Consortium
Manufacturing Institute
Mason Contractors Association of America
Merit Contractors Association of Canada
NACE International
National Association of Minority Contractors
National Association of Women in Construction
National Insulation Association
National Ready Mixed Concrete Association
National Technical Honor Society
National Utility Contractors Association
NAWIC Education Foundation
North American Technician Excellence
Painting & Decorating Contractors of America
Portland Cement Association
Skills USA
Steel Erectors Association of America
U.S. Army Corps of Engineers
University of Florida, M. E. Rinker School of Building Construction
Women Construction Owners & Executives, USA

Contents

Module Eight

Loaders

Covers the uses of wheel and track loaders, as well as operator maintenance, loader safety, and operating procedures. Includes procedures for using loaders in excavation, grading, and demolition work. (Module ID 22205-13; 17.5 Hours)

Module Nine

Scrapers

Describes the types of scrapers used in site preparation, as well as the safe practices associated with the operation of scrapers. Covers operator inspection and maintenance requirements, along with startup, shutdown, and operating techniques. (Module ID 22204-13; 17.5 Hours)

Glossary

Index

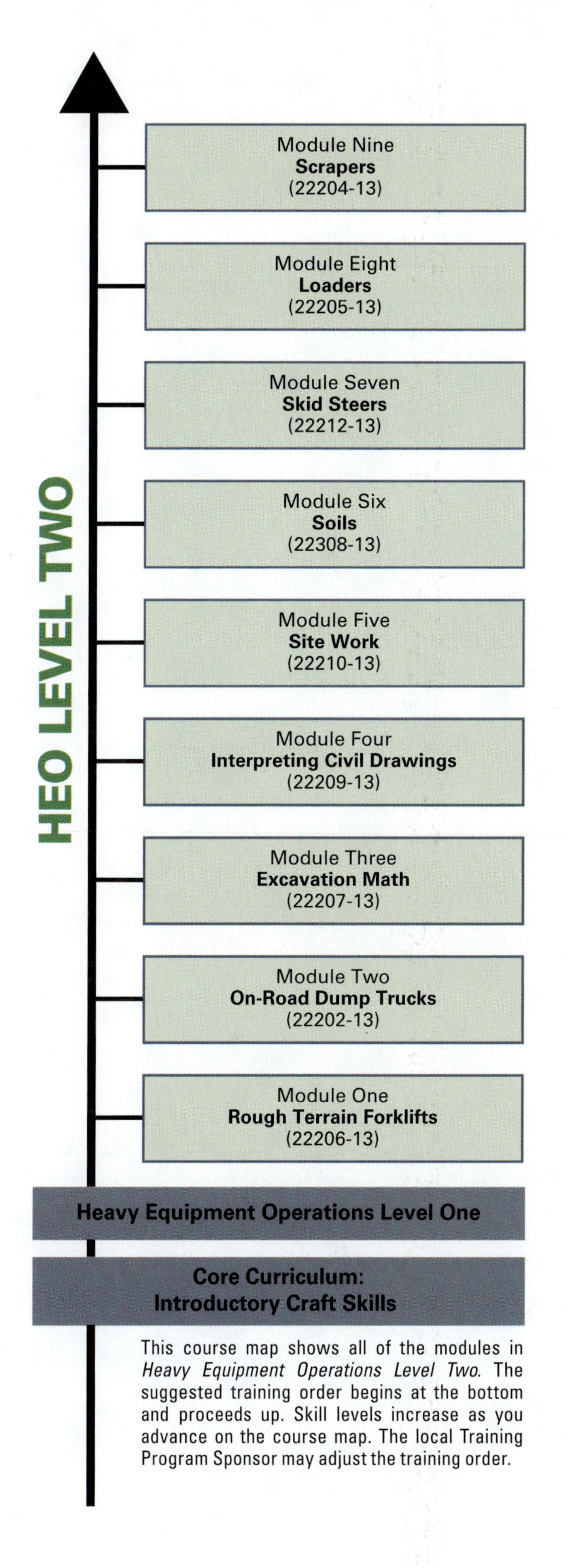

This course map shows all of the modules in *Heavy Equipment Operations Level Two*. The suggested training order begins at the bottom and proceeds up. Skill levels increase as you advance on the course map. The local Training Program Sponsor may adjust the training order.

22206-13

Rough-Terrain Forklifts

Overview

Forklifts, also called lift trucks, are motorized vehicles equipped primarily with a fork assembly that allows the machine to lift and carry materials. Forklifts are used inside and outside of buildings. Those used inside are built low to the ground and are usually equipped with slick tires. The tires used on inside forklifts may be solid rubber, or inflated. Those forklifts used outside are usually built higher off the ground and are equipped with treaded tires that have more traction in dirt or other natural surfaces. Outdoor forklifts normally have inflated rubber tires. The forklifts found on most construction sites are usually referred to as rough-terrain forklifts.

Module One

Trainees with successful module completions may be eligible for credentialing through NCCER's National Registry. To learn more, go to **www.nccer.org** or contact us at **1.888.622.3720**. Our website has information on the latest product releases and training, as well as online versions of our *Cornerstone* newsletter and Pearson's product catalog.

Your feedback is welcome. You may email your comments to **curriculum@nccer.org,** send general comments and inquiries to **info@nccer.org**, or fill in the User Update form at the back of this module.

This information is general in nature and intended for training purposes only. Actual performance of activities described in this manual requires compliance with all applicable operating, service, maintenance, and safety procedures under the direction of qualified personnel. References in this manual to patented or proprietary devices do not constitute a recommendation of their use.

22206-13
ROUGH-TERRAIN FORKLIFTS

Objectives

When you have completed this module, you will be able to do the following:

1. Identify and describe the components of a rough-terrain forklift.
   a. Identify and describe chassis components.
   b. Identify and describe the controls.
   c. Identify and describe the instrumentation.
   d. Identify and describe the attachments.
2. Describe the prestart inspection requirements for a rough-terrain forklift.
   a. Describe prestart inspection procedures.
   b. Describe preventive maintenance requirements.
3. Describe the startup and operating procedures for a rough-terrain forklift.
   a. State rough-terrain forklift-related safety guidelines.
   b. Describe startup, warm-up, and shutdown procedures.
   c. Describe basic maneuvers and operations.
   d. Describe related work activities.

Performance Tasks

Under the supervision of your instructor, you should be able to do the following:

1. Complete a proper prestart inspection and maintenance on a rough-terrain forklift.
2. Perform proper startup, warm-up, and shutdown procedures.
3. Execute basic maneuvers with a rough-terrain forklift.
4. Interpret a forklift load chart.
5. Perform basic lifting operations with a rough-terrain forklift.
6. Demonstrate proper parking of a rough-terrain forklift.

Trade Terms

Crab steering
Four-wheel steering
Fulcrum
Oblique steering
Powered industrial trucks
Power hop
Telehandler
Tines

Industry Recognized Credentials

If you are training through an NCCER-accredited sponsor, you may be eligible for credentials from NCCER's Registry. The ID number for this module is 22206-13. Note that this module may have been used in other NCCER curricula and may apply to other level completions. Contact NCCER's Registry at 888.622.3720 or go to **www.nccer.org** for more information.

Contents

Topics to be presented in this module include:

Figures

Figures (*continued*)

SECTION ONE

1.0.0 ROUGH-TERRAIN FORKLIFT COMPONENTS

Objective 1

Identify and describe the components of a rough-terrain forklift.

a. Identify and describe chassis components.
b. Identify and describe the controls.
c. Identify and describe the instrumentation.
d. Identify and describe the attachments.

Trade Terms

Crab steering: A steering mode where all wheels may move in the same direction, allowing the machine to move sideways on a diagonal, also known as oblique steering.

Four-wheel steering: A steering mode where the front and rear wheels may move in opposite directions, allowing for very tight turns, also known as independent steering or circle steering.

Oblique steering: A steering mode where all wheels may move in the same direction, allowing the machine to move sideways on a diagonal. Also known as crab steering.

Powered industrial trucks: An OSHA term for several types of light equipment that include forklifts.

Telehandler: A type of powered industrial truck characterized by a boom with several extendable sections known as a telescoping boom. Another name for a shooting boom forklift.

Tines: A prong of an implement such as a fork. For forklifts, tines are often called forks.

The Occupational Safety and Health Administration (OSHA) uses the term **powered industrial trucks** for forklifts. OSHA uses the American Society of Mechanical Engineers (ASME) definition that describes a powered industrial truck as being "a mobile, power-propelled truck used to carry, push, pull, lift, stack, or tier material." Powered industrial trucks are classified by their manufacturers and the Industrial Truck Association (ITA) according to the individual characteristics of the trucks. Those classifications are as follows:

- *Class 1* – Electric motor, sit-down rider, counterbalanced trucks (on solid or pneumatic tires)
- *Class 2* – Electric motor, narrow-aisle, trucks (on solid tires)
- *Class 3* – Electric motor, hand trucks or hand/rider trucks (on solid tires)
- *Class 4* – Internal combustion engine trucks (on solid tires)
- *Class 5* – Internal combustion engine trucks (on pneumatic tires)
- *Class 6* – Electric and internal combustion engine trucks (on solid and pneumatic tires)
- *Class 7* – Rough-terrain forklift trucks (on pneumatic tires)

Manual lift hand trucks moved by humans are classified as Class 8 fork trucks. This module focuses only on the Class 7 rough-terrain forklift trucks. They are referred to as either rough-terrain forklifts, or simply forklifts. The module also focuses only on forklifts powered by combustion engines. OSHA and the ASME further separate forklifts into the following groups, based on where they are being used:

- General Industry
- Shipyards
- Marine Terminals
- Longshoring
- Construction

The rough-terrain forklifts grouped under the construction banner are frequently used in logging, agriculture, and general construction. In those environments, the surfaces over which they are driven are often uneven with foreign objects possibly present.

Generally, rough-terrain forklifts are further divided into two broad categories: fixed-mast and telescoping-boom forklifts. A telescoping boom forklift is normally called a **telehandler**. *Figure 1* shows examples of a rough-terrain fixed-mast forklift and of rough-terrain telehandlers that are often seen on construction sites.

1.1.0 Chassis Components

The frame used in any vehicle is often called its chassis. The frame or chassis is the backbone of the vehicle. The engine, transmission, axles, operator cab, and the main body parts are attached directly to the chassis.

1.1.1 Axles and Steering

Within the fixed-mass forklift and telehandler categories, the forklifts are further differentiated by their drive train, steering, and capacity. Both types have either two-wheel or four-wheel drive. A two-wheel-drive forklift has a drive train that transmits power to the axle and wheels closest to the **tines**, or forks. That axle and those wheels

FIXED-MAST ROUGH-TERRAIN FORKLIFT

ROUGH-TERRAIN TELEHANDLER FORKLIFTS

22206-12_F01.EPS

Figure 1 Examples of rough-terrain forklifts.

In The Beginning...

The Pennsylvania Railroad put battery-powered platform trucks to work for the first time, using them to move luggage in an Altoona train station in 1906. Between the years of 1917 and 1920, several companies entered the market. Clark built this unit in 1917 and placed it in service at a Michigan plant. It was manually loaded and unloaded, but others quickly saw the value in it. About 75 units were built for sale in 1919. Both Towmotor and Yale & Towne entered the market with their own products around the same time. According to Clark, about 350,000 of their units are presently in use worldwide.

22206-12_SA01.EPS

are considered to be the front wheels. The front wheels are often larger than the rear wheels. Two-wheel-drive forklifts typically have rear-wheel steering. The rear wheels move when the steering wheel is turned. Rear-wheel steering provides greater maneuverability. This is especially useful when backing up with a load. Some rear-wheel-drive forklifts may have only one rear wheel. These three-wheeled forklifts are good for general use, but they lack power and stability on rough-terrain.

A four-wheel-drive forklift has a drive train that transmits power to both the front and rear axles. These forklifts are well suited for rough-terrain and other conditions that require additional traction. When four-wheel drive is engaged, power is transmitted to all four wheels.

Rough-terrain forklifts may also have different steering modes, as illustrated in *Figure 2*. Different steering modes enable the forklift to maneuver in smaller spaces. Some models can be operated in the following steering modes:

- The rear wheels can be locked so only the front wheels steer. This is called two-wheel steering.
- All wheels may move in the same direction. This is called **crab steering** or **oblique steering**.
- The front and rear wheels may move in opposite directions. This is called **four-wheel steering**, circle steering, or independent steering. Some manufacturers may also refer to it as round steering.

1.1.2 Fixed-Mast Forklifts

The mast is the upright structure mounted to the front of the forklift chassis. It is a frame on which the fork assembly travels. A typical fixed-mast forklift has a mast that tilts 20 degrees forward and 10 degrees rearward. These types of forklifts are suitable for placing loads vertically and traveling with the load. However, their reach is limited to how close the machine can be driven to the pickup or landing point. For example, a fixed-mast forklift cannot place a load beyond the leading edge of a roof because of its limited reach. For this reason, telescoping-boom forklifts or telehandlers are frequently used in construction. *Figure 3* shows a typical fixed-mast rough-terrain forklift. *Figure 4* shows the specifications for the same forklift.

On a fixed-mast forklift, a fork carriage rides up and down the mast. Most carriages are driven by chain drives. Some forklifts are equipped with one or more hydraulic cylinders that adjust the overall length of the mast to allow the fork carriage to be lifted higher than a forklift without the hydraulic cylinder(s).

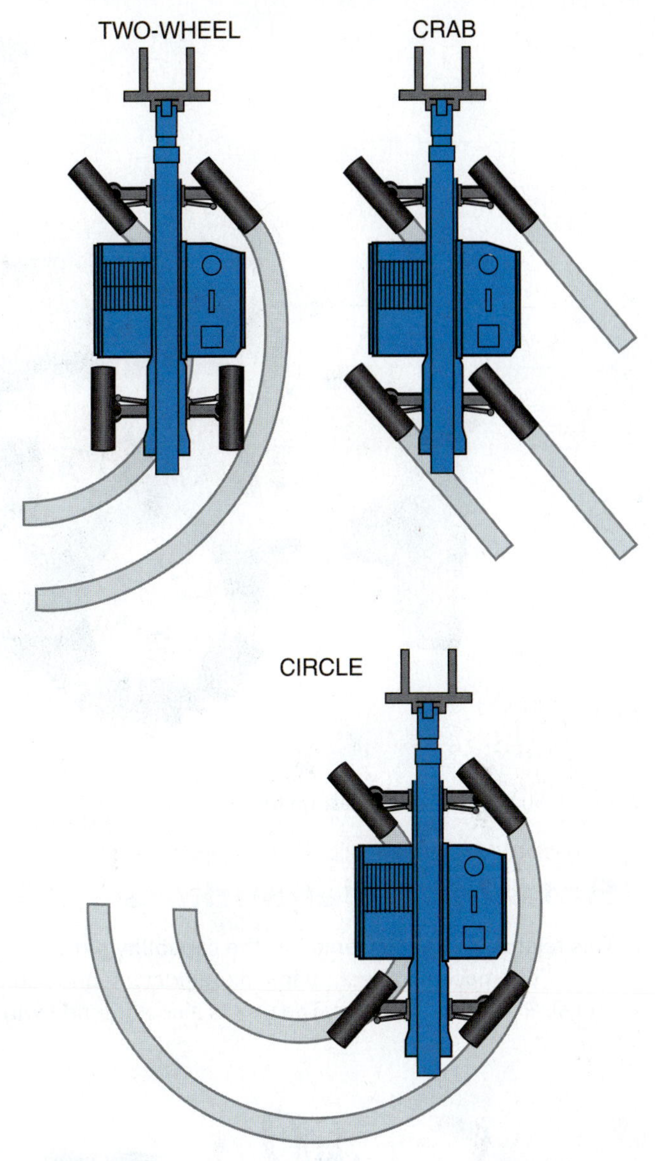

Figure 2 Steering modes for forklifts.

> **CAUTION**
> Fixed-mast forklift operators must be careful and watch the top of the mast when passing under any low passageways. If the top of the mast comes in contact with a solid overhead object, the mast and the overhead object may be damaged.

The forklift's forks are anchored on a heavy horizontally mounted rod built into the fork carriage. On most fixed-mast forklifts, the forks must be manually adjusted along the anchor rod to whatever width of separation is desired. On some higher-end fixed-mast forklifts, a hydraulically

Figure 3 Fixed-mast rough-terrain forklift.

Rotating Telehandlers

This telehandler from Genie has the capability of rotating the entire cab and boom assembly 360 degrees. This allows the machine to set up in a fixed position and rotate around a vertical axis to quickly handle repetitive lifts and similar work activities. The unit is also equipped with four independently controlled outriggers.

22206-12_SA02.EPS

S160 ROUGH TERRAIN FORKLIFT
LIFT CAPACITY OF 16,000 LBS @ 24" LOAD CENTER

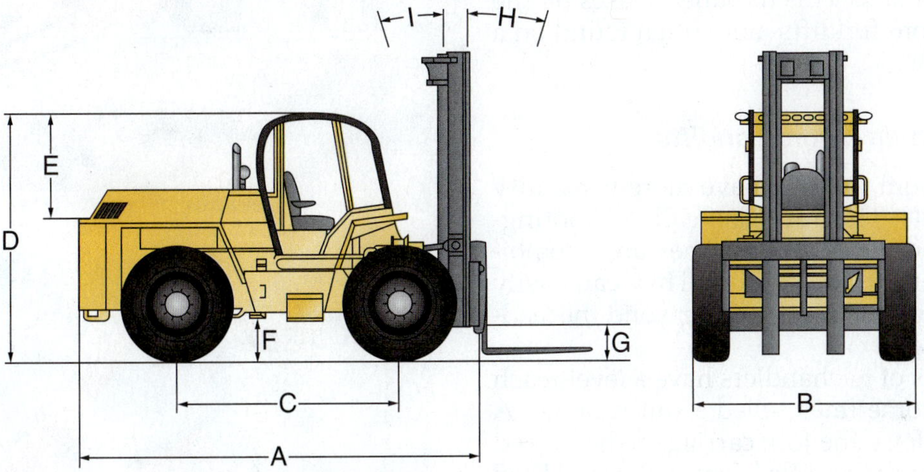

I H E D F G C A B

Model	Ref	S160
Lift capacities @ 24 inch load center		16,000 lbs
Overall length	A	188"/4 775 mm
Overall width	B	100"/2 540 mm
Wheelbase	C	100"/2 540 mm
Height to top of overhead guard	D	106"/2 692 mm
Distance from seat to overhead guard	E	43"/1 092 mm
Outside turning radius		150"/3 810 mm
Frame underclearance	F	21"/533 mm
Mast underclearance	G	12"/305 mm
Carriage sideshift		6"/152 mm
Mast tilt - forward	H	15°
Mast tilt - rearward	I	12°
Axle(s)	Dana 213 Series	
Travel speed (standard)	16 mph	
Gradeability	25%	
Tire size type & size	Front/Rear (pneumatic) 15.5 × 25 – 16 ply L3	
Electrical System	Alternator – 95A; Battery 1,000 CCA (12 volt system)	
Engine model	Dieselmax 444 Series intercooled turbocharged	
Cylinders	4	
Engine output (gross power)	114 hp @ 2,200 rpm	
Peak torque	325 ft·lbs @ 1,300 rpm	
Transmission manufacturer & type	Dana T20000 Automatic Powershift	

Because of our constant dedication to product improvement, all specifications are subject to change without notice.

22206-12_F04.EPS

Figure 4 Fixed-mast rough-terrain forklift specifications.

driven device on the carriage allows the operator to spread or narrow the forks without getting off the operator seat. This module focuses on the telescoping-boom forklifts more often found on a construction site.

1.1.3 Telescoping-Boom Forklifts

Telescoping-boom forklifts have more versatility in placing loads. They are also called shooting-boom forklifts or telehandlers. They are a combination of a crane and a forklift. They can easily place loads on upper floors, well beyond the leading edge (*Figure 5*).

Some models of telehandlers have a level-reach fork carriage, sometimes called a squirt boom. A squirt boom allows the fork carriage to be moved forward or backward while keeping the load level and at the same elevation (*Figure 6*). Squirt booms can also extend as the load is being raised, again keeping the load level and rising in a straight vertical plane.

This module describes the operations of telehandlers similar to the SkyTrak and JLG rough-terrain telehandlers often used on construction sites. These machines are typically used for light and medium work, including staging of materials and equipment.

22206-12_F05.EPS

Figure 5 Telescoping-boom forklift.

22206-12_F06.EPS

Figure 6 Example of squirt boom in action.

NOTE

Always read the operator manual before operating equipment. Follow all safety and startup procedures.

Most telescoping forklifts, or telehandlers, have similar parts. To help balance the machine, the engine is mounted along one side of the chassis, and the operator cab is mounted on the opposite side. *Figure 7* shows the engine and operator cab locations on a SkyTrak telehandler's right and left sides.

The boom assembly is centered on and mounted above the main chassis. The boom is attached at a pivot point toward the rear of the chassis. Large hydraulic cylinders lift and lower the boom as needed. Some telehandlers have a single section boom, but most have booms made up of three or more telescoping sections. Hydraulic cylinders and chain drives extend or retract the boom sections. At the end of the last boom section is an attachment tilt cylinder that allows the operator to level or tilt (up or down) an attachment such as a

Figure 7 Locations of engine and operator cab on a telehandler.

fork assembly. *Figure 8* shows the basic parts of a three-section telehandler boom.

1.1.4 Engine Area

The engine on most telehandlers is mounted on the side of the chassis opposite the operator cab. The engine on later telehandler models has a cover that can be raised to allow a person standing on the ground to inspect and service the engine. Obviously; larger telehandlers may require the operator or service person to use a short ladder or climb up onto the machine to reach the engine area. However, most can be reached while standing on the ground. To locate specific engine components, such as filters, dipsticks, and fill ports, refer to the operator manual for the machine being used. Most telehandlers are powered by diesel engines.

1.1.5 Transmissions

A telehandler's engine provides power for all the machine's operations. The power from the engine is transmitted through a transmission to the forklift's drive axle(s). The engine and transmission systems are similar to those found in most tractors. Most modern forklifts use an electronically controlled hydrostatic transmission. It provides infinitely variable speed within the speed range of the machine. Many have a load-sensing feature that automatically adjusts the speed and power to changing load conditions.

Some forklifts are driven only by a single axle (two-wheel drive) while others are driven by both axles (four-wheel drive). As noted earlier, most forklifts are steered by their rear axle and wheels. Because the forklifts are steered primarily by their rear wheels, it is critical that they not be put into situations where the load is either too heavy or extended too far out in front. This relieves too much weight from the rear wheels, and the forklift may become difficult or impossible to steer.

1.1.6 Loader Hydraulic System Components

The engine of a forklift also powers the hydraulic system that provides power to the machine's steering and lifting devices. The engine supplies power to one or more pumps that build pressure in the hydraulic system. The output pressure of the hydraulic pumps varies with the engine speed. Therefore, hydraulic components operate faster when the engine speed is increased. In some cases, a hydraulic component may not move at all if the machine is only operating at a low idle. The operational controls open and close valves to the hydraulic lines connected to the machine's steering and lifting hydraulic cylinders and pistons. When the hydraulic fluid is applied to the lifting cylinders, the pistons extend and raise the boom. When the flow of hydraulic fluid is reversed, the cylinders retract their pistons and the boom lowers. The hydraulic cylinder(s) located nearest the attachment end of the last boom section tilts the attachment (fork assembly or other) up or down. Ensuring that the hydraulic cylinders and hoses are not damaged

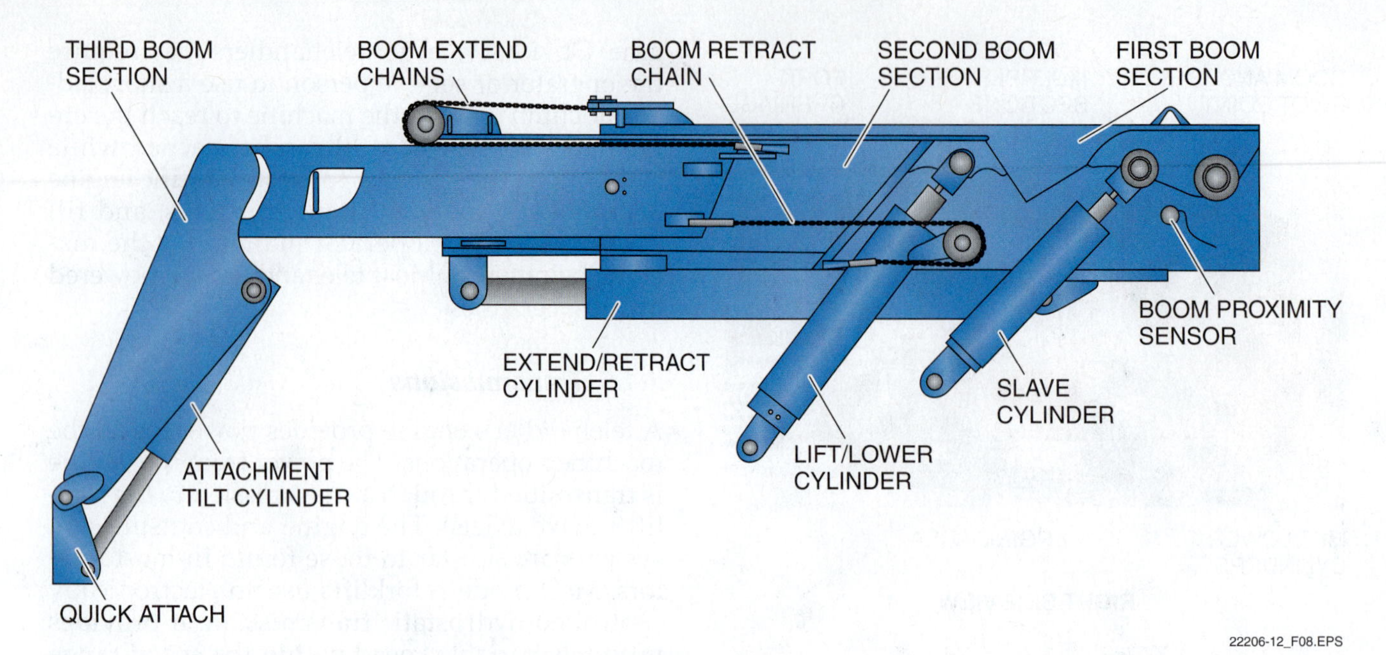

Figure 8 Components of a telehandler boom.

is critically important to the safe operation of the forklift. The same goes for all the hydraulic lines connected to the hydraulic devices.

1.1.7 Operator Cab

The operator cab is normally on the left side of most telehandlers. The cabs are normally installed low along the side of the chassis so that the operator can better see the movements of the boom. Every forklift is somewhat different, but most share similar locations for things like the steering wheel, brake pedal, accelerator pedal, and the joysticks that control the boom and the installed attachment. *Figure 9* shows the general layout of a SkyTrack telehandler's operator cab.

The operator's cab of a forklift serves several purposes. The cab provides some operator protection in case the forklift was to overturn through the rollover protective structure (ROPS). The ROPS identifies a structural integrity and cab safety model that heavy equipment such as rough-terrain forklifts must incorporate into the design to protect the operator. It may provide a controlled environment in which the operator can work comfortably on some models. The cab also serves as the central hub for forklift operations. An operator must understand the controls and instruments before operating a forklift.

Big or small, the operator cabs on most rough-terrain forklifts are laid out in a similar manner. The steering wheel is in the center of the cab. The operator's seat is at the back of the cab and centered on the steering wheel. A dash panel with indicators and some controls is below the steering wheel. The brake and accelerator pedals are near the floor and under the steering wheel. A controls console, usually on the right of the seat, houses most of the controls used on the forklift. Most modern operator cabs also have good heating and cooling systems, as well as adjustable and comfortable seating for the operator.

1.1.8 Outriggers

Telehandlers are equipped with outriggers for stability when it is picking up or dropping off a load (*Figure 10*). To use the outriggers, the forklift must be stopped and put into neutral with its brakes set. When the outriggers are extended, they must be set on firm terrain. After they are properly set, a boom and outrigger interlock allows the operator to extend the boom to its full extension. Without the outriggers in position, boom extension is intentionally limited for safety. The outriggers can be individually controlled to level the telehandler. The specifications indicate the range of correction. For example, the Skytrak Model 6036 can adjust the angle of the forklift up to 10 degrees. The load charts for the telehandler indicate how much load it can handle with and without the outriggers in use.

1.2.0 Controls

All the controls for a forklift or telehandler are located within reach of the operator inside the operator cab. These include the controls for acceleration, braking, steering, and the movement of the boom and any installed attachments. If the machine is

Figure 9 Operator cab of a telehandler.

Figure 10 Telehandler with outriggers spread and set.

equipped with outriggers, they are also controlled from the operator's position in the cab.

1.2.1 Disconnect Switches

Some forklifts are equipped with disconnect switches located outside of the cab. These switches disconnect critical functions so that the machine cannot be operated. Refer to the operator manual to determine where these disconnect switches are located, and what they control.

The disconnect switches must be activated before the machine can be started. Turn them on before mounting the machine. They offer an additional level of safety and security. Unauthorized users are usually unaware of, or cannot locate, these switches and will not be able to operate the equipment. Some of the newer machines may even require a security code to be entered into the machine's computer system before the machine will start.

Some models have a fuel shutoff valve or switch that must be turned to the On position before the machine can be operated. The fuel shutoff valve or switch physically prevents fuel from flowing from the fuel tank into the supply lines. This prevents unwanted fuel flow during idle periods or when transporting the machine, significantly reducing the potential for fuel leaks.

Some machines may also have a battery disconnect switch. When the battery disconnect switch is turned off, the entire electrical system is disabled. This switch should be turned off when the machine is idle overnight or longer to prevent a short circuit or active components from draining the battery. Before mounting the machine, check that the switch is in the On position.

1.2.2 Operator Seat and Steering Wheel Adjustments

Every forklift is slightly different when it comes to adjusting the position of the operator's seat and the steering wheel. Not all forklifts or telehandlers have adjustable steering wheels and seats. The seat in *Figure 11* can be moved up or down, and forward or backward. The operator needs to review the operator manual to see what adjustments can be made on the machine being used. When the seat is in the proper position, the operator should be able to reach all the foot pedals comfortably. When any pedal is fully depressed, there should still be a slight bend at the knee.

22206-12_F11.EPS

Figure 11 Telehandler operator seat.

After the operator gets the seat and steering wheel positioned to the most comfortable position, he or she needs to make sure that the seat belts are in good condition and usable. After that, the operator can move on to the other controls.

1.2.3 Engine Start Switch

The ignition switch functions can vary widely between makes and models of forklifts. Some only activate the starter and ignition system. Others may activate fuel pumps, fuel valves, and starting aids. In some cases, the starter and starting aids are engaged by other manual controls. The engine start switch on most forklifts is located somewhere near the dash and steering wheel. Most engine start switches have the following four positions:

- *Off* – Turning the key to this position stops the engine. It also disconnects power to electrical circuits in the cab. However, several lights remain active when the key is in the Off position, including the hazard warning light, the interior light, and the parking lights.
- *Auxiliary* – If available, turning the key to this position only applies power to accessory power outlets located on top of the front console, or to devices such as radios that may be connected to accessory power. Power is not supplied to the instrument panel or other electrical controls, other than a battery gauge or indicator.

CAUTION

Leaving the key in the Auxiliary position without the engine running will drain the battery.

- *On* – Turning the key to the On position activates all of the electrical circuits except the starter motor circuit. When the key is first turned to the On position, it may initiate a momentary instrument panel and indicator bulb check.
- *Start* – The key is turned to the Start position to activate the starter, which starts the engine. This position is spring-loaded to return to the On position when the key is released. If the engine fails to start, the key must be returned to the Off position before the starter can be activated again. To reduce battery load during starting, the ignition switch of some forklifts may be configured to shut off power to accessories and lights when the key is in the start position.

CAUTION

Activate the starter for a maximum of 30 seconds. If the machine does not start, turn the key to the off position and wait 2 minutes before activating the starter again. Overuse of the starter will cause it to overheat.

1.2.4 Vehicle Movement Controls

Forklift movement is controlled in a manner similar to that of cars and trucks. The throttle and brakes are operated by foot pedals. A steering wheel is used to turn the vehicle. Rotating the steering wheel counterclockwise guides the machine to the left. Rotating the steering wheel clockwise guides the machine to the right. The steering on modern forklifts is aided by computer systems built into the machine. The computer system(s) detects the ground speed of the machine and then adjusts the reaction speed of the steering to ensure that the steering is smooth and controlled. This helps prevent oversteering when the forklift is moving too fast to safely execute the maneuver.

Some steering wheels are equipped with a knob. This gives the operator greater leverage and faster control of the wheel when steering with one hand. This is important when the operator is using one hand to steer and the other to operate the boom and forks with the joystick.

One of the things that a forklift operator must always watch when steering is the swing of the forks and the rear of the machine. Since the forklift can be moved with the boom elevated, the operator must also watch where the top of the boom is in relationship to anything overhead.

As noted earlier, the power of a loader's engine is sent through the machine's transmission to the drive axle(s). A hydrostatic transmission is controlled by a forward/neutral/reverse (F-N-R) control that is normally mounted on the steering wheel shaft. *Figure 12* shows a typical F-N-R transmission control mounted to the left of the steering wheel. The selected transmission direction may be indicated on one of the instrument panels in the cab, depending on the model. Models without a hydrostatic transmission likely have a manual gear shift with multiple gears from which to choose.

Forklifts have transmissions with multiple speeds. On most forklifts, the first two or three speeds can be used for either forward or reverse movements. The highest gear is reserved for forward motion only. Do not skip gears when downshifting. For which gears to use, refer to the operator manual for the forklift being used.

CAUTION

Always come to a complete stop before changing from Forward to Reverse, or vice versa. Changing travel direction while the machine is moving can damage some machines. Although it is possible to change gears while in motion on machines with a hydrostatic transmission, it is not recommended. Changing directions suddenly can also dislodge the load.

1.2.5 Boom and Lift Attachment Controls

The boom and forks on a forklift can be controlled with levers or with a joystick. The forks can be moved in several directions. The forks can be tilted up and down and the boom can be raised and lowered. Some forks can be moved side to side and others can be tilted or angled from side to side. Telescopic booms can be extended and retracted.

Older forklifts are controlled with a series of levers. Each lever controls one aspect of the motion of the forks. For example, some forklifts have three levers. One lever moves the forks up and down, while another tilts them forward and backward. The third lever moves the forks from side to side. Because different makes and models have different controls, review the operator manual to become familiar with the controls before operating any forklift. This module describes joystick controls for fork and boom movement.

The joystick is used in conjunction with switches, triggers, or buttons to control the boom and fork movement. The six ways to move the fork and boom are as follows:

- Boom raise
- Boom lower
- Boom extend
- Boom retract
- Tilt forward
- Tilt back

The speed of the movement is usually determined by both how far the joystick is moved and the engine speed. Increase the engine speed and then move the joystick slowly until the forks are moving at the desired speed. The maximum speed of movement occurs when the joystick is moved to its limit in some direction, and the engine speed is high. When the joystick is released it returns to the central or neutral position and movement stops. Avoid sudden movement with the forks, as that can dislodge the load or cause accidents.

A typical control arrangement is shown in *Figure 13*. It is important to note that joystick control layouts can vary dramatically. Even though joy-

Figure 12 F-N-R transmission control on steering column.

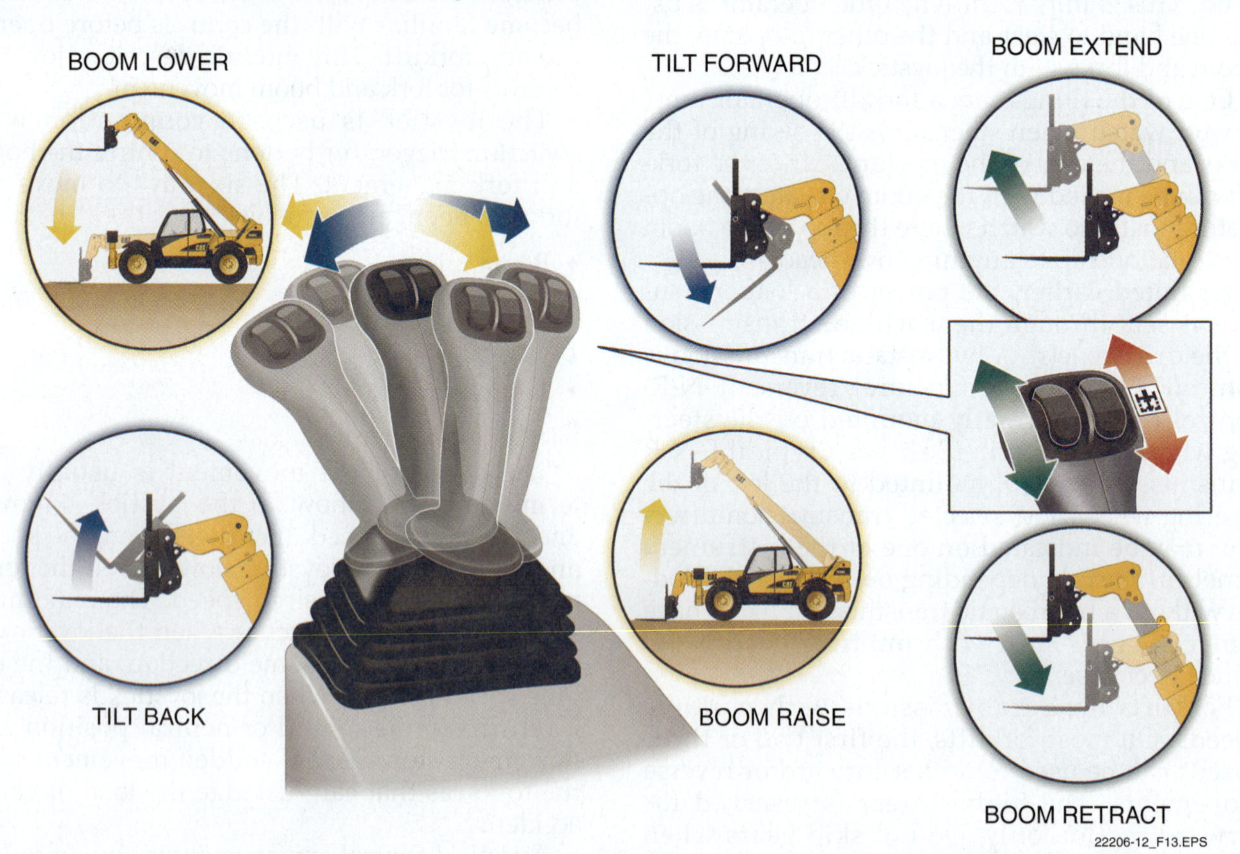

Figure 13 Forklift boom controls.

sticks may appear to be the same, they may have different button and switch assignments from one model to another. Make sure to review and clearly understand the controls shown in the operator manual of the machine being used. Never make assumptions about joystick controller actions.

For the joystick shown, right and left movement of the joystick tilts the forks forward and back. Backward and forward movement of the joystick raises and lowers the boom. On this joystick, there are two sliding switches shown on top. The sliding switch on the left retracts and extends the telescopic boom. The switch on the right raises and lowers the stabilizers. For these two switch-actuated movements, the speed at which the boom extends and retracts depends only on the engine speed.

With a joystick, multiple movements can be made simultaneously. Moving the joystick diagonally tilts the forks while raising or lowering the boom. The switch can also be activated to retract or extend the boom while the joystick is off-center. With practice, an operator will be able to move the boom and forks smoothly in several directions at once.

The carriage on some forklifts can be shifted from side to side or tilted up to 10 degrees to either side. The tilt feature allows the operator to position the forks to pick up a load that is not level. This is particularly useful on rough terrain. The side-shift feature allows the forks to move side to side horizontally while the boom remains stationary. The operator can precisely position the forks under a load without repositioning the forklift.

1.2.6 Control Switches

Each model of forklift has its own set of control switches (*Figure 14*). In most cases, the switches are located on the dash in front of the operator, off to the operator's right or left side. They activate the features and functions of the forklift. Some switches are used in conjunction with other controls, including the steering mode select and the quick coupler. The following are some examples of control switches:

- *Service brake or parking brake* – A control that locks down the machine's brakes.
- *Quick coupler* – Allows the operator to disconnect from one attachment and hook onto another attachment.
- *Steering mode select* – Allows the operator to change the speed of the machine's steering. This option is especially important when working in close quarters.

Figure 14 Control switches.

- *Fog lights* – Allows the operator to have some additional lighting when needed.
- *Left stabilizer* – Allows the operator to operate only the left stabilizer.
- *Right stabilizer* – Allows the operator to operate only the right stabilizer.
- *Work lights* – Allows the operator to control the lights on the front or rear of the operator cab, and any lights that may be mounted on the boom.
- *Rotating beacon* – Some jobs require forklifts to have a rotating beacon on and operating when the machine is in motion.

1.2.7 Special Features

There are several features that override the normal functions of a forklift. These include the transmission neutralizer and the differential lock. These features are an important part of operating the forklift. They offer added control of the machine in specialized situations.

The transmission neutralizer switch is a two-position switch. With the switch activated, the transmission is disengaged (neutralized) when the service brake is applied. With the switch turned off, the transmission remains engaged when the service brakes are applied. It can be considered a clutch that is interlocked with the brake pedal.

The differential lock control overrides the normal operation of the front axle differential. The differential normally allows the drive wheels to rotate at different speeds when turning the forklift. Locking the differential disables this feature.

Torque is then transmitted equally to both wheels, even though one wheel may not have good traction. This helps maintain traction when ground conditions are soft or slippery.

If one or both wheels start to spin, release pressure on the accelerator until the wheels stop spinning. Engage the differential lock and increase pressure on the accelerator. Once clear of the area, release pressure on the accelerator, and release the switch to disengage the differential lock. When using the differential lock, the forklift should be driven straight ahead or in reverse; only very mild and shallow turns should be made.

CAUTION

Limit steering maneuvers while the differential lock is engaged. Steering maneuvers with the differential lock engaged can damage the machine.

1.2.8 Operator Comfort and Other Controls

There are several controls designed to adjust the seat and controls for maximum operator comfort. Adjust the seat, armrests, mirrors, and cab climate controls while the machine is warming up or parked. Do not move or operate the machine before adjusting the seat and mirrors for comfort and visibility. The seat belt should be worn any time the engine is running.

WARNING!

The seat belt must be worn at all times. Failure to use a seat belt may result in serious personal injury or death.

Seat and mirror adjustments affect safety as well as operator comfort. Operators who are adjusting the seat or mirrors after beginning operations are not giving their full attention to machine operations. If an operator cannot reach all of the controls, the machine may be hard to control and cause injury or property damage. Properly setting the climate controls reduces operator fatigue and increases alertness. Failure to properly set the climate controls can also reduce visibility if the windshield is fogged.

If the machine will be used at night, the operator must become familiar with all of the light switches. Once it is dark, it is too late to look for the light switches. Make sure that all lights are functioning properly. Adjust the lights so that the work area is properly illuminated.

1.3.0 Instruments

Many of the instruments on a forklift are very similar to those found on a car or truck driven on the street. Among other things, the instruments tell the operator the status of the vehicle's engine and cooling system. They also let the operator know when something is wrong with the vehicle. Forklifts also have instruments that tell the operator how level the machine is, or where the boom is positioned.

An operator must pay attention to the instrument panel of any machine being operated. There are several warning lights and indicators that must also be monitored. An operator can seriously damage the equipment by ignoring the instrument panel.

The instrument panel can vary dramatically on different makes and models of forklifts. Generally, they include indicators for the engine coolant temperature, transmission oil temperature, fuel level, and service hour meter (*Figure 15*). Other types of forklifts may have different gauges, such as the one shown in *Figure 16*. Refer to the operator manual for the specific gauges on the machine you are operating.

1.3.1 Fuel Level Gauge or Indicator

A fuel gauge indicates the amount of fuel in the forklift's fuel tank. On diesel engine forklifts, the gauge may contain a low-fuel warning zone. On forklifts with analog gauges (those with needle

Figure 15 Instrument panel.

22206-12_F16.EPS

Figure 16 Instrument panel with LCD display.

movements), the warning zone is usually yellow. On models with a digital readout, a low fuel warning light may turn on when the fuel is too low. Avoid running out of fuel on diesel engine forklifts because the fuel lines and injectors must be bled of air before the engine can be restarted.

1.3.2 Engine Coolant Temperature Gauge or Indicator

The engine coolant temperature gauge indicates the temperature of the coolant flowing through the engine cooling system. Refer to the operator manual to determine the correct operating range for normal forklift operations. Temperature gauges normally read left to right, with cold on the left and hot on the right. If the gauge is in the white zone, the coolant temperature is in the normal range. Most gauges have a section that is red. If the needle is in the red zone, the coolant temperature is excessive. Some machines may also activate warning lights if the engine overheats.

> **CAUTION**
>
> Operating equipment when temperature gauges are in the red zone may severely damage it. Stop operations, determine the cause of the problem, and resolve it before continuing to operate the equipment.

If the engine coolant temperature gets too high, stop immediately. Get out of the cab and investigate the problem. There are three primary causes of the engine overheating: low coolant level, a nonfunctional cooling fan, or blocked air flow through the radiator. An operator needs to check them all before turning the machine in for service. These three issues can often be resolved easily on site with a fan belt, water, and/or an antifreeze solution.

Never try to open the cap on a radiator while the coolant is still hot. While waiting for the coolant to cool, check the radiator fins to ensure that they are not fouled. If they are fouled or clogged, clean them as necessary.

> **WARNING!**
>
> Engine coolant may be extremely hot and under pressure. Check the operator manual and follow the procedure to safely check and fill engine coolant. Accessing the coolant system while pressures and temperatures are high can result in severe burns or scalding.

Next, check to see if the fan belt is loose or broken. Replace it if necessary. If the radiator fins are clear and the fan blade and belt are good, check the radiator and its hoses for leaks. If no leaks can be seen, and the radiator has cooled, open the radiator cap and check the level of coolant inside the radiator. Add coolant and/or an antifreeze solution as needed. Restart the engine and monitor the temperature gauge. If the temperature goes up again, stop operations and take the machine out of service. If the issue is not related to these three common maintenance issues, then an internal problem, such as a thermostat or water pump, is likely.

1.3.3 Transmission Oil Temperature Gauge or Indicator

The transmission oil temperature gauge indicates the temperature of the oil flowing through the transmission. This gauge also reads left to right in increasing temperature. It has a red zone that indicates excessive temperatures. When the weather is colder, allow the transmission oil to warm up sufficiently before operating the machine. Some models may not provide a gauge showing the present transmission oil temperature. Instead, only a warning light may be provided, or a digital display may need to be changed to a different display mode to view the temperature. Most rough-terrain models have a transmission fluid cooler, constructed similar to the radiator and located in the same area.

1.3.4 Hydraulic Oil Temperature Gauge or Indicator

As noted earlier, forklifts use hydraulic cylinders to lift and lower the boom, and to control the tilt of the attachment at the end of the boom. Hydraulic system pressure is also used to help steer the forklift. Any time the fluid in a hydraulic system is being used to perform work, its temperature increases. The hydraulic fluid returning to the hydraulic system's reservoir is filtered and cooled before it is reused. Sensors in the hydraulic system show the temperature of the hydraulic fluid. Some forklifts may not have a gauge or indicator for the hydraulic oil temperature, but many will. Others may have only a warning light when the fluid temperature rises too high. When the weather is colder, the forklift operator may need to operate the boom and fork assembly several times to build up some heat in the hydraulic fluid. Like the transmission, the hydraulic system also has a radiator-like cooling coil, usually in the same area as the radiator.

1.3.5 Service Hour Meter

The service hour meter on any machine is important because it indicates the total hours of operation. It indicates the period of time the machine has been running over the life of the machine. It cannot be reset, since the total number of hours needs to be tracked for a variety of reasons.

The number of hours a machine has been in service determines when the machine needs to have its next periodic maintenance performed. The operator must keep track of the hours since the last service using an hour meter and a service log. Periodic maintenance is covered in more detail later.

1.3.6 Speedometer/Tachometer

Some machines also have a tachometer and speedometer. A tachometer indicates the engine speed in revolutions per minute (rpm). A speedometer shows the machine's ground speed. Typically they can be set for either miles per hour (mph) or kilometers per hour (kph).

1.3.7 Other Indicators

Some forklifts have a series of lights above the steering wheel or on the dash behind the steering wheel. These lights show the operator when various features are activated under normal operating conditions, or they may alert the operator to a problem. Indicator lights show the status of a component or system, while warning lights indicate a dangerous or undesirable condition. Two warning lights, for example, are the alternator light and oil pressure light. If the alternator light is on, the battery is not receiving a charge. If the oil pressure light is on, the oil pressure is too low. Stop the machine and resolve the problem before continuing operations. Typical indicator lights show the activation of the following features:

- Left and right turn signals
- High-beam headlights
- Circular steering
- Crab steering

> **CAUTION**
>
> A flashing warning light requires immediate attention by the operator. If a warning light illuminates, stop working and investigate the problem. Ignoring such indicators may cause serious damage to the equipment.

1.3.8 Frame Level Indicator and Longitudinal Stability Indicator

Just above the front windshield or front opening of most forklifts is an indicator that shows the level of the forklift's frame (*Figure 17*). This is normally called the frame level indicator. When its pointer moves too far to the left or right of zero (0), the forklift has the potential of rollover.

Just below and to the right of the frame level indicator is a load stability indicator that is actually mounted on the side of the boom. Load stability is an important consideration in forklift operations. The longitudinal stability indicator is located on the right side of the front windshield or front opening in the operator cab. This way, the operator can easily monitor the indicator and the load at the same time. This indicator determines the forward stability of the load. It produces visual (and possibly audible signals) to indicate the limits of forward stability of the machine. The indicator alerts the operator to make adjustments to the machine or how it is being operated. Carefully monitor this indicator as the load is being raised to maintain load stability. If the machine senses that a dangerous load stability level is being approached, an audible alert may sound. Such issues often include audible warnings, since the operator's eyes are typically focused on the load or the surroundings.

For example, if the forklift is equipped with a digital readout system, the digital system may show the machine's forward stability went from 20 to 80 percent of its range while the boom was being extended. When the stability reaches 85 percent of its range, an indicator light comes on and an alarm may sound intermittently. At this point, the operator must not increase the outreach of the boom any further. In addition, the boom may need to be retracted some before the load is lowered. If the stability reaches 100 percent of the maximum allowed, red lights illuminate and the

VIEW FROM OPERATOR SEAT

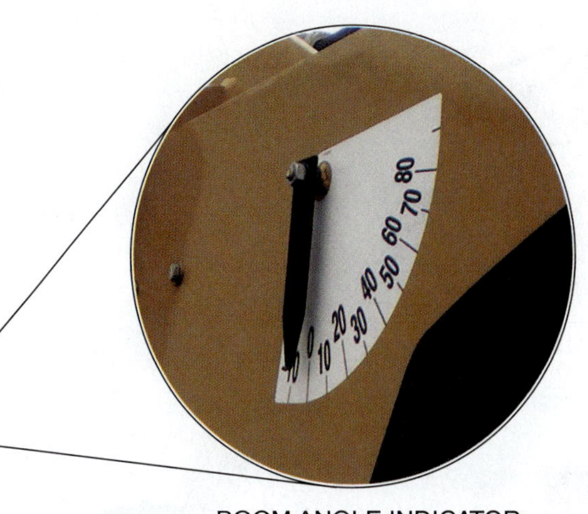

BOOM ANGLE INDICATOR

22206-12_F17.EPS

Figure 17 Frame level and load stability indicators.

alarm sounds continuously. The operator must immediately lower the load or retract the boom to avoid an accident.

> **WARNING!** Extending the boom or lowering a raised boom increases the outreach of the load and reduces forward stability. If you are approaching the machine's stability limit, retract the boom before lowering it. Otherwise the machine could tip over and cause personal injury or death.

If the forklift being used has a digital system, it must be tested daily. The load stability indicator can be adjusted for operations on tires or with the stabilizers lowered. The visual and audible alarms may operate momentarily if the machine is carrying a load close to its maximum capacity, especially when traveling over rough terrain.

1.4.0 Attachments

The main attachment on a forklift is the forks. However, on many types of forklifts, the forks can be detached and replaced with different forks or other attachments. The length and configuration of the forks can be changed to maximize productivity when handling certain materials. Some forks are designed to handle specific materials like pallets, cubes of brick and block, or sheet metal. Forks are also available in different lengths.

> **CAUTION** The lifting capacity of the forklift changes any time a different set of forks or another attachment is used. Load charts must be rechecked for the different weights. Also, ensure that any new set of forks or attachment is securely fastened before use.

In addition to lifting and moving, forklifts can be used for other operations. Different attachments, such as sweepers, buckets, or hoppers, can be fitted onto some forklifts. A forklift can be used as a raised workstation when an access platform is attached. These attachments greatly expand the capabilities of the forklift and increase its usefulness on the construction site. Use the proper attachment to lift loads safely. Although a sling can be attached to the forks, it is safer to use the boom and hook attachment to lift loads from the top. Make sure that the attachment is certified by the manufacturer for the intended lift.

> **WARNING!** Only use attachments that are designed for the machine. Using attachments that were not designed for the specific model forklift can cause a serious accident and/or equipment failure.

Some forklift attachments are designed for use in specific industries such as farming or waste handling. A bale handler is attached to the forklift shown in *Figure 18*. The forklift can easily move bales of scrap even though they are not palletized. The attachment is hooked up to the forklift's hydraulic system. The bale grabber is operated using auxiliary switches on the forklift controls.

1.4.1 Forks

Forks are available in various sizes to handle different types of materials, including standard forks, cube forks, and lumber forks. Standard forks are 48 inches long, 2 inches thick at the head, and 4 inches wide. Cube forks (*Figure 19*) have narrower, 2-inch wide tines. They are used in sets of four or more, rather than two. Lumber

22206-12_F18.EPS

Figure 18 Forklift with bale-handling attachment.

22206-12_F19.EPS

Figure 19 Cube forks.

forks are normally longer and wider than the others, at 60 inches long and 7 inches wide.

Some carriages have fixed forks. On other models, the space between the forks can be adjusted. To manually adjust the forks, lift the tip of the fork and tilt it upward. Slide it along the carriage until it reaches the desired position. Lower the tip of the fork to lock it into place.

WARNING!

Manually adjusting the forks poses a severe pinch hazard. Read the operator manual and follow all safety precautions. Do not drop forks into place. Lower them carefully to avoid crushing your fingers or hands.

The forklift's hydraulic system can be used to power an attachment installed over the forks. For example, some forks are designed to swing from side to side. The carriage is connected to the forklift's hydraulic system. The auxiliary hydraulic switches are used to rotate the carriage up to 45 degrees to either side. This provides better maneuverability and allows loads to be lifted or placed when the forklift cannot be directly perpendicular to the load. On rough terrain, this is a common problem.

1.4.2 Booms and Hooks

Some loads are not palletized or easily lifted from the bottom. These loads must be lifted from the top using slings and a lifting hook. The lifting hook and boom extension allow the forklift to operate like a crane. This arrangement is often used in construction to place trusses or to set pipe. The operator should always remain at the controls while a load is suspended from either the boom extension or the lifting hook.

Boom extensions can be either fixed angle (no angle adjustments) or adjustable angle. *Figure 20* shows a Caldwell fixed angle telescoping (extension) boom with its associated load chart. Note that the boom of the attachment must be extended manually and a pin is placed through aligned holes at the desired point. It is not operated hydraulically. This is typical for this attachment. If the boom extension is used on a telehandler, the telehandler boom can still extend or retract.

The forks of the forklift are inserted into the boom slots of the boom extension's main body. After the forks are fully engaged with the extension, one or more chains or straps are used to secure the extension to the mass of the forklift.

The boom of such extension booms can be extended by one-foot segments. With this Caldwell boom fully retracted (Model FB-30), the boom is capable of lifting 3,000 pounds (assuming that the forklift has that capability). Note that the load chart shown in *Figure 20* is for the extension boom only, and not the forklift itself. For this discus-

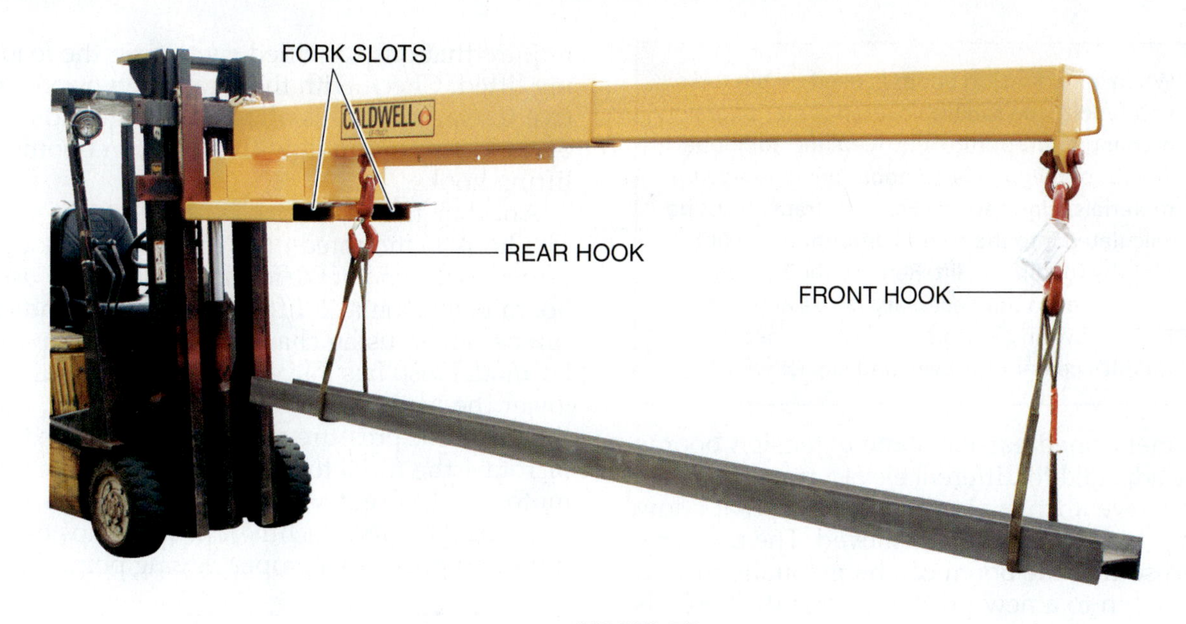

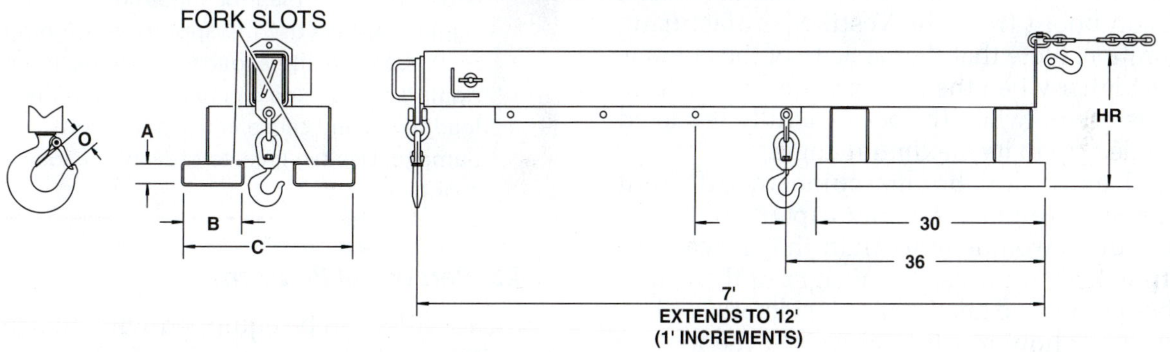

NOTE: All dimensions on drawings shown in inches unless stated.

SPECIFICATIONS

Model Number	Dimensions (in.)					Maximum Capacity at Hook Position (lb.)							Weight (lb.)
	A	B	C	HR	0	3'–6'	7'	8'	9'	10'	11'	12'	
FB–30	2½	7½	22	16	1.00	3,000	3,000	2,600	2,200	1,900	1,600	1,500	340
FB–40				16	1.09	4,000	3,200	2,600	2,200	1,900	1,600	1,500	340
FB–60	2½	7½	22	17	1.36	6,000	5,000	4,200	3,500	3,000	2,700	2,500	390
FB–80				18	1.61	8,000	7,000	5,700	4,800	4,100	3,600	3,100	520

WARNING
Capacities are for boom only and are dependent on fork lift used. Check with forklift manufacturer for load capacities before use.

22206-12_F20.EPS

Figure 20 Extension boom with load chart.

sion, assume that the forklift can lift the weights shown. When the boom is extended out to 7 feet, the boom is still capable of lifting 3,000 pounds, but extend it another foot and the lifting capacity is reduced to 2,600 pounds. Extending the boom out to 10 feet results in a maximum capacity of 1,900 pounds. When the boom is fully extended to 12 feet, it can only lift 1,500 pounds.

WARNING!

When an extension boom is used on the forks of a forklift, the forklift's overall center of gravity is changed proportionally. Also, the additional weight of the extension boom and any rigging materials (clevises, hooks, and straps) must be calculated into the overall amount of weight that the forklift can lift. Refer to the operator manual, or to the necessary tables and charts, to see how an extension boom changes the forklift's center of gravity and capacity.

As mentioned earlier, some extension booms can be adjusted to different elevation angles. Such booms have anchor pins at their base that allow the angle of the boom to be moved. The pin is removed so that the boom can be manually moved up or down to a new position. After the boom is reset, the pin is replaced and the boom is secured in its new location. *Figure 21* shows an adjustable extension boom from the Vestil Manufacturing Corporation. Note that the capacity of the boom is always higher when the load is closest to the forklift. This occurs when the boom is fully retracted and angled up to its maximum angle.

Vestil extension booms come in different lengths and weights. A boom capable of lifting 8,000 pounds weighs more than the one capable of lifting 4,000 pounds. The weight of the extension boom must be taken into consideration when determining how much total weight the forklift can actually lift.

When the Vestil 8K boom is lowered to 0 degrees and fully retracted (at the 36-inch point), the boom can lift up to 8,000 pounds at its outer hook, assuming that the forklift is capable of lifting that much weight. The further the boom is extended, the less weight the boom can handle at its outer hook. If the boom is fully extended and then moved to a higher elevation angle, the amount it can lift changes proportionally. When the boom is elevated, anything being lifted by its outer hook comes closer to the boom's base. The closer the load is to the boom's base, the more load the boom can lift. Refer to the load charts in the owner manual of the boom being used.

WARNING!

Do not exceed the rated lift capacity of the forklift, the extension boom, or any rigging component used to lift the load.

Any time that a forklift is equipped with an attachment that allows it to be used as a crane, such as the extension booms, OSHA guidelines may require that only certified riggers rig the load being lifted. Check with the local safety representative to clarify who is allowed to rig loads lifted by forklifts equipped with extension booms and lifting hooks.

Another lifting device is shown in *Figure 22*. The hook is mounted on a truss boom that is attached to the end of the forklift boom. Only use the boom extension and lifting hook with loads that can be rigged using chains or slings. This particular model also has a hydraulic winch to raise and lower the load without moving the boom. It uses the hydraulic pressure source normally used to tilt or rotate the forks to power the hydraulic winch motor. A different style of lifting hook is shown in *Figure 23*. This attachment simply slips over the forks and provides a proper rigging point.

WARNING!

Never rig directly to a set of forks. The rigging could slip off or the fork edges could cut the rigging. Always use the appropriate rigging methods to secure a load to the hook. Improper rigging can fail unexpectedly, causing the load to fall and cause injury or property damage. Use tag lines to prevent the load from swinging.

1.4.3 Personnel Platform

Some forklifts can be equipped with an articulating personnel platform (*Figure 24*). This provides an elevated workstation for two workers and their tools and materials. The platform uses the forklift's hydraulic system and is operated with controls located on the platform itself. Once the machine is positioned and the stabilizers lowered, the occupant controls the movement of the platform. However, unlike boom lifts designed specifically for this purpose, the forklift cannot be driven from the platform.

As an additional safeguard, other machine controls are locked out with a key switch. The lockout switch disables the stabilizers, frame level, transmission control, and quick coupler. This prevents the machine from being moved while the platform is raised. Moving the machine while workers are on the platform is dangerous and must be prevented.

WARNING!

Always use the proper fall prevention and arrest equipment when working on an elevated workstation. Follow all safety procedures for operating the access platform. Improper operation of the platform could result in injury or death.

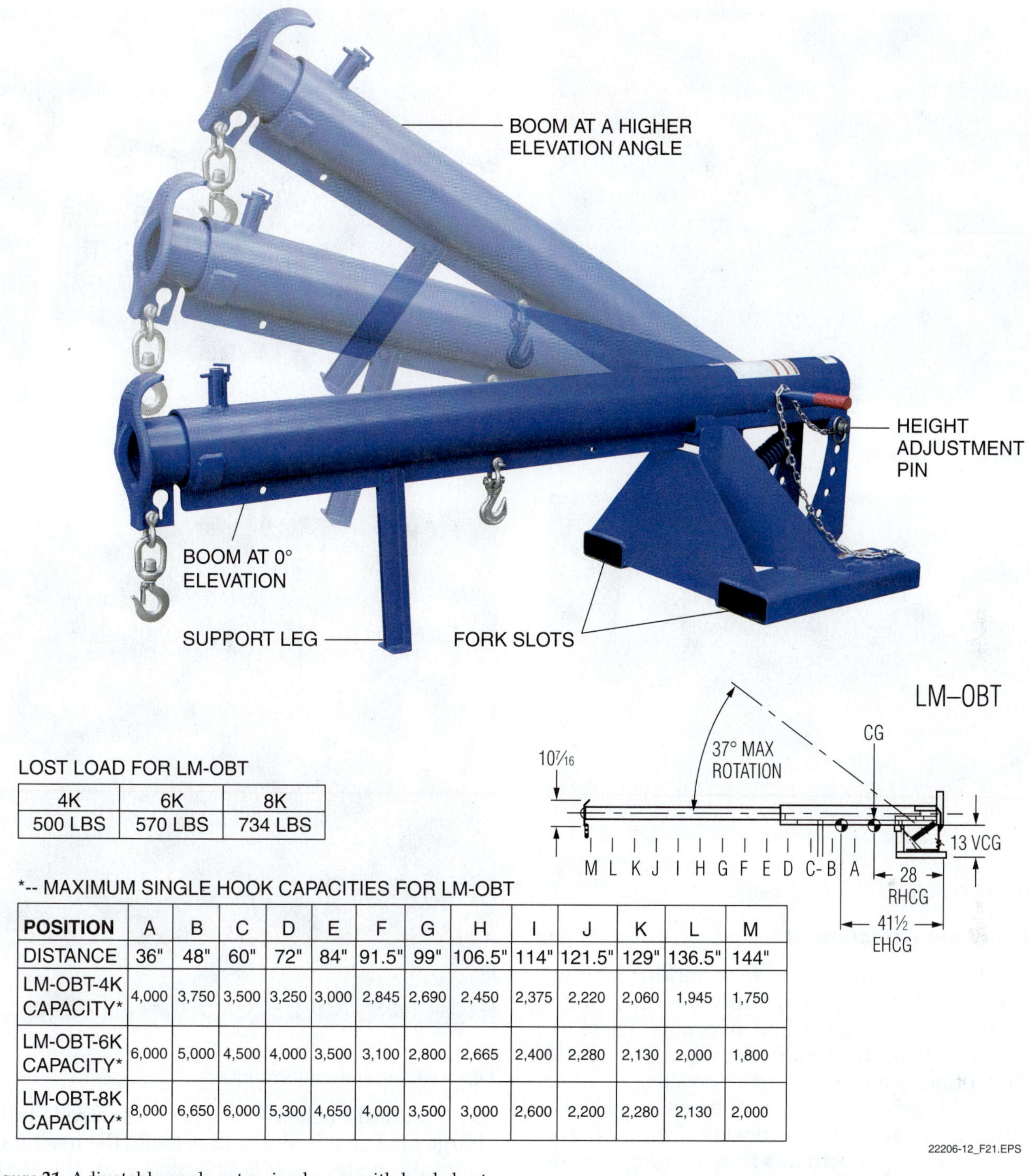

LOST LOAD FOR LM-OBT

4K	6K	8K
500 LBS	570 LBS	734 LBS

*-- MAXIMUM SINGLE HOOK CAPACITIES FOR LM-OBT

POSITION	A	B	C	D	E	F	G	H	I	J	K	L	M
DISTANCE	36"	48"	60"	72"	84"	91.5"	99"	106.5"	114"	121.5"	129"	136.5"	144"
LM-OBT-4K CAPACITY*	4,000	3,750	3,500	3,250	3,000	2,845	2,690	2,450	2,375	2,220	2,060	1,945	1,750
LM-OBT-6K CAPACITY*	6,000	5,000	4,500	4,000	3,500	3,100	2,800	2,665	2,400	2,280	2,130	2,000	1,800
LM-OBT-8K CAPACITY*	8,000	6,650	6,000	5,300	4,650	4,000	3,500	3,000	2,600	2,200	2,280	2,130	2,000

Figure 21 Adjustable angle extension boom with load chart.

Review the operator manual or other literature before operating the access platform. Follow all safety precautions to minimize hazards from slips and falls. Keep the platform clear of power lines, and do not allow the platform to contact any structure. An operator should not move the forklift unless the access platform is lowered to 18 inches or less off the ground, and is unoccupied.

Note that not every personnel platform is articulating. A simple personnel basket can be mounted and secured to the forks for easy aerial tasks that require very limited movement of the basket itself. Remember to always use proper fall arrest and/or fall prevention PPE while in a personnel basket.

22206-12_F22.EPS

Figure 22 Lifting hook.

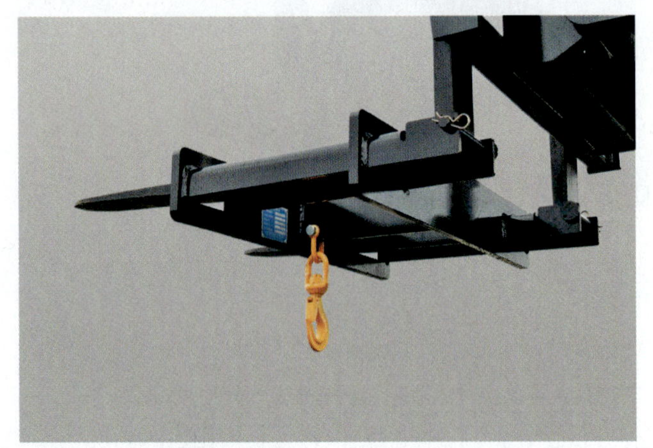

22206-12_F23.EPS

Figure 23 Fork-mounted lifting hook.

1.4.4 Other Attachments

Other types of attachments for a forklift include concrete buckets, hoppers, utility buckets, multipurpose buckets, augers, and sweepers. These attachments enable the forklift to be used for a wide variety of tasks on the construction site.

Several types of buckets can be used with the forklift. General-purpose buckets (*Figure 25*) are used for a variety of activities from digging to loading loose material, such as dirt or gravel. There are several bucket sizes for different types of materials. Light material buckets are generally larger than the standard bucket. Light material buckets are used for low-density materials such as wood chips.

A multipurpose bucket allows the bucket to function several different ways. It is also known as a clamshell or a grappler (*Figure 26*). It can be used as a standard bucket, a clamp, or a controlled-discharge bucket. It is also connected to the machine's hydraulic system and is operated with the auxiliary switches on the joystick controls.

Concrete hoppers and buckets (*Figure 27*) allow the forklift to carry loose or wet materials. Hoppers can be manually dumped. They can lift ½ or 1 cubic yard of concrete over columns or other forms. The concrete bucket has a clamshell gate on the bottom that is opened to place the concrete.

22206-12_F24.EPS

Figure 24 Personnel platform.

22206-12_F25.EPS

Figure 25 General-purpose bucket.

22206-12_F26.EPS

Figure 26 Clamshell, or grappler, attachment.

> **WARNING!**
> When moving a concrete hopper, the contents can easily shift as the machine moves. This could result in an imbalance that may cause the machine to tip over, especially when the hopper is high in the air.

The function and operation of an attachment is described in the manufacturer's literature and the operator manual. Always review the instructions before using any equipment or attachments. Improper operation can result in personal injury, machine damage, or property damage.

22206-12_F27.EPS

Figure 27 Concrete hopper for forklift use.

> **WARNING!**
> Only use attachments that are compatible with the machine. Using nonstandard attachments could damage the equipment or cause serious personal injury.

1.0.0 Section Review

1. The specific steering mode that allows the front and back wheels to move in different directions is known as _____.
   a. crab steering
   b. locked steering
   c. oblique steering
   d. four-wheel steering

2. The maximum amount of time that the starter should be engaged is _____.
   a. 15 seconds
   b. 20 seconds
   c. 30 seconds
   d. 45 seconds

3. The service hour meter is reset after each maintenance action is performed.
   a. True
   b. False

4. Cube forks are different from standard forks in that the tines are _____.
   a. wider
   b. narrower
   c. longer
   d. shorter

Section Two

2.0.0 Inspection and Maintenance

Objective 2

Describe the prestart inspection requirements for a rough-terrain forklift.

a. Describe prestart inspection procedures.
b. Describe preventive maintenance requirements.

Performance Task 1

Complete proper prestart inspection and maintenance for a rough-terrain forklift.

Trade Term

Power hop: Action in heavy equipment that uses pneumatic tires to create a bouncing motion between the fore and aft axles. Once started, the oscillation back and forth usually continues until the operator either stops or slows down significantly to change the dynamics.

Preventive maintenance is an organized effort to regularly lubricate and service the machine in order to avoid poor performance and breakdowns at critical times. Performing preventive maintenance on a forklift keeps it operating more efficiently and safely, and may help avoid costly and untimely failures.

The preventive maintenance requirements for forklifts are essential, but most tasks are quite simple, if you have the right tools and equipment. Preventive maintenance should become a habit, and be performed on a regular basis. Inspect and lubricate the machine on a daily basis. Be aware of hours of service and have the machine serviced at the appropriate intervals.

Many maintenance activities are performed after certain intervals of service. Others, however, must be done daily. Daily maintenance is typically done before the machine is started at the beginning of the workday. This group of activities done at the beginning of the day is usually referred to as a prestart inspection. The operator manual includes lists of inspections and servicing activities required for each time interval. *Appendix A* is an example of typical periodic maintenance requirements.

Maintenance time intervals for most machines are established by the Society of Automotive Engineers (SAE) and adopted by most equipment manufacturers. Instructions for preventive maintenance are usually in the operator manual for each piece of equipment. Typical time intervals are: 10 hours (daily); 50 hours (weekly); 100 hours, 250 hours, 500 hours (quarterly); and 1,000 hours (semi-annually). Note that the comparison of operating hours to days, weeks, and months is based on a typical 40 to 50 hour workweek. Forklifts may operate around the clock at some sites, although rough-terrain models are less likely to operate this way. When they do, an inspection that is typically done semi-annually can quickly become one that is required monthly, based on the operating hours accumulated.

2.1.0 Prestart Inspections

The first thing to do each day before beginning work is to conduct the prestart, or daily, inspection. Some companies provide an inspection checklist that must be completed daily. A model inspection checklist for forklifts that was developed by OSHA is included in *Appendix B*. The prestart inspection obviously should be done before starting the engine. It will identify any potential problems that could cause a breakdown and indicate whether the machine can be operated safely.

Note that the prestart inspection should not be the only time that the condition of the equipment is considered. Operators should always be attentive to the condition of the forklift while it is in use, and after it is shut down as well.

The prestart inspection is sometimes called a walk around. The operator should walk completely around the machine checking various items. Many manufacturers provide a diagram of the suggested walk-around path, noting items to check as the machine is circled. The following items are typically checked and serviced on a prestart inspection:

> **WARNING!**
> Do not check for hydraulic leaks with your bare hands. Use cardboard or a similar material. Pressurized fluids can cause severe injuries to unprotected skin. Long-term exposure to these fluids can cause cancer or other chronic diseases.

- Look around and under the machine for leaks, damaged components, or missing bolts or pins. Fluids that can leak include battery electrolyte, fuel, hydraulic fluid, engine oil, and engine coolant.
- Inspect the cooling system for leaks or faulty hoses. Remove any debris from the radiator. If

a leak is observed, locate the source of the leak and fix it. If leaks are suspected, check fluid levels more frequently.

- Inspect all attachments and implements for wear and damage. Make sure there is no damage that would create unsafe operating conditions or cause an equipment breakdown.
- Inspect all hydraulic cylinders and actuators. Wipe surfaces clean of any debris that could damage the seals.
- Inspect the chassis and main structure for cracks in welds or parent metals.
- Carefully inspect any and all sensors, limit switches, and other electrical devices. These devices provide the information that initiates alarms or status indicator lights. If they have been damaged or dislodged, they may not function which creates an unsafe condition.
- Check to ensure safety decals are in place, as well as placards that provide operating information, load capacity charts, and travel limits for moving components.
- Inspect and clean steps, walkways, and handholds.
- Inspect the engine compartment and remove any debris. Clean access doors.
- Inspect the ROPS for obvious damage.
- Inspect the hydraulic system for leaks, faulty hoses, or loose clamps.
- Inspect the lights and replace any broken bulbs or lenses.
- Inspect the axles, differentials, wheel brakes, and transmission for leaks.
- Inspect tires for damage and replace any missing valve caps. Check that the tires are inflated to the correct pressure and with the correct substance.
- Check the condition and adjustment of drive belts on the engine.
- Inspect the operator's compartment and remove any trash.
- Inspect the windows for visibility and clean them if needed.
- Adjust the mirrors.

Most manufacturers require that daily maintenance be performed on specific parts or systems. These items are usually identified as 10-hour maintenance items. Checking the engine oil level, using the dipstick, is a common daily requirement. Although dipstick markings vary, the indication of a Safe level and a Full level are marked in some manner. The oil level should be between those markings, but not above the Full level.

Another common 10-hour inspection item for forklifts is the hydraulic fluid level. Most machines have a vertical sight glass to view the fluid level. The sight glass usually has markings to indicate the safe operating level. If the fluid level is low, locate the fill cap and fill the reservoir to the proper level.

CAUTION

Operating the forklift with a low hydraulic fluid level could cause a sudden failure of hydraulically operated components, resulting in equipment or load damage.

2.1.1 Tires

Forklift tires can be inflated with one of several substances including air, nitrogen, or a liquid mixture. Tires inflated with liquid ballast provide added machine stability. The liquid may be a mixture of water, corrosion inhibitors, and calcium chloride; the latter ingredient provides antifreeze protection.

CAUTION

Using a liquid ballast product in tires that lack corrosion-inhibiting ingredients (such as plain water) can cause severe and rapid corrosion of the wheel, from the inside out. Ensure that only products designed for this purpose are used.

Heavy equipment sometimes suffers from a condition known as **power hop**. Rough-terrain forklifts, as well as tractors, loaders, and other heavy equipment using inflatable tires have the potential for power hop to occur. Heavy equipment with a short wheelbase tend to experience its ill effects more so than longer machines. The machine may bounce or hop during operation, causing a poor ride, load instability, a loss of traction, and excessive wear and tear on the equipment. Operators usually have to slow down and operate in a lower gear to reduce the effects. The conditions that are likely to cause power hop are: dry soil, especially when it is a loose mixture over a hard sub-layer; pulling soil implements that require a great deal of torque; and improperly adjusted or connected implements.

Liquid ballast has been used in heavy equipment tires for many years. Ballasting helps prevent slipping by increasing traction, and also may improve fuel economy. A liquid ballast remains near the bottom of the tire as it turns, concentrating weight in the traction area. Adjusting the percentage of inner tire area filled with liquid helps to resolve power hop.

Power hop can often be controlled by stiffening front tires using higher inflation pressures and/or adding more liquid ballast. Rear tires are softened by reducing the inflation pressure and/or removing liquid ballast in some cases. The combination of changes to front and rear tires can significantly reduce power hop.

The operator needs to know if liquid ballast is being used, and what the appropriate tire pressure should be. Regardless of what is used inside the tires, check the pressure daily to make sure it meets the recommended pressure in the operator manual. While checking the inflation pressure, also inspect the tires for significant damage and excessive wear. Tires that have suffered cuts or tears that expose the underlying cords should be considered unsafe and be taken out of service. Also check to ensure that all the lug nuts are present and tight on the wheels.

2.1.2 Fuel Service

Most rough-terrain forklifts operate on diesel fuel, but some require gasoline. Although no engine benefits when the fuel supply is allowed to run out, diesel engines are especially sensitive to damage or operating problems. It is essential that diesel engines not be allowed to run out of fuel while in operation.

It is best to refuel the machine at the end of the day. Leaving the tank only partially filled overnight can lead to water condensing in the fuel tank as temperatures change overnight. To prevent water from leaving the tank and entering the fuel lines, a fuel/water separator (*Figure 28*) is typically provided. Since water is heavier than diesel fuel, it collects in the bottom of the separator. To eliminate the water, a drain cock on the bottom is opened, and the water is allowed to drain into a proper container until only fuel is clearly draining. Draining the fuel/water separator may be identified as a 10-hour or a 50-hour maintenance requirement.

2.2.0 Preventive Maintenance

When servicing a forklift, follow the manufacturer's recommendations and service chart. Any special servicing for a particular forklift model is highlighted in the manual. Normally, the service chart recommends specific intervals, based on hours of operation, for such things as changing oil, filters, and coolant.

Since inspection items identified as 10-hour requirements are considered daily or prestart inspection tasks, preventive maintenance tasks are typically those that are conducted at longer intervals.

Figure 28 Fuel/water separator.

The first step in the maintenance process is to prepare the forklift. The following steps should be taken:

- Park the forklift on firm, level ground and apply the parking brake.
- Shift the transmission into Neutral.
- Lower the forks or other attachment to the ground; the ground should support any attachment, rather than relying on the hydraulic system to hold it up.
- Operate the engine at idle speed for three to five minutes.
- Shut the engine down and take the key out of the ignition.
- Turn off the electrical master switch (*Figure 29*) and the fuel system switch (if equipped).
- Firmly chock the wheels to prevent any movement.

2.2.1 Weekly (50-Hour) Maintenance

There are a number of items that are typically part of a 50-hour maintenance program. However, it is essential that the manufacturer's recommenda-

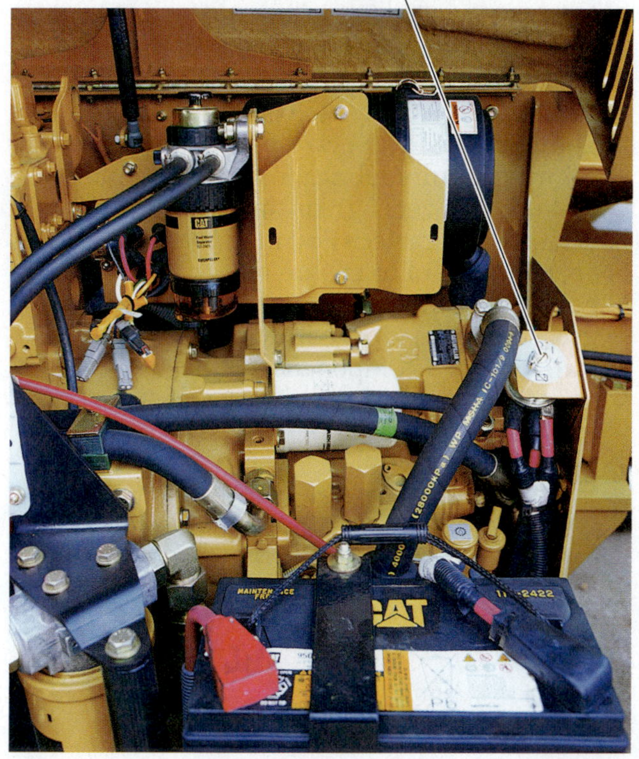

22206-12_F29.EPS

Figure 29 Electrical master switch.

tions for the specific model of forklift in use be followed.

It is important to note that new forklifts may start out with a special maintenance schedule. For example, replacement of the axle oil, wheel-end oil, and the engine oil is often required after the first 50 hours of operation. These components, including the new engine, typically produce significantly more wear to metals during the first 50 hours of operation than they will in the future. Replacing the lubricant in these components after this short interval removes these extra metal bits and flakes, preventing them from causing damage or plating other surfaces. Fluid filters are also changed the first time after a short interval. If a new forklift is being driven, be sure to follow the required maintenance schedule to ensure the warranty remains valid.

The following are common examples of 50-hour maintenance tasks.

- *Batteries* – The battery should be inspected every 50 hours of operation. During this inspection, the battery case should be inspected for any signs of damage from an external source, casing cracks, or buckling. Sealed batteries require only a visual inspection.

WARNING!

Eye protection should be worn at all times when working near batteries. The electrolyte solution is a strong acid that can cause severe burns or blindness.

- *Engine coolant* – The engine coolant may be checked one of two ways, depending on the engine and forklift manufacturer. Some engines may have a transparent overflow reservoir (*Figure 30*). The tank allows the level to be checked easily, without removing any caps. Water and/or antifreeze products can also be added here. If the engine is not equipped this way, then the radiator cap must be removed.

WARNING!

The radiator may be under pressure, and the coolant may be at or near the boiling point. When the cap is removed and the radiator returns to atmospheric pressure, steam and/or extremely hot water may be expelled. Use extreme caution and the proper PPE (safety glasses, heavy rubber gloves) when removing the cap from a radiator.

Radiator caps have three positions. When it is in the closed position, it forms a tight seal that allows pressure to build in the radiator. The pressure is controlled by a pressure relief valve in the cap. If the pressure exceeds the relief setting, pressure will be relieved. By allow-

22206-12_F30.EPS

Figure 30 Coolant fill and overflow tank.

ing pressure to build in the cooling system, the boiling point of the coolant will be increased. Cooling systems under pressure can operate somewhat above 212°F (100°C) without boiling over.

The second position of the radiator cap allows pressure to escape, but the cap is still locked to the radiator fill neck. This prevents it from being blown off by the pressure behind it as a safety measure. This position is usually reached with a quarter- to half-turn counterclockwise. With slight downward pressure, the cap can again be turned counter-clockwise for complete removal. The coolant level should then be clearly visible. If coolant needs to be added, be sure to maintain the appropriate ratio of antifreeze and water.

- *Boom wear pads* – The sectional booms are equipped with wear pads that provide a smooth contact surface for the boom sections as they slide in and out. If they are not maintained, the mounting hardware may eventually make contact with and damage the boom section(s). The wear pads may also require periodic lubrication, even though some are made of slick materials. Some pads may be made with silicone or Teflon® to provide an inherently slick surface. As the name implies, the wear pads experience wear, protecting the metal boom surfaces from being damaged. Wear pads can be in locations that are difficult to access for lubrication. It is not unusual for a telehandler to have 30 to 50 individual wear pads. The manufacturer may provide a dimension or other criteria to determine if the wear pads need to be replaced. The pad shown in *Figure 31* is considered functional until the top is worn down beyond the beveled edge shown on either end. Worn out, missing, or poorly lubricated wear pads are usually evidenced by chattering as the boom moves in and out.
- *Lubrication* – The forklift manual provides a lubrication schedule, which generally points out the exact locations that require attention. It is not unusual for some locations to require attention during the 50-hour inspection, while others may require lubrication at longer intervals. Note that several grease products, each with different characteristics, may be required. It is very important to ensure that correct grease is being applied at each point. Using the wrong grease may lead to component failure, especially if it is a component that operates at high speeds or is exposed to significant stress.

NOTE

In cold weather, it may be preferable to lubricate pivot points at the end of a work shift when the mechanism is warm. Warm the grease gun before using it for better grease flow and penetration.

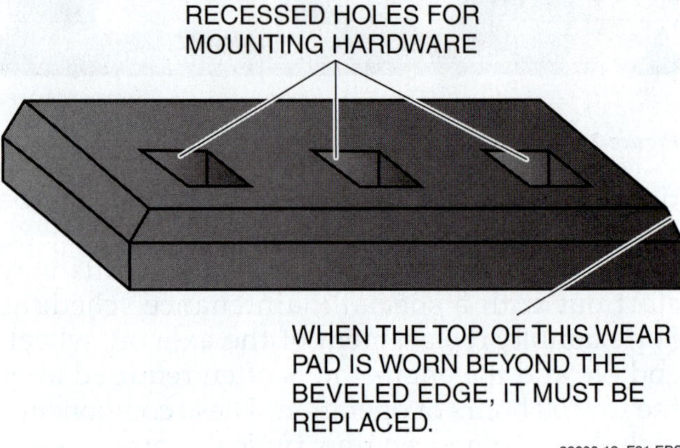
RECESSED HOLES FOR
MOUNTING HARDWARE
WHEN THE TOP OF THIS WEAR
PAD IS WORN BEYOND THE
BEVELED EDGE, IT MUST BE
REPLACED.
22206-12_F31.EPS

Figure 31 Wear pad inspection.

Hydraulic System Assassins

There are a number of things that can destroy hydraulic systems and components. Aeration creates tiny air bubbles that collapse violently and can erode metal surfaces over time. Hydraulic fluid in this condition appears milky and makes crackling noises as the pressure rises. Cavitation has much the same effect, with air bubbles expanding and exploding as they approach the low-pressure zone of the pump inlet. However, contamination is credited with causing roughly 70 percent of all system failures. Particles of dirt and debris also erode metal surfaces, increasing the level of contamination and wear. These problems can be avoided through periodic fluid sampling and testing.

2.2.2 Long Interval Maintenance

A number of tasks are required at longer intervals. Generally, the tasks become more complex and are done in the maintenance shop. Tasks that are typically done at longer intervals include:

- 250 hours:
  - Replacement of engine oil and oil filter
  - Sampling and testing of engine oil and hydraulic fluid
  - Checking axle and wheel end lubricant levels
  - Replacement of air filter elements
  - Inspecting and lubricating the boom drive chain
- 500 hours:
  - Replacement of the fuel filter
  - Verification of wheel lug nut torque values
- 1,000 hours:
  - Changing the transmission fluid and filter
  - Changing the hydraulic fluid and filter
  - Changing the differential oil
  - Checking and adjusting the boom chain tension
  - Inspecting and/or testing brake systems
  - Changing axle and wheel end lubricants
- 1,500 to 2,000 hours:
  - Changing the engine coolant

Remember that the above list is only a single example of a maintenance schedule. Make sure that the schedule for the specific forklift model in use is followed.

Hydraulic fluid maintenance is addressed in all maintenance schedules. However, the hydraulic fluid should be changed whenever it becomes dirty or breaks down due to overheating. Continuous operation of the hydraulic system in hot environments can heat the hydraulic fluid to the boiling point and cause it to break down. In dusty areas, the fluid cooler may become clogged or blocked, causing the fluid to remain hot. As a result, the hydraulic fluid condition should be observed consistently, and replaced if it shows signs of overheating and breakdown. Filters should also be replaced whenever the fluid is changed.

2.2.3 Preventive Maintenance Records

Accurate, up-to-date maintenance records are essential for knowing the history of your equipment. Each machine should have a record that describes any inspection or service that is to be performed and the corresponding time intervals. Typically, an operator manual and some sort of inspection sheet are kept with the equipment at all times. Actions taken, along with the date, are recorded on the log. Various operators can then share information about fluids that have been added, for example, so that patterns in fluid loss are noted.

The operator manual usually has detailed instructions for performing periodic maintenance. If you find any problems with your machine that you are not authorized to fix, inform the foreman or field mechanic before operating the machine.

2.0.0 Section Review

1. The hydraulic fluid level is typically checked by _____.
   a. removing a dipstick
   b. observing a vertical sight glass
   c. observing a floating indicator arm
   d. removing the fill cap and observing the level

2. Which of the following fluids is most likely to be replaced on a new forklift after 50 hours of service?
   a. engine oil
   b. brake fluid
   c. hydraulic fluid
   d. transmission fluid

SECTION THREE

3.0.0 STARTUP AND OPERATING PROCEDURES

Objective 3

Describe the startup and operating procedures for a rough-terrain forklift.

a. State rough-terrain forklift-related safety guidelines.
b. Describe startup, warm-up, and shutdown procedures.
c. Describe basic maneuvers and operations.
d. Describe related work activities.

Performance Tasks 2 through 6

Perform proper startup, warm-up, and shutdown procedures.

Interpret a forklift load chart.

Execute basic maneuvers with a rough-terrain forklift.

Perform basic lifting operations with a rough-terrain forklift.

Demonstrate proper parking of a rough-terrain forklift.

Trade Term

Fulcrum: A point or structure on which a lever sits and pivots.

Now that all the basic rough-terrain forklift components have been covered along with the operator-performed inspections and preventive maintenance activities, the next step is to actually start and operate the machine. Before that can happen, operators must fully understand the safety issues associated with this versatile piece of equipment. The operator manual for a given forklift contains safety information about that specific machine. Forklifts and telehandlers in general are similar in their operation. However, they all have slight differences. Some may be a bit clumsier to operate than others are. The type of steering modes available and the terrain also change how a machine maneuvers.

3.1.0 Safety Guidelines

Safe forklift operation is the responsibility of the operator. Operators must develop safe working habits and recognize hazardous conditions to protect themselves and others from injury or death. Always be aware of unsafe conditions to protect the load and the forklift from damage. Become familiar with the operation and function of all controls and instruments before operating the equipment. Read and fully understand the operator manual. Operators must be properly trained and certified.

3.1.1 Operator Training and Certification

Not everyone is allowed to operate a forklift. Due to liability and insurance guidelines, only trained and qualified personnel are allowed to operate forklifts on the job. Operators receive a license certifying them to operate specific forklifts. They also have to be re-evaluated at different times over the years. OSHA recommends that operators be recertified every three years. Since operators often move from one job site to another or from one company to another, they may be required to requalify at each new location. The following explanation of operator training requirements was extracted from OSHA's *Occupational Safety & Health Administration (OSHA) Directive CPL 02-01-028*, which deals with Compliance Assistance for the Powered Industrial Truck Operator Training Standards:

> The training requirement found in *29 CFR 1910.178(l)* for operators of powered industrial trucks and the same requirement for operators of powered industrial trucks in the construction *[1926.602(d)]* and maritime *[1915.120, 1910.16(a)(2)(x), 1910.16(b)(2)(xiv), 1917.1(a)(2)(xiv), 1918.l(b)(10)]* industries specify that the employer must develop a complete training program. OSHA requires that operators of powered industrial trucks be trained in the operation of such vehicles before they are allowed to operate them independently. The training must consist of instruction (both classroom-type and practical training) in proper vehicle operation, the hazards of operating the vehicle in the workplace, and the requirements of the OSHA standard for powered industrial trucks. Operators who have completed training must then be evaluated while they operate the vehicle in the workplace. Operators must also be periodically evaluated (at least once every three years) to ensure that their skills remain at a high level and must receive refresher training whenever there is a demonstrated need. To maximize the effectiveness of the training, OSHA will not require training that is duplicative of other training the employee has previously received if the operator has been evaluated and found competent to operate the truck safely. Finally, the training provisions require that the employer certify that the training and evaluations have been conducted.

Forklift-Related Fatality

A 39-year-old supply motorman was helping a forklift operator place crib blocks on a mine flat car. The motorman was using a shovel to remove ice and snow from the bottom of a crib block while it was raised on the forklift. The load slipped and crushed the motorman. He later died from his injuries.

The Bottom Line: Always maintain a safe distance from the load, and do not approach a load that is unstable for any reason. Ice and snow represent significant hazards in the work place, changing many situations from acceptably safe to unpredictable.

Individual companies may have their own guidelines that specify how they comply with these OSHA guidelines. It is in the individual's best interest that he or she maintains copies of any operator certifications received. In some cases, it may keep the operator from having to recertify at a new job site.

3.1.2 Operator Safety

There are a number of things workers can do to protect themselves and those around them from getting hurt on the job.

Know and follow your employer's safety rules. Your employer or supervisor will provide you with the requirements for proper dress and safety equipment. The following are recommended safety procedures for all occasions:

- Operate the machine from the operator's cab only.
- Wear the seat belt and tighten it firmly. The effectiveness of the ROPS depends heavily on the operator remaining in the seat in the event of an accident.
- Mount and dismount the equipment carefully.
- Wear a hard hat, safety glasses, safety boots, and gloves when operating the equipment.
- Do not wear loose clothing or jewelry that could catch on controls or moving parts.
- Keep the windshield, windows, and mirrors clean at all times.
- Never operate equipment under the influence of alcohol or drugs.
- Never smoke while refueling.
- Do not use a cell phone and avoid other sources of static electricity while refueling. Cell phones should also never be used while operating the machine.
- Never remove protective guards or panels.
- Never attempt to search for leaks with your bare hands. Hydraulic and cooling systems operate at high pressure. Fluids under high pressure can cause serious injury.
- Always lower the forks or other attachments to the ground before performing any service or when leaving the forklift unattended.

3.1.3 Safety of Co-Workers and the Public

You are not only responsible for your personal safety, but also for the safety of other people who may be working around you. Sometimes, you may be working in areas that are very close to pedestrians or motor vehicles. In these areas, take time to be aware of what is going on around you. Create a safe work zone using cones, tape, or other barriers. Remember, it is often difficult to hear when operating a forklift. Use a spotter and a radio in crowded conditions.

The main safety points when working around other people include the following:

- Walk around the equipment to make sure that everyone is clear of the equipment before starting and moving it.
- Always look in the direction of travel.
- Do not drive the forklift up to anyone standing in front of an object or load.
- Make sure that personnel are clear of the rear area before turning (*Figure 32*).
- Know and understand the traffic rules for the area you will be operating in.
- Use a spotter when landing an elevated load or when you do not have a clear view of the landing area.
- Exercise particular care at blind spots, crossings, and other locations where there is traffic or where pedestrians may step into the travel

Did You Know?

Each year in the United States nearly 100 workers are killed and another 20,000 are seriously injured in forklift-related incidents. The most frequent type of accident is a forklift striking a pedestrian. This accounts for 25 percent of all forklift accidents. However, these figures include all forklifts. Rough-terrain models have a lower incidence of accidents, especially in striking pedestrians, since they are generally used outdoors where visibility is better. The tight confines and blind corners of a warehouse significantly increase the accident rate.

22206-12_F32.EPS

Figure 32 Checking the rear before turning.

path. Sound the horn to communicate your presence at such locations.
- Do not swing loads over the heads of workers. Make sure you have a clear area to maneuver.
- Travel with the forks no higher than 12 to 18 inches above the ground. Always drop the forks or attachment to the ground when parking the machine.
- Do not allow workers to ride in the cab or on the forks.

When working around pedestrians or in other public areas, create a safe work zone for forklift operations. Use barrels, cones, tape, or barricades to keep others out of your work area. This protects both you and bystanders.

3.1.4 Equipment Safety

The forklift has been designed with certain safety features to protect you as well as the equipment. For example, it has guards, canopies, shields, rollover protection, and seat belts. Know your equipment's safety devices and be sure they are in working order.

Forklift overturns are the leading cause of deadly forklift accidents. They represent 25 percent of all forklift-related deaths. In too many cases, the seat belt was not in use. Know the weight of the load and the limits of your machine. Don't take risks and overload the machine. Be especially careful moving suspended loads in windy conditions.

Use the following guidelines to keep your equipment in good working order:

- Perform prestart inspection and lubrication daily (*Figure 33*).
- Look and listen to make sure the equipment is functioning normally. Stop the equipment if it is malfunctioning. Correct or report trouble immediately.
- Always travel with the forks or load low to the ground.
- Never exceed the manufacturer's limits for speed, lifting, or operating on inclines.
- Know the weight of all loads before attempting to lift them. Review the appropriate load chart. Do not exceed the rated capacity of the forklift. It is important to remember that a forklift can rarely lift its maximum load capacity. Especially when a mast is raised and tilted forward, or a telescoping boom is extended, the actual lifting capacity is usually a small percentage of the maximum load capacity.
- Always lower the forks, engage the parking brake, turn off the engine, and secure the controls before leaving the equipment.
- Never park on an incline.
- Maintain a safe distance between your forklift and other equipment that may be on the job site.

Know your equipment. Learn the purpose and use of all gauges and controls as well as your equipment's limitations. Never operate your machine if it is not in good working order.

3.1.5 Spill Containment and Cleanup

As noted in earlier sections of this text, forklifts can be powered by batteries or combustible fuels. Fuels must be contained during refueling activities. Most companies have designated refueling areas specifically built to contain any fuel spills that may occur.

Forklifts also use one or more hydraulic systems to steer the forklift and to extend, retract, tilt, lift, and lower the forklift mast and forklift assembly. With so many of the forklift operational controls relying on the hydraulic system, it is critical that hydraulic fluid be contained without spills or leaks. Anyone who has ever worked around any hydraulic system knows that it is almost impossible to stop all leaks. Leaks need to be cleaned and repaired whenever reasonably possible. Hydraulic fluid spills, on the other hand, are much more of a problem. If a hydraulic hose ruptures under pressure, gallons of hydraulic fluid are spilled. To protect the environment, companies must maintain an oil-spill kit that is easily acces-

22206-12_F33.EPS

Figure 33 Check all fluids daily.

sible to the forklift operator. A typical oil spill kit may include the following:

- A large (20-gallon or more) salvage drum/container to store spill kit materials until needed and for disposal after materials are used
- An emergency response guide for spills of petroleum-based products
- Personal protective equipment (goggles and unlined rubber gloves)
- One or more 3" by 10' absorbent socks for containment
- Half dozen 3" by 4' absorbent socks for containment
- About 20 absorbent pads for cleanup
- About 50 wiper towels or rags for cleanup
- 3 to 5 pounds of powdery absorbent material
- Several disposable bags with ties

NOTE

Although a flat-tipped shovel may not fit into a spill kit, having one handy would be a good idea for spill cleanups.

3.1.6 Hand Signals Used with Forklifts

Forklift operators often work alone. They operate their forklift controls based on what they can see out of the forklift operator's area. When they are placing or picking up a load and cannot see exactly everything they need to see, they must use a spotter or signal person. A spotter uses electronic communication devices, voice commands, or hand signals to tell the forklift operator what moves are needed. Most companies have published examples of the hand signals they want their people to use when performing forklift operations. *Figure 34* shows a few basic hand signals used with forklifts.

3.2.0 Startup, Warm-Up, and Shutdown Procedures

Before starting forklift operations, make sure that you are familiar with the load and the area of operations. Check the area for both vertical and horizontal clearances. Make sure that the path is clear of electrical power lines and other obstacles.

The following suggestions can help improve operating efficiency:

- Observe all safety rules and regulations.
- Determine the weight of the load, review the load charts, and plan operations before starting.
- Use a spotter if you cannot see the area where the load will be placed.

RAISE THE TINES
With forearm vertical, forefinger pointing up, move hand in small horizontal circle.

LOWER THE TINES
With arm extended, palm down, lower arm vertically.

TILT MAST BACK
With forearm vertical, thumb extended, jerk thumb over shoulder.

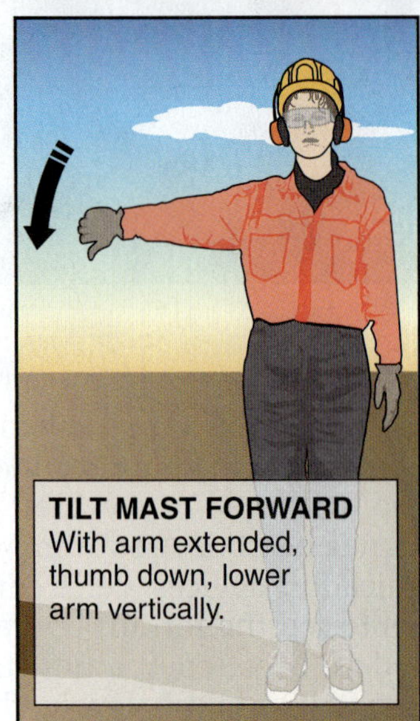

TILT MAST FORWARD
With arm extended, thumb down, lower arm vertically.

MOVE TINES IN DIRECTION FINGER POINTS
With arm extended, palm down, point forefinger in direction of movement.

DOG EVERYTHING
Clasp hands in front of body.

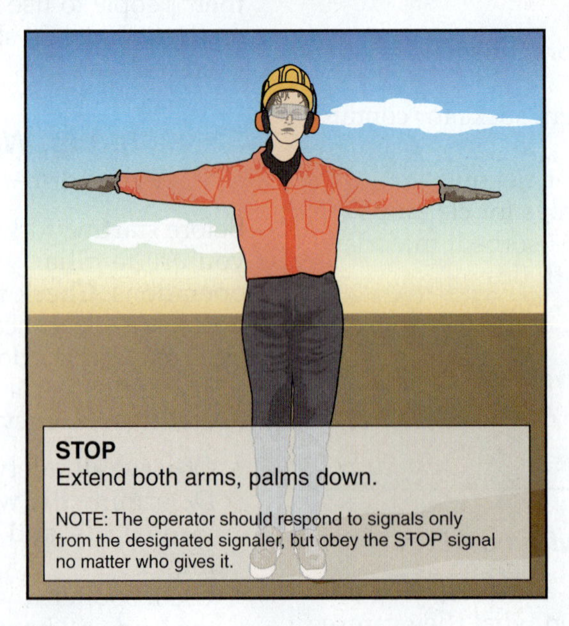

STOP
Extend both arms, palms down.
NOTE: The operator should respond to signals only from the designated signaler, but obey the STOP signal no matter who gives it.

Figure 34 Basic hand signals for forklifts.

3.2.1 Preparing to Work

Preparing to work involves getting organized in the cab, fastening the seat belt, and starting the machine. Mount your equipment using the grab rails and foot rests. Getting in and out of equipment can be dangerous. Always face the machine and maintain three points of contact when mounting and dismounting a machine. That means you should have three out of four of your hands and feet on the equipment. That can be two hands and one foot or one hand and two feet.

> **WARNING!**
> OSHA requires that approved seat belts and a ROPS be installed on virtually all heavy equipment. Old equipment must be retrofitted and the seat belt must be used at all times. Do not use heavy equipment that is not equipped with these safety devices.

Adjust the seat to a comfortable operating position. The seat should be adjusted to allow full pedal travel with your back firmly against the seat back. This will permit the application of maximum force on the brake pedals. The knees should still be slightly bent when each pedal is pushed to its maximum position. Make sure you can see clearly and reach all the controls.

The startup and shutdown of an engine is very important. Proper startup lengthens the life of the engine and other components. A slow warm up is essential for proper operation of the machine under load. Similarly, the machine must be shut down properly to cool the hot fluids circulating through the system. These fluids must cool so that they can cool the metal parts of the engine before it is switched off.

3.2.2 Startup

There may be specific startup procedures for the piece of equipment you are operating. But in general, the startup procedure should follow this sequence:

Step 1 Be sure the transmission control is in neutral.

Step 2 Engage the parking brake (*Figure 35*). This is done with either a lever or a knob, depending on the forklift make and model.

> **NOTE**
> When the parking brake is engaged, an indicator light on the dash should light up or flash. If it does not, stop and correct the problem before operating the equipment.

Step 3 Depress the throttle control slightly.

Step 4 Turn the ignition switch to the start position. The engine should turn over. Never operate the starter for more than 30 seconds at a time. If the engine fails to start, wait two to five minutes before cranking again. The lesser amount of time typically applies in cold weather.

Step 5 As soon as the engine starts, release the key; it should return to the On position.

Adjust the engine speed to one-third to one-half throttle for the warm-up period. Keep the engine speed low until the oil pressure shows on the gauge. The oil pressure light should initially light and then go out. If the oil pressure light does not turn off within 10 seconds, stop the engine, investigate, and correct the problem.

If the machine you are using has a diesel engine, there are special procedures for starting the engine in cold temperatures. Many diesel engines have glow plugs that heat up the engine for ignition. Some units are also equipped with ether starting aids. Review the operator manual so that you fully understand the procedures for using these aids. Follow the manufacturer's instructions for starting the machine in cold weather.

22206-12_F35.EPS

Figure 35 Parking brake.

> **WARNING!**
> Ether is a highly flammable liquid. It mixes with the air in the carburetor. As the pistons compress the gases, ignition takes place. The flame from the ether produces enough heat to overcome the compression heat lost through the cold metal of the engine. Ether should only be used when temperatures are below 0°F (-17.7°C) and when the engine is cold.

3.2.3 Warm Up

Warm up a cold engine for at least five minutes. Warm up the machine for a longer period in colder temperatures. While allowing time for the engine to warm up, complete the following tasks:

Step 1 Check all the gauges and instruments to make sure they are working properly. Watch the temperature gauge as the coolant temperature climbs.

Step 2 Shift the transmission into forward and rotate the gear control (if any) to low range.

Step 3 Release the parking brake and depress the service brakes to ensure they work as the forklift begins to roll.

Step 4 Check the steering for proper operation.

Step 5 After a few minutes, manipulate the controls to be sure all components are operating properly.

> **CAUTION**
> Do not pick up a load if the hydraulics are sluggish. Allow the machine to warm up until the hydraulic components function normally. The hydraulics can fail if not warmed up completely.

Step 6 Shift the gears to neutral and lock.

Step 7 Reset the parking brake while the machine continues to warm up.

Step 8 Make a final visual check for leaks, unusual noises, or vibrations.

If there are any problems that have no obvious cause, shut down the machine and investigate or get a mechanic to look at the problem.

3.2.4 Shutdown

Shutdown should also follow a specific procedure. Proper shutdown reduces engine wear and possible damage to the machine.

Step 1 Find a dry, level spot to park the forklift. Stop the forklift by decreasing the engine speed, depressing the clutch, and bring the machine to a full stop.

Step 2 Place the transmission in neutral and engage the parking brakes.

Step 3 Release the service brake and make sure that the parking brake is holding the machine.

Step 4 Make sure that the boom is fully retracted. Lower the forks or other attachment so that they are resting on the ground.

Step 5 Place the speed control in low idle and let the engine run for approximately five minutes.

> **CAUTION**
> Failure to allow the machine to cool down can cause excessive temperatures in the engine or systems. The engine oil, hydraulic fluid, or transmission fluid may overheat or even boil.

Step 6 Turn the engine start switch to the Off position.

Step 7 Release hydraulic pressures by moving the control levers until all movement stops.

Step 8 Turn the engine switch to Off and remove the key. If you must park on an incline, chock the wheels.

Some machines have security panels or vandalism caps for added security. The panels cover the controls and can be locked when the machine is not in use. Lock the cab door, and secure and lock the engine enclosure. Always engage any security systems when leaving the forklift unattended.

3.3.0 Basic Maneuvering

To maneuver the forklift, you must be able to move it forward, backward, and turn.

3.3.1 Moving Forward

The first basic maneuver is learning to drive forward. To move forward, follow these steps:

Step 1 Before starting to move, use the joystick to raise the forks to roughly 15 inches above the ground. This is the travel position.

Step 2 Put the shift lever in low forward. Release the parking brake, and press the accelerator pedal to start moving the forklift.

Step 3 Steer the machine using the steering wheel.

Step 4 Once underway, shift to a higher gear to drive on a smooth road. To shift from a lower to a higher gear, rotate the collar of the shift lever.

WARNING!

Always travel with the forks low to the ground (12 to 18 inches). Be aware of the tips of the forks and avoid hitting things with them. On very rough terrain, an even higher travel position may be required.

3.3.2 Moving Backward

To back up or reverse direction, always come to a complete stop first. Then move the shift lever to reverse. Once in reverse gear, you can apply some acceleration and begin to move backwards.

Depending on the steering modes available or in use, backing up can be challenging. The operator should always be looking in the direction of travel, and using the proper steering inputs while looking backwards requires some practice.

3.3.3 Steering and Turning

How a forklift is steered depends on the make and model. However, most rough-terrain models have a steering wheel. Some forklift steering wheels can be operated with one hand. This allows the operator to use the other hand to control the forks and boom. The steering wheels on a forklift operate in the same manner as steering wheels on cars and trucks. Moving the wheel to the right turns the forklift to the right. Turning the wheel to the left moves the wheels to the left.

Some forklifts have different steering modes, including two-wheel, four-wheel, and crab steering. When the machine is in a two-wheel steering mode, only the front wheels may move; if the unit is front-wheel drive, then the rear wheels will likely move. Use two-wheel steering when you are traveling, especially on a smooth road. Use four-wheel steering for work activities and close-quarter operations. When the machine is in four-wheel steering mode, the front and back wheels turn in opposite directions. This allows the machine to make tight turns in a circle. When working in a confined area, crab steering is probably the best choice. In crab steering, the front and back wheels turn in the same direction. The machine will travel forward and to one side, or backward and to one side. The work area and tasks to be done determine which steering mode is the most advantageous.

CAUTION

Always straighten the wheels before switching modes. You can damage the steering if the wheels are not centered before operating the machine in two-wheel steer.

NOTE

It is possible for the steering to go out of synchronization if the correct procedures are not followed when changing steering modes. Review the operator manual before changing steering modes. Follow the procedure in the operator manual to synchronize the wheels if they become unsynchronized.

If the forklift is equipped with a differential lock, it should be disengaged before making any turns. The differential lock ensures that an equal amount of torque is sent to the drive wheels. As a result, they will attempt to rotate at the same speed. The differential lock is best used when in slippery or muddy conditions, and one drive wheel may be spinning while the other is nearly or completely motionless. Locking the differential rebalances the torque to each drive wheel and helps get the forklift moving. When turning the forklift, the drive wheels on the inside of the turn must rotate more slowly than those on the outside of the turn. Turning with the differential lock engaged does not allow this to occur, causing a clumsy turn and unnecessary stress on the drive train.

3.3.4 Leveling the Forklift

Keeping the forklift level is a vital part of safe operations. The forklift must be level and balanced before significant loads are lifted. If the forklift is not level when a load is lifted, it could tip over. Remember that an overturned forklift is the leading cause of death in forklift operations.

Most rough-terrain forklifts have some type of level indicator. On many telehandlers, the level indicator is located on the frame above the front windshield as shown in *Figure 36*. An air bubble or small bead is contained within a small-arced sight glass. When the bubble or bead is centered in the middle of the sight glass, the forklift frame is level. A scale on the side of the glass is marked zero in the center and increases to either side. The numbers represent the number of degrees off level.

22206-12_F36.EPS

Figure 36 Forklift level indicator.

Note that a forklift must always be leveled before a load is raised. Although leveling can be done with a load on the forks, the forks should be very near the ground before any leveling is attempted.

WARNING!

Never attempt to level a forklift with a load raised in the air. Using any method of leveling with the load raised could result in a serious imbalance leading to serious personal injury and/or property damage.

There are two ways to level the forklift. One is to lower and adjust the outriggers until the machine is level. When the forklift must be leveled while the machine is on its wheels, use the frame-leveling control to level the machine (if the machine is so equipped). This feature rotates the cab and boom assembly on the chassis to level the machine. The frame level controls cannot be used after the outriggers are lowered.

The frame-leveling controls are usually located on the console near the joystick. The default position for the three-position switch is center or hold. Depressing the switch to either side rotates the frame. Pressing the right side of the switch lowers the right side of the machine. Pressing the left side of the switch lowers the left side.

Use the frame-leveling controls when a load must be lifted from an uneven surface. The controls tilt the frame 10 degrees to the left or right. Lower the boom before using the level controls. The boom attachment should be close to, but not touching, the ground. Depress the appropriate switch until the machine is level. When you release the switch, it returns to the hold position and the machine stops rotating.

CAUTION

Do not operate the frame-leveling controls when the boom is raised. Doing so could damage the machine.

The outriggers can also be used to level the machine. The outrigger controls are used to lower and raise the left and right outriggers. The outriggers are controlled with switches located on the instrument panel or the joystick. Lowering the outriggers firmly to the ground provides a firm and level base for lifting. When the stabilizers are lowered, the forklift can handle heavier loads and the boom can be fully extended.

WARNING!

Check that the ground will support the stabilizers before they are lowered. If the soil collapses from the weight of the machine, the forklift could overturn and cause injury or death. Make sure that all personnel stand clear when operating the stabilizers. Moving stabilizers poses a significant crushing hazard.

Before lowering the stabilizers, check that the area is free of obstructions and that the ground will support the stabilizers. Lower and retract the boom. The boom should be close to, but not on, the ground. Use the frame-leveling controls to level the machine. Run the engine at sufficient speed to supply enough hydraulic pressure to the stabilizers. Check that all personnel are clear of the area. Depress or slide the switch for the right outrigger to lower it. Release it when the outrigger has reached the desired position. Repeat the procedure for the left outrigger. When both outriggers are lowered, the front wheels of the forklift should be off the ground (*Figure 37*). Adjust the positions of the outriggers until the frame is level as indicated in the sight glass.

Before raising and retracting the outriggers, fully retract and lower the boom. Check that all personnel are clear of the area before retracting the outriggers. Make sure that both outriggers are fully raised and retracted before moving the forklift.

3.4.0 Work Activities

Operation of the forklift is fairly straightforward, but it requires attention to detail. With proper planning, you will have no trouble operating the forklift. The basic work activities performed with a forklift are described in this section. One thing that is different with these machines is the terrain. The terrain is very predictable in a warehouse en-

22206-12_F37.EPS

Figure 37 Outriggers in use; front wheels raised.

vironment. The environment where rough-terrain models are used presents an entirely different set of hazards and surprises.

> **NOTE**
> The controls on specific forklifts may be different from those described in the procedures. Check your operator manual for information about the controls and limitations of your equipment.

3.4.1 Load Charts

The most important factor to consider when operating any forklift is its lifting capacity. Each forklift is designed with an intended capacity that must never be exceeded. Exceeding the capacity jeopardizes the equipment, the load, and the safety of everyone near the equipment. Capacity information is provided in the load charts, which are included in the operator manual or posted in the forklift. Be sure to read and follow the load chart.

The ability of a forklift to lift a load without tipping is called the rated capacity. The capacity of a forklift varies depending on the angle and height of the boom. Load charts, like the ones shown in *Figure 38*, provide data on the maximum capacity of the forklift. A forklift operator must be able to read load charts and make sure that the load does not exceed the capacity.

Different load charts are used for different machine configurations. For example, the capacity changes if the outriggers are lowered or raised. Using different attachments also requires that you use a different load chart. Make sure that you are using the correct load chart for the lift you are performing.

Lifting a load, regardless of how it is done, is a matter of leverage. Almost everyone has played on a playground seesaw. When nobody is on the seesaw, the board can be balanced on the pivot point, which is the **fulcrum** for the seesaw. When equal weights are placed on each end of the board, the board can still be balanced. With a forklift, the rear end provides a given amount of weight that acts as a counterweight to any load the forklift tries to lift. The front axle, aided by the wheels, is the fulcrum, or pivot point. Anything heavy placed on the forks is the load. If the forklift's rear weight is enough of a counterweight, the forklift can lift the load. As with any lifting device, the further the load gets away from the fulcrum horizontally, the more likely it is to pivot due to the added leverage. If it is too heavy and placed far enough from the fulcrum, the rear wheels of the forklift rise off the ground.

In summary, the closer the load is to the front axle horizontally, the more stable the lift. When the boom's elevation is kept at zero degrees, and the load is extended out further away from the axle, the forklift's balance is more unstable.

Understanding a concept known as the stability triangle helps operators to better evaluate a lift situation. The stability triangle is represented by three points on the chassis of the forklift (*Figure 39*). The center of the two front wheels and the center of the rear axle form the three points. The center of the triangle is the center of gravity (CG) of the forklift with no load, the forks down, and the boom retracted. Adding a fourth point—straight up from the CG to the boom—creates a pyramid. Imagine that there is a weight hanging from the boom at this point. As long as the weight is hanging inside the lines of the triangle below, the forklift remains stable. As the boom is raised, the CG rises within the pyramid. As a result, the triangle begins to shrink in size. The smaller the triangle, the more difficult it becomes to keep the imaginary weight hanging within it. If the forklift is not level, especially if it is tilting left or right, the imbalance can easily cause the machine to tip over as the load rises.

Forklifts are often selected based on their maximum load capacity. However, most can only achieve the maximum capacity under ideal conditions, on ideal terrain, and without any horizontal extension of the boom or forward-tilting of the forks. A 6,000-pound forklift can quickly become a 1,500-pound forklift as the position of the load in relation to the fulcrum changes.

Figure 38 shows load charts for a telehandler with and without outriggers being used. With the outriggers retracted (not in use or non-existent), the machine on the right can lift up to 8,000

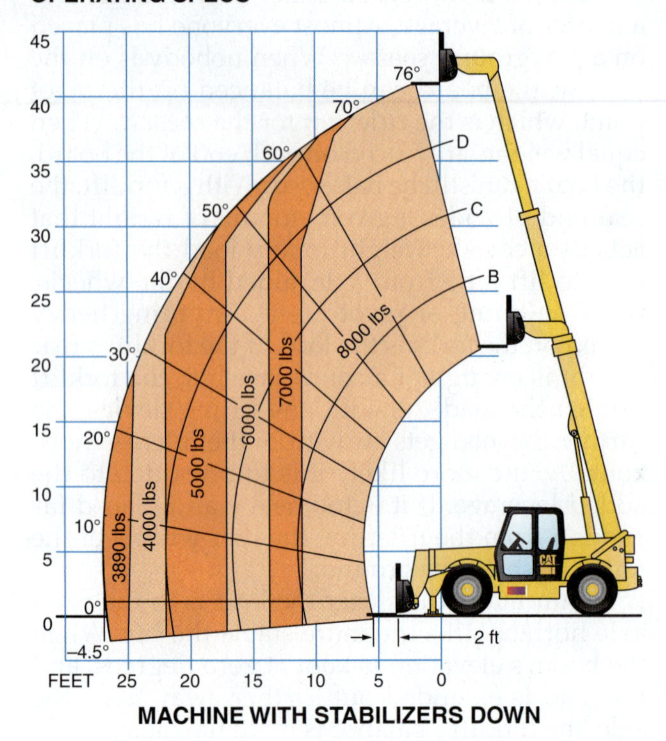

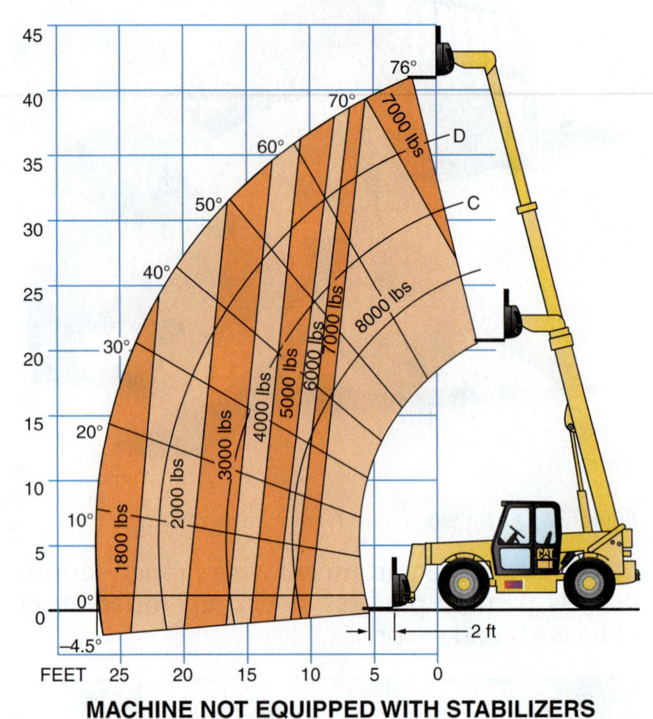

Figure 38 Load charts.

pounds when its boom is fully retracted (closest in to the axle) and lowered to 0 degrees. It can also lift that same load to a height of nearly 40 feet as long as the fork tips remain within 6 feet forward of the front wheels, which occurs at a boom angle of 70 degrees. If the boom is kept at zero degrees, with the fork tips extended out to 15 feet, you will see that the machine can only lift 4,000 pounds. If the fork tips are extended all the way out to 25 feet, while being kept at zero degrees, the machine can then lift only 1,800 pounds.

Now, still operating without the outriggers, see what the forklift can do when the boom is elevated to some angle above zero degrees. If the boom is fully retracted and then raised, the machine can lift an 8,000-pound load up to 70 degrees, which would put the load up to nearly 40 feet above the ground. If the boom is raised beyond 70 degrees, the amount of load the machine can lift is reduced to 7,000 pounds because the height of the load becomes unstable when lifted almost directly over the machine's front axle.

If the same machine, not using its stabilizers, has its boom extended and raised at the same time, the load chart shows that a different set of limitations must be applied. If the boom is extended to 15 feet and elevated to 20 degrees, the machine's lifting capacity is reduced to 4,000 pounds. If the boom is kept at 20 degrees, but extended out to 25 feet, you can see that the machine's lifting capacity is further reduced to only 1,800 pounds.

There are times when the machine operator has plenty of room to use the machine's stabilizers or outriggers. The chart on the left of *Figure 38* shows that the machine can lift heavier loads further away from its axle when the outriggers are used. This particular machine can lift up to 3,890 pounds when its boom is at zero degrees elevation and fully extended out to 25 feet. When the machine has its stabilizers in use and its boom retracted, it can safely elevate its boom with the full 8,000 pounds all the way up to the 76-degree point. The stabilizers allow for the increased lifting capacity.

The point of this discussion about load charts is that the machine operator must clearly understand how much of a load is to be lifted and what boom extensions and elevation angles are needed to move and position the load. Before beginning any forklift work, make sure that you clearly understand the rated capacities of that given machine. However, it is equally important to remember that, regardless of the operator's expertise in interpreting load charts, it is of no value at all if the actual weight of the load is not first determined. Ignoring or trying to guess the weight of a significant load can start a tragic series of events.

Load capacities for forklifts are calculated based on the assumption that the load's CG is 24 inches up from the forks and 24 inches forward

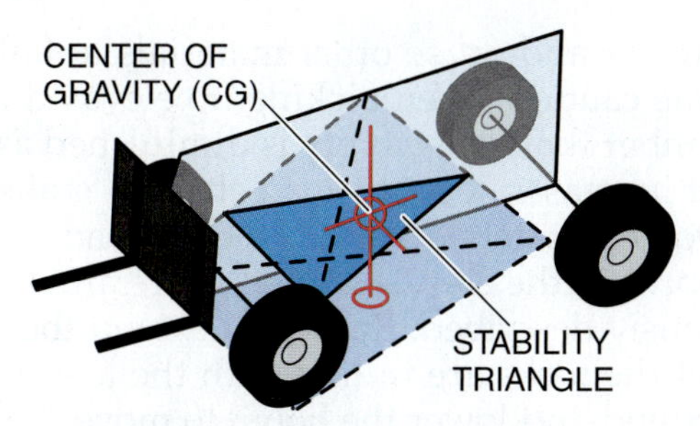

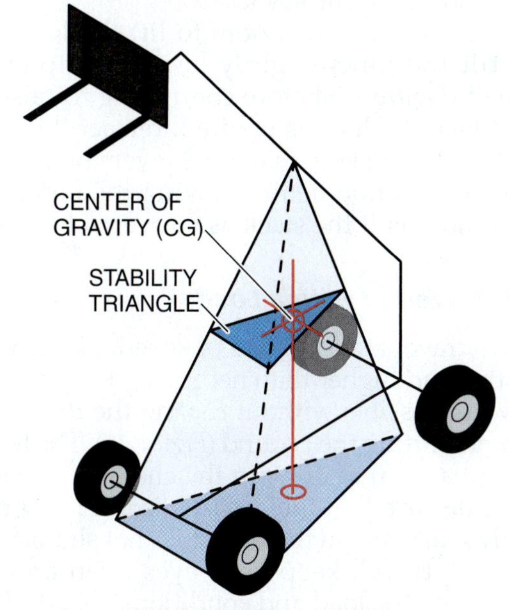

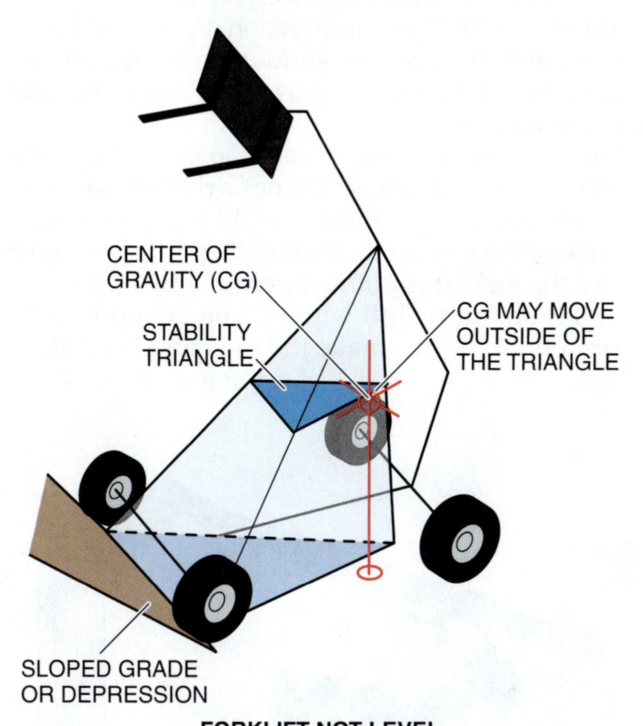

Figure 39 The stability triangle changes due to the terrain and boom position.

of the back of the forks. Attachments can make a dramatic difference in determining the capacity. Operators must consider the added weight of the attachment at a minimum. However, they must also consider how the attachment changes the CG of the load. A good example of this is seen in the use of lifting hooks/boom extensions and fork-mounted lifting hooks. Fork-mounted lifting hooks, such as the one shown in *Figure 23*, do not usually change the center of gravity at all. Only the added weight of the attachment must be considered. However, with a lifting hook such as the one shown in *Figure 22* is used, the CG is moved far forward of the forks. Operators must then consider the location of the load CG on the load chart and disregard the location of the forks. Of course, the load capacity of the attachment itself must always be considered as well. The load capacity of the attachment and the forklift at a given set of conditions must be compared, and the lowest value of the two becomes the maximum capacity for the lift conditions.

3.4.2 Picking Up a Load

To pick up a load, first check the position of the forks. They should be centered on the carriage. If they must be adjusted, check the operator manual for the proper procedure. Usually, there is a pin at the top of each fork. When the fork is lifted, you can slide it along the upper backing plate until it reaches the desired position. Lock them into place.

Travel to the area at a safe rate of speed. Use the foot throttle and the steering wheel to maneuver. Always keep the forks 12 to 18 inches above the ground and tilted slightly back.

Before picking up the load, make sure it is stable. If it looks like the load might shift when picked up, secure the load with strapping or other tie-downs. The center of gravity is critical to load stability. If necessary, make a trial lift and adjust the load.

Approach the load so that the forks straddle the load evenly. It is important that the weight of the load is distributed evenly across the forks. Overloading one fork can damage it or cause the forklift or the load to overturn. In some cases, it may be advisable to measure and mark the load's center of gravity. When lifting critical, sensitive, or unusual equipment, check the equipment manufacturers documentation for weight and balance information.

Drive up to the load with the forks straight and level (*Figure 40*). Approach the load so that the forklift is square and level to the intended load. If necessary, level the forklift using the frame level controls or the stabilizers. If the load is on a pallet, make sure that the forks will clear the pallet board. If the load is on blocks, you may need to get out and check if the forks will clear.

Move forward slowly until the leading edge of the load rests against the back of both forks. If you cannot see the forks engage the load, ask someone to signal for you. This can avoid expensive damage and injury. Raise the carriage slowly until the forks contact the load. Continue raising the carriage until the load safely clears the ground. Tilt the mast fully rearward to cradle the load. This minimizes the possibility of the load slipping in transport.

> **WARNING!**
> Set the parking brake and turn off the engine before exiting the cab. Do not leave the forklift running unattended.

> **CAUTION**
> Oil, ice, mud, grease, rain, or snow can cause the forks to become very slippery. Clean the forks whenever they become wet or fouled. Wet metal on wet metal is extremely slippery. Be extremely careful when picking up metal loads with wet forks.

3.4.3 Picking Up an Elevated Load

Picking up a load from an elevated position is more difficult than picking a load off the ground. First, it is more difficult to see an elevated load. Use a spotter if necessary. Second, the load can fall farther and cause other materials to fall. Use extreme caution when picking up elevated loads. Remember that the stability is diminished as load height increases. Use the load chart to make sure that you can safely lift and place the load.

Approach the elevated load square and level as previously described. Raise and extend the boom so that the forks are in line with the load (*Figure 41*). Extend and lower the boom to move the forks under the load. The forks will remain level. Be careful not to hit the stack.

Carefully raise the boom to lift the load, and then tilt the forks slightly backward to cradle the load (*Figure 42*) before continuing to raise the load. Once the load is cradled, retract the boom slowly until the load is clear. Lower the boom to the travel position before moving the forklift. Be careful not to hit the stack as you retract the load.

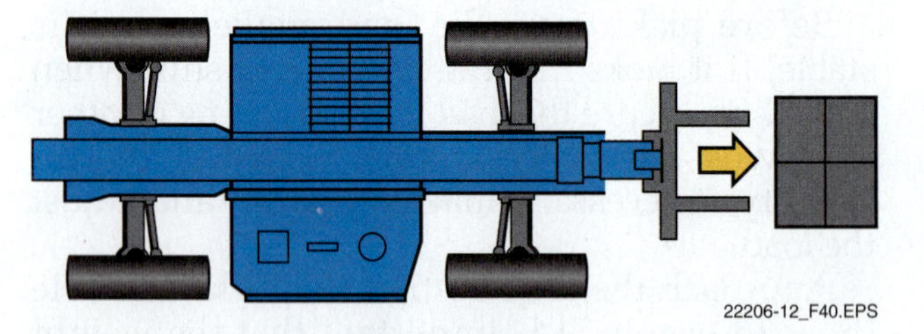

Figure 40 Approach the load squarely.

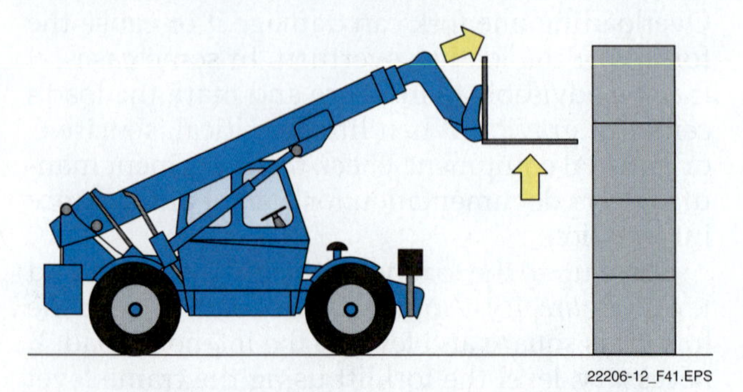

Figure 41 Raise the boom to align the forks with the load.

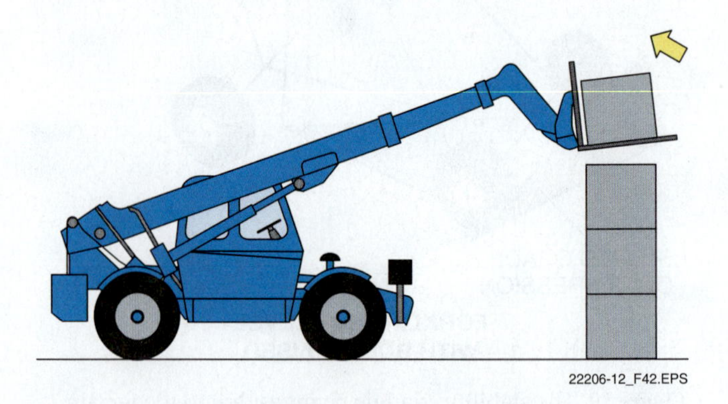

Figure 42 Tilt the forks back to cradle the load.

3.4.4 Traveling with a Load

Always travel at a safe rate of speed. Never travel with the load higher than necessary. Keep the load as low as possible without risking the forks or attachment hitting the ground (*Figure 43*). The terrain has a great deal to do with the choice of traveling height. Be sure that the carriage is level and tilted slightly rearward so the load does not slip off.

As you travel, keep your eyes open and stay alert. Watch the load and conditions ahead of you. Alert others to your presence. Avoid sudden stops and abrupt changes in direction. Be careful when downshifting because sudden deceleration can cause the load to shift. Be aware of front and rear swing when turning.

If you have to drive on a steep slope, keep the load as low as possible. Do not drive across steep slopes, because the forklift could easily overturn. Drive or back up and down a steep hill, but make sure the forklift is pointed uphill with a load (*Figure 44*) and downhill without one. If you have to turn on an incline, make the turn wide and slow. This minimizes the risk of tipping over.

22206-12_F43.EPS

Figure 43 Keep the load low.

22206-12_F44.EPS

Figure 44 Traveling uphill with a load.

Driving over bumps and holes with a load can cause you to lose a load. If the machine bumps, the load can bounce off. Avoid driving over bumps and holes. If impossible to avoid, drive over them slowly and keep the load low.

3.4.5 Placing a Load

Position the forklift at the landing point so the load can be placed where needed. Remember that it is easier to reposition the forklift than to reposition the load after it is placed. If the load is not palletized, place blocking material where the load is to be placed. The blocking will create space under the load so that you can remove the forks. Make sure that the blocking will adequately support the load and be sure that everyone is clear of the area.

The area under the load must be clear of obstruction and must be able to support the weight of the load. If you cannot see the placement, use a spotter to guide you.

When the forklift is in position, tilt the forks forward to the horizontal position. Lower the load slowly. You can usually feel when the load is resting on the ground or blocking. When the load has been placed, lower the forks a little more to clear the underside of the load. Back carefully away from the load to disengage the forks.

NOTE

Do not lower the forks too much after placing the load. You can lower the forks too far and dig up the ground under the load when backing up. Lower the forks just enough to clear the bottom of the load but still remain off the ground.

3.4.6 Placing an Elevated Load

You must take extra precautions when placing elevated loads. It is extremely important to level the machine before lifting the load. Failure to do so can cause the machine, the load, or the existing stack to tip over.

One of the biggest safety hazards for elevated load placement is poor visibility (*Figure 45*). There may be workers in the area who cannot be seen. The landing point itself may not be visible. Your depth perception decreases as the height of the load increases. To be safe, use a signal person to help you spot the load. Use tag lines for long loads.

Drive the forklift as close as possible to the landing point with the load kept low. Set the parking brake. Lower the outriggers if necessary. The outriggers offer a significant advantage in stability with a raised load. Raise the load slowly and carefully while maintaining a slight rearward tilt to keep the load cradled against the back of the forks. Do not tilt the load forward until the load is over the landing point and ready to be set down.

If the forklift's rear wheels start to lift off the ground, stop immediately but not abruptly. Lower the load. Slowly reposition it, break it down into smaller loads, or use the outriggers. If the surface conditions are poor at the unloading site, it may be necessary to reinforce the surface to provide more stability.

3.4.7 Unloading a Flatbed Truck

The following steps are used to unload material from a flatbed truck:

Step 1 Position the forklift at either side of the truck bed.

22206-12_F45.EPS

Figure 45 Place elevated loads carefully.

Step 2 Manipulate the control levers in order to obtain the appropriate fork height and angle.

Step 3 Drive forks into the opening of the pallet or under the loose material. Use care not to damage any material.

Step 4 Adjust the controls as required to lift the material slightly off the bed.

Step 5 Tilt the forks back slightly to keep the pallet or other material from sliding off the front of the forks.

Step 6 Retract the boom or back the forklift away from the truck.

Step 7 Lower the forks to the travel position and move to the stockpile area.

Step 8 Position the forklift so that the material can be placed in the desired area.

Step 9 Lower or raise the boom until the material is set on the required surface.

Step 10 Adjust the forks with the boom lever in order to relieve the pressure under the pallet.

Step 11 Back the forklift away from the pallet.

Step 12 Repeat the cycle until the truck is unloaded.

3.4.8 Using Special Attachments

In addition to the various types of forks, there are several special attachments that expand the forklift's operational capability. The three main attachments are the hook, the boom extension, and the bucket. Always read the operator manual to make sure you follow the proper procedure for securing attachments to the forklift.

Some forklifts have a coupler system that allows the operator to easily change attachments. *Figure 46* shows the main features of the coupler. The coupler is activated with switches located on the instrument panel or on the joystick. On some models the coupler is controlled with the joystick after another switch is activated to change the joystick mode.

Before detaching the forks from the forklift, you need to disengage the hydraulics. Typically, there is a diverter valve on the hydraulic hoses that must be closed. Check the operator manual for the correct procedures. If the attachments are not secured properly, they could fail, causing property damage and significant injury.

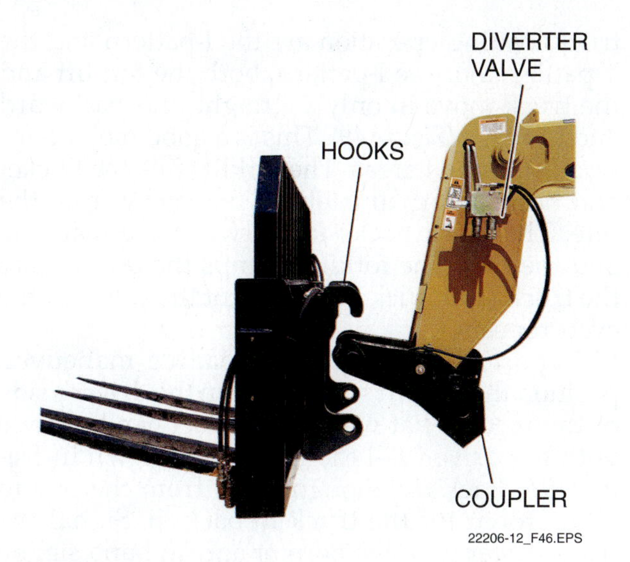

Figure 46 Quick coupler.

> **WARNING!**
> The hydraulic system is under pressure. Follow the safety procedures listed in the operator manual for relieving pressure before disconnecting hydraulic hoses. The release of fluids under pressure can cause significant injury.

To attach the coupler to the forks or other attachments, position the coupler in line with the attachment. Tilt the coupler forward so that it is below the levels of the hooks. Move the forklift forward or extend the boom until the coupler contacts the carriage. Tilt the coupler back until the lower part of the carriage contacts the coupler, and then secure the attachment to the coupler.

The boom extension and lifting hook allow the forklift to lift objects from above. These attachments feature a sturdy hook mounted on the carriage or the end of a boom extension (*Figure 47*). Loads must be securely rigged using approved lifting equipment. Each attachment has different lifting capacities. Operators must be aware that lifting attachments may have lower capacities than the forklift itself. Be sure to use the correct load chart when planning the lift, and consider the limitations of the attachment first. Once the capacity of the attachment is known, the operator can then determine if the forklift or the attachment represents the weakest link. Do not exceed the lifting capacity of the equipment or the attachment.

> **WARNING!**
> Only use approved chains, slings, hooks, and other rigging. Nonstandard rigging can fail and cause property damage and personal injury.

Figure 47 Forklift with boom extension and lifting hook.

Position the hook or lifting point directly above the load before lifting. If it is not directly above, the load could swing when it is lifted. Secure the load to the hook. Using shorter slings also reduces swinging (*Figure 48*). Use tag lines to control load swing and placement. Once the load is secured, use the boom controls to lift and position the load.

> **CAUTION**
> In extremely cold temperatures, the load can freeze to the ground. Free the load before attempting to lift it. Lifting a frozen load can cause a jolt that could affect the stability of the machine or dislodge the load.

3.4.9 Using a Bucket

Loading trucks, bins, and other containers can be done using a forklift with a bucket attachment. Usually, material is loaded from a forklift by tak-

Figure 48 Shorter rigging reduces swinging.

ing the material from a stockpile. The procedure for carrying out a loading operation from a stockpile is as follows:

Step 1 Travel to the work area with the bucket in the travel position.

Step 2 Position the bucket parallel to and just skimming the ground.

Step 3 Drive the bucket straight into the stockpile.

Step 4 Tilt the bucket backwards to fill it.

Step 5 Work the tilt control lever back and forth to move material to the back of the bucket. This is called bumping. When the bucket is full, move the tilt control lever to the tilt-back position.

Step 6 Shift the gears to reverse and back the forklift away from the stockpile.

Step 7 Place the bucket in the travel position and move the forklift to the truck.

Step 8 Center the forklift with the truck bed and raise the bucket high enough to clear the side of the truck.

Step 9 Move the bucket over the truck bed and shift the bucket control lever forward to dump the bucket.

Step 10 Pull the bucket control lever to retract the empty bucket, and back the forklift away from the truck as soon as the bucket is empty.

Step 11 Lower the bucket to the travel position and return to the stockpile.

Step 12 Repeat the cycle until the truck is loaded.

As the truck fills, the material needs to be pushed across the truck bed to even the load. As the leading edge of the bucket passes the sideboard of the truck, roll the bucket down quickly. Dump the material in the middle of the bed. The load is then pushed across the truck as the bucket is raised. By raising the bucket and backing up slowly, the material is distributed evenly across the bed.

CAUTION

Avoid hitting the side of the truck with the boom or bucket when you are unloading.

While there are many ways to maneuver a forklift, the two most common patterns for a truck loading operation are the I-pattern and the Y-pattern. For the I-pattern, both the forklift and the truck move in only a straight line, backward and forward (*Figure 49*). This is a good method for small, cramped areas. The forklift fills the bucket and backs approximately 20 feet away from the pile. The truck backs up between the machine and the pile. The forklift dumps the bucket into the truck. The truck moves out of the way and the cycle repeats.

To perform this I-pattern loading maneuver, position the forklift so that it is on the driver's side of the truck. That way, eye contact can be made with the driver. Fill the bucket, as shown in *Figure 49A*. Back far enough away from the pile to allow room for the truck to back in. Signal the truck driver with the horn or agreed hand signal. The truck backs up to a predetermined position, as shown in *Figure 49B*. Move the forklift forward and center it on the truck bed. Raise the bucket to clear the side of the truck and place it over the truck bed. Move the boom and bucket control lever to dump the bucket. At the same time, raise the forklift boom to make sure the bucket clears the truck bed. When the bucket is empty, move the boom and bucket control from side to side to shake out the last of the material. Back the forklift away from the truck and signal the truck driver to move. When the truck is out of the way, lower the bucket and position the forklift to return for another bucket of material.

The other loading pattern is the Y-pattern (*Figure 50*). This method is used when larger open areas are available. The dump truck remains stationary and as close as possible to the pile. The forklift does all the moving in a Y-shaped pattern.

To perform the Y-pattern, position the truck so that eye contact can be made with the driver. Fill the bucket with material. While backing up, turn the forklift to the right or left, depending on the position of the truck. Shift to a forward gear and turn the forklift while approaching the truck slowly. Stop when the forklift is lined up with the truck bed. Dump the bucket in the same way done for the I-pattern. When the bucket is empty, back away from the truck while turning toward the pile. Drive forward into the pile to repeat the pattern. Repeat the cycle until the truck is full.

3.4.10 Clamshell Bucket

The clamshell bucket has a hydraulically operated clamshell design. The bucket can perform four basic functions. It can be used as a clamshell bucket, scraper, dozer, and a regular forklift bucket.

The clamshell can be used for removing stumps and large rocks, as well as picking up de-

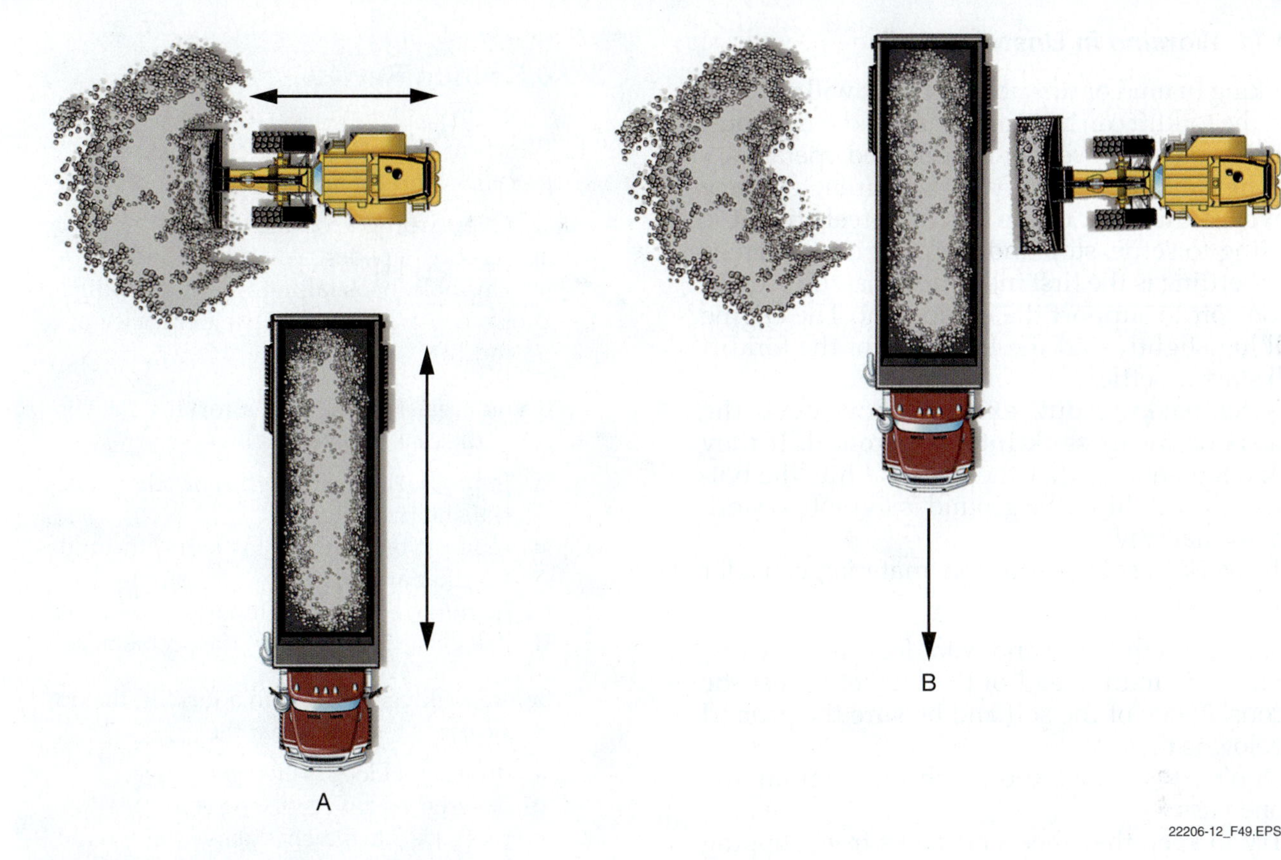

Figure 49 I-pattern for loading.

Figure 50 Y-pattern for loading.

bris and brush. To do this, the operator must open the bowl and position the bucket over the material to be loaded, then lower the bucket, and close the bowl to fill. Material can then be transported to the truck or stockpile for dumping. Use of the clamshell configuration gives the forklift added height for dumping and better handling of sticky material such as wet clay soils.

Using the bucket as a scraper requires the operator to open the bowl and use the backside of the bucket to cut material. When the material is filling the back of the bucket, close the bowl over the material, and raise the bucket for transporting to the dumpsite.

For use as a dozer or pusher, the operator must open the bowl fully and use the back of the bucket as a blade. Level cutting is maintained by the bucket lift control. The material can then be pushed into an area to create a stockpile. The multipurpose bucket is useful for roughing-in access roads.

With the added height of the multipurpose bucket, loading trucks becomes easier because the boom can remain higher and stay away from the side of the truck. The clamshell configuration also makes it easier to dump sticky material because the bucket does not compact the material.

3.4.11 Working in Unstable Soils

Working in mud or unstable soils that will not support the forklift can be aggravating and dangerous. This is a problem even for experienced operators.

When entering a soft or wet area, go very slowly. If the front of the machine feels like it is starting to settle, stop and back out immediately. That settling is the first indication that the ground is too soft to support the equipment. The engine will lug slightly and the front end of the forklift will start to settle.

After backing out, examine how deep the wheels or tracks sank into the ground. If they sink deep enough that the material hits the bottom of the machine, the ground is too soft to work in a normal way.

To work in soft or unstable material, consider this approach:

- Start from the edge and work forward slowly.
- Push the mud ahead of the bucket to test the consistency of the soil and be sure the ground below is firm.
- Don't try to move too much material on any one pass.
- Try to keep the wheels or tracks from slipping and digging in.

Partially stable material can also be a hazard because an operator may drive in and out over relatively firm ground many times, while it slowly gets softer because the weight of the wheels or tracks pumps more water to the surface. If this happens, the wheels or tracks will sink a little more each time until the forklift finally gets mired in the hole. Then the machine must be pulled out.

To keep this from happening, do not run in the same track each time entering or leaving an area. Move over slightly in one direction or the other so the same tracks are not pushed deeper into the unstable material each time.

3.0.0 Section Review

1. Which of the following accidents causes 25 percent of forklift fatalities?
   a. Overturning a forklift
   b. Striking a pedestrian with a forklift
   c. Unstable loads falling from the forklift
   d. Exceeding the safe lifting capacity of a forklift

2. If you are asked to drive a forklift and discover the seat belt strap is broken, _____.
   a. improvise by using a rope or other material and drive it
   b. repair it by securely fastening the ends together and drive it
   c. do not operate it and report the problem
   d. drive it as is and report the problem

3. Before making a turn with a forklift, the operator needs to ensure that the _____.
   a. differential lock is engaged
   b. differential lock is disengaged
   c. forks are 3 to 5 inches above the ground
   d. boom is at no more than a 10-degree angle

4. The two common driving patterns for loading a truck using a bucket attachment are the _____.
   a. I- and the Y-patterns
   b. A- and the B-patterns
   c. V- and the Y-patterns
   d. I- and the Z-patterns

SUMMARY

Rough-terrain forklifts are used primarily for lifting and loading material on job sites. The forks are used to lift palletized loads and other bundled material. The forklift can be fitted with several attachments that enable it to lift and place loose or bulky material. These attachments include a lifting hook, a boom extension, a hopper, and a bucket.

Vehicle movement is controlled with the steering wheel, accelerator, and brake pedals. The forks and boom are controlled with either levers or a joystick and switches. Study the operator manual to become familiar with the machine you will be operating.

Safety considerations when operating a forklift include keeping the forklift in good working condition, obeying all safety rules, being aware of other people and equipment in the same area where you are operating, and not taking chances. Perform inspections and maintenance daily to keep the forklift in good working order. One of the primary safety considerations for forklift operations is the rated capacity of the forklift. Know the weight of the load and the capacity of the machine. You must be able to read and interpret a load chart. Exceeding the rated capacity of the machine can cause injury and death.

Always position the forklift so that it is square and level to the load. Pick a load up slowly and tilt the forks backward to cradle the load. Lower the load before traveling to the unloading area. Use extra caution when picking up elevated loads.

Review Questions

1. Two-wheel drive forklifts usually use the _____.
   a. rear wheels to both drive and steer
   b. front wheels to both drive and steer
   c. rear wheels to drive and the front wheels to steer
   d. front wheels to drive and the rear wheels to steer

2. The steering mode that allows all four wheels to turn and point in the same direction at the same time is called _____.
   a. articulated steering
   b. dynamic steering
   c. oblique steering
   d. circle steering

3. A hydrostatic transmission provides _____.
   a. only manual shifting, using a clutch
   b. highway speeds for a rough-terrain forklift
   c. more reverse gears than a common transmission
   d. infinitely variable speeds

4. A forklift fitted with an F-N-R control typically has a _____.
   a. squirt boom
   b. pivoting boom
   c. set of outriggers
   d. hydrostatic transmission

5. With a joystick in use, the operator can make multiple movements occur at the same time, such as raising the boom and tipping the forks.
   a. True
   b. False

6. The typical means of adjusting the length of a boom extension for lifting is _____.
   a. manually
   b. electrically
   c. hydraulically
   d. engine power take-off

7. Per the chart shown in *Figure 1*, the lifting capacity of the FB-40 boom attachment at an 11-foot extension is _____.
   a. 1,500 pounds
   b. 1,600 pounds
   c. 3,000 pounds
   d. 4,800 pounds

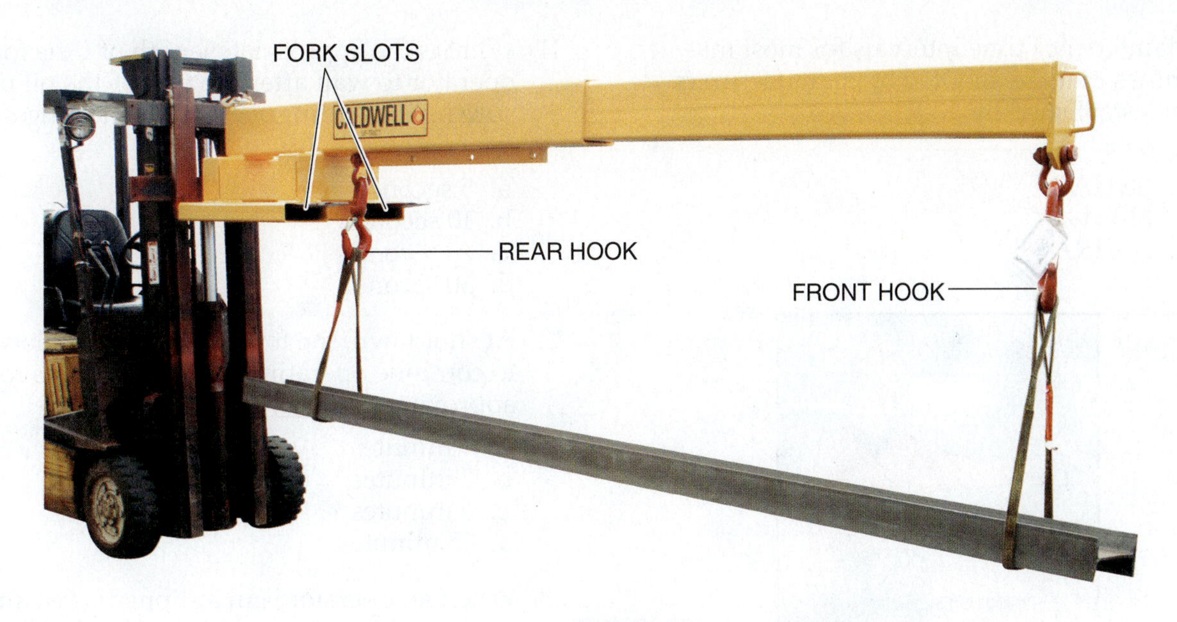

MODEL FB
FIXED TYPE FORK LIFT BOOM

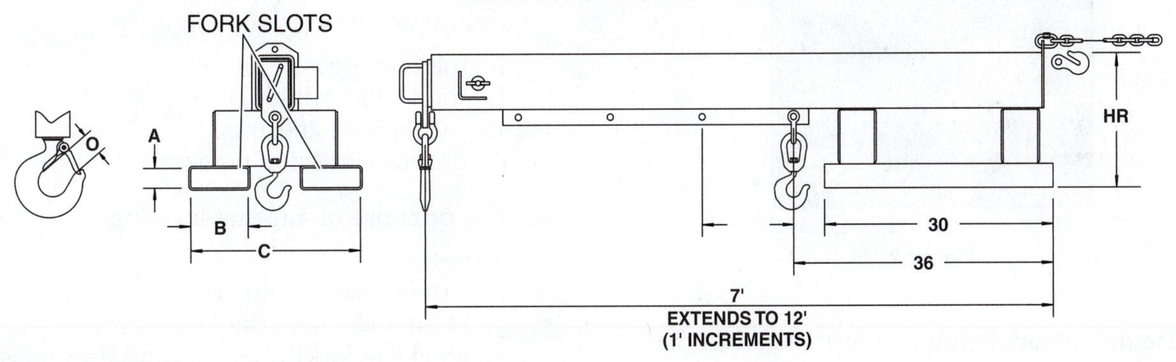

NOTE: All dimensions on drawings shown in inches unless stated.

SPECIFICATIONS

Model Number	Dimensions (in.)					Maximum Capacity at Hook Position (lbs.)							Weight (lbs.)
	A	B	C	HR	0	3'–6'	7'	8'	9'	10'	11'	12'	
FB–30	2½	7½	22	16	1.00	3,000	3,000	2,600	2,200	1,900	1,600	1,500	340
FB–40				16	1.09	4,000	3,200	2,600	2,200	1,900	1,600	1,500	340
FB–60	2½	7½	22	17	1.36	6,000	5,000	4,200	3,500	3,000	2,700	2,500	390
FB–80				18	1.61	8,000	7,000	5,700	4,800	4,100	3,600	3,100	520

Capacities are for boom only and are dependent on fork lift used. Check with forklift manufacturer for load capacities before use.

22206-12_RQ01.EPS

Figure 1

8. Maintenance time intervals for most machines that are adopted by manufacturers are established by _____.
   a. SAE
   b. OSHA
   c. NAMA
   d. ASHRAE

22206-12_RQ02.EPS

Figure 2

9. The movement being communicated to the operator in *Figure 2* is _____.
   a. lower the forks
   b. retract the boom
   c. tilt the mast forward
   d. raise one or more outriggers

10. When the machine is equipped with a ROPS and was built after 2008, the seat belt is not necessary.
    a. True
    b. False

11. What is the appropriate length of time for an operator to wait after startup for the oil pressure light to extinguish before shutting down the engine?
    a. 5 seconds
    b. 10 seconds
    c. 20 seconds
    d. 30 seconds

12. At shutdown, the forklift should be allowed to continue operating to cool the fluids for approximately _____.
    a. 1 minute
    b. 2 minutes
    c. 5 minutes
    d. 15 minutes

13. When an operator is in a slippery area and one wheel is spinning while the other is motionless, the component or feature that should be engaged to help is _____.
    a. the outriggers
    b. the differential lock
    c. two-wheel steering
    d. the transmission neutralizer

14. The purpose of a frame-leveling control is to _____.
    a. extend the outriggers
    b. raise and lower the outriggers
    c. level the forklift by rotating the chassis on its wheels and axles
    d. automatically sense an out-of-level condition and adjust the outriggers

15. If the forklift's rear wheels start to lift off the ground with a load, _____.
    a. stop raising the load and slowly lower it
    b. raise the outriggers higher on both sides
    c. drop the load to the ground immediately
    d. back a little farther away from the load and try again

Trade Terms Introduced in This Module

Crab steering: A steering mode where all wheels may move in the same direction, allowing the machine to move sideways on a diagonal; also known as oblique steering.

Four-wheel steering: A steering mode where the front and rear wheels may move in opposite directions, allowing for very tight turns; also known as independent steering or circle steering.

Fulcrum: A point or structure on which a lever sits and pivots.

Oblique steering: A steering mode where all wheels may move in the same direction, allowing the machine to move sideways on a diagonal; also known as crab steering.

Powered industrial trucks: An OSHA term for several types of light equipment that include forklifts.

Power hop: Action in heavy equipment that uses pneumatic tires to create a bouncing motion between the fore and aft axles. Once started, the oscillation back and forth usually continues until the operator either stops or slows down significantly to change the dynamics.

Telehandler: A type of powered industrial truck characterized by a boom with several extendable sections known as a telescoping boom; another name for a shooting boom forklift.

Tines: A prong of an implement such as a fork. For forklifts, tines are often called forks.

Appendix A

Typical Periodic Maintenance Requirements

Maintenance Interval Schedule

Note: All safety information, warnings, and instructions must be read and understood before you perform any operation or maintenance procedure.

Before each consecutive interval is performed, all of the maintenance requirements from the previous interval must also be performed.

The normal oil change interval is every 500 service hours. If you operate the engine under severe conditions or if the oil is not Caterpillar oil, the oil must be changed at shorter intervals. Refer to the Operation and Maintenance Manual, "Engine Oil and Filter – Change" for further information. Severe conditions include the following factors: high temperatures, continuous high loads, and extremely dusty conditions.

Refer to the Operation and Maintenance Manual, "S.O.S. Oil Analysis" in order to determine if the oil change interval should be decreased. Refer to your Caterpillar dealer for detailed information regarding the optimum oil change interval.

The normal interval for inspecting and adjusting the clearance between the wear pads and the boom is 500 service hours. If the machine is working with excessively abrasive material then the clearance may need to be adjusted at shorter intervals. Refer to the Operation and Maintenance Manual, "Boom Wear Pad Clearance – Inspect/Adjust" for further information.

When Required

Battery – Recycle
Battery or Battery Cable – Inspect/Replace
Boom Telescoping Cylinder Air – Purge
Boom and Frame – Inspect
Engine Air Filter Primary Element – Clean/Replace
Engine Air Filter Secondary Element – Replace
Engine Air Filter Service Indicator – Inspect
Engine Air Precleaner – Clean
Fuel System – Prime
Fuel System Primary Filter – Replace
Fuel System Secondary Filter – Replace
Fuel Tank Cap and Strainer – Clean
Fuses and Relays – Replace
Oil Filter – Inspect
Radiator Core – Clean
Radiator Screen – Clean
Transmission Neutralizer Pressure Switch – Adjust
Window Washer Reservoir – Fill
Window Wiper – Inspect/Replace

Every 10 Service Hours or Daily

Backup Alarm – Test
Boom Retracting and Boom Lowering with Electric Power – Check
Braking System – Test
Cooling System Coolant Level – Check
Cooling System Pressure Cap – Clean/Replace
Engine Oil Level – Check
Fuel System Water Separator – Drain
Fuel Tank Water and Sediment – Drain
Indicators and Gauges – Test
Seat Belt – Inspect
Tire Inflation – Check
Transmission Oil Level – Check
Wheel Nut Torque – Check
Windows – Clean

Every 50 Service Hours or 2 Weeks

Axle Support – Lubricate
Bearing (Pivot) for Axle Drive Shaft – Lubricate
Boom Cylinder Pin – Lubricate
Boom Pivot Shaft – Lubricate
Brake Control Linkage – Lubricate
Carriage Cylinder Bearing – Lubricate
Carriage Pivot Pin - Lubricate
Compensating Cylinder Bearing – Lubricate
Cylinder Pin (Grapple Bucket) – Lubricate
Cylinder Pin and Pivot Pin (Bale Handler) – Lubricate
Cylinder Pin and Pivot Pin (Multipurpose Bucket) – Lubricate
Cylinder Pin and Pivot Pin (Utility Fork) – Lubricate
Fork Leveling Cylinder Pin – Lubricate
Frame Leveling Cylinder Pin – Lubricate
Pulley for Boom Extension Chain – Lubricate
Pulley for Boom Retraction Chain – Lubricate
Quick Coupler – Lubricate
Stabilizer and Cylinder Bearings – Lubricate

Initial 250 Service Hours (or after rebuild)

Boom Wear Pad Clearance – Inspect/Adjust
Service Brake – Adjust

Initial 250 Service Hours (or at first oil change)

Hydraulic System Oil Filter – Replace

22206-12_A01.EPS

Every 250 Service Hours or 3 Months

Axle Breathers – Clean/Replace
Belts – Inspect/Adjust/Replace
Boom Chain Tension – Check/Adjust
Differential Oil Level – Check
Drive Shaft Spline – Lubricate
Drive Shaft Universal Joint – Lubricate
Engine Oil Sample – Obtain
Final Drive Oil Level – Check
Hydraulic System Oil Level – Check
Longitudinal Stability Indicator – Test
Transfer Gear Oil Level – Check
Transmission Breather – Clean

Every 500 Service Hours

Differential and Final Drive Oil Sample – Obtain

Every 500 Service Hours or 6 Months

Boom Wear Pad Clearance – Inspect/Adjust
Engine Oil and Filter – Change
Fuel System Primary Filter – Replace
Fuel System Secondary Filter – Replace
Fuel Tank Cap and Strainer – Clean
Hydraulic System Oil Filter – Replace
Hydraulic System Oil Sample – Obtain
Hydraulic Tank Breather – Clean
Service Brake – Adjust
Transmission Oil Sample – Obtain

Every 1,000 Service Hours or 1 Year

Differential Oil – Change
Engine Valve Lash – Check
Final Drive Oil – Change
Rollover Protective Structure (ROPS) and Falling Object Protective Structure (FOPS) – Inspect
Transfer Gear Oil – Change
Transmission Oil – Change
Transmission Oil Filter – Replace

Every 2,000 Service Hours or 2 Years

Fuel Injection Timing – Check
Hydraulic System Oil – Change

Every 3 Years After Date of Installation or Every 5 Years After Date of Manufacture

Seal Belt – Replace

Every 3,000 Service Hours or 3 Years

Boom Chain – Inspect/Lubricate
Cooling System Coolant Extender (ELC) – Add
Cooling System Water Temperature Regulator – Replace
Engine Mounts – Inspect

Every 6,000 Service Hours or 6 Years

Cooling System Coolant (ELC) – Change

22206-12_A02.EPS

Appendix B

OSHA Inspection Checklist

Operator's Daily Checklist – Internal Combustion Engine Industrial Truck – Gas/LPG/Diesel Truck

Record of Fuel Added

Date		Operator		Fuel	
Truck #		Model #		Engine Oil	
Department		Serial #		Radiator Coolant	
Shift		Hour Meter		Hydraulic Oil	

SAFETY AND OPERATIONAL CHECKS (PRIOR TO EACH SHIFT)
Have a **qualified** mechanic correct all problems.

Engine Off Checks	OK	Maintenance
Leaks – Fuel, Hydraulic Oil, Engine Oil or Radiator Coolant		
Tires – Condition and Pressure		
Forks, Top Clip Retaining Pin and Heel – Check Condition		
Load Backrest – Securely Attached		
Hydraulic Hoses, Mast Chains, Cables and Stops – Check Visually		
Overhead Guard – Attached		
Finger Guards – Attached		
Propane Tank (LP Gas Truck) – Rust Corrosion, Damage		
Safety Warnings – Attached (Refer to Parts Manual for Location)		
Battery – Check Water/Electrolyte Level and Charge		
All Engine Belts – Check Visually		
Hydraulic Fluid Level – Check Level		
Engine Oil Level – Dipstick		
Transmission Fluid Level – Dipstick		
Engine Air Cleaner – Squeeze Rubber Dirt Trap or Check the Restriction Alarm (if equipped)		
Fuel Sedimentor (Diesel)		
Radiator Coolant – Check Level		
Operator's Manual – In Container		
Nameplate – Attached and Information Matches Model, Serial Number and Attachments		
Seat Belt – Functioning Smoothly		
Hood Latch – Adjusted and Securely Fastened		
Brake Fluid – Check Level		
Engine On Checks – Unusual Noises Must Be Investigated Immediately	OK	Maintenance
Accelerator or Direction Control Pedal – Functioning Smoothly		
Service Brake – Functioning Smoothly		
Parking Brake – Functioning Smoothly		
Steering Operation – Functioning Smoothly		
Drive Control – Forward/Reverse – Functioning Smoothly		
Tilt Control – Forward and Back – Functioning Smoothly		
Hoist and Lowering Control – Functioning Smoothly		
Attachment Control – Operation		
Horn and Lights – Functioning		
Cab (if equipped) – Heater, Defroster, Wipers – Functioning		
Gauges: Ammeter, Engine Oil Pressure, Hour Meter, Fuel Level, Temperature, Instrument, Monitors – Functioning		

22206-12_A03.EPS

Figure Credits

Courtesy of JLG Industries, Inc., Module opener, Figures 1B, 5–10, 12, 19, 22, 23, and 37

Courtesy of Sellick Equipment Limited, Figures 1A, 3, and 4

Courtesy of CLARK Material Handling Co., SA01

Courtesy of Terex Aerial Work Platforms, SA02

Reprinted courtesy of Caterpillar Inc., Figures 11, 13–16, 18, 24–26, 28–30, 32, 33, 35, 36, 38, 43–48, and Appendix A

Topaz Publications, Inc., Figure 17

Courtesy of The Caldwell Group, Inc., Figure 20

Vestil Manufacturing Corporation, Figure 21

Courtesy of Gar-Bro Manufacturing Co., Figure 27

US Department of Labor, Appendix B

Section Review Answers

Answer	Section Reference	Objectives
Section One		
1 d	1.1.1	1a
2 c	1.2.3	1b
3 b	1.3.5	1c
4 b	1.4.1	1d
Section Two		
1 b	2.1.0	2a
2 a	2.2.1	2b
Section Three		
1 a	3.1.4	3a
2 c	3.2.1	3b
3 b	3.3.3	3c
4 a	3.4.9	3d

NCCER CURRICULA — USER UPDATE

NCCER makes every effort to keep its textbooks up-to-date and free of technical errors. We appreciate your help in this process. If you find an error, a typographical mistake, or an inaccuracy in NCCER's curricula, please fill out this form (or a photocopy), or complete the online form at **www.nccer.org/olf**. Be sure to include the exact module ID number, page number, a detailed description, and your recommended correction. Your input will be brought to the attention of the Authoring Team. Thank you for your assistance.

Instructors – If you have an idea for improving this textbook, or have found that additional materials were necessary to teach this module effectively, please let us know so that we may present your suggestions to the Authoring Team.

NCCER Product Development and Revision

13614 Progress Blvd., Alachua, FL 32615

Email: curriculum@nccer.org
Online: www.nccer.org/olf

❑ Trainee Guide ❑ AIG ❑ Exam ❑ PowerPoints Other ______________________________

Craft / Level: Copyright Date:

Module ID Number / Title:

Section Number(s):

Description:

Recommended Correction:

Your Name:

Address:

Email: Phone:

22202-13

On-Road Dump Trucks

Overview

A skilled operator can recognize several types of dump trucks and choose the best one for the task at hand. Safe operators have a working knowledge of all of the instruments and controls, perform daily inspections and maintenance, and know the rules of the road for both public roads and on-site haul roads.

Module Two

Trainees with successful module completions may be eligible for credentialing through NCCER's National Registry. To learn more, go to **www.nccer.org** or contact us at **1.888.622.3720**. Our website has information on the latest product releases and training, as well as online versions of our *Cornerstone* newsletter and Pearson's product catalog.

Your feedback is welcome. You may email your comments to **curriculum@nccer.org**, send general comments and inquiries to **info@nccer.org**, or fill in the User Update form at the back of this module.

This information is general in nature and intended for training purposes only. Actual performance of activities described in this manual requires compliance with all applicable operating, service, maintenance, and safety procedures under the direction of qualified personnel. References in this manual to patented or proprietary devices do not constitute a recommendation of their use.

22202-13
On-Road Dump Trucks

Objectives

When you have completed this module, you will be able to do the following:

1. Identify the types of on-road dump trucks.
   a. Identify and describe standard dump trucks.
   b. Identify and describe special dump trucks and trailers.
2. Identify and describe the instruments and specialized control systems found on an on-road dump truck.
   a. Identify and describe instruments.
   b. Identify and describe control systems.
3. Describe the operator inspection and maintenance requirements for an on-road dump truck.
   a. Describe inspection, startup, and shutdown procedures.
   b. Identify preventive maintenance procedures that must be performed.
4. Describe safe on-road driving practices for on-road dump trucks.
   a. State the normal driving practices associated with dump truck operation.
   b. Describe how to handle a dump truck in an emergency.
5. Describe the procedures for operating a dump truck on the job.
   a. State the safety practices associated with dump truck operation on a job site.
   b. Describe proper loading, dumping, and snow plowing procedures.

Performance Tasks

Under the supervision of your instructor, you should be able to do the following:

1. Complete proper prestart inspection and maintenance for a dump truck.
2. Perform the proper startup, warm-up, and shutdown procedures on a dump truck.
3. Carry out basic operations with a dump truck:
   - Dump a load in a designated spot, and tailgate-spread the load.
   - Back up with a trailer attached.
   - Perform tailgate adjustment, as applicable.

Trade Terms

Auxiliary axle
Cab guard
Clutch
Engine retarder (engine brake)
Governor
Hoist
Lug
Tag axle
Tandem-axle

Industry Recognized Credentials

If you are training through an NCCER-accredited sponsor, you may be eligible for credentials from NCCER's Registry. The ID number for this module is 22202-13. Note that this module may have been used in other NCCER curricula and may apply to other level completions. Contact NCCER's Registry at 888.622.3720 or go to **www.nccer.org** for more information.

Contents

Topics to be presented in this module include:

Figures

SECTION ONE

1.0.0 TYPES OF ON-ROAD DUMP TRUCKS

Objective 1

Identify the types of on-road dump trucks.

a. Identify and describe standard dump trucks.
b. Identify and describe special dump trucks.

Trade Terms

Auxiliary axle: An additional axle that is mounted behind or in front of the truck's drive axles and is used to increase the safe weight capacity of the truck.

Cab guard: Protects the truck cab from falling rocks and load shift.

Hoist: Mechanism used to raise and lower the dump bed.

Tandem-axle: Usually a double-axle drive unit.

Dump trucks are among the most widely used vehicles in construction. They are built on a heavy-duty chassis and have a large bed that is used to move large amounts of loose material, such as sand, soil, gravel, or asphalt. Dump trucks can be quickly loaded using a chute or heavy equipment, such as a loader or excavator (*Figure 1*). The beds can be raised and tilted back with a **hoist** that is mounted on the truck, so the driver can quickly unload the cargo without help.

Dump trucks are divided into two major categories: on-road trucks and off-road trucks. This module covers on-road dump trucks. Off-road trucks are covered in another module. On-road dump trucks vary in size and load capacity and may be operated on public roads. Off-road trucks are usually much larger than on-road vehicles and are prohibited by law from routine operation on public roads because of their great size and weight.

Dump trucks, like all motor vehicles, are rated by the amount of weight their axles may safely carry. The truck's weight capacity can be found in the operator manual. It is the driver's responsibility to ensure that the maximum load limit for the vehicle is not exceeded. Operating an overweight vehicle can damage the vehicle and increase the risk of an accident. Moreover, states

22202-12_F01.EPS

Figure 1 Excavator loading a dump truck.

have regulations governing the maximum loads for highways and bridges. Drivers who violate those regulations are subject to fines. State and federal load limits vary, and may not be same as the truck's load limit. Always observe the lowest limit.

Roads and bridges are rated for the weight they can safely support. As trucks became heavier, the federal government introduced the Federal Bridge Gross Weight Formula (also called the Federal Bridge formula, or simply bridge formula) to protect the nation's bridges and roadways from damage due to excessive weight. The bridge formula specifies the maximum weight that each axle on a truck can legally carry. Although the federal government sets the maximum weight, it allows each state to make its own laws, which may be even stricter than the federal standards.

Weigh Stations

Anyone who travels on the Interstate Highway System has seen the weigh stations at intervals along the highway. Commercial vehicles using the highways are required to pass through these stations. The more modern weigh stations can record the weight while the vehicle is moving through the station. Others require the vehicle to stop on a scale. The primary purpose of these stations is to collect road use taxes, which are based on vehicle weight. Another purpose is to determine if any vehicle exceeds the maximum weight specified by the state for the type of vehicle. In addition to these permanent weigh stations, police and Department of Transportation (DOT) agents also have portable scales that can be set up at any location desired.

It is therefore important to know the laws for the states in which you work.

> **WARNING!**
> Never operate an overweight dump truck. Not only can it damage the vehicle, but it is also a safety hazard. Carrying too heavy a load negatively affects the maneuverability of the truck and stresses the vehicle's brakes, suspension, axles, and tires. If you do not know the weight limits, check the vehicle's operator manual.

The on-road dump truck designation does not mean that all trucks of this type are operated on public roads, since on-road dump trucks may also be driven off-road. The on-road designation means that the truck falls within the legal size limits for vehicles driven on public roads and has the safety equipment required for such service. When on-road vehicles are driven on public roads, they must be in good working order and follow all local laws and regulations. This includes proper licensing and vehicle registration, state and local safety inspections, and traffic laws. Before driving a vehicle on a public road, make sure the vehicle is roadworthy and that the proper documentation is available to prove that both the driver and the vehicle are permitted to be on the road.

When a dump truck is driven on a public road—even for a short distance—the operator must have a commercial driver's license (CDL). The CDL program sets forth minimum national standards that drivers of commercial vehicles must meet. States may choose to add to those standards, so it is the driver's job to know and follow local laws. When a truck is taken onto a public road, it is the driver's responsibility to ensure that the vehicle meets all state and local requirements. The driver must be licensed and trained to operate the vehicle safely. If you are stopped by the police or are involved in an accident, you will be held responsible for illegal operation of a vehicle on a public road.

> ## Size Matters
> In 1982, the federal government increased the maximum width of vehicles traveling on public roads from 96 inches to 102 inches (8' to 8'-6"). Some states still limit the legal width of on-road vehicles to 96 inches. The legal length of on-road vehicles varies greatly from state to state.

1.1.0 Standard On-Road Dump Trucks

Standard dump trucks are the most commonly used dump trucks. These vehicles are used to haul light, medium, and heavy loads such as soil, stone, and gravel from one site to another. One common characteristic on all standard dump trucks have in common is rigid-frame construction. This means that the cab and the dump body are mounted on a common frame and operate together.

Figure 2 shows a typical dump truck. Dump trucks are generally equipped with a load cover that is used to prevent debris from flying off the load while the truck is moving. Using the cover is especially important when the truck is driven at highway speeds. It is required by law in most states. Standard rigid-frame dump trucks are equipped with a **cab guard**, which protects the operator from falling objects during loading.

The key difference in dump trucks of different sizes is the number of axles (*Figure 3*). Increasing the number of axles increases the amount of weight that a truck can safely carry. However, the truck is not permitted to exceed the maximum load allowed on a road or bridge, regardless of the number of axles.

Standard dump trucks include the following rear axle configurations:

- *Single-axle* – Single-axle trucks have one rear axle and are used to haul light loads. They normally have six wheels—two on the front axle and four on the rear axle.
- *Dual-axle* – Dual-axle trucks have two rear axles and are used to haul heavy loads. They normally have 10 wheels—two on the front axle and eight on the rear axles. (Some states refer to any truck

Figure 2 Typical dump truck.

(A)

(B)

22202-12_F03.EPS

Figure 3 Dump truck axles.

with dual-axle or above as **tandem-axle** trucks, while some refer to a tandem-axle as a dual-axle only.)

- *Tri-axle* – Tri-axle trucks have three rear axles and are used to haul heavier and larger loads than dual-axle vehicles. They normally have 12 wheels—two on the front axle and ten on the rear axles.
- *Quad-axle* – Quad-axle trucks have four rear axles that are used to haul the heaviest and largest loads. These trucks are rarely used.

A close look at *Figure 3B* shows that some of the tires at the rear of the truck are not touching the ground. These axles are called **auxiliary axles**, or secondary axles. (Primary axles are drive axles with the tires always touching the ground.) Auxiliary axles, also known as lift axles or airlift axles, are lowered manually or by a drive system that is operated from the cab. Auxiliary axles are used to distribute the weight when the truck is carrying a load. When the tires on the auxiliary axle are touching the ground, they support part of the weight of the load, thus easing the strain on the other axles and permitting the truck to carry more weight than it can with just the primary axles. The extra axles also distribute the weight, which can allow the truck to meet the weight-per-axle restrictions imposed by the Federal Bridge Formula.

1.2.0 Special Dump Trucks and Trailers

In addition to the standard dump truck, there are dump trucks and trailers made for special hauling situations. The following sections describe some of these trucks and trailers.

1.2.1 Transfer Dump Truck

Transfer dump trucks (*Figure 4*) standard dump trucks that tow a second dump bed on a trailer. When the bed of the standard dump truck is empty, the dump bed on the trailer is slid into the bed of the truck and then emptied using the truck's hoist mechanism.

1.2.2 Pup Trailers

Pup trailers are similar to the trailers used in transfer dump trucks. See *Figure 5*. The chief difference is that the trailer has its own hoist mechanism, so it can be emptied without using a standard dump truck.

1.2.3 Bottom Dump Trucks

Bottom (belly) dump trucks (*Figure 6*) are trailers that are towed behind a truck tractor. As the name implies, the load is dumped through doors in the bottom of the trailer. These trailers are used when the load needs to be dumped in a row. To dump the load, the driver drives forward while opening the bottom doors and the load is deposited in a row. A good use for this type of dumper would be to dump materials, including asphalt, gravel, and dirt, in a row on a roadbed. The material handler can follow the trailer to quickly spread and compact the asphalt.

1.2.4 Side Dump Truck

Side dump trucks are trailers that are towed behind a truck tractor. This type of trailer dumps its load by tilting the trailer body on its side. A side dumper can only be used when the dumpsite is long enough to permit the trailer access to it. The advantage to this truck is that it can be unloaded quickly.

22202-12_F04.EPS

Figure 4 Dump truck with transfer trailer.

22202-12_F05.EPS

Figure 5 Pup trailer.

22202-12_F06.EPS

Figure 6 Bottom dump truck.

Roll-Off Dump Trucks

Some dump trucks are designed to leave their dump body at the job site and return to pick it up when it has been filled. Once reattached to the truck, the bed can be raised and dumped like that of any other dump truck. Trucks like this are convenient for demolition sites.

22202-12_SA01.EPS

GOING GREEN

Diesel Engine Emission Controls

Diesel engines provide the power and efficiency needed to drive heavy equipment. A major drawback of past diesel engines has been the level of soot and nitrous oxide (NO_x) emitted through their exhaust. In order to reduce these emissions, the federal government's Clean Air Act of 1990 included regulations for reducing diesel engine emissions. The regulations were phased-in over time. One early element of these regulations was the requirement for diesel engines to use ultra-low sulfur diesel fuel (USDF). On-road vehicles manufactured since 2010 are required to meet Tier IV exhaust emission standards by using and by implementing a system for treating the engine exhaust. Most engine manufacturers adopted the selective catalytic reduction (SCR) exhaust treatment system. This system injects a solution known as diesel emission fluid (DEF) into the engine exhaust system. DEF is an ammonia-based solution known as urea. The urea is mixed with air in a mixing valve before being injected. Urea is stored in a tank with a capacity of 10 to 30 gallons. The photo below shows a DEF tank on a Tier IV-compliant dump truck.

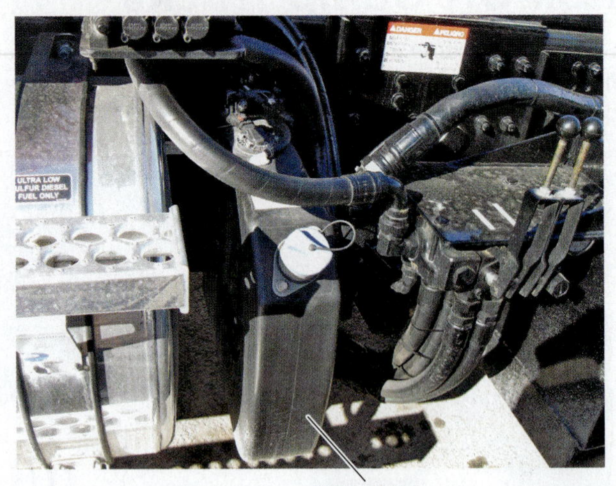

22202-12_SA02.EPS

Additional Resources

Truck Driver's Guide to CDL, First Edition. New York, NY: Prentice Hall Press.

1.0.0 Section Review

1. A dual-axle dump truck typically has _____.
   a. 4 wheels
   b. 6 wheels
   c. 8 wheels
   d. 10 wheels

2. The truck that tows a second dump bed that is slid onto the empty bed of the truck for dumping is a _____.
   a. transfer dump truck
   b. pup trailer
   c. bottom dumper
   d. side dumper

SECTION TWO

2.0.0 INSTRUMENTS AND CONTROL SYSTEMS

Objective 2

Identify and describe the instruments and specialized control systems found on an on-road dump truck.

a. Identify and describe instruments.
b. Identify and describe control systems.

Trade Terms

Clutch: Device used to disengage the transmission from the engine.

Engine retarder (Engine brake): An alternate braking system activated from the cab that slows down the vehicle by reducing engine power.

Governor: Device for automatic control of speed, pressure, or temperature.

Tag axle: Auxiliary axle that is mounted behind the truck's drive wheels. It may be called a pusher axle if it is placed in front of the drive axle.

Dump trucks are designed so the driver can perform most duties from the cab. A typical cab configuration is shown in *Figure 7*. The layout and operation of the vehicle's instruments and controls vary among manufacturers, models, and uses of the truck. Many instruments and controls are optional, so they are not on all trucks. Since each truck is slightly different, the operator must study the manual for the particular vehicle. This should be the first step when assigned to a new truck. The operator manual should be considered part of the equipment and of no less importance than other components such as the seat belt.

2.1.0 Instruments

The instrument panel of a dump truck (*Figure 8*) contains many of the same instruments found in a car, including an odometer, tachometer, and various temperature and pressure gauges. The panel to the driver's right (*Figure 9*) contains instruments and controls that are unique to the dump truck. Of special note are the trailer brake control and parking brake control. Keep in mind that the arrangement shown is specific to one truck model. Other trucks contain many, if not all, of the same instruments and controls, but the

22202-12_F07.EPS

Figure 7 Example of a cab layout.

22202-12_F08.EPS

Figure 8 Example of a dump truck instrument panel.

22202-12_F09.EPS

Figure 9 Truck controls and indicators.

arrangement is likely to be different. Specialized instruments that may be found on dump trucks include the following:

- *Transmission temperature gauge* – The transmission temperature gauge (*Figure 10*) indicates the temperature of the transmission oil and can alert the driver when the transmission is overheating.
- *Brake system air pressure gauge* – The brake system air pressure gauge indicates the pressure level in the air brake system. If the system pressure falls below the normal level, a light on the gauge provides an indication to the driver. An audible alarm also sounds.
- *Air filter status gauge (not shown)* – The air filter status gauge provides an indication of the air filter condition. It indicates in the red area if the air filter becomes blocked and needs to be replaced.
- *Auxiliary axle indicator* – Auxiliary axle gauges show the status of the auxiliary axles.

NOTE

Warning lights are used instead of gauges in some trucks.

2.2.0 Control Systems

Because of its size and the nature of the work it is used for, every dump truck has specialized control systems that are needed for safe and efficient operation of the vehicle. The controls for these systems are located in the cab of the truck. Most of them are on the driver's instrument panel. *Figure 11* shows the controls and indicators related to the vehicle control systems, which are described in the sections that follow.

2.2.1 Differential Lock

Many dump trucks that are operated in snowy or muddy work environments are equipped with a differential lock feature. To understand a differential lock, it is necessary to understand the function of a differential.

A differential is part of a drive axle assembly. It permits the wheels on the axle to turn at two different speeds and allows vehicles to turn corners smoothly. That is, the outside wheel turns faster than the inside one, because the outside wheel needs to cover more distance than the inside wheel in the same time period. When the drive wheels of a vehicle are stuck on ice or in snow or mud with one wheel spinning and the other not moving, that is a differential at work.

The differential lock is used to lock the differential so that both wheels turn at the same speed. This allows the truck to move as long as one of the drive wheels has some traction. However, since the drive wheels will be locked, normal steering control of the vehicle is limited. Before using this feature, ensure that the wheels on the steering axle are pointing forward.

WARNING!

Never attempt normal operation of a vehicle with differential lock engaged. It can cause an accident and will damage driveline components.

2.2.2 Power Takeoff

Some dump trucks are equipped with a power takeoff (PTO) unit. A PTO unit is a mechanical link to an engine or transmission to which a

22202-12_F10.EPS

Figure 10 Close-up of driver gauges.

22202-12_F11.EPS

Figure 11 Truck system controls.

cable, belt, or shaft may be connected in order to power another device. PTO units are usually located on the front of the truck. Sometimes they are used to control the hydraulics for operating the dump bed hoist. The control for the PTO is inside the cab, either on the dashboard or on the floorboard. Operation is different depending on the type of transmission (standard or automatic) and the truck model. Refer to the manufacturer's instructions for details about PTO operation.

2.2.3 Auxiliary Axles

Auxiliary axles (*Figure 12*) are sometimes called secondary axles and are used only when the truck is loaded. They permit the truck to safely carry more weight than the vehicle would without the auxiliary axles by dividing the load weight over more axles. In dump trucks, these axles are usually mounted at the back of the truck, either in front of or behind the drive wheels. When the axle is in front of the drive wheels, it is called an auxiliary or pusher axle. When it is behind the drive wheels, it is called a **tag axle**.

Most auxiliary axles are pneumatically raised or lowered using a switch located in the cab. When the truck is equipped with auxiliary axles, it is the driver's responsibility to know how and when to use them; improper use cannot only damage the vehicle but can also cause an accident. Remember, in the eyes of the law, the truck operator is responsible for the safe and proper operation of the vehicle.

When a dump truck is empty, the operator can raise these axles to improve the vehicle's fuel mileage and maneuverability as well as to reduce toll expenses, since many toll roads base fees on the number of axles on the road. When the vehicle is loaded, auxiliary axles are lowered to spread the load weight over more axles, thus reducing the stress on the other axles and the truck's suspension.

22202-12_F12.EPS

Figure 12 Auxiliary axle.

CAUTION

Never lower the auxiliary axle when the vehicle is traveling at higher than the manufacturer's recommended speed. This action can severely damage the auxiliary axle tires.

2.2.4 Air Brake System

Most large trucks use air brake systems. The air brake system is a separate endorsement on the CDL test. If you have not taken the air brake part of the test, do not operate a vehicle with air brakes on a public road. An air braking system is very different from the brakes systems used in a personal vehicle. The personal vehicle probably uses hydraulic brakes to create stopping power, but trucks use air pressure to create stopping power.

In an air brake system (*Figure 13*), an air compressor powered by the truck's engine supplies air to the reservoir tanks. The reservoir has a safety relief valve and the air compressor has a **governor**, so the pressure from either one cannot get too high. The compressed air is stored in the reservoir tanks until the driver pushes on the brake pedal. Then the air is released into the brake chambers where it creates a mechanical force to slow the turning of the wheels.

The normal pressure in the reservoir tanks is usually between 60 and 100 pounds per square inch (psi). The driver must monitor the air pressure gauge while driving to be sure there is adequate pressure at all times. Many trucks have an alarm or buzzer that will sound when the air pressure is low. Make a habit of glancing at the air pressure gauge to check the reading while driving and after applying the brakes.

Air brake systems must be treated differently than other braking systems. First, it takes time for the air to move from the reservoir tanks to the brake chambers. Stopping distances are therefore much longer than they are with other systems. Second, frequent use of the brakes drains the air from the reservoir, reducing stopping power, so avoid pumping the brakes or driving with your foot resting on the brake. Third, the reservoirs in air brake systems gather condensation that can take up room in the tank, displacing air and lowering the availability of compressed air. Further, the water in the tank can freeze and damage the brake system, so the air reservoirs need to be drained periodically, usually after each shift.

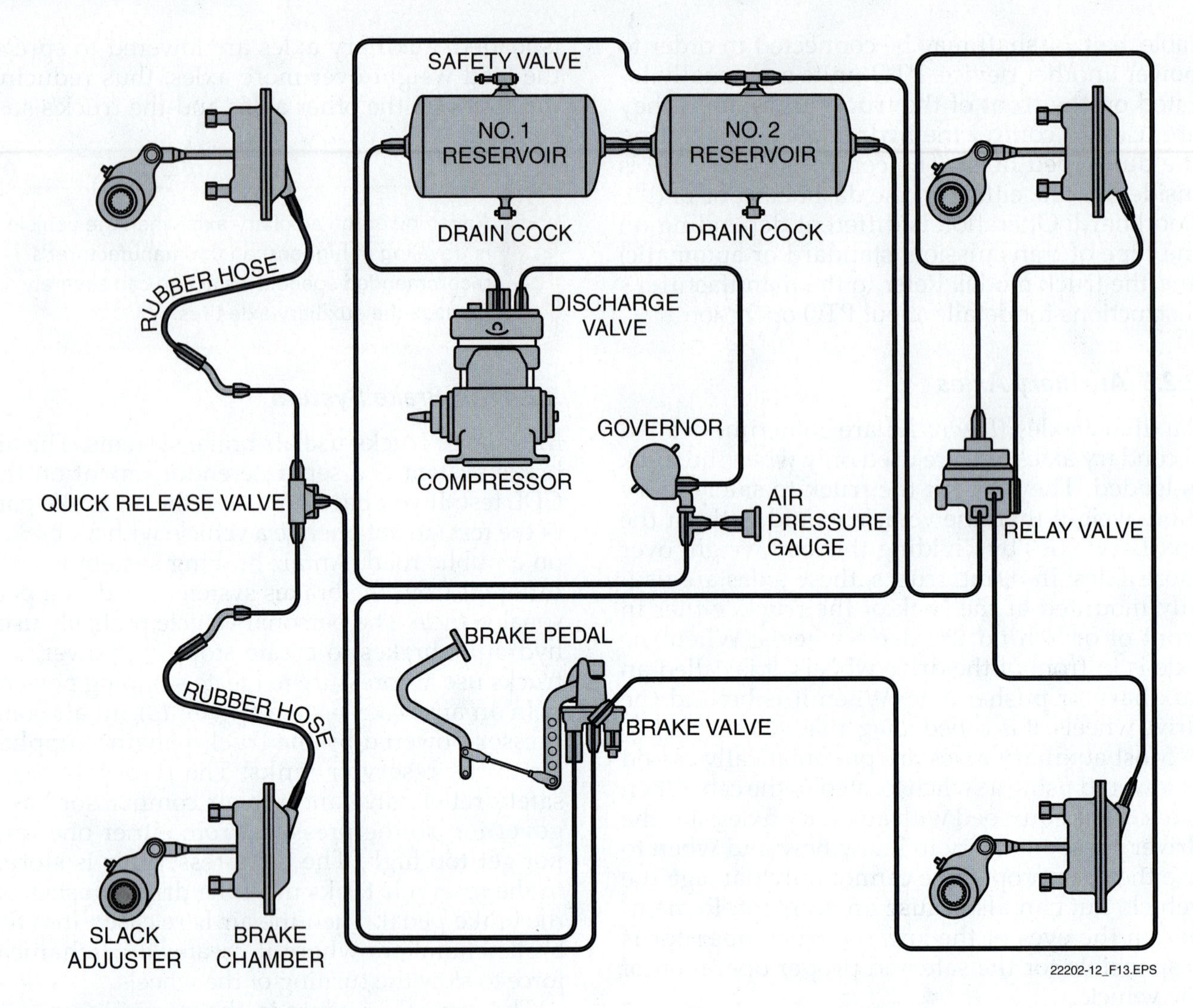

Figure 13 Air brake system diagram.

> **NOTE**
> Your instructor will help you learn how to use the air brakes in your vehicle. When you are on the job and using a different vehicle, practice using the brakes at low speed and in a safe area until you are sure of your abilities.

2.2.5 Hoist Mechanism

The hoist mechanism is used to raise and lower the dump bed. *Figure 14* shows two types of hoist mechanisms. Depending on the type of hoist control used, it may be located on the floor near the driver's seat or on the dashboard control panel. *Figure 15* shows three examples of hoist controls. The control shown in *Figure 15C* is the control for a roll-off bed truck.

Generally, the dump body can only be raised when the transmission is in neutral, in first gear, or below a certain ground speed. On some truck models, moving the dump body control automatically shifts the transmission into neutral. A dump body control such as the one in *Figure 15A* may have three or four positions. A four-position control includes Raise, Hold, Float, and Lower. The control positions are as follows:

- *Raise (1)* – Pull the lever up to raise the dump body and empty the load.
- *Hold (2)* – Move the lever down to the Hold position and the dump body will not move.
- *Float (3)* – Push the lever down to the Float position and the dump body seeks its own level. This is the primary and default position.
- *Lower (4)* – Push the lever all the way down to lower the dump body. When the lever is released, it returns to the Float position.

A three-position control (*Figure 15A*) would include Raise, Float, and Lower positions.

22202-12_F14.EPS

Figure 14 Hoist mechanisms.

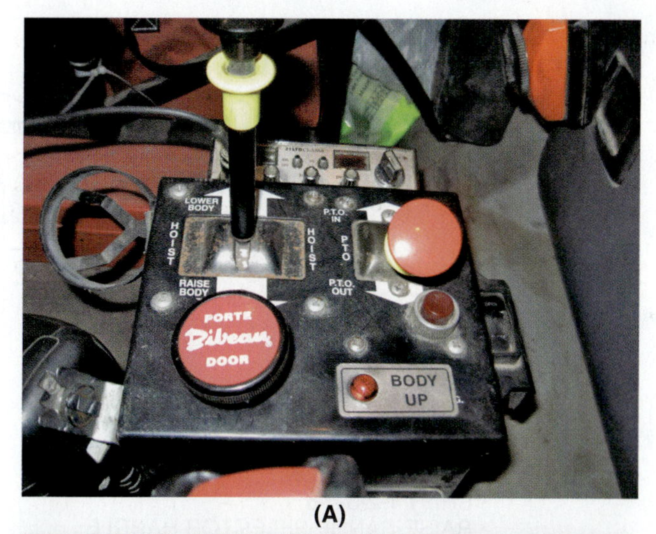

(A)

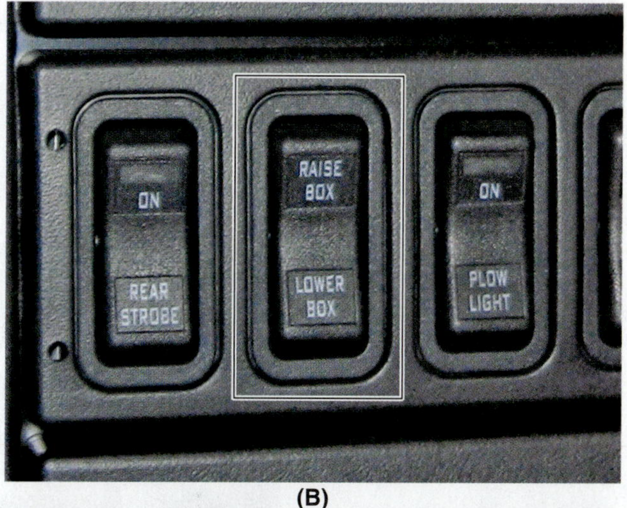

(B)

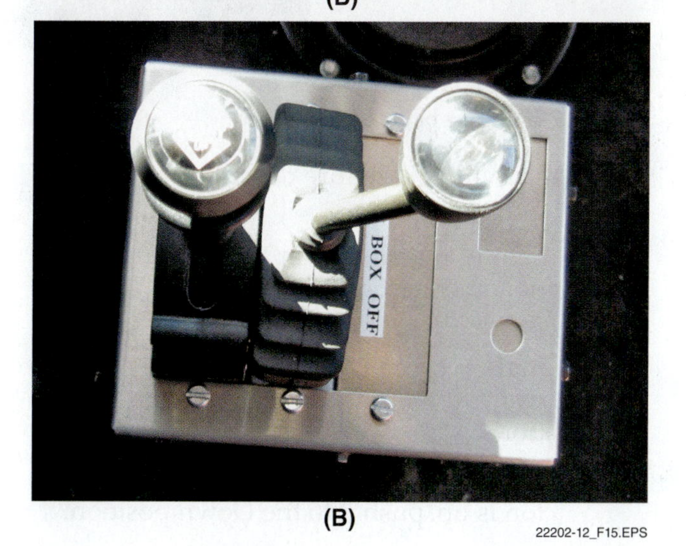

(B)

22202-12_F15.EPS

Figure 15 Examples of hoist controls.

2.2.6 Manual Transmission

A dump truck may be equipped with either a manual or automatic transmission. Automatic transmissions have become commonplace, but many trucks have a manual transmission. If a truck has a manual transmission, the driver needs to operate the **clutch** and gearshift in order to drive the truck. Regardless of the type of transmission, follow the manufacturer's shifting procedure. The steps in shifting are the same although the pattern may be different. *Figure 16* shows the typical gearshift patterns for both the Roadranger® and Mack Maxitorque® transmissions, two common types of manual transmission.

The gear range control switch is located on the gearshift. This switch changes the transmission gears from low (first to fifth gear) to high (sixth to tenth gear). To use low gears, the switch is set to the Down position. Low gears provide the truck with more power than the higher gears, so always start your truck from a stop in first gear. Lower gears are also used to climb and descend hills and to slow the speed of the truck so that it can stop. Shifting to low gears before stopping helps to spare the brakes from wear and possible overheating.

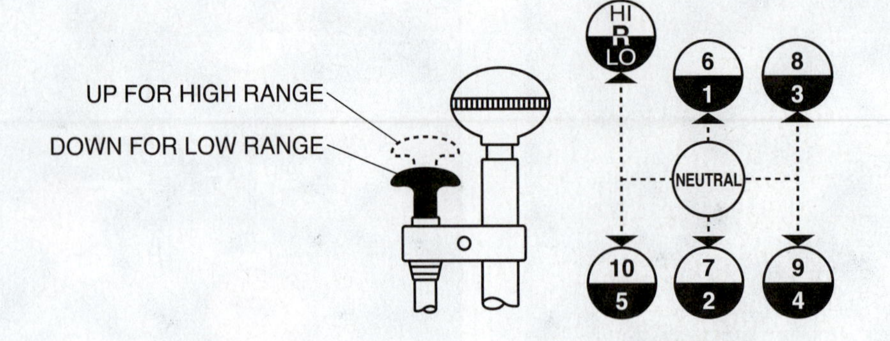

UPSHIFT

- START WITH RANGE SELECTOR IN DOWN POSITION
- SHIFT 1-2-3-4-5
- RAISE RANGE SELECTOR HANDLE
- SHIFT 6-7-8-9-10

DOWNSHIFT

- SHIFT 10-9-8-7-6
- PRESS RANGE SELECTOR HANDLE DOWN
- SHIFT 5-4-3-2-1

22202-12_F16.EPS

Figure 16 Examples of gearshift patterns.

The following procedure describes how to upshift:

Step 1 Set the gearshift lever to the neutral position and start the engine.

Step 2 Wait for the air system to reach normal pressure (usually 60 to 100 psi), but check the manufacturer's recommendations.

Step 3 Check the range control button. If the button is up, push it to the Down position.

Step 4 Push the clutch pedal to the floor with your left foot. Shift the gearshift to first gear, and then slowly lift your foot off the clutch while slowly pressing the accelerator with your right foot until the truck moves.

> **CAUTION**
>
> When operating a clutch, always push the clutch to the floor before shifting and then slowly let out the clutch to drive. Never rest your foot on the clutch pedal while the engine is running; this is called riding the clutch and can cause premature wear of the clutch. Unless you are operating the clutch, your left foot should rest on the floor of the truck.

Step 5 Shift progressively through the gears at manufacturer's recommended rpm until you reach fifth gear. Then shift the range selector to high range by pulling the selector knob up and shift into sixth gear. As the lever passes through the neutral position, the transmission automatically shifts from low range to high range.

Step 6 Shift progressively through the upper gears at manufacturer's recommended rpm.

The following describes the procedure for downshifting:

Step 1 Move the shift lever from tenth through each successive lower gear to sixth. When in sixth gear, locate the range control button with your hand.

Step 2 While in sixth gear, push the range control button down, and move the lever to fifth gear. As the lever passes through the neutral position, the transmission automatically shifts from high range to low range.

Step 3 Shift downward through each of the remaining gears.

> **CAUTION**
>
> Never shift from high range to low range at high speeds. Never make range shifts with the vehicle moving in reverse gear. Always leave the vehicle parked in low gear.

Automatic transmissions have four, five, or six forward speeds, along with reverse and neutral. Unlike a car, the truck transmission lacks a Park position because the parking brake is used instead.

2.2.7 Engine Retarder

Dump trucks are very heavy, especially when loaded, and on-road trucks are designed to operate at highway speeds. For that reason, they are difficult to stop. Using brakes to stop the vehicle causes a great amount of wear on the braking system. In addition, when a truck is driving downhill, constantly applying the brakes can cause them to overheat and fail. For this reason, on-road trucks are often equipped with an **engine retarder (engine brake)**, which works by using the power coming from the engine to slow the truck. *Figure 9* shows an engine brake control at the upper left edge of the panel. The engine retarder is activated with a switch that is located in the cab of the truck. Engine braking is often blamed for the familiar machine gun-like noise that is associated with big trucks. Because the loud noise bothers nearby residents, engine braking is prohibited in many areas (see *Figure 17*).

22202-12_F17.EPS

Figure 17 Engine braking prohibited sign.

2.0.0 Section Review

1. If the light on the air brake indicator on the instrument panel is lit, it means that air pressure is normal.
   a. True
   b. False

2. In a truck with air brakes, it is a good idea to _____.
   a. pump the brakes, rather than applying firm pressure
   b. apply the brake slowly because air brakes react instantly
   c. allow for longer stopping distance
   d. disable the governor

SECTION THREE

3.0.0 INSPECTION AND MAINTENANCE

Objective 3

Describe the operator inspection and maintenance requirements for an on-road dump truck.

a. Describe inspection, startup, and shutdown procedures.
b. Identify preventive maintenance procedures that must be performed.

Performance Task 1

Complete proper prestart inspection and maintenance for a dump truck.

In order to operate efficiently and remain in good repair, dump trucks require periodic maintenance. The operator may not perform this maintenance, but is responsible to know the maintenance schedule and to make sure the maintenance is performed when required. Dump trucks must also be inspected by the driver before and after each shift to ensure that the truck is safe to operate. In addition, there are specific procedures for starting up and shutting down a dump truck.

3.1.0 Inspection, Startup, and Shutdown

The operator is responsible for performing a daily walk-around inspection. Don't underestimate the importance of these inspections. Taking a few minutes to perform some simple tasks can extend the life of a vehicle and help keep the driver and others safe.

The vehicle must be inspected before and after each shift. Federal regulations may require written documentation of these inspections. At a minimum, an employer should have a written checklist that must be completed before the truck is used. *Figure 18* is an example of such a checklist. A visual inspection often identifies problems that affect safety or operating performance. For example, a tire that has struck a curb or other obstacle sometimes develops a bubble on the sidewall that is readily seen under close inspection. If it is left undetected, the tire can blow out while the truck is in use. A defective brake light can result in a traffic stop, along with a lengthy delay and an expensive traffic citation.

3.1.1 Startup

A broad-based pre-operational inspection may be mandated, but at minimum the following systems must be checked to ensure that they are working properly:

- Service and parking brakes
- Steering mechanism
- Headlights, parking lights, brake lights, and all reflectors
- Tires
- Horn
- Windshield
- Windshield wipers
- Rearview mirrors
- Coupling devices (if used)

Make sure any problems found during the inspection are repaired before using the truck. Once the inspection is complete, follow this procedure to start up the truck:

- Set the parking brake (*Figure 19*) and ensure that the gearshift is in neutral, or in Park if the transmission has one. Adjust all mirrors and the seat, then fasten the seat belt.

> **NOTE**
> Only small trucks have a transmission with a Park position.

- Start the engine and listen for any unusual noises. Investigate as needed.
- Check the readings of all indicators and gauges.
- Check the foot pedals for proper operation.
- Check the operation of the windshield wipers.
- If the vehicle is equipped with communications, check its operation.
- Ask a co-worker to help check that the headlights, turn signals, reverse lights, backup alarm, and brake lights are operational.
- Ask a co-worker to act as a safety spotter and check the operation of the hoist.
- Check the tailgate latches.
- Check the load cover (tarp) and its mechanism.

3.1.2 Shutdown

The shutdown procedure is shorter than the startup procedure, but it is essential for proper cooling of all parts. Following these basic steps, as well as any special procedures in the manufacturer's recommendations, increases the life of the engine:

Step 1 Stop the truck, preferably on a level location well out of the flow of traffic.

HEAVY DUMP TRUCK
Preventive Maintenance Checklist

Comments:

____

____

____

Note: The items below should be inspected during a typical preventive maintenance check. Additional checklist items may be required depending on equipment or circumstances.

UNDERHOOD

- ☐ Motor oil, power steering
- ☐ Coolant level, hoses
- ☐ Fuel line leaks
- ☐ Belt tensions
- ☐ Fuel level
- ☐ Batteries
- ☐ Windshield washer

INTERIOR

- ☐ Brakes
- ☐ Steering
- ☐ Horn & safety devices
- ☐ Wiper blades & control
- ☐ Heater
- ☐ Seats & seat belts
- ☐ Clutch

EXTERIOR

- ☐ Stop lights
- ☐ Head, tail, direction lights
- ☐ Clear, spot, warning lights
- ☐ Cab, body, glass
- ☐ Reflectors
- ☐ Coupling devices
- ☐ Hydraulic lines
- ☐ Tires, wheels, lug bolts
- ☐ Hydraulic reservoirs
- ☐ Springs – steering mechanism
- ☐ Drive line, universal joints
- ☐ Drain air reservoirs

GENERAL

- ☐ Exhaust system
- ☐ Engine
- ☐ Fire extinguisher
- ☐ Emergency triangle
- ☐ First aid kit

22202-12_F18.EPS

Figure 18 Example of a driver inspection checklist.

22202-12_F19.EPS

Figure 19 Parking brake.

22202-12_F20.EPS

Figure 20 Compressed-air reservoir bleed-off valve.

Step 2 Place the gearshift in neutral position.

Step 3 Set the parking brake. Chock the wheels when necessary.

Step 4 If the truck has a diesel engine, run it at idle for two to three minutes before shutting the engine down. This permits the engine to cool and allows for gradual and uniform cooling of all engine parts. Turn the key to the Off position.

Step 5 Perform a walk-around inspection and note leaks and broken or missing parts.

Step 6 Purge the air reservoirs of fluid by following the manufacturer's recommendation and your employer's policy. Moisture can build up in the reservoir and reduce the efficiency of the brakes. *Figure 20* shows the bleed valve on a compressed-air reservoir.

> **NOTE**
>
> Most air brake systems now have air dryers to remove moisture from the compressed air that feeds the brake system. However, some moisture can get through and build up in the air tanks. This is a serious concern in cold weather, when the moisture can freeze and potentially disable the brakes. So, even if air dryers are present, it is still necessary to bleed off the moisture as part of the daily routine.

Step 7 Lock the cab and remove the keys.

Step 8 Report any deficiencies to the proper authority at your company.

> **NOTE**
>
> Safely parking and securing the vehicle is as important as driving safely. Apply the service brakes to stop the truck. Move the gearshift to neutral and engage the parking brakes. Park on a level surface if possible. Otherwise, chock the wheels.

3.2.0 Preventive Maintenance Checks

A professional truck driver always makes sure that a vehicle is in good working order whether it is operated off-road or on a public road. The following requirements cover a typical check procedure that should be performed periodically to ensure that all systems work properly:

- Check under the truck for fluid leaks, loose wires or parts, and other damage.
- Examine the windshield and mirrors for dirt, cracks, scratches, or debris that obstructs the view. Clean or repair as needed.
- Open the engine compartment and look for leaking fluid and worn wiring, insulation, or hoses. Check for loose electrical connections. Repair if necessary.
- Inspect drive belts for serviceability and tightness (*Figure 21*).
- Check the level of oil, coolant, power steering, and transmission fluid (automatic transmission). Add fluids as required. *Figure 22* shows examples of the service points for these checks on a dump truck.
- If the truck is equipped with windshield washer fluid, check its level and top it off as

22202-12_F21.EPS

Figure 21 Engine drive belts.

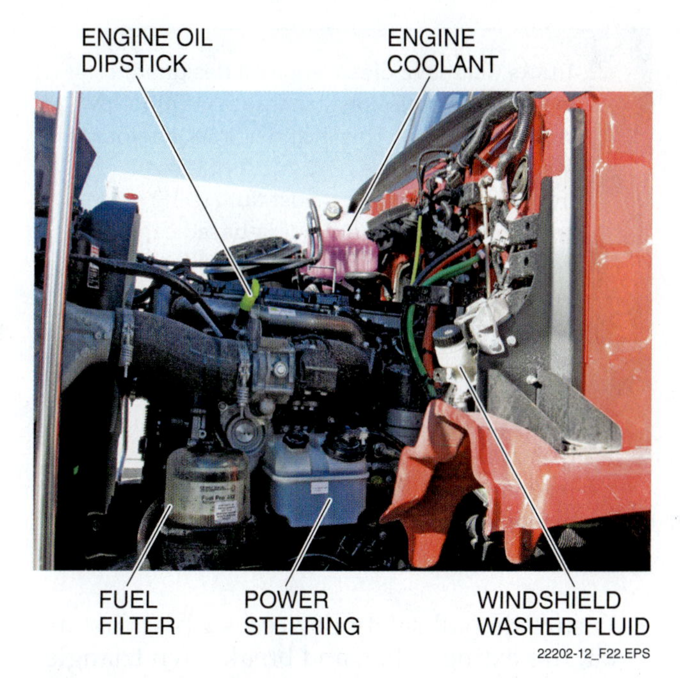

22202-12_F22.EPS

Figure 22 Engine fluid check points.

needed (*Figure 23*). This is a critical check, especially if the truck is driven on snowy or slushy roads that have been treated with sand or salt. Without washer fluid, visibility through the windshield can become severely reduced.

- Check battery connections for tightness and corrosion, and clean or tighten as required.
- Examine the air filter and clean or replace it as needed.

> **NOTE**
> The frequency at which some maintenance procedures are performed depends on the working conditions. In a dusty environment, for example air filters may need to be changed more frequently than would be required in a clean environment. This is usually spelled out on the manufacturer's maintenance schedule.

- Check the pressure and condition of each tire. Check the manufacturer's recommendations for tire inflation. This information is generally located on a label attached to the door frame.

> **CAUTION**
> Tire pressure decreases as the surrounding air temperature decreases. Tires that are correctly inflated in a warm shop become underinflated as the truck is operated in freezing weather. This can result in excessive tire wear. Check the tire pressure at the temperature in which the truck will operate.

- Clean all grease fittings and lubricate according to the manufacturer's recommendation. In cold areas of the country, this step is performed at the end of a shift so the equipment is warm and more readily receives the lubricant.
- Inspect the braking system according to the manufacturer's instructions and adjust as necessary. If the truck is equipped with air brakes, ensure that the air reservoirs have been drained of fluid. In cold areas of the country, reservoirs are bled at the end of a shift to prevent the condensation from freezing and damaging the brake system.
- Look in the dump bed and clean it if necessary. Ensure that the tailgate is operational and all pins, hooks, and chains are in place for safe and effective operation.
- In the cab, remove any trash or other items that can obstruct your view through the windows and windshield or your ability to reach the truck's controls.

> **NOTE**
>
> Trucks that have diesel engines designed to meet Tier IV emission standards use ultra-low sulfur diesel fuel. This includes truck engines manufactured since 2010. Most of these trucks have a diesel exhaust fluid (DEF) tank containing urea, an ammonia-based chemical. The DEF is used to treat engine exhaust in order to reduce particulate matter and nitrous oxide emissions. The DEF tank must be checked regularly and refilled as needed. The fluid must also be periodically replaced in accordance with the manufacturer's instructions. The vehicle control system should be equipped with a means to alert the operator when the DEF is low or when is has become contaminated.

22202-12_F23.EPS

Figure 23 Windshield washer fluid reservoir.

- Ensure that all safety devices, such as first aid kit, fire extinguisher, and breakdown triangles are available and operational. See *Figure 24*.

Perform these preventive maintenance tasks intelligently. When a truck is operated in a dusty environment or on short hauls, perform lubrications more frequently. If the truck is pulling to one side excessively, stop and look at the tires and check the tire pressure with a gauge. When you sense a problem, check it out even if it was checked during the pre-operational check.

22202-12_F24.EPS

Figure 24 Safety devices.

NOTE

It is normal for tires that are warm from use to have a higher air pressure than tires that are cold, because air expands as it is heated.

Balancing Act

Changing the oil and oil filter on a big truck engine can be expensive, so it should not be done too soon. On the other hand, running an engine on used-up oil can result in engine damage. It's a balancing act. Modern vehicles have on-board computers that keep track of maintenance needs. The dashboard display tells the driver how much life is left in the engine oil, for example. The same information can be transmitted to a central location so fleet maintenance managers can keep track of maintenance needs and schedule maintenance actions on a just-in-time basis.

Additional Resources

www.osha.gov/dts/shib/shib091806.html discusses proper blocking of the truck bed while it is raised for maintenance.

3.0.0 Section Review

1. The person responsible for performing daily truck inspections is the _____.
   a. driver
   b. foreman
   c. OSHA inspector
   d. mechanic

2. Tire pressure increases as the temperature of the surrounding air decreases.
   a. True
   b. False

SECTION FOUR

4.0.0 SAFE DRIVING

Objective 4

Describe safe on-road driving practices for on-road dump trucks.

a. State the normal driving practices associated with dump truck operation.
b. Describe how to handle a dump truck in an emergency.

Performance Task 2

Perform the proper startup, warm-up, and shutdown procedures on a dump truck.

Trade Terms

Lug: Effect produced when engine is operating in too high a transmission gear. Engine rotation is jerky, and the engine sounds heavy and labored.

Dump trucks can be used to tow dump trailers, equipment trailers, and other heavy equipment. Because of its great size and weight, a dump truck can be outfitted with a snowplow and can then be used to clear snow from roads, parking lots, and work areas. In addition, it may be fitted with a spreader (*Figure 25*) that scatters sand, salt, or other material onto icy surfaces.

Because of the dump truck's usefulness, large size, and power, the job of a dump truck operator is complex. These trucks can be a danger to operators, workers, and the public when used improperly or unsafely. A dump truck operator needs to know how to maintain safe control of the vehicle at all times. This includes not only driving the truck, but also safe loading and unloading.

Professional truck drivers have a responsibility to their employer to use a vehicle properly. This includes performing pre-operation inspection and servicing before operating the truck. It also means allowing the truck to warm up before driving it, especially in cold weather. When driving the truck, observe all gauges and indicators. Look, listen, and feel for defects, strange noises, and changes in engine or braking power. Allow diesel engines to idle for a few minutes before shutdown to allow the engine to cool slowly.

22202-12_F25.EPS

Figure 25 Spreader for salt and sand.

A large part of operating a dump truck on a public highway is looking out for other vehicles and pedestrians. Always look ahead and think ahead. When fueling the truck, always ground the fuel nozzle to the fuel tank filler neck to prevent sparks. Do not re-enter the vehicle during fueling to avoid creating static electricity, which can be an ignition source. Never fill the tanks while smoking or when near an open flame. If you spill fuel, immediately clean it up. When refueling has been completed, replace and tighten the fuel tank cap.

4.1.0 Driving Procedures

Learn to shift gears smoothly, and shift as needed to keep the engine operating at the manufacturer's suggested rpm. Use lower gears for low-speed operation, such as on a work site. Engines **lug** (hesitate) when driven at too low a speed for the gear in use. This is hard on an engine and can cause the engine to stall unexpectedly. Use high gears for highway speeds. Lower the speed and gear on rough roads and avoid driving over curbs, rocks, and other obstacles. Drive the vehicle at moderate speeds, especially when it is fully loaded. To avoid excessive tire wear, accelerate slowly to avoid spinning the tires and allow ample stopping time to avoid sliding the tires.

When entering the cab, adjust the seat and mirrors, and then fasten the seat belt. Before starting the engine, ensure that the parking brake is set and that the gearshift is in neutral (or Park if applicable). Follow the vehicle starting procedure recommended by the manufacturer.

At the end of the shift, safely park the vehicle in a firm, level spot out of the way of other traffic. If it is necessary to park on an incline, position

the vehicle across the slope and then follow the vehicle shutdown procedure.

Even with all the safety features that have been built into the design of dump trucks, there are still dangers. Drive defensively; watch for unsafe situations created by other drivers and be extremely careful in the operation of the equipment. To help prevent accidents on the road, observe the following safety rules at all times:

- Become acquainted with the operation and maintenance manuals for the truck.
- Know the location and functions of all the controls, gauges, and warning devices.
- Always wear the seat belt. Also, make sure any passengers wear their seat belts. Do not move the equipment until everyone is buckled in.
- Pay close attention at intersections and wait before accelerating when a red light turns green.
- Never accelerate through a yellow signal.
- If it is necessary to pass a vehicle, pass only in designated passing zones and only after checking blind spots for clearance.
- Obey all state and local traffic regulations at all times. Make sure your CDL is up to date. Never take a truck on the road without an up to date CDL, even if ordered to do so.
- Obey all state and local laws regarding the use of cell phones and other communications equipment. Do not use a hand-held device while driving.
- Always drive at a speed that permits full control of the vehicle and allows for factors such as road, weather, and traffic conditions.
- Do not have loose articles in the cab that might obstruct your view or keep you from maintaining complete control of the vehicle.
- Perform a pre-operational inspection before using the truck.
- If you get drowsy, pull over and take a break. Get out and walk around or change drivers.

To be a safe dump truck operator you must know and practice safe operating principles and procedures on a continual basis. Several areas of operation require particularly close attention, including climbing and descending hills, runaway vehicles, recovering from a skid, and negotiating curves.

4.1.1 Climbing and Descending Hills

It is important not to depend solely on the brakes when descending steep hills. Brakes can overheat and fail to work. Before entering a downhill grade, apply the brakes to reduce speed, and then shift the transmission into a lower gear. Using a lower gear gives the same effect as using the brakes and helps avoid overheating the brakes. A good rule for driving downhill is to use one gear lower than would be used to drive up the same grade.

It is risky to change gears while climbing or descending steep hills. The safest procedure is to select the proper gear before starting to climb or descend. If it is necessary to go to a lower gear during the climb, be sure to shift before the engine lugs to a point where it might stall.

When starting out on a road with long, steep hills, it may be a good idea to come to a complete stop and inspect the vehicle before proceeding. Only do this if it can be done safely. Check the tires for proper inflation and brakes for proper travel and operation. If everything is working correctly, start down the hill in the lowest gear. Do not use the accelerator. Shift gears as needed to match speed gains as the truck picks up momentum.

Starting a loaded truck on a positive grade without any rollback can be difficult. It may be necessary to release the parking brake as the truck is accelerating in first gear in order to prevent rollback. If the steepness of the grade is causing the truck to slow, downshift when the engine rpm reaches the shift point for the next lower gear. Then continue to downshift to match the power demands for the grade until there is no longer a loss of power.

The following are safety tips for descending a steep downhill grade:

- Reduce speed with minimal use of the foot brake.
- Check the speedometer and tachometer frequently. Never exceed the recommended engine speed in any gear.
- Take extra care when weather and road conditions are unfavorable. Never exceed the advised truck speed for any downgrade.
- Keep the truck under complete control at all times.

4.1.2 Taking Curves

Taking a curve at high speed with a full load is dangerous. Curves must be taken more slowly when a truck is carrying a load than when the truck is empty. When rounding a curve, centrifugal force pushes the truck toward the outer edge of the curve. The only resistance is the friction of the tires on the road. An increase in load raises the center of gravity, increases the force that causes the truck to slide sideways, and also increase the risk of tipping over. If traction is good, the truck tilts slightly, compressing the outside springs and tires, and the truck stays on the road. If the outward force is greater than the vehicle can handle, the truck tips over or goes into a skid. The safest action is to reduce speed and downshift before going into the curve, and accelerate gently as the vehicle is coming out of the curve.

4.1.3 Backing Safely

Backing is a hazardous maneuver under any circumstance. The driver is always responsible for knowing what is behind the truck. Do not use a backing maneuver any time you can drive forward. Whenever possible, use a spotter to help guide the truck and reduce the chance of injury or property damage. The rearview mirrors show only what is to the side of the truck. Anything that is low or immediately behind the vehicle is not visible to the driver. It is a good practice to get out of the cab and look behind the vehicle before backing.

The following tips should help in safe backing:

- Back slowly. Be sure there is sufficient clearance when backing into narrow spaces.
- Remain properly seated when backing the vehicle, using the mirrors or a spotter.
- Avoid long backing runs. It is much safer to turn around and drive forward.
- Avoid backing downhill.

4.2.0 Emergency Procedures

Under certain conditions, the driver can lose control of the truck. It is important to know what to do in such circumstances. The sections that follow describe how to handle a truck in various emergency situations.

4.2.1 Runaway Vehicle

If brakes fail to hold the vehicle and it starts to run out of control down a hill, the last resort is to ditch the vehicle. Running it off the road against a bank at a gradual angle will slow and stop the vehicle. This must be done promptly before the runaway vehicle has gained too much speed. Proper action in such an emergency may prevent a much more serious accident. Remaining belted in the cab is the safest position.

Runaway Truck Ramps

Brake failure due to overheating or loss of pressure can cause a truck to run out of control on a steep downgrade. Many states, especially mountainous states, have ramps designed to bring a runaway truck to a safe stop on long, steep downgrades. The ramp is usually made from sand or gravel, which will rapidly retard the speed of the truck. The ramps have a steep upward grade, which further helps in retarding speed.

4.2.2 Controlling a Skid

Statistics show that about one-fourth of all vehicle accidents involve a skid. The reason a skid occurs at all is that the tires have lost their traction on the road surface. When this happens, you begin to lose control of the vehicle's direction. Most skid problems are caused by slippery road conditions like rain, snow, ice, or wet leaves.

If the vehicle should start to skid, stay calm. Avoid hard braking, since slamming on the brakes locks the wheels, causes further loss of traction, and increases the skid. The secret is to make small steering corrections. Steer in the direction of the skid; as the vehicle begins to correct, straighten the front wheels carefully and be ready to correct in the other direction. Some skids need more than one correction to regain control. Avoid over-steering; turning the steering wheel too far will whip the rear end of the truck into a skid in the opposite direction. Keep the clutch engaged (clutch pedal not depressed) and the transmission in gear (or automatic transmission selector lever in drive). Holding the vehicle in gear helps reduce speed and provides the most control. Avoid lifting your foot from the accelerator suddenly, as this action can worsen the skid.

It is better to try to prevent a skid than to have to recover from one. Adjusting speed to the conditions in the road reduces the chance of a skid. Driving within your sight distance and maintaining an adequate distance from the vehicle in front reduces the need for sudden stops and the possibility of a braking skid.

Additional Resources

www.driving-truck-school.com/dump_truck_driving_tips.html—Dump Truck Driving Tips.

4.0.0 Section Review

1. The best way to maintain a safe speed on a downhill grade is to _____.
   a. keep steady pressure on the brakes
   b. pump the brakes continuously
   c. downshift as necessary
   d. select the proper gear before starting down

2. In order to regain control in a skid, you should _____.
   a. leave the steering wheel in a fixed position
   b. steer in the direction of the skid
   c. push in the clutch
   d. slam on the brakes

SECTION FIVE

5.0.0 SAFE OPERATION ON THE JOB

Objective 5

Describe the procedures for operating a dump truck on a job site.

a. State the safety practices associated with dump truck operation on a job site.
b. Describe proper loading, dumping, and snow plowing procedures.

Performance Task 3

Carry out basic operations with a dump truck:

- Dump a load in a designated spot, and tailgate-spread the load.
- Back up with a trailer attached.
- Perform tailgate adjustment, as applicable.

Because the dump truck is one of the most frequently used pieces of heavy equipment in the earthmoving and construction business, many accidents occur each year involving dump truck operation, both on the road and at the job site. Most accidents can be avoided by following required safety procedures and using common sense. Prevention of accidents involving dump trucks depends primarily on the person operating the equipment. Manufacturers design many features into their equipment that make driving safer and easier; however, avoiding unsafe situations and using forethought, good judgment, and skill is solely up to the operator.

This section describes procedures and actions that can be used to ensure safe operation of a dump truck while on the job.

5.1.0 Working Safely

While working on a job, it will be necessary to coordinate your vehicle's movements with those of other vehicles. Usually one person at a site, known as a signal person or spotter, is appointed to direct the movement of vehicles in congested areas. Find out who this is and then watch for and obey all signals. When the signal person is not in view, exercise caution when moving the truck. Look and think ahead to avoid situations that could disrupt the smooth flow of traffic. Yield to vehicles approaching from the right, but never take the right of way without first assessing conditions. When operating the vehicle near workers who are on foot, keep them in sight.

A driver behind the wheel of a loaded dump truck is responsible for guiding several tons of material around a work site or on a public road. Avoid actions that can weaken your skills or judgment on the job. Never use drugs or alcohol on the job. They can impair alertness, judgment, and coordination, and may be grounds for termination. When required to take prescription or over-the-counter medications, seek medical advice about whether you can safely operate machinery. Only you can judge your physical and mental condition; don't take chances with your life or someone else's.

Never perform any unsafe maneuver or operations, even if told to do so by another person. To be qualified, you must understand the manufacturer's written instructions; have training, including actual operation of the vehicle; be properly licensed; and know the safety rules for the job site. Current federal regulations state that at the time of your initial assignment and then at least annually, your employer must instruct you in the safe operation and servicing of equipment that you will use. Most employers have rules about operation and maintenance of equipment; make sure you know them. Finally, your employer must authorize you to operate a vehicle.

The key to staying safe is to develop good habits while learning a new skill. This way safety becomes second nature. The following are some good habits to practice:

- Read the operator manual thoroughly. Know the machine.
- Always follow the job site's safety precautions.
- Always use a seat belt. Use other personal protective equipment (PPE) as required by the job site.
- Do not haul people in the dump bed or on the running board.
- Never get under the dump bed when it is raised unless it is securely blocked using an approved method.
- Keep windshield, windows, and mirrors clean at all times.
- Use a proper three-point mounting and dismounting technique when entering or leaving the vehicle. Never enter or leave the truck by grabbing the steering wheel.
- Exit the cab facing the vehicle, and keep a firm grip on the handholds until you are on the ground.
- Never use cell phones or other devices that can distract you while driving.
- Never get under the truck unless the wheels are securely blocked.

- Clean slippery materials off your shoes. Clean shoes help prevent slipping on steps or having your feet slip from the clutch and brake pedals.
- Set the parking brake and block the wheels when parking trucks.
- Always observe local laws regarding the weight and height limits of vehicles.

5.1.1 Pinch Points

Pinch points result from the motion between two or more mechanical parts of the equipment. Being caught in a pinch point can cause injury or even death. On dump trucks, pinch points exist in several areas. These are shown in *Figure 26* and marked with "PP". The following are the main pinch points:

- Between the cab guard and the back of the cab
- Between the dump bed and the frame
- Between the tailgate and the sides or bottom of the dump bed
- Between the rear wheels and the side of the frame
- Between the cowling of the hood and the truck body
- Around exposed rotating machinery (such as belts and gears) that are accessible to the hands

It is the operator's responsibility to work safely and make sure co-workers also follow proper safety procedures. Follow these tips to work safely around pinch points:

- In general, do not allow workers to be located at a pinch point area, either on the machine or around the machine, while it is in operation.
- Do not stand or allow others to stand or work under a raised dump bed. If work must be performed on the truck with the dump bed raised, the dump bed should be empty and suitable blocking placed between the bed and the frame according to the manufacturer's instructions. Manufactured devices designed for this purpose are readily available and should be used. Blocking with wood beams, I-beams and similar methods is not considered safe.
- Keep fingers out of the area between the tailgate, tailgate latch, and the sides of the dump bed. Always stand clear when the bed is being lowered because the gate will slam shut as the bed comes down.
- Do not stand on the frame behind the cab as the bed is being lowered. The hydraulic action to lower the bed may be too fast to allow you to get out of the way. Stand on the ground and to the side if it is necessary to operate the dump controls at the back of the cab.

Accident Case History

In July of 2008, a property owner discovered a landscape contractor crushed to death between the cab and the dump body of a dump truck, a space only 7 inches wide. Investigators speculated that the dump bed had gotten stuck and the contractor had climbed up onto the chassis to free it by kicking the linkage between the dump lever and the dump body. In doing so, he caused the dump body to fall. The takeaway message from this incident is that no one should be under a raised dump body unless the dump body is securely blocked or cribbed.

Source: New York State Fatality Assessment and Control Evaluation (NY FACE)

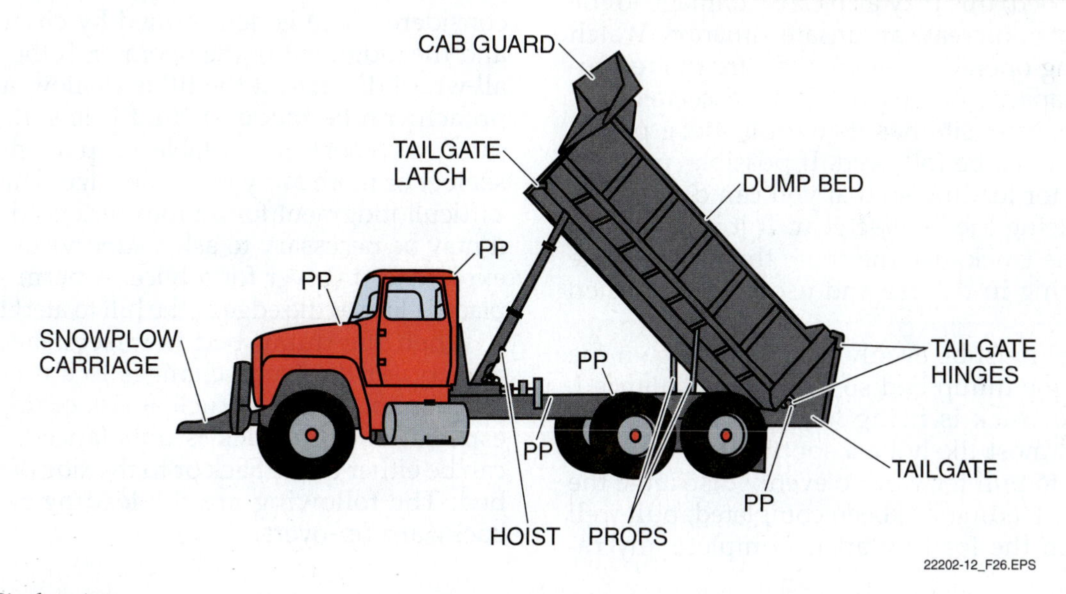

Figure 26 Pinch points.

- Do not reach into the area between the rear tires and the frame unless the truck has been shut down and the wheels are properly chocked. Any movement by the wheels forward or backward can crush hands and feet, causing serious injury.

5.2.0 On the Job

The most common use of a dump truck is to haul large loads of material in its bed. However, they are also widely used for plowing snow and spreading sand or chemicals to melt snow and ice. An improperly loaded dump bed can make the vehicle harder to handle. Follow proper loading procedures to ensure a balanced load. Never perform any unsafe maneuver or operation, even if told to do so by another person.

Use the proper vehicle for the type of material being hauled. The truck should be in good, safe operating condition. Make sure the frame is sound, the bed is firmly attached, and the tailgate is solid and can be opened easily and locked securely. These items need to be checked before any material is placed in the bed. Avoid the temptation to use a questionable vehicle.

5.2.1 Safe Loading Practices

Overloading a truck causes serious problems for dumping, as well as for on-road driving. For example, the material may be too heavy for the hydraulic hoist to raise the bed. If that happens, the material may have to be removed using an excavator or other machine, which could damage the dump bed. Another problem is the force of excess weight against the sides and tailgate of the dump bed; this may also cause damage to the equipment and create an unsafe situation. Watch the loading operation and make sure the recommended capacity of the truck is not exceeded.

Each loading site has its own loading procedure that must be followed. If possible, position the truck for loading so that you can drive away without using the reverse gear. If it is necessary to back the truck, use the route that reduces the time moving in reverse and use a spotter when possible.

Material must be evenly distributed down the center of the dump bed so the load is balanced. When the truck is being loaded with a chute, there will most likely be a spotter who will signal when to pull forward to evenly distribute the load. Once loading has been completed, pull well away from the loading area. Complete any required paperwork before leaving the loading site. Remember to secure the load cover if it is needed.

WARNING!

Never drive with an unbalanced load. It can cause you to lose control of your vehicle and may cause an accident. Ask the workers at the loading site for help to balance the load or contact a supervisor.

5.2.2 Safe Dumping Practices

The truck is at the most risk for tipping over while the load is being dumped, especially if the load is unbalanced. Even when the cargo has been loaded properly, many factors can affect the unloading of the truck. Stay alert for problems during the dumping procedure and be ready to stop the procedure at any time the conditions warrant.

Most dump areas using rear-dump trucks require turning and backing to the edge of the fill. If possible, the fill should be arranged so that any turn is made near the dump spot. The turn spot should be wide enough so that reverse gear is used only once. Turns in reverse should be toward the driver's left to give the driver maximum visibility. The truck should be level or facing uphill for dumping. Use the low range of the reverse gear when the load is heavy, when the ground is soft or rough, or when complicated steering is required.

When dumping off the edge of a fill, back so that both rear wheels are the same distance from the edge, rather than at an angle. Check the stability of the ground at the dump area before driving on it. If one wheel sinks in the ground deeper than the other, it may not be possible to dump the truck safely or to pull out with a load.

The distance from the edge of a fill that can be considered safe is determined by circumstances and the judgment of the operator. If the truck has all-wheel drive, or if the fill is shallow, a close approach can be made. If the fill is soft, slippery, sandy, or otherwise unstable, keep the rear wheels six feet or more away from the edge. This can be a difficult judgment for an inexperienced driver, so it may be necessary to ask a supervisor or a more experienced driver for advice. A berm should be placed close to the edge of the hill to alert the driver.

When the dump bed is raised, the center of gravity for the truck changes to a much higher position, placing the truck at risk of tipping over, especially if the truck is unbalanced. Tip-overs can be either to the back or to the side of the dump bed. The following are the leading reasons for backward tip-overs:

- *Top-heavy load* – The rear portion of the load is dumped, leaving the rest of the load stuck at the front of the bed, as shown in *Figure 27.*
- *Material stuck together* – Either the material is frozen together or to the bed of the truck.
- *Uneven loading* – This can be either front-to-back or side-to-side.
- *Uneven dumping* – This is similar to uneven loading, but is caused by trying to dump part of a load in a particular spot.

When material sticks to the front part of the bed in top-heavy loads, it places a great strain on the hydraulics and causes an unstable condition for the whole truck. The best action to take is to lower the bed to a level position and use a shovel or other hand tool to dislodge the load. Never stand in the dump bed and try to work the material loose while the bed is raised.

When part of a load remains stuck in the bed of the truck, a hazardous condition is created that could cause tipping or damage to the dump bed. This situation is sometimes referred to as a split load. A split load can be caused by frozen or sticky material that attaches to the bottom and one side of the bed. Compacting the material in the bed when it is loaded may also cause the material to stick on a side and in the corners.

In a split side load, the material splits down the center of the bed and leaves part of the load completely to one side of the bed, as shown in *Figure 28.* This shifts the vertical center of gravity and can tip the unit over to one side. Uneven loading, excessive rear spring deflection, or under-inflated tires can cause this condition. To correct a split load, lower the bed and use a shovel or other hand tool to loosen the load.

Another problem encountered while dumping is allowing the material to pile up against a hinged tailgate. In this case, the upper body of the load usually empties and the load piles up at the back. A load stuck in this position may unbalance the truck so that the front wheels lift off the ground. Until the material is freed, the truck cannot move because there is no steering capability. If the load cannot be dumped, the bed should be lowered and the tailgate opening adjusted or the tailgate removed.

The following actions will help prevent tip-over due to materials problems:

- Make sure the bed is empty and clean before loading. Keep the bed washed down with proper fluids when required.
- Test the hoist mechanism with an empty bed to ensure proper operation.
- Adjust the tailgate for loading according to the manufacturer's recommendation.
- Do not dump with the truck on a slope.
- Raise the bed slowly.
- Load the bed evenly. If the bed is loaded unevenly, lower the bed and even out the load with a shovel. Do not do this while the bed is raised.

It is the operator's responsibility to ensure that the equipment is working properly and that the material is being handled safely. Never perform any unsafe maneuver or operations, even if told to do so by another person.

5.2.3 Using Trucks With Bottom Dump Trailers

In some parts of North America, bottom dumps are used for hauling and spreading, especially for asphalt and aggregate material. These trucks come in many different designs and sizes, but

Figure 27 Top-heavy load.

Figure 28 Split-side load.

they all dump their load through gates in the bottom of the dump bed. Under certain conditions, there are advantages to using bottom dumps. For example, they can place material in a windrow, which can be worked more easily than dumped piles.

There are precautions that need to be taken with bottom dump trailers:

- Follow the directions of a spotter when starting the dump. If the material is asphalt or aggregate that needs to be placed at exact locations, do not rely on your sense of distance to judge when the dump should start. Being off by half a truck length could cause problems for other equipment.
- Do not allow the material to flow out of the bottom of the bed faster than you can move to maintain an even row. Allowing the material to be dumped in one spot may cause the trailer to become stuck on the pile.
- Never dump while turning. The motion of the bottom of the trailer bed against the side of the dumped material could cause the doors or mechanism to jam.

5.2.4 Snow Plowing With a Dump Truck

In some areas, dump trucks are used to plow snow and spread sand or salt over icy surfaces. See *Figure 29*. These trucks are seasonally equipped with snowplows, spreaders, or both. Some snowplows are fixed at one level and angle and can be adjusted only by manually moving and resetting the blade. Most snowplow systems use hydraulics to allow the operator to adjust the level and angle of the plow from controls in the cab. Other snowplow systems are very sophisticated and are partially automatic, so that the plow rides over some obstructions without operator intervention. Study the manufacturer's instructions for the equipment used on your truck.

When working in cold, snowy weather, take additional precautions. First, always dress suitably. Wear appropriate protective clothing; especially warm gloves, hats, and socks. Fingers, ears, and toes are the first extremities to freeze. Wear moisture-repellent clothing and dress in layers. Layers of lightweight clothing provide more warmth than a single heavyweight garment. Dressing in layers also allows you to remove clothing as you warm up from working.

Know the symptoms of cold exposure and frostbite. Prolonged exposure to cold weather can cause drowsiness and affect judgment, so take breaks in a warm area when necessary. If that is not possible, exercise, such as jogging in place, will help. Keep warm liquids handy, and avoid drinking alcohol.

Sunlight is intensified when reflected from snow and can cause eye fatigue or temporary blindness. Wear good quality sunglasses to protect your eyes from the strong light. When sunglasses are not needed, remove them to maximize your vision.

Snow presents an additional hazard, since it covers many obstacles. Some obstructions will dislocate or damage the plow and perhaps damage the truck. Always use extreme caution when plowing around the following:

- Bridge and pavement expansion joints
- Headwalls of culverts
- Cattle guards
- Signposts and guardrails
- Hard-packed snow or ice
- Road shoulders
- Raised pavement markers, curbs, and islands
- Fire hydrants

22202-12_F29.EPS

Figure 29 Dump truck with snow plows.

Operation of the snowplow should be done with extreme caution. Working in poor environmental conditions means that snowplow operators need to make every effort to protect themselves and their equipment from accident and injury. Only experience will allow you to develop a feel for plowing different kinds of snow under different conditions. For the mechanical operation of the plow, follow the manufacturer's instructions.

Before starting to plow, perform the normal pre-operational inspection and then check to see that the plow is operational. When driving, use extreme caution because the blade has a low clearance even when it is raised. Drive according to road, snow, and visibility conditions. Be alert for hazards. At the area to be plowed, adjust the blade to the prescribed height. If that height is not specified, set the blade for approximately one-inch clearance. Set the blade angle as shallow as conditions allow. See *Figure 30*. The shallower the angle, the wider the plow lane is. A light, dry snow requires a shallow angle, but wet, heavy snow requires an acute angle.

> **NOTE**
> Be sure that the plow angle is not set so that it fails to clear a path for the truck wheels.

When plowing a roadway, plow in the direction of traffic flow so that the snow is removed to the shoulder of the road. For two-lane roads, plow from the center line out to shoulders and make sure the blade overlaps the center line slightly on the first pass. Plow to the low side of ramps or curves when possible. If plowed to the high side, the snow melts across the highway and ices over. Be alert for vehicles approaching from behind. Rear-end collisions are common. Plow away from wind whenever possible and practical. Clear snow past intersections and raise the blade before making a turn. Use caution when making a turnaround.

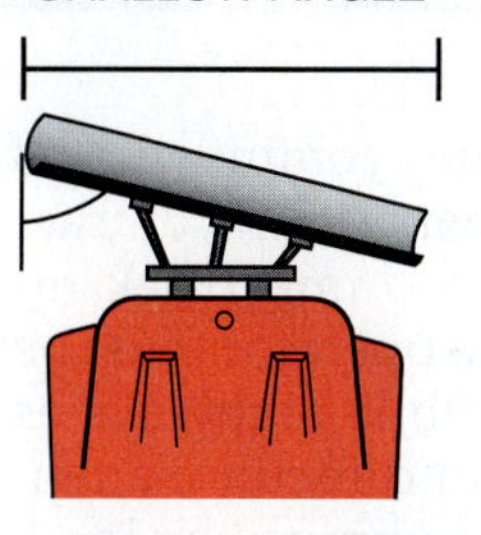

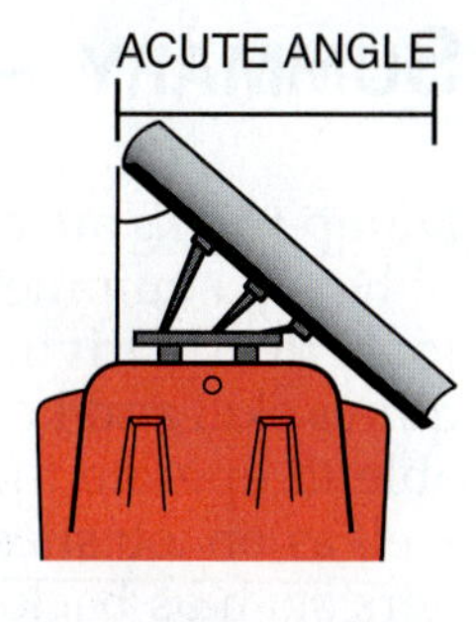

22202-12_F30.EPS

Figure 30 Snow plow angles.

Additional Resources

Excavating and Grading Handbook Revised. Nicholas E. Capachi. Carlsbad, CA: Craftsman Book Company.

5.0.0 Section Review

1. If a vehicle approaches your truck from the right at a job site, you should keep moving because you have the right of way.
   a. True
   b. False

2. The way to deal with a split load is to _____.
   a. pop the clutch to jar the material loose
   b. drop the bed suddenly to jar the material loose
   c. lower the bed and use a shovel to loosen the material
   d. adjust the tailgate to provide a wider opening

3. When making a turn while plowing snow, the blade should be _____.
   a. lifted
   b. kept on the ground
   c. turned in the opposite direction
   d. pitched downward

SUMMARY

Dump trucks are among the most common items of heavy equipment in the construction contractor's fleet. Driving an on-road dump truck requires skill and knowledge. The operator must be able to operate the truck on public highways as well as on job sites and perform difficult maneuvers such as backing up and dumping the load. The driver must be able to do these tasks safely and efficiently.

An on-road dump truck consists of a dump bed and cab mounted on a rigid frame. The engine can either be gasoline or diesel powered. The transmission can be automatic or manual. Manual transmissions usually have 10 or more gears, while automatic transmissions have 4 to 6 forward speeds. The configuration of the axles can either be single, dual, or multiple in the rear, with single or double tires mounted on each side. The truck's dump bed is operated by a hydraulic hoist mounted on the frame. Most trucks come with an air brake system, which the operator must know how to use and maintain in proper working order. Operating a truck with air brakes requires a special CDL endorsement.

A dump truck operator is responsible for its preventive maintenance and safety. This includes performing daily walk-around inspections to make sure the truck is operating properly and safely. Proper preventive maintenance will prolong the life of the truck and must be performed per the manufacturer's schedule. Do not drive a truck that is not operating properly or is unsafe. Follow the procedures in the operator manual for proper inspection and operation activities.

Operating a dump truck on a highway requires specialized knowledge and the skill to properly use the transmission and brakes to negotiate hills. It is especially important to know how to manage the truck in an emergency such as a skid or loss of brakes.

Dump trucks may tip over if loaded improperly or when dumping material that is frozen or stuck together. The tipping usually occurs when the dump bed is being hoisted and the material is being unloaded. Tipping can also take place if a large load causes the center of gravity to be too high and the truck is moving too fast around a curve.

Dump trucks are often used for snow removal. A snowplow can be attached to the front of the truck and a spreader can be loaded into the bed or mounted on the back of the truck. These attachments are only mounted on the truck when needed and are removed and stored at other times.

Review Questions

1. If you don't know the maximum weight capacity of a truck, you should _____.
   a. ask another dump truck driver
   b. ask your immediate supervisor
   c. look it up in the truck's manual
   d. load the bed until the rear tires compress

2. A standard dump truck with a single rear axle is used to haul _____.
   a. only light loads
   b. light and heavy loads
   c. medium loads
   d. the heaviest loads

3. A good use for a bottom dump truck is _____.
   a. moving soil to a dumping site
   b. hauling away excavated soil
   c. depositing sand for a foundation
   d. dumping asphalt for a road bed

4. Understanding the controls of your dump truck is critical to your safety and that of others. When you are assigned to drive an unfamiliar dump truck, what should your first step be?
   a. Drive the truck around the parking lot until you are comfortable with it.
   b. Ask an experienced co-worker to show you how the truck controls work.
   c. Study the manufacturer's manual before you even start the truck's engine.
   d. Slowly drive the truck on public roads—experience is the best teacher.

5. The differential lock is helpful if the truck is stuck in mud and one tire has traction.
   a. True
   b. False

6. A power takeoff unit is generally located _____.
   a. at the rear of the truck
   b. at the front of the truck
   c. in the cab
   d. in the dump bed

7. When operating a dump truck equipped with air brakes, you should _____.
   a. use them like other braking systems
   b. rest your foot on the brake pedal to maintain pressure
   c. monitor the system air pressure while driving
   d. pump the brakes to avoid locking them

8. In general, a dump body can only be raised when the transmission is in neutral.
   a. True
   b. False

9. To shift gears on a manual transmission, you need to operate the gearshift and _____.
   a. clutch
   b. differential
   c. choke
   d. power takeoff

10. It is a good idea to shift the transmission into low gear before stopping in order to save wear on the _____.
    a. clutch
    b. brakes
    c. retarder
    d. PTO

11. Preventive maintenance such as replacing the engine air filter should be performed _____.
    a. every day at the beginning of the shift
    b. in exact accordance with the manufacturer's schedule
    c. when conditions such as air quality require it
    d. whenever the truck is not being used

12. Before entering a curve with a fully loaded dump truck, you should _____.
    a. reduce speed and downshift
    b. increase speed and upshift
    c. shift the transmission to the next higher gear
    d. shift the transmission into neutral and coast

13. When backing a dump truck, you should first_____.
    a. look over your shoulder just as you do in an automobile
    b. use your mirrors so you can see the area immediately behind the truck
    c. look straight ahead and follow the directions of the spotter
    d. get out of the truck before backing to see what is behind it

14. When is an employer required to provide instruction on the safe operation of your equipment?
    a. When you have an accident
    b. At least once a year
    c. At monthly tailgate meetings
    d. Only when you are first hired

15. When leaving the truck cab, you should exit _____.
    a. while holding onto the steering wheel
    b. facing the vehicle and using the handholds
    c. from the passenger side of the truck
    d. facing front and quickly jumping

16. In a correctly loaded dump truck, the load is _____.
    a. balanced by placing it down the center of the bed
    b. piled as high as the sides of the truck allow
    c. piled as high as possible
    d. compacted to use the most bed space possible

17. When dumping at the edge of a fill, it is best to _____.
    a. back the truck so that it is at an angle to the edge
    b. keep the back wheels the same distance from the edge
    c. dump the load at the fill edge and then shovel it in
    d. always get the truck as close to the edge as possible

18. When dumping, turns in reverse should be toward the driver's left for _____.
    a. maximum turning radius
    b. maximum visibility
    c. greater stability
    d. better traction

19. When it is necessary to work in cold snowy weather, you should wear layers of clothing rather than a single heavy outer garment.
    a. True
    b. False

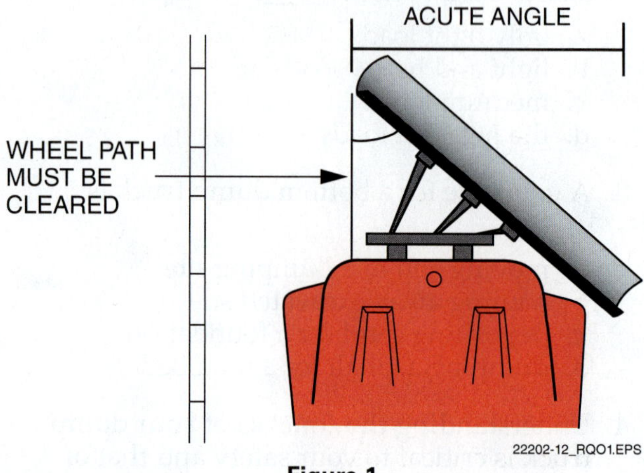

Figure 1

20. The plow shown in *Figure 1* is set to _____.
    a. plow heavy wet snow
    b. plow dry light snow
    c. clear ice from the road
    d. clear snow or ice

Trade Terms Introduced in This Module

Auxiliary axle: An additional axle that is mounted behind or in front of the truck's drive axles and is used to increase the safe weight capacity of the truck.

Cab guard: Protects the truck cab from falling rocks and load shift.

Clutch: Device used to disengage the transmission from the engine.

Engine retarder (engine break): An alternate braking system activated from the cab that slows down the vehicle by reducing engine power.

Governor: Device for automatic control of speed, pressure, or temperature.

Hoist: Mechanism used to raise and lower the dump bed.

Lug: Effect produced when engine is operating in too high a transmission gear. Engine rotation is jerky, and the engine sounds heavy and labored.

Tag axle: Auxiliary axle that is mounted behind the truck's drive wheels. It may be called a pusher axle if it is placed in front of the drive axle.

Tandem-axle: Usually a double-axle drive unit.

Additional Resources

This module presents thorough resources for task training. The following resource material is suggested for further study.

Excavating and Grading Handbook Revised. Nicholas E. Capachi. Carlsbad, CA: Craftsman Book Company.

Truck Driver's Guide to CDL, First Edition. New York, NY: Prentice Hall Press.

www.driving-truck-school.com/dump_truck_driving_tips.html—Dump Truck Driving Tips.

www.osha.gov/dts/shib/shib091806.html discusses proper blocking of the truck bed while it is raised for maintenance.

Figure Credits

Courtesy of Wardlaw Trucking, Module opener, Figure 4

Reprinted courtesy of Caterpillar Inc., Figures 1 and 14B

Topaz Publications, Inc., Figures 2, 3, 7, 8, SA02, 9–12, 14A, 15, 16B, 17, 19–23, 24C, 25, 29

Courtesy of CBI Manufacturing, Figure 5

Courtesy of Robert Lafrenière, Figure 6

Courtesy of **Trucks.com**, SA01

Courtesy of Amerex Corporation, Figure 24A

Courtesy of Peterson Mfg., Co., Figure 24B

Section Review Answers

Answer	Section Reference	Objective
Section One		
1 d	1.1.0	1a
2 a	1.2.1	1b
Section Two		
1 b	2.1.0	2a
2 c	2.2.4	2b
Section Three		
1 a	3.1.0	3a
2 b	3.2.0	3b
Section Four		
1 d	4.1.1	4a
2 b	4.2.2	4b
Section Five		
1 b	5.1.0	5a
2 c	5.2.2	5b
3 a	5.2.4	5b

NCCER CURRICULA — USER UPDATE

NCCER makes every effort to keep its textbooks up-to-date and free of technical errors. We appreciate your help in this process. If you find an error, a typographical mistake, or an inaccuracy in NCCER's curricula, please fill out this form (or a photocopy), or complete the online form at **www.nccer.org/olf**. Be sure to include the exact module ID number, page number, a detailed description, and your recommended correction. Your input will be brought to the attention of the Authoring Team. Thank you for your assistance.

Instructors – If you have an idea for improving this textbook, or have found that additional materials were necessary to teach this module effectively, please let us know so that we may present your suggestions to the Authoring Team.

NCCER Product Development and Revision

13614 Progress Blvd., Alachua, FL 32615

Email: curriculum@nccer.org

Online: www.nccer.org/olf

❑ Trainee Guide ❑ AIG ❑ Exam ❑ PowerPoints Other ____________________

Craft / Level: Copyright Date:

Module ID Number / Title:

Section Number(s):

Description:

Recommended Correction:

Your Name:

Address:

Email: Phone:

22207-13

Excavation Math

Overview

Math is an everyday reality in excavation work. Operators must be able to determine a volume of soil to remove, calculate load weights, and calculate the area of a cut or fill.

Module Three

Trainees with successful module completions may be eligible for credentialing through NCCER's National Registry. To learn more, go to **www.nccer.org** or contact us at **1.888.622.3720**. Our website has information on the latest product releases and training, as well as online versions of our *Cornerstone* newsletter and Pearson's product catalog.

Your feedback is welcome. You may email your comments to **curriculum@nccer.org**, send general comments and inquiries to **info@nccer.org**, or fill in the User Update form at the back of this module.

This information is general in nature and intended for training purposes only. Actual performance of activities described in this manual requires compliance with all applicable operating, service, maintenance, and safety procedures under the direction of qualified personnel. References in this manual to patented or proprietary devices do not constitute a recommendation of their use.

22207-13
Excavation Math

Objectives

When you have completed this module, you will be able to do the following:

1. Explain how to use formulas.
   a. Explain the sequence of operations in solving a problem using a formula.
   b. Explain how squares and square roots are derived.
   c. Define angles and identify the types of angles.
2. Explain how math is used to solve right triangle problems.
   a. Explain how to determine the length of a slope.
   b. Explain how a building is laid out using right triangle math.
3. Define area and explain why determining the area of a space is required.
   a. Determine the area of squares and rectangles.
   b. Determine the area of a triangle.
   c. Determine the area of a trapezoid.
   d. Determine the area of a circle.
4. Define volume and explain the purpose of calculating volume.
   a. Calculate the volume of a cube.
   b. Calculate the volume of a prism.
   c. Calculate the volume of a cylinder.
   d. Describe the estimating process used to determine the volume and weight of simple and complex excavations.

Performance Task

Under the supervision of the instructor, you should be able to do the following:

1. Using information provided by the instructor, calculate the volume and weight of a given excavation project.

Trade Terms

Average
Constant
Hypotenuse
Parallel
Parallelogram
Quadrilateral
Squared
Variable

Required Trainee Materials

1. Pencil and paper
2. Ruler
3. Calculator

Industry Recognized Credentials

If you are training through an NCCER-accredited sponsor, you may be eligible for credentials from NCCER's Registry. The ID number for this module is 22207-13. Note that this module may have been used in other NCCER curricula and may apply to other level completions. Contact NCCER's Registry at 888.622.3720 or go to **www.nccer.org** for more information.

Contents

Topics to be presented in this module include:

Figures and Tables

69D

SECTION ONE

1.0.0 WORKING WITH FORMULAS AND EQUATIONS

Objective 1

Explain how to use formulas.

a. Explain the sequence of operations in solving a problem using a formula.
b. Explain how squares and square roots are derived.
c. Define angles and identify the types of angles.

Trade Terms

Constant: A value in an equation that is always the same; for example pi is always 3.14.

Variable: A value in an equation that depends on the factors being considered; for example, the lengths of the sides of a triangle may vary from one triangle to another.

Squared: Multiplied by itself.

During excavations, heavy equipment operators need to understand how to calculate the volume of material to be removed from or brought into a site. Once the volume of material is determined, it is necessary to calculate its weight. Knowing the volume and weight of excavation material is necessary for many reasons, but the most important is safety. Because heavy equipment is rated by weight and volume, it is important to make sure that a load does not exceed the capacity of the machine being used. Overloading a vehicle makes it harder to operate and places workers in danger of an accident. In addition, overloading a vehicle can damage it, requiring costly repairs.

In studying this module, resist the desire to skip practice problems. Math is just like anything else—skill comes with practice. Initially, calculations may seem difficult, but practice will bring steady improvement.

Excavations require determining the volume of a cut or fill area. To determine the volume of any excavation, you will first calculate the area of the shape.

Math is governed by formulas, which are equations made up of letters and symbols. The letters represent values and symbols (mathematical signs such as + and ×) define what to do. Study the following formula for calculating the area of a circle:

$$a = \pi r^2$$

In this case, the letter *a* means area, π means pi (pronounced *pie*), and the letter *r* means radius. To read this formula, you would say, "Area is equal to pi times the radius squared." When you see an expression such as πr^2 in a formula, it means that you must multiply the values. In this case, you multiply the radius squared by π.

Some values in formulas are **variables**. This means that the value is not a single number, so any number can replace a variable. In this sample formula, r is a variable. No matter what size a circle is, and thus no matter what its radius is, its area can be determined with this formula.

Some values, however, are **constant**. Pi is a constant. Its value is always 3.14159, often rounded off to 3.14. Whenever you see the word pi or its symbol (π) in a formula, you know to replace it with 3.14.

Equations are collections of numbers, symbols, and mathematical operators connected by equal signs (=). Everything on the left of the equal sign must match the right side. Consider the following equation for calculating the area of a rectangle:

$$\text{Area} = l \times w$$

In this case, the letter *l* means length and the letter *w* means width. The formula means multiply the length by the width. It can also be written as lw, without the multiplication sign. No multiplication sign is required when the intended relationship between symbols and letters is clear. For example, 2l means two times l.

Complicated equations must be solved by performing the indicated operations in a prescribed order: parentheses, exponents, multiply and divide, and add and subtract (PEMDAS). Always move from left to right when performing multiplication and division, and addition and subtraction. For example, the following equation can result in a number of answers if the PEMDAS order is not followed:

$$(3 + 3) \times 2 - 6 \div 3 + 1 = ?$$

Step 1 Parentheses:

$$\underline{(3 + 3)} \times 2 - 6 \div 3 + 1 = ?$$

Step 2 Multiply and divide:

$$\underline{6 \times 2} - 6 \div 3 + 1 = ?$$

$$12 - \underline{6 \div 3} + 1 = ?$$

Step 3 Add and subtract:

$$\underline{12 - 2} + 1 = ?$$

$$\underline{10 + 1} = ?$$

$$\text{Result} = 11$$

When none of the numbers are grouped within parentheses, the process is as follows:

$$3 + 3 \times 2 - 6 \div 3 + 1 = ?$$

Step 1 Multiply and divide:

$$3 + \underline{3 \times 2} - 6 \div 3 + 1 = ?$$

$$3 + 6 - \underline{6 \div 3} + 1 = ?$$

Step 2 Add and subtract:

$$\underline{3 + 6} - 2 + 1 = ?$$

$$\underline{9 - 2} + 1 = ?$$

$$\underline{7 + 1} = ?$$

$$\text{Result} = 8$$

> **NOTE**
> When solving problems in this module, round to the nearest tenth as you progress.

1.1.0 Squares and Square Roots

The formula for the area of a circle is one of the formulas that involve squared numbers. You will also work with formulas in which you must find the square root of a given number.

> ## Converting Fractions to Decimals
>
> When dimensions are stated in inches and fractions of inches, it is easier to work with them if you convert the dimensions to decimal inches. Adding fractions can be difficult, and often results in errors. Decimals are much easier to add. With some practice, you can do most of the conversions in your head.

A square is the product of a number or quantity multiplied by itself. For example, the square of 6 means 6 × 6. To denote a number as squared, simply place the exponent 2 above and to the right of the base number. An exponent is a small figure or symbol placed above and to the right of another figure or symbol to show how many times the latter is to be multiplied by itself. For example:

$$6^2 = 6 \times 6 = 36$$

$$6^2 = 6 \times 6 \times 6 = 216$$

The square root of a number is the divisor which, when multiplied by itself (squared), gives the number as a product. Extracting the square root refers to a process of finding the equal factors which, when multiplied together, return the original number. The process is identified by the radical symbol [√]. This symbol is a shorthand way of stating that the equal factors of the number under the radical sign are to be determined. Finding the square root is necessary in many calculations, including those involving right triangles.

For example, $\sqrt{16}$ is read as the square root of 16. The number consists of the two equal factors 4 and 4. Thus, when 4 is squared, it is equal to 16. Again, squaring a number simply means multiplying the number by itself.

The number 16 is a perfect square. Numbers that are perfect squares have whole numbers as the square roots. For example, the square roots of perfect squares 4, 25, 36, 121, and 324 are the whole numbers 2, 5, 6, 11, and 18, respectively.

Squares and square roots can be calculated by hand, but the process is very time consuming and subject to error. Most people find squares and square roots of numbers using a scientific calculator like the one shown in *Figure 1*. To find the square of a number, the calculator's square key [x^2] is used. When pressed, it takes the number shown in the display and multiplies it by itself. For example, to square the number 4.235, you would enter 4.235, press the [x^2] key, then read 17.935225 on the display.

Similarly, to find the square root of a number, the calculator's square root key [√] or [$\sqrt{}_x$] is used. When pressed, it calculates the square root of the number shown in the display. For example, to find the square root of the number 17.935225, enter 17.935225; press the [√] or [$\sqrt{}_x$] key, then read 4.235 on the display.

> **NOTE**
> On some calculators, the [√] or [$\sqrt{}_x$] key must be pressed before entering the number.

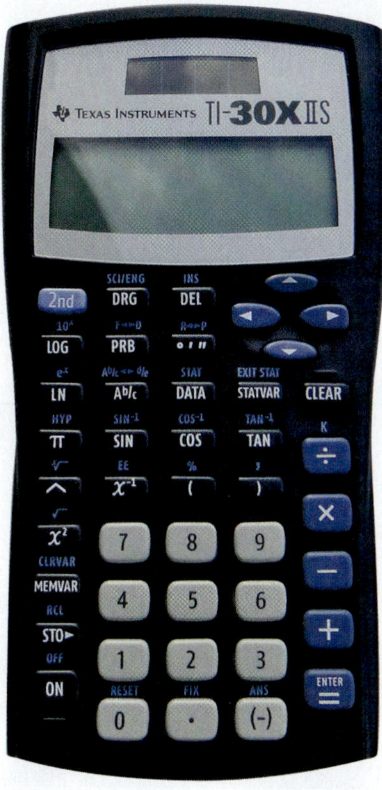

Figure 1 Scientific calculator.

1.2.0 Using Formulas to Solve Problems

To solve problems using formulas, enter the measurements into the equation that makes up the formula. For example, let's say you need to know the area of a round access cover located in one area of a job site. See *Figure 2*. You would measure across the cover and determine that its diameter is four feet. Its radius (r) is half the size of the diameter, or two feet. To determine the area of the round access cover, this number can be plugged into the formula for the area of a circle (area = πr^2) in place of the letter r. Look at the formula with the numbers that represent the variable (r) and the constant (π) plugged into the equation:

$$a = \pi r^2$$
$$a = (3.14)(2^2)$$

Notice that 3.14 and 2^2 have been placed inside parentheses. When numbers are enclosed in parentheses, calculations inside each set of parentheses must be finished before completing any other calculations. In this case, you will first find 2^2 by multiplying 2×2:

$$a = (3.14)(4)$$
$$a = 12.56 \text{ square feet}$$

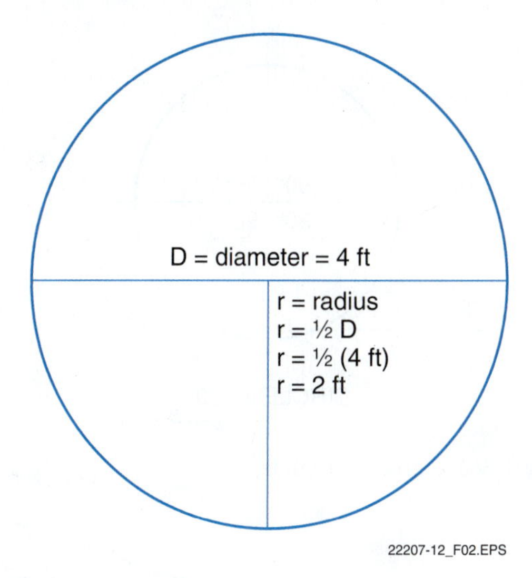

Figure 2 Access cover.

The parentheses in this equation are a grouping symbol. Grouping symbols are just like punctuation in writing. Imagine how hard it would be to read this module if there were no periods or commas to tell you where to stop or pause. Grouping symbols help make sense of an equation, just like punctuation helps make sense of a sentence. Grouping symbols identify numbers that belong together and which functions to perform first. It is important to pay attention to how terms are grouped in a formula and to do the calculations in the right order. In more complex problems, additional types of grouping symbols, such as square brackets [ ] and braces { }, are used.

1.3.0 Angles

Angles are measured in degrees and can be related to a circle. As shown in *Figure 3*, a circle contains 360 degrees. It can be evenly divided into four parts that each contain 90 degrees. An angle is made when two straight lines meet (*Figure 4*). The point where they meet is called a vertex (point B in *Figure 4*). The two lines are the sides of the angle. These lines are called the rays of the angle. The angle is the amount of opening that exists between the rays. It is measured in degrees. Two ways are commonly used to identify angles. One is to assign a letter to the angle, such as angle D shown in *Figure 4*. This is written: ∠D. The other way is to name the two end points of the rays and put the vertex letter between them; for example, ∠ABC. When showing the angle measure in degrees, it should be written inside the angle, if possible. If the angle is too small to show the measurement, it may be put outside of the angle and an arrow drawn to the inside.

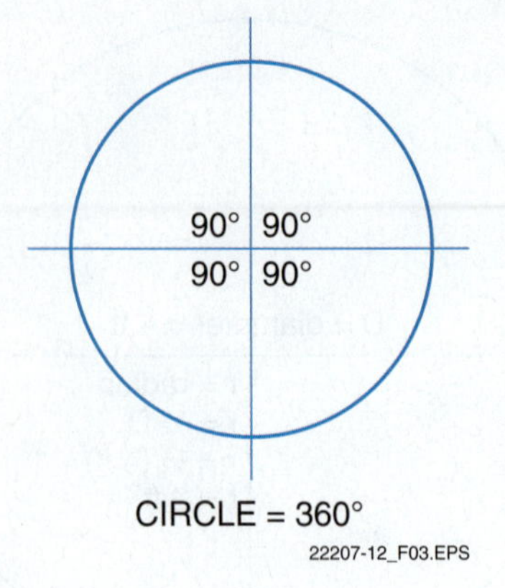

Figure 3 360 degrees in a circle.

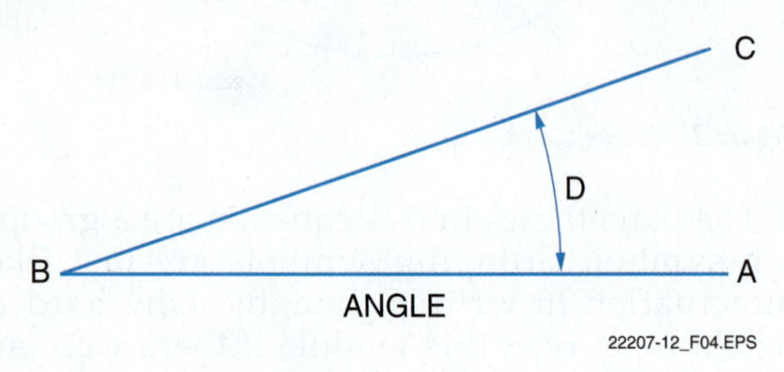

Figure 4 An angle.

There are several kinds of angles:

- *Right angle* – A right angle has rays that are perpendicular to one another (*Figure 5A*). The measure of this angle is always 90 degrees. The right angle is used often in construction, so remember that it is always 90 degrees.
- *Straight angle* – A straight angle (*Figure 5B*) does not look like an angle at all. The rays of a straight angle lie in a straight line, and the angle measures 180 degrees.
- *Acute angle* – An acute angle has less than 90 degrees (*Figure 5C*).
- *Obtuse angle* – An obtuse angle is greater than 90 degrees, but less than 180 degrees (*Figure 5D*).
- *Adjacent angles* – When three or more rays meet at the same vertex, the angles formed are adjacent (next to) one another. In *Figure 6A*, the angles ∠ABC and ∠CBD are adjacent angles. The ray BC is common to both angles.

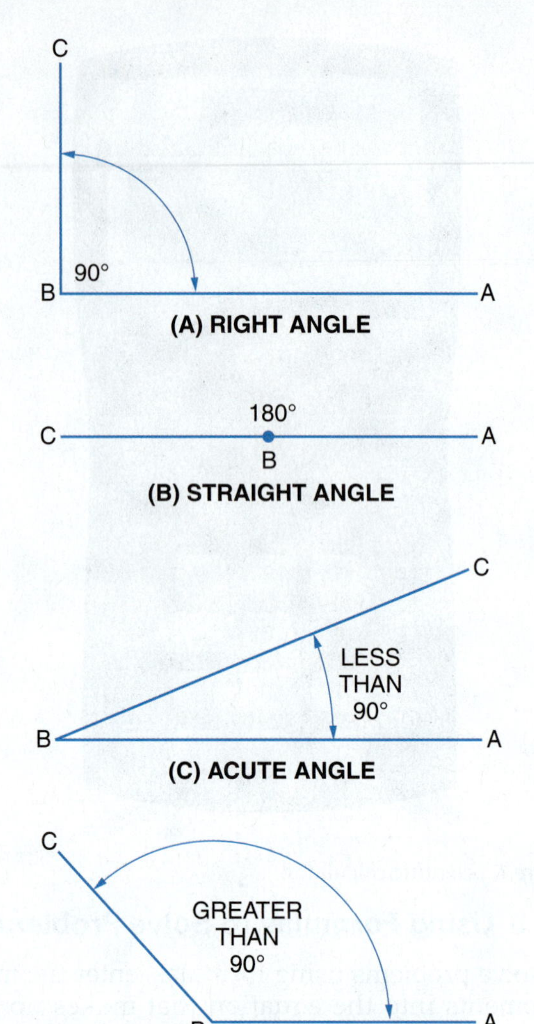

Figure 5 Right, straight, acute, and obtuse angles.

- *Complementary angles* – Two adjacent angles that have a combined total measure of 90 degrees. In *Figure 6B*, ∠DEF is complementary to ∠FEG.
- *Supplementary angles* – Two adjacent angles that have a combined total measure of 180 degrees. In *Figure 6C*, ∠HIJ is supplementary to ∠JIK.

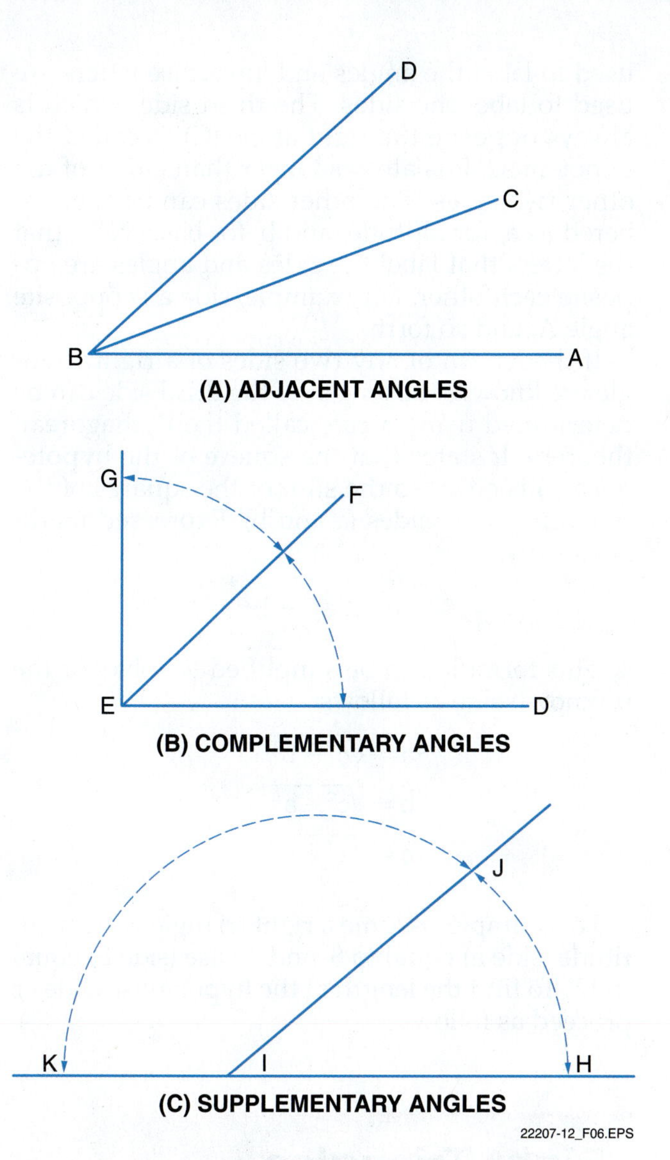

Figure 6 Adjacent, complementary, and supplementary angles.

Additional Resources

Applied Construction Math: A Novel Approach, Alachua, FL: NCCER.

1.0.0 Section Review

1. When solving an equation, the first step is to _____.
   a. multiply
   b. divide
   c. add
   d. subtract

2. The cube of 4 is _____.
   a. 12
   b. 16
   c. 48
   d. 64

3. An angle containing more than 90 degrees is known as a(n) _____.
   a. obtuse angle
   b. acute angle
   c. right angle
   d. adjacent angle

SECTION TWO

2.0.0 WORKING WITH RIGHT TRIANGLES

Objective 2

Explain how math is used to solve right triangle problems.

a. Explain how to determine the length of a slope.
b. Explain how a building is laid out using right triangle math.

Trade Terms

Hypotenuse: The long dimension of a right triangle and always the side opposite the right angle.

The right triangle is perhaps the most used shape in construction. Any vertical object or structure, such as a column or the face of a building, is part of a right triangle. If you draw an imaginary line from a point on the ground to the top of the structure, such as the top of a column or the roof of a building, that line forms the **hypotenuse** of a right triangle. The base of the triangle extends from the bottom of the pole or structure to the point on the ground that was the starting point for your line.

2.1.0 Right Triangle Calculations

Because the right triangle has one right angle, the other two angles are acute angles. They are also complementary angles, the sum of which equals 90 degrees. The right triangle has two sides perpendicular to each other, thus forming the right angle. To aid in writing equations, the sides and angles of a right triangle are labeled as shown in *Figure 7*. Normally, capital (uppercase) letters are used to label the angles and lowercase letters are used to label the sides. The third side, which is always opposite the right angle (C), is called the hypotenuse. It is always longer than either of the other two sides. The other sides can be remembered as a, for altitude, and b, for base. Note that the letters that label the sides and angles are opposite each other. For example, side a is opposite angle A, and so forth.

If the length of any two sides of a right triangle are known, the length of the third side can be determined using a rule called the Pythagorean theorem. It states that the square of the hypotenuse (c) is equal to the sum of the squares of the remaining two sides (a and b). Expressed mathematically:

$$c^2 = a^2 + b^2$$

This formula can be simplified to solve for the unknown side as follows:

$$a = \sqrt{c^2 - b^2}$$

$$b = \sqrt{c^2 - a^2}$$

$$c = \sqrt{a^2 + b^2}$$

For example, assume a right triangle with an altitude (side a) equal to 8' and a base (side b) equal to 12'. To find the length of the hypotenuse (side c), proceed as follows:

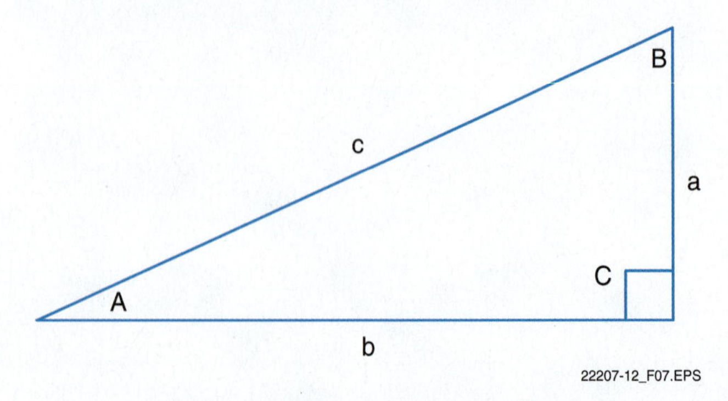

Figure 7 Labeling of angles and sides of a right triangle.

Right Triangles

Pythagoras was a Greek philosopher and mathematician who lived about 2,500 years ago. He is credited with developing the math for solving right triangle problems. Fifty years before Pythagoras, Thales, another Greek mathematician, figured out how to measure the height of the Egyptian pyramids using a technique that later became known as trigonometry. Trigonometry recognized that there is a relationship between the size of an angle and the lengths of the sides of a right triangle.

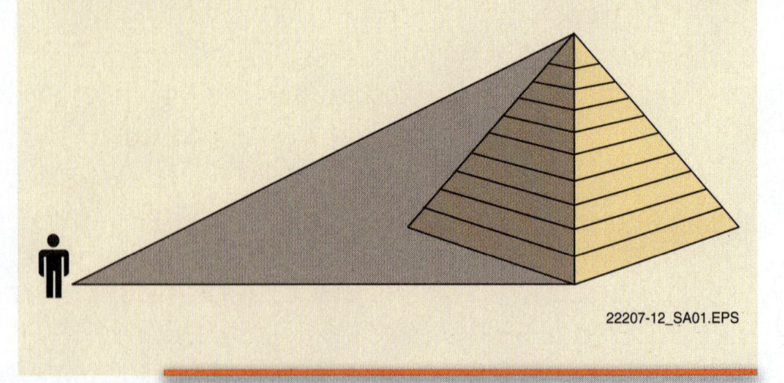

$$c = \sqrt{a^2 + b^2}$$
$$c = \sqrt{8^2 + 12^2}$$
$$c = \sqrt{64 + 144}$$
$$c = \sqrt{208}$$
$$c = 14.422'$$

To determine the actual length of the hypotenuse using this formula, it is necessary to calculate the square root of the sum of the sides squared. Fortunately, this is easy to do using a scientific calculator. On many calculators, you simply key in the number and press the square root [√] key. On some calculators, the square root does not have a separate key. Instead, the square root function is the inverse of the [x^2] key, so you have to press [INV] or [2nd F], depending on the calculator, followed by [x^2], to obtain the square root.

For an example of the Pythagorean theorem, let's say you know the length of a support cable attached to the top of a telephone pole. You wish to know the height of the pole, but measuring it from the ground would not be practical. You can, however, easily measure the distance between the end of the cable attached to the ground and the bottom of the pole—which forms the base of a right triangle.

If the length of the cable (the hypotenuse, or *c* from the equation) is 25' and the distance from the cable to the base of the pole is 10', the c and b parts of formula are known:

$$c^2 = a^2 + b^2$$
$$25^2 = a^2 + 10^2$$
$$625 = a^2 + 100$$

Subtracting 100 from both sides of the equation yields the following:

$$525 = a^2$$

The square root of 525, therefore, is the height of the pole (roughly 22.9').

2.2.0 Laying Out and Checking 90-Degree Angles Using the 3-4-5 Rule

The 3-4-5 rule describes a simple method for laying out or checking 90-degree angles (right angles) and requires only the use of a tape measure. The rule is based on the Pythagorean theorem and has been used in building construction for centuries. The numbers 3-4-5 represent dimensions in feet that describe the sides of a right triangle. Right triangles that are multiples of the 3-4-5 triangle, such as 9-12-15, 12-16-20, 15-20-25, and 30-40-50, are commonly used. The specific multiple used is determined by the relative distances involved in the job being laid out or checked.

An example of the 3-4-5 rule using the multiples 15-20-25 is shown in *Figure 8*. In order to square or check a corner as shown in the example, first measure and mark 15' down the line in one direction, then measure and mark 20' down the line in the other direction. The distance measured between the 15' and 20' points must be exactly 25' to ensure that the angle is a perfect right angle.

The specific multiple used is determined mainly by the relative distances involved in the job being laid out or checked. It is best to use the highest multiple that is practical. When smaller multiples are used, any error made in measurement results in a much greater angular error.

Figure 9 shows an example of the 3-4-5 rule involving the multiple 48-64-80 (multiple of 16). In order to square or check a corner as shown in the

The 3-4-5 Rule

The 3-4-5 rule has been used in construction for centuries. It is a simple method for laying out and checking 90-degree angles (right angles). It requires only the use of a tape measure. Because they are easy to check, right triangles based on the 3-4-5 rule are commonly used in construction. Shown below is an example of the 3-4-5 rule using the multiples of 15-20-25. In order to check or square a corner, first measure and mark 15' down the line in one direction Then measure and mark 20' down the line in the other direction. The distance between the 15' and 20' points must be 25' for a perfect right triangle.

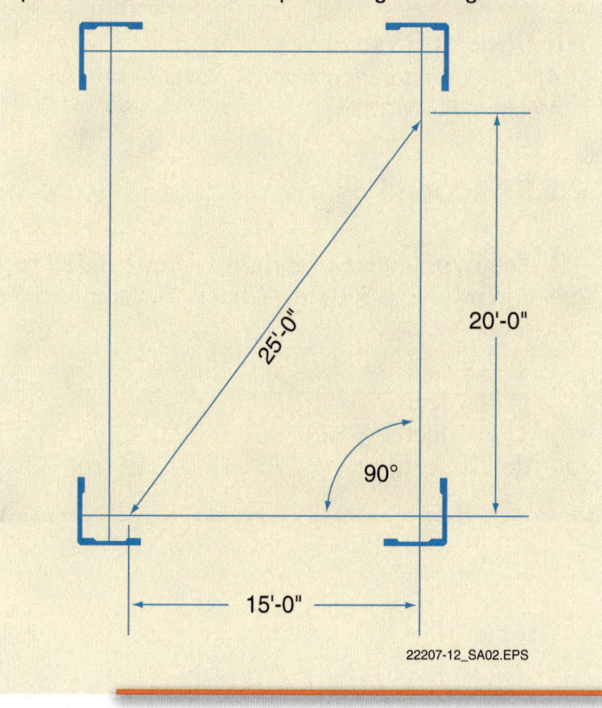

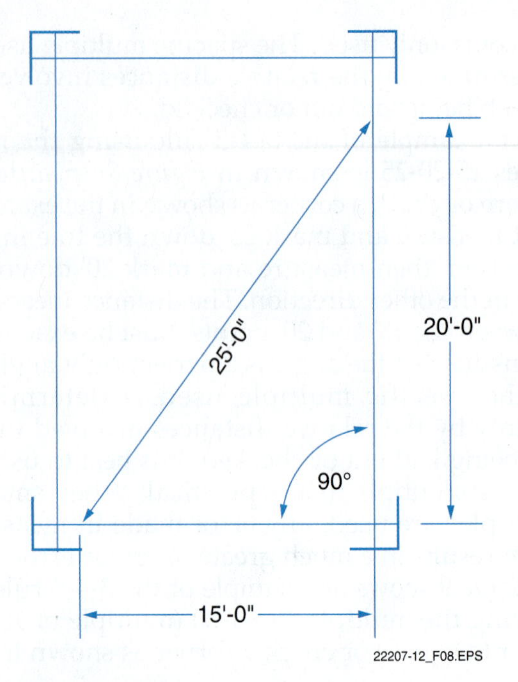

Figure 8 The 3-4-5 rule.

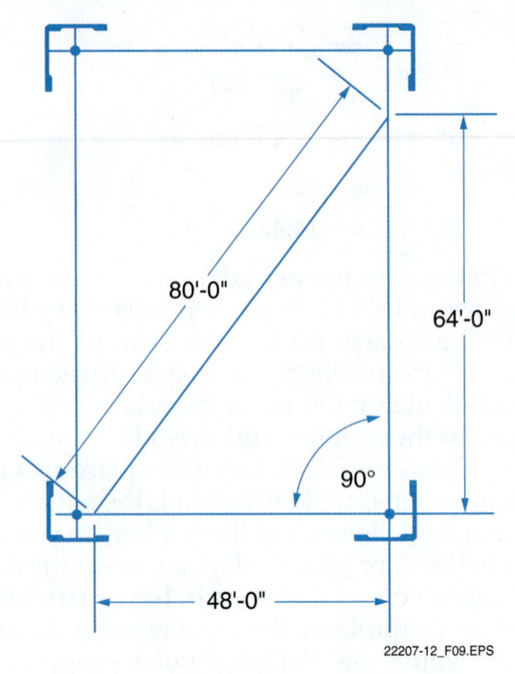

Figure 9 Checking a layout using the 3-4-5 rule.

example, first measure 48' down the line in one direction, then 64' down the line in the other direction. The distance measured between the 48' and 64' points must be exactly 80' if the angle is to be a perfect right angle. If the measurement is not exactly 80', the angle is not 90 degrees. This means that the direction of one of the lines or the corner point must be adjusted until a right angle exists.

It cannot be emphasized enough that exact measurements are necessary to get the desired results when using the 3-4-5 method of laying out or checking a 90-degree angle. Any error in the measurements of the distances will result in not establishing a right angle. If an existing 90-degree angle is being checked, inaccurate measurements may cause an unnecessary adjustment.

Additional Resources

Applied Construction Math: A Novel Approach. Alachua, FL: NCCER.

2.0.0 Section Review

1. The hypotenuse of a right triangle is 720 feet and one leg is 530 feet. What is the length of the other leg?
   a. 320 feet
   b. 487 feet
   c. 647 feet
   d. 720 feet

2. Using the 3-4-5 method, measurements that would be used to verify the squareness of the building lines for a 24' × 32' building are _____.
   a. 21-28-35
   b. 24-32-40
   c. 24-32-48
   d. 20-40-60

Section Three

3.0.0 Calculating Area

Objective 3

Define area and explain why determining the area of a space is required.

a. Determine the area of squares and rectangles.
b. Determine the area of a triangle.
c. Determine the area of a trapezoid.
d. Determine the area of a circle.

Trade Terms

Average: The middle point between two numbers or the mean of two or more numbers. It is calculated by adding all numbers together, and then dividing the sum by the quantity of numbers added. For example, the average (or mean) of 3, 7, 11 is 7 (3 + 7 + 11 = 21; 21 ÷ 3 = 7).

Parallel: Two lines that are always the same distance apart even if they go on into infinity (forever is called infinity in mathematics).

Parallelogram: A two-dimensional shape that has two sets of parallel lines.

Quadrilateral: A four-sided, closed shape with four angles whose sum is 360 degrees.

Area is the measurement of the amount of space on a flat surface. Houses and apartments are advertised using area—area determines how much floor space is available. For example, a 12' × 12' room has 144 square feet of floor space, and that is its area.

Area is measured in square units, such as square inches, feet, yards, and miles (in the metric system, square centimeters, meters, and kilometers). This unit of measure is often written with the abbreviation inches^2, feet^2, meters^2, or in^2, ft^2, m^2.

A square inch is the area in a shape that is 1 inch long and 1 inch wide. Shapes that are measured in square units are called two-dimensional because they have only two measurements. *Figure 10A* shows a block that has an area of 1 square inch. The block is 1 inch long and 1 inch wide so it is a 1-inch square and has an area of 1 square inch. *Figure 10B* shows a larger block. That block is 2 inches long and 1 inch wide and contains two 1-inch squares, so it has an area of 2 square inches.

You have probably already guessed that *Figure 10C* shows a block that has an area of 4 square inches, because it is 4 inches long and 1 inch wide and contains four 1-inch squares.

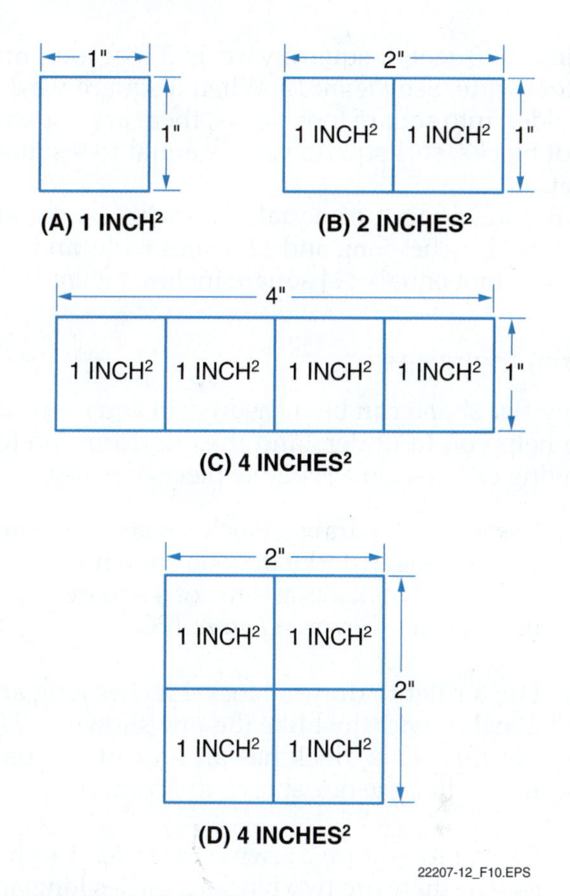

Figure 10 Square inches.

See *Figure 10D*. The block still contains four 1-inch squares so it has an area of 4 square inches just like *Figure 10C*, but this block is 2 inches long and 2 inches wide. (Calculating the area of three-dimensional objects is covered in later sections.) In *Figure 10D*, the shape of the surface does not matter—it is still the same area as the shape in *Figure 10C*.

Just as square inch is the area on the surface of a shape that is 1 inch long and 1 inch wide, a square foot is the area on the surface of a shape that is 1 foot long and 1 foot wide. See *Figure 11*.

When a shape is 1 yard long and 1 yard wide, it has the area of 1 square yard. Since 1 yard is

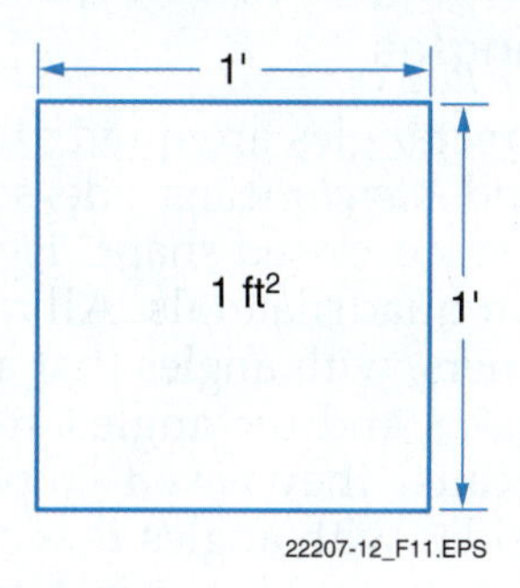

Figure 11 One square foot.

equal to 3 feet, a square yard is 3 feet long and 3 feet wide. See *Figure 12*. When a square yard is divided into square foot blocks, there are 9 square foot blocks, so 1 square yard is equal to 9 square feet.

Because one foot equals 12 inches, a square foot is 12 inches long and 12 inches wide, and one square foot equals 144 square inches (*Figure 13*).

Area Exercises

Any flat shape can be measured in square units. To help you to understand this, perform the following exercise on a separate piece of paper.

1. Use a ruler to draw a block 4 inches long and 1 inch wide just like the one shown in *Figure 10C*. This block has an area of 4 square inches like the one shown in *Figure 10C*.

2. Use a ruler to draw a block 2 inches long and 2 inches wide just like the one shown in *Figure 10D*. This block has an area of 4 square inches like the one shown in *Figure 10D*.

3. Divide the figure drawn in exercise 1 in half so that there are two blocks 2 inches long and 1 inch wide. Since the whole block has an area of 4 square inches, half of the block must have an area of 2 square inches.

4. Divide the figure drawn in exercise 2 in half by drawing a line from its upper right corner to its lower left corner. Since the whole block has an area of 4 square inches, half of the block must have an area of 2 square inches.

5. Compare the lower half of the figure drawn in exercise 4 to the right half of the one drawn in exercise 3. Both have the same area of 2 square inches, but the shapes are quite different.

3.1.0 Calculating Area of Squares and Rectangles

Squares and rectangles are **quadrilaterals**. *Quad* means four and *lateral* means side, so a quadrilateral is a four-sided closed shape. *Figure 14* shows some common quadrilaterals. All quadrilaterals have four corners, with angles that add up to 360 degrees. Squares and rectangles are also **parallelograms** because they have two pairs of opposite **parallel** sides with angles that add up to 360 degrees. Of the shapes shown in *Figure 14*, all but the trapezoid are parallelograms.

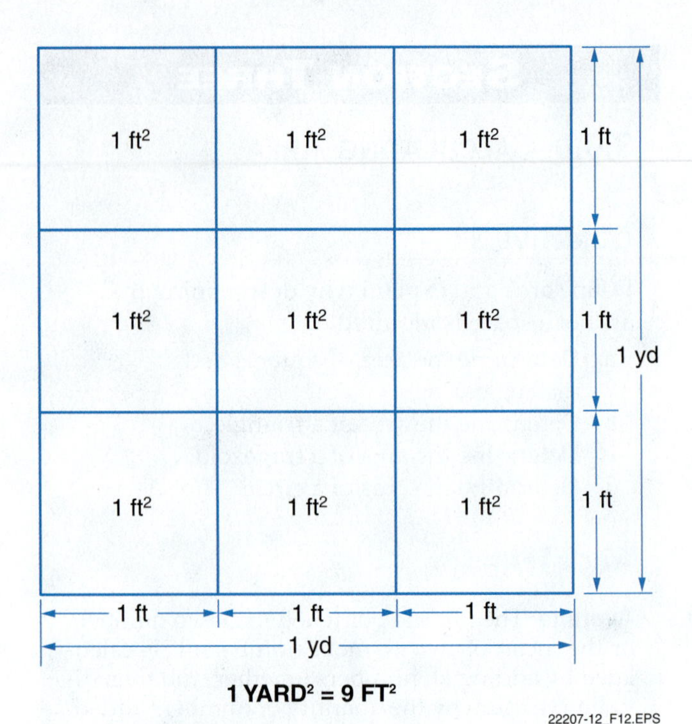

Figure 12 One square yard.

A rectangle (*Figure 15*) is a four-sided shape joined so that four 90-degree angles are formed. The sum of the four angles in any rectangle is 360 degrees. A rectangle has two pairs of equal sides. The longer side is called the length and is designated with the letter *l*, while the shorter side is called the width and is designated with the letter *w*. The area of a rectangle is calculated by multiplying the length times width (l × w or lw).

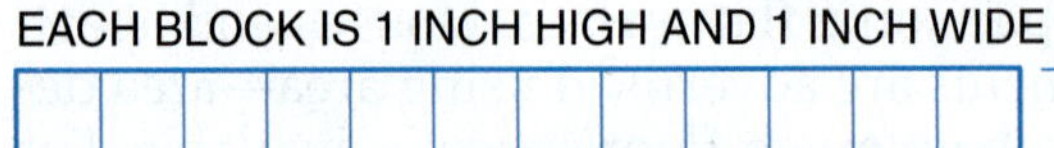

EACH BLOCK IS 1 INCH HIGH AND 1 INCH WIDE

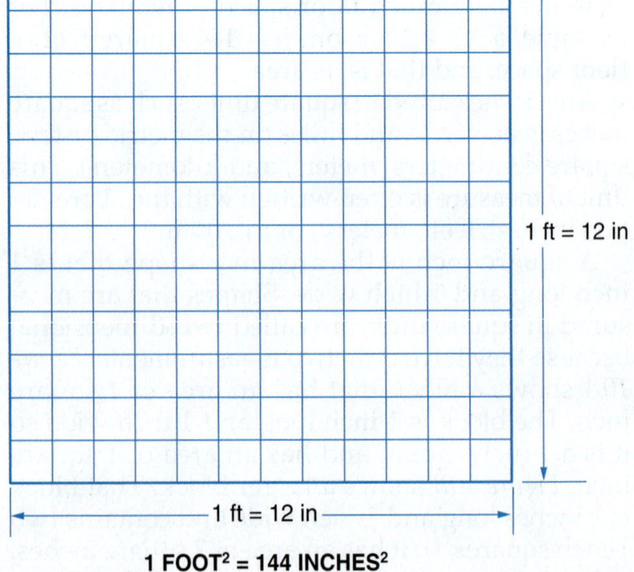

Figure 13 144 square inches.

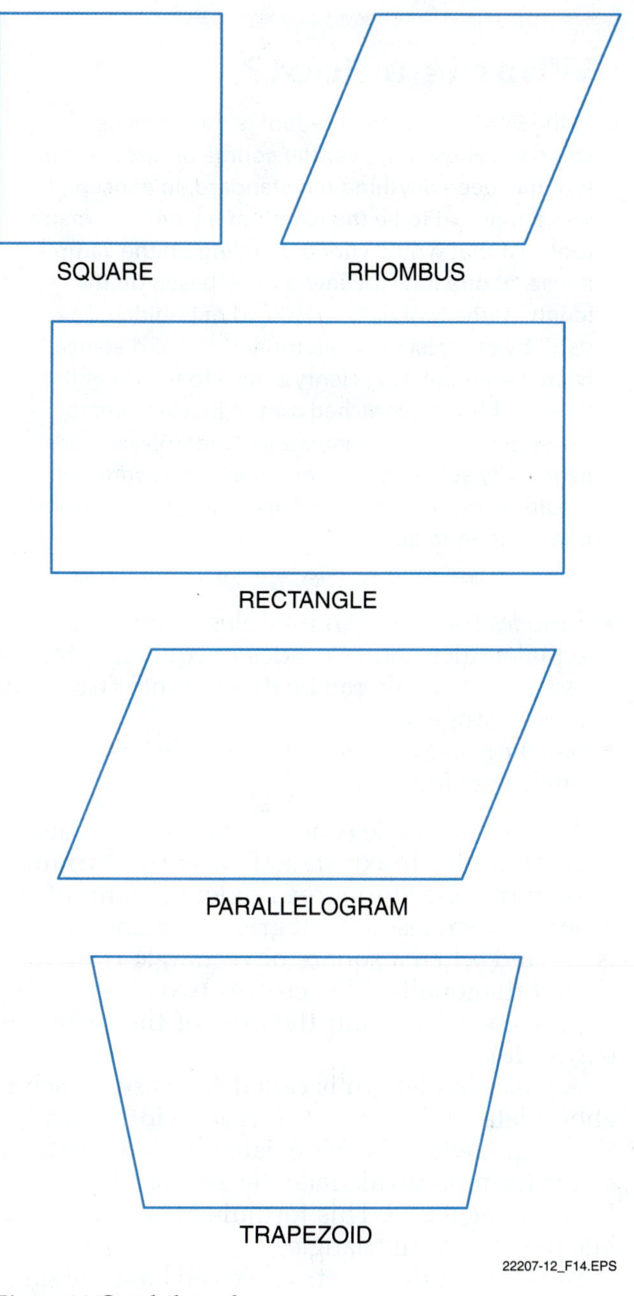

Figure 14 Quadrilaterals.

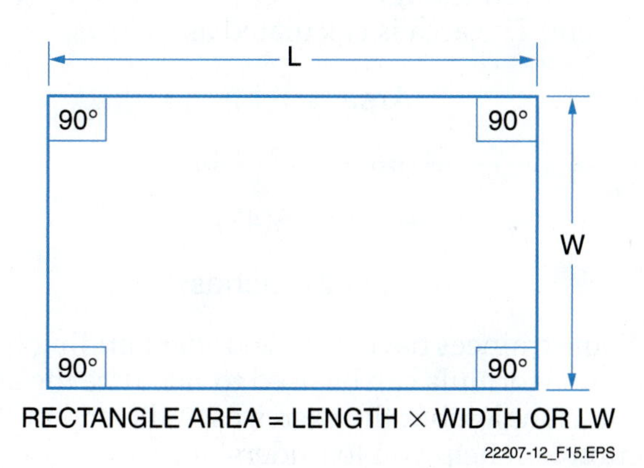

Figure 15 A rectangle.

A rectangle with a length of 3 inches and width of 6 inches has an area of 18 square inches. It is calculated as follows:

$$\text{Area} = l \times w$$

$$\text{Area} = 3 \text{ inches} \times 6 \text{ inches}$$

$$\text{Area} = 18 \text{ inches}^2$$

When you are calculating the area of any object, the numbers multiplied together must be in the same units. For example, a rectangle with a length of 3 inches and a width of 1 foot has an area of 36 square inches, because 1 foot equals 12 inches. It is calculated as follows:

$$\text{Area} = 18 \text{ inches}^2$$

$$\text{Area} = 3 \text{ inches} \times 1 \text{ foot}$$

$$\text{Area} = 3 \text{ inches} \times 12 \text{ inches}$$

(convert 1 foot to 12 inches)

$$\text{Area} = 36 \text{ inches}^2$$

A square (*Figure 16*) is similar to a rectangle. It has four sides that are joined to form four 90-degree angles. The sum of these angles is 360 degrees—just like a rectangle—but all sides of the square are the same length. The sides of the square are labeled with a single letter. In this module, the sides of a square will be labeled with the letter *e*. The area of a square is calculated by multiplying two sides together ($e \times e$ or e^2).

A square that is 3 inches long has an area of 9 square inches, which is calculated as follows:

$$\text{Area} = e^2$$

$$\text{Area} = 3 \text{ inches} \times 3 \text{ inches}$$

$$\text{Area} = 9 \text{ inches}^2$$

A square that is 12 inches long has an area of 144 square inches, which is calculated as follows:

$$\text{Area} = e^2$$

$$\text{Area} = 12 \text{ inches} \times 12 \text{ inches}$$

$$\text{Area} = 144 \text{ inches}^2$$

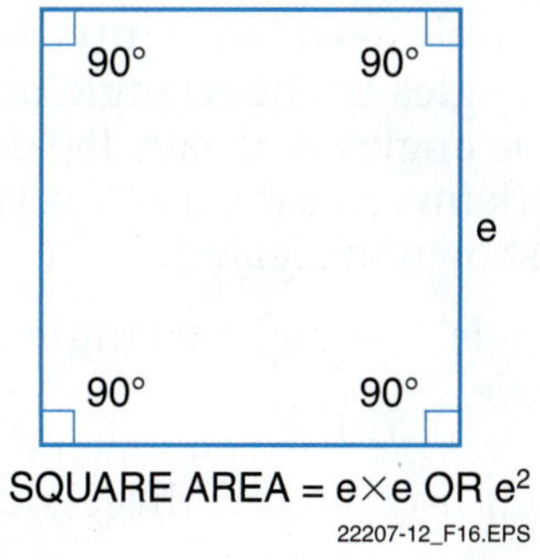

90° 90°

e

90° 90°

SQUARE AREA = e×e OR e^2

22207-12_F16.EPS

Figure 16 A square.

The numbers quickly become large when working with inches, so it is better to work with smaller numbers by converting inches into feet or even yards. In the example above, 12 inches was the value used to arrive at the answer of 144 square feet. By converting the 12 inches to one foot, the area can be calculated as follows:

$$\text{Area} = e^2$$
$$\text{Area} = 1 \text{ foot} \times 1 \text{ foot}$$
$$\text{Area} = 1 \text{ foot}^2$$

These two calculations show that 1 square foot is the same as 144 square inches. If it is necessary to convert square feet to square inches, multiply the square feet by 144 as follows:

$$1 \text{ foot}^2 \times 144 = 144 \text{ inches}^2$$
$$2 \text{ feet}^2 \times 144 = 288 \text{ inches}^2$$
$$3 \text{ feet}^2 \times 144 = 432 \text{ inches}^2$$

To convert square inches to square feet, divide the square inches by 144 as follows:

$$144 \text{ inches}^2 \div 144 = 1 \text{ foot}^2$$
$$432 \text{ inches}^2 \div 144 = 3 \text{ feet}^2$$
$$729 \text{ inches}^2 \div 144 = 5\tfrac{1}{2} \text{ feet}^2$$

Complete the following three exercises by drawing the shape described and then calculating the area of each shape. Hint: always convert the measurements into the same units before calculating the area.

1. The building plans call for a building that is 40 feet long and 30 feet wide.

2. Outside the building, the parking pad is 15 feet by 15 feet.

3. The driveway leading up to the building is 12 yards long and 12 feet wide.

3.2.0 Calculating the Area of a Triangle

A triangle is a three-sided figure that has three angles. The angles in the triangle may vary, but the sum of the angles is always 180 degrees. Construction work involves the use of several types of triangles as shown in *Figure 17*:

- *Right triangle* – A right triangle has one 90-degree angle.
- *Equilateral triangle* – An equilateral triangle has three equal angles and three sides of equal length.
- *Isosceles triangle* – An isosceles triangle has two equal angles and two sides of equal length. An isosceles triangle can be divided into two equal right triangles.
- *Scalene triangle* – A scalene triangle has no equal angles or side lengths.

The right triangle is one of the most frequently used triangles in construction. A right triangle must have one 90-degree angle; the sum of the other two angles is 90 degrees. A right triangle is created when a square or rectangle is divided in half diagonally. This creates two identical triangles, each with half the area of the rectangle (*Figure 18*).

A triangle's length is called the base, which is abbreviated with the letter *b*. Its width is called the height, which is abbreviated with the letter *h*, so the formula to calculate the area of a triangle is ½bh. See *Figure 19*. This formula is used to calculate the area of all triangles.

In *Figure 20*, the four triangles all have the same base (7 inches) and the same height (6 inches), so the area of each triangle is the same (21 square inches), even though their appearances are very different. The area is calculated as follows:

$$\text{Area} = \tfrac{1}{2}\text{bh}$$
$$\text{Area} = \tfrac{1}{2}(7 \times 6)$$
$$\text{Area} = \tfrac{1}{2}(42)$$
$$\text{Area} = 21 \text{ inches}^2$$

Some trainees have trouble understanding how the same formula can be used to calculate the area of all triangles, but the exercises at the end of this section will help you to understand.

What is a Foot?

In the English system, the foot is the distance standard. However, over the course of history, the foot has been anything but standard. In concept, it was supposed to be the length of an average man's foot, but that would vary depending on the sample. It was, at one time, believed to be based on the length of the foot of King Henry I of England. The yard, by contrast was determined as the distance from the end of King Henry's nose to the tip of the thumb of his outstretched arm. When seeking to develop a common, regulated standard, scientists eventually settled on the meter as a measure that would be more precise and less likely to vary from one location to another.

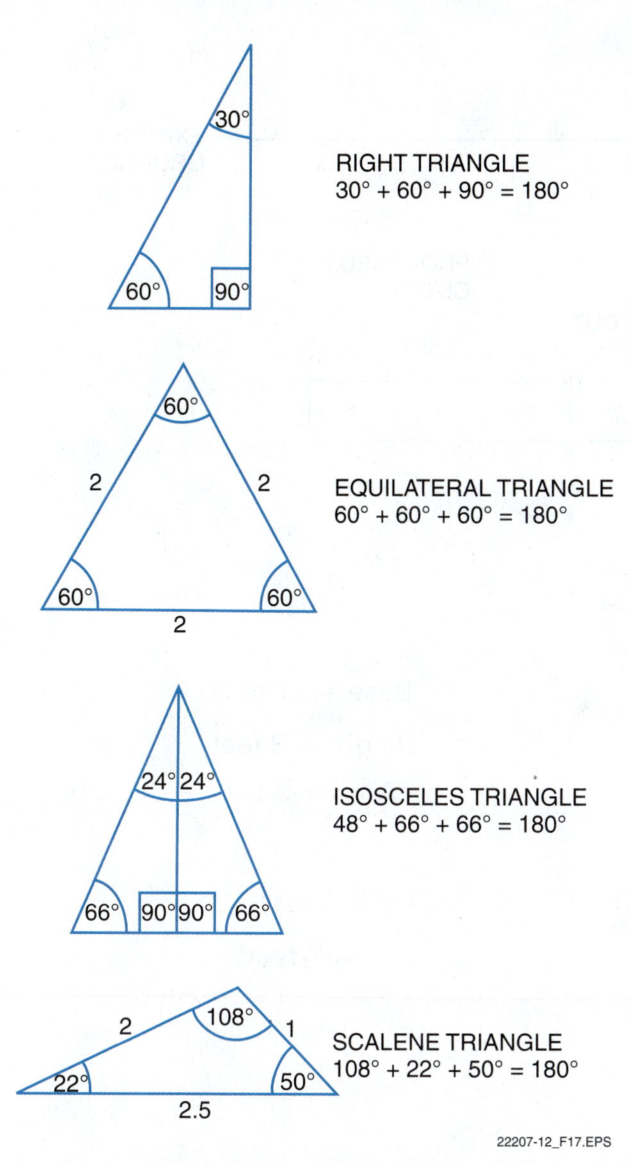

Figure 17 The sum of a triangle's three angles is always 180 degrees.

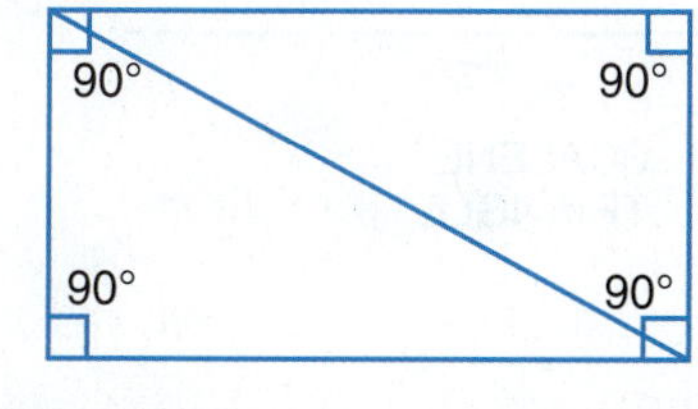

Figure 18 One rectangle, two triangles.

Triangles are related to excavation. See *Figure 21*. This highway job requires the removal of part of the existing ground to make an inslope and backslope. The slope stake directs a cut of 3 feet and a grade of 3:1. This means the ground must drop 1 foot every 3 feet from the stake until the cut depth is 3 feet. In this example, assume that

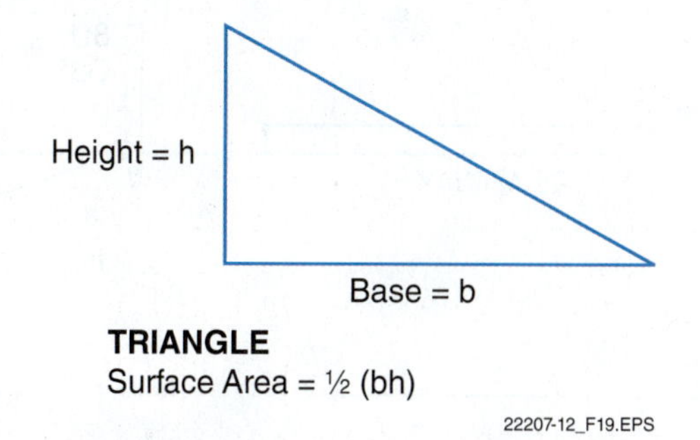

Figure 19 Triangle area.

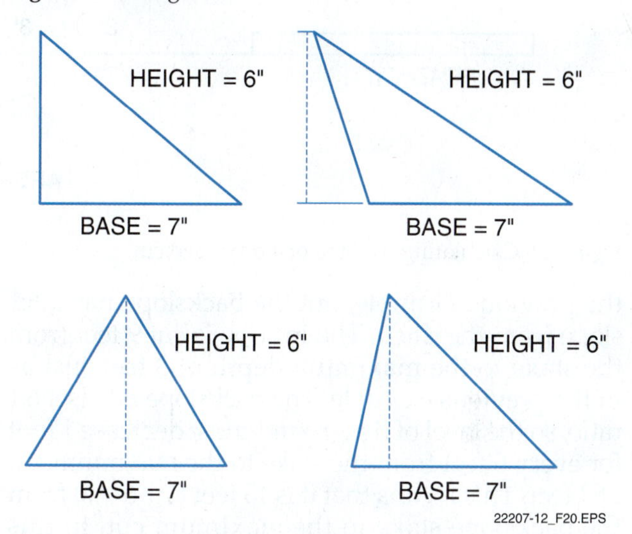

Figure 20 Different triangles, same area.

the backslope will be cut the same way. Look carefully at *Figure 21*. The proposed cut forms an inverted isosceles triangle.

Think about the stake directions. The maximum cut is 3 feet, so the height of the triangle must be 3 feet. A 3:1 slope means the ground must drop 1 foot every 3 feet from the stake. So to the point of maximum cut, the horizontal distance is 9 feet (3 feet + 3 feet + 3 feet). Since the backslope is being cut the same as the inslope, that side is 9 feet, too. So the base of the triangle is 18 feet. You now have all of the information needed to calculate the area of the proposed cut, which is as follows:

$$\text{Base} = 18 \text{ feet}$$
$$\text{Height} = 3 \text{ feet}$$
$$\text{Area} = \tfrac{1}{2}bh$$
$$= \tfrac{1}{2}(18 \times 3)$$
$$= \tfrac{1}{2}(54)$$
$$= 27 \text{ feet}^2$$

This process can be used to calculate the area of any cut that forms a triangle. See *Figure 22*. On this job, the inslope needs to be cut just like the one in

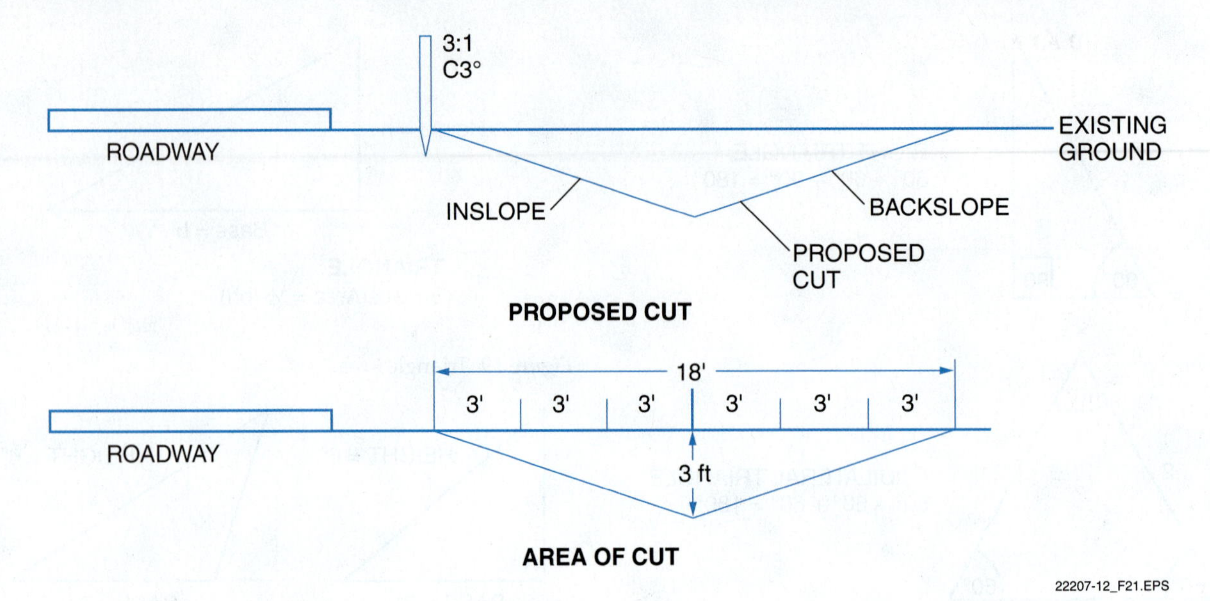

Figure 21 Calculating the area of a proposed cut.

the previous example, but the backslope has a 6:1 slope from the ditch. The inslope is cut 9 feet from the stake to the maximum depth of 3 feet, just as in the previous example. The backslope needs a 6:1 ratio, so the level of the ground must decrease 1 foot for every 6 feet from the stake to the maximum cut of 3 feet. This means that it is 18 feet (6 + 6 + 6) from the backslope stake to the maximum cut. In this case, the cut forms an inverted scalene triangle.

The base of the triangle is 27 feet and the height is 3 feet. You now have all of the information you need to calculate the area of the proposed cut, which is as follows:

$$\text{Base} = 27 \text{ feet}$$

$$\text{Height} = 3 \text{ feet}$$

$$\text{Area} = \tfrac{1}{2}\text{bh}$$

$$= \tfrac{1}{2}(27 \times 3)$$

$$= \tfrac{1}{2}(81)$$

$$= 40\tfrac{1}{2} \text{ feet}^2$$

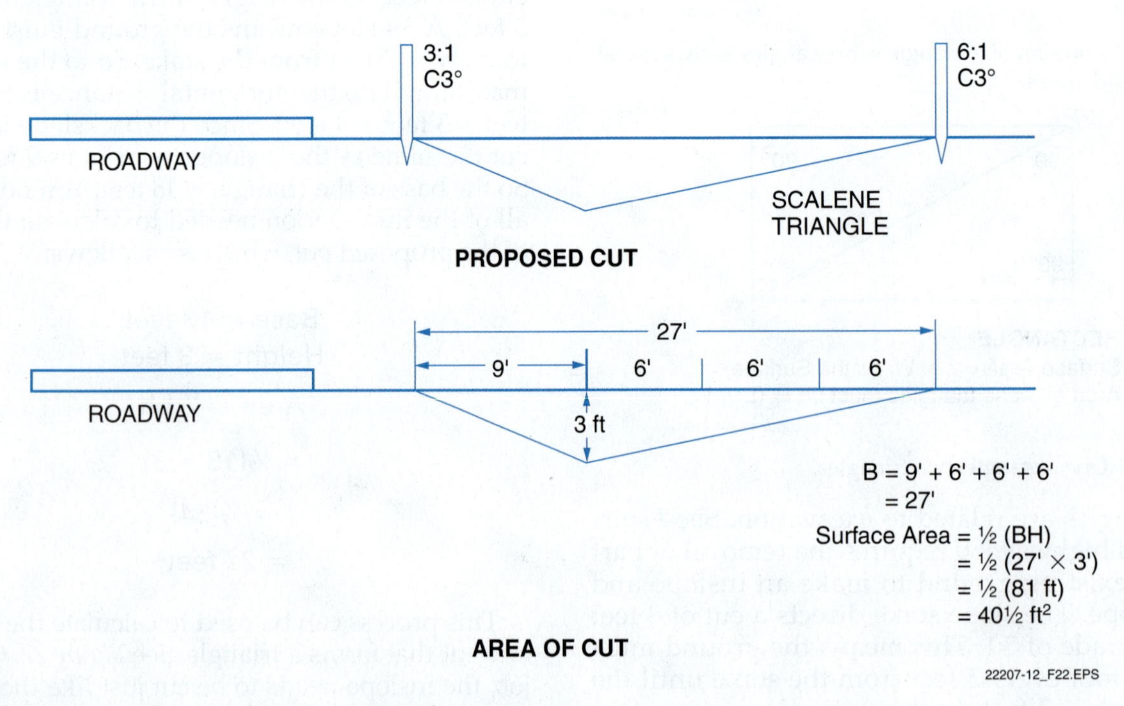

Figure 22 Calculating the area of another proposed cut.

3.2.1 Triangle Area Exercises

The following exercises will provide a better understanding of how to calculate the area of a triangle. Read each question carefully and draw figures as necessary.

1. Find the following for *Figure 23*:

   a. Length =
   b. Width =
   c. Area =

2. Find the following for *Figure 24*:

   Triangle 1:

   a. Base =
   b. Height =
   c. Area =

   Triangle 2:

   d. Base =
   e. Height =
   f. Area =

   Triangles 1 and 2:

   g. Total area =

3. Find the following for *Figure 25*:

   Triangle 1:

   a. Base =
   b. Height =
   c. Area =

Triangle 2:

   d. base =
   e. height =
   f. Area =

   Triangle 3:

   g. Base =
   h. Height =
   i. Area =

   Triangles 1, 2, and 3:

   j. Total area =

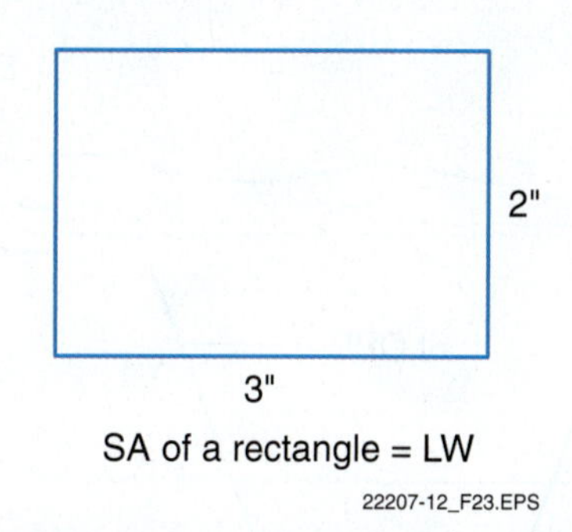

Figure 23 Triangle Exercise 1.

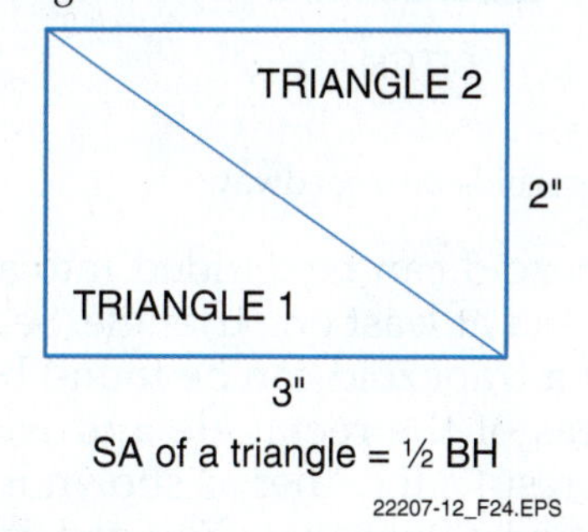

Figure 24 Triangle Exercise 2.

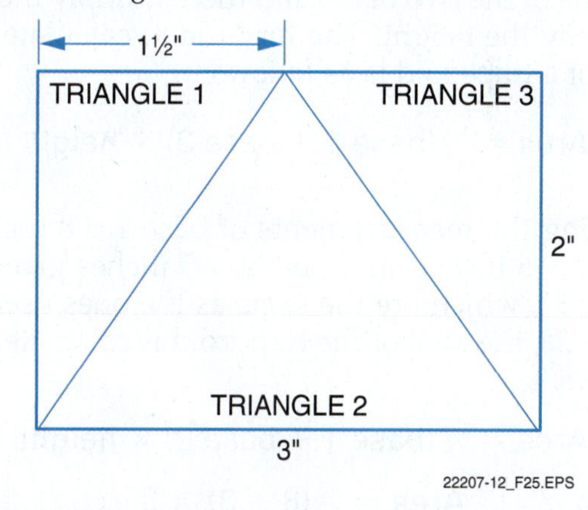

Figure 25 Triangle Exercise 3.

3.3.0 Calculating the Area of a Trapezoid

A trapezoid is also a quadrilateral. It has four sides with two parallel sides. The angles formed by the sides add up to 360 degrees. *Figure 26* shows some examples of trapezoids. Note that the two parallel lines are called base 1 and base 2. *Figure 27* shows the cross-section of a roadway with the shoulder slope, ditch, and backslope. The shape made by the two slopes and ditch is a trapezoid.

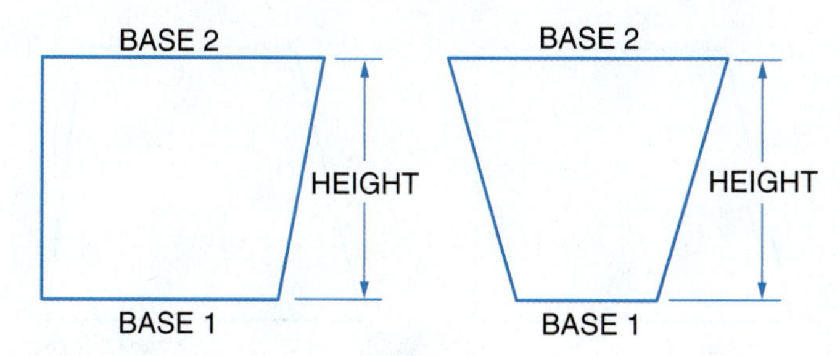

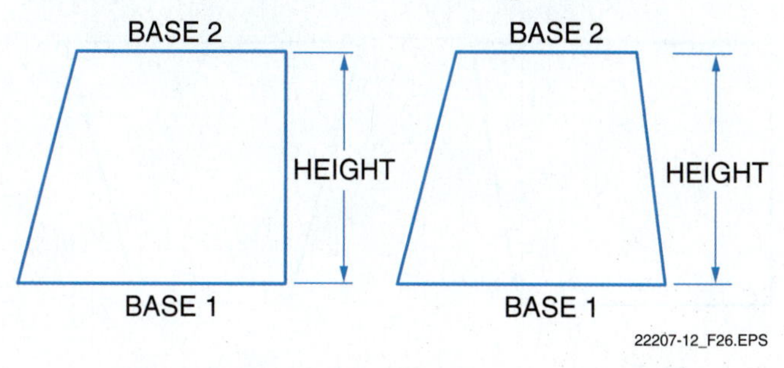

Figure 26 Trapezoids.

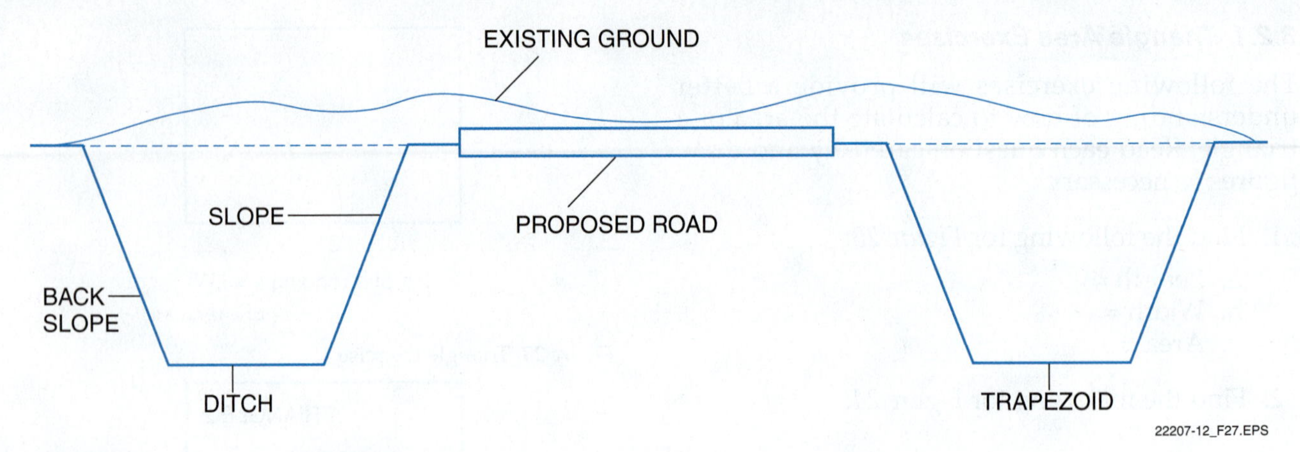

Figure 27 Trapezoids on a roadway.

Any trapezoid can be divided into a rectangle (or square) and at least one triangle. See *Figure 28.* The area of a trapezoid can be found by calculating the areas of the rectangle and triangle and adding the results together as shown in *Figure 29,* but there is an easier way. You can **average** the lengths of the two bases and then multiply the average by the height. The formula to calculate the area of a trapezoid is as follows:

$$\text{Area} = \frac{1}{2}(\text{base 1} + \text{base 2}) \times \text{height}$$

Using the measurements of base 1 = 8 inches, base 2 = 6 inches, and height = 5 inches given in *Figure 30,* which are the same as the ones used in *Figure 28,* the area of the trapezoid is calculated as follows:

$$\text{Area} = \frac{1}{2}(\text{base 1} + \text{base 2}) \times \text{height}$$

$$\text{Area} = \frac{1}{2}(8 + 3) \times 5$$

$$\text{Area} = \frac{1}{2}(11) \times 5$$

$$\text{Area} = 5\frac{1}{2} \times 5$$

$$\text{Area} = 27\frac{1}{2} \text{ inches}^2$$

It may help to remember the trapezoid formula by relating it to the formula for a rectangle. The rectangle in *Figure 31* has a length of 3 inches and a width of 2 inches, so it has an area of 6 square inches. Using the trapezoid formula for area, you need to use the length as base 1 and base 2, and calculate the average of the bases as follows:

$$\frac{1}{2}(3 + 3) = 3$$

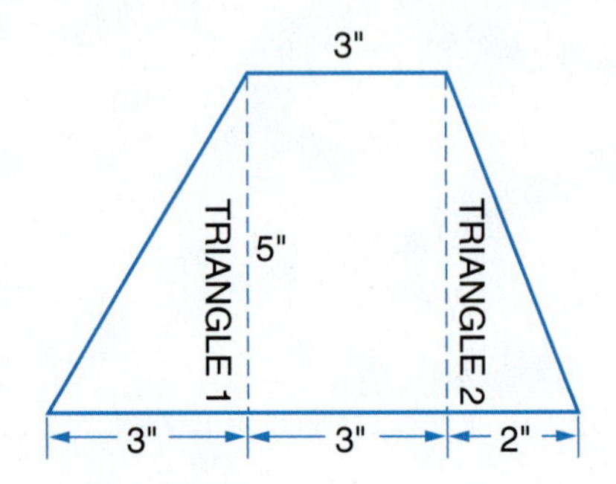

RECTANGLE
SA = LW
SA = 3" × 5"
= 15 in²

TRIANGLE 1
SA = ½ (BH)
SA = ½ (3" × 5")
= ½ (15")
= 7½ in²

TRIANGLE 2
SA = ½ (BH)
SA = ½ (2" × 5")
= ½ (10")
= 5 in²

TRAPEZOID
SA = SA Rectangle + SA Triangle 1 + SA Triangle 2
SA = 15 in² + 7½ in² + 5 in²
= 27½ in²

22207-12_F29.EPS

Figure 29 Trapezoid area using triangles and rectangles.

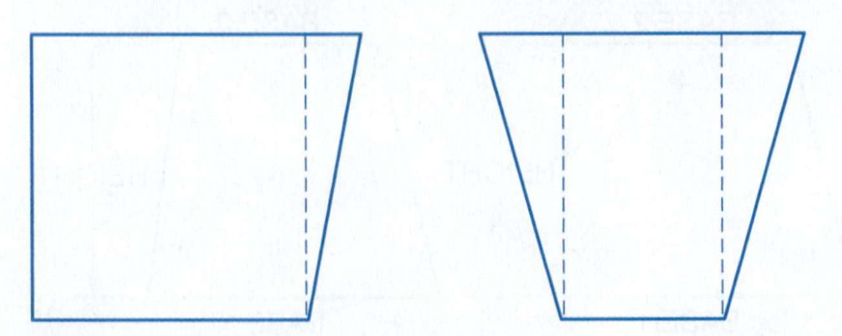
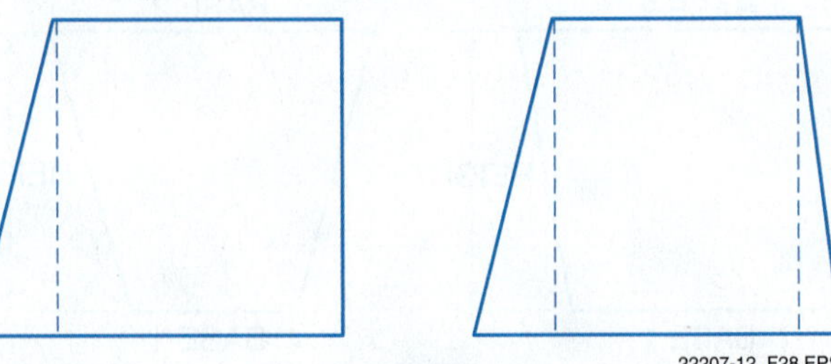

Figure 28 Trapezoids contain triangles and rectangles.

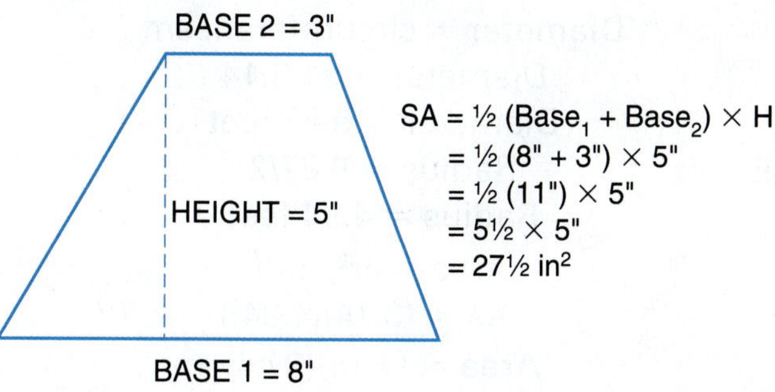

Figure 30 Trapezoid area using formula.

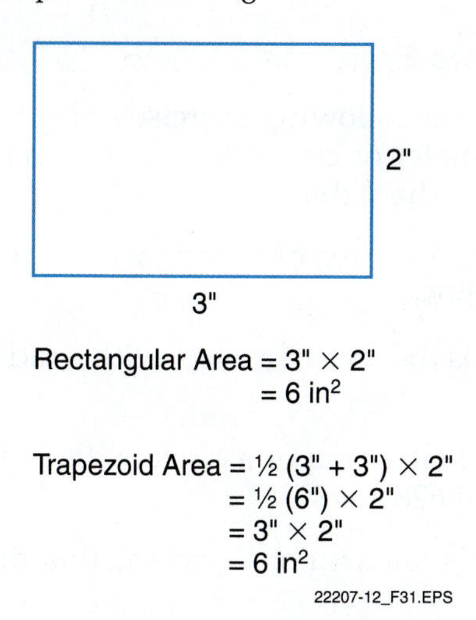

Figure 31 Rectangle and trapezoid relationship.

Then use the width as the height and calculate the area as follows:

$$3 \times 2 = 6 \text{ inches}^2$$

3.3.1 Trapezoid Exercises

Refer to *Figure 32* and complete the following exercises. Remember to convert units of measure as required.

Scenario: A crew is working on a highway job and needs to cut a shoulder slope, ditch, and backslope into existing ground. The cross-section plan is shown in *Figure 32*. The cut is 3 feet, the ditch is 2 feet wide, and the slope for the shoulder slope and backslope is 1:1 to the edges of the ditch. (Hint: the plan specifies that a 3-foot cut be made from the shoulder into the existing ground; 1:1 means that for every one foot of travel there is a one-foot drop in elevation.)

1. Base 1 =
2. Base 2 =
3. Height =
4. Area =

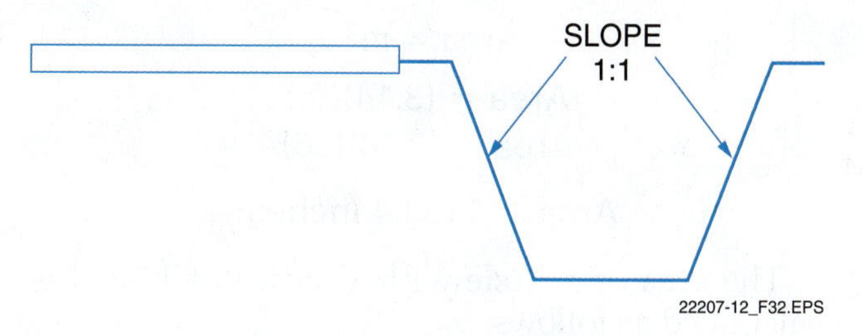

Figure 32 Trapezoid exercise cross-section plan.

3.4.0 Calculating the Area of a Circle

A circle is a single curved line that connects with itself (*Figure 33*). A circle also has these properties:

- All points on a circle are the same distance (equidistant) from the point at the center.
- The distance from the center to any point on the curved line, called the radius (r), is always the same.
- The shortest distance from any point on the curve through the center to a point directly opposite is called the diameter (d). The diameter is therefore equal to twice the radius (d = 2r).
- The distance around the outside of the circle is called the circumference. It can be determined by using the equation: circumference = πd, where π is a constant approximately equal to 3.14 and d is the diameter.
- A circle is divided into 360 parts, with each part called a degree. Therefore, one degree = $\frac{1}{360}$ of a circle. The degree is the unit of measurement commonly used in construction for measuring the size of angles.
- The total measure of all the angles formed by all consecutive radii equals 360 degrees.

The formula to calculate the area of a circle is πr^2. The π symbol represents pi, which is a constant of 3.14, and the r means radius. The area of a circle with a radius of 6 inches is calculated as follows:

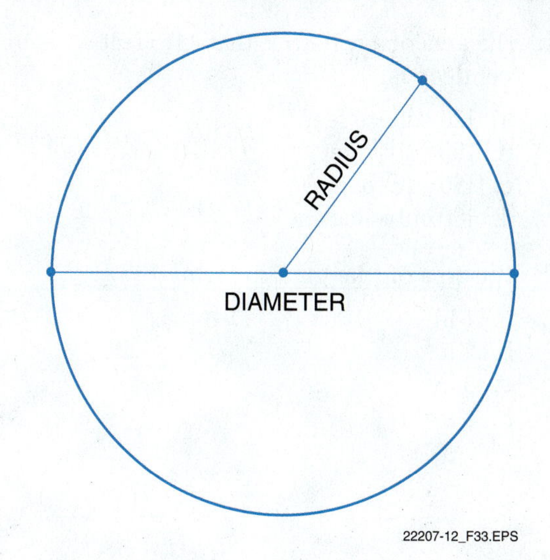

Figure 33 A circle.

$$\text{Area} = \pi r^2$$
$$\text{Area} = (3.14)(6^2)$$
$$\text{Area} = (3.14)(36)$$
$$\text{Area} = 113.04 \text{ inches}^2$$

The area of a circle with a radius of 10 feet is calculated as follows:

$$\text{Area} = \pi r^2$$
$$\text{Area} = (3.14)(10^2)$$
$$\text{Area} = (3.14)(100)$$
$$\text{Area} = 314 \text{ feet}^2$$

The area of a circle with a diameter of 10 feet is calculated as follows:

$$\text{Area} = \pi r^2, \text{and radius} = d \div 2,$$
$$\text{so } r = 10 \div 2 = 5 \text{ feet}$$
$$\text{Area} = (3.14)(5^2)$$
$$\text{Area} = (3.14)(25)$$
$$\text{Area} = 78\tfrac{1}{2} \text{ feet}^2$$

In earthwork, it will sometimes be easier to get the circumference of a circle than the diameter or radius—such as when a large depression needs to be filled. The formula for finding the diameter of a circle with the circumference is: $c \div \pi$, where c is circumference and π is 3.14. The area of a circle with a circumference of 31 feet is calculated as follows:

$$\text{Diameter} = \text{circumference}/\pi$$
$$\text{Diameter} = 31/3.14$$
$$\text{Diameter} = 9.87 \text{ feet}$$
$$\text{Radius} = 9.87/2$$
$$\text{Radius} = 4.94 \text{ feet}$$
$$\text{Area} = \pi r^2, \text{ and}$$
$$\text{Area} = (3.14)(4.94^2)$$
$$\text{Area} = (3.14)(24.4)$$
$$\text{Area} = 76.62 \text{ feet}^2$$

3.4.1 Circle Exercises

Complete the following exercises. Hint: when the circle diameter or circumference is given, be sure to calculate the radius.

1. What is the area of a circle with a diameter of 10 inches?
2. What is the area of a circle with a radius of 100 feet?
3. What is the area of a circle with a circumference of 628 feet?
4. What is the area of a circle with a radius of 7 feet?
5. What is the area of a circle with a diameter of 12 inches?

Additional Resources

Applied Construction Math: A Novel Approach. Alachua, FL: NCCER.

3.0.0 Section Review

1. The area of a rectangle that is 15 feet wide by 12 feet deep is _____.
   a. 120 square feet
   b. 150 cubic feet
   c. 180 square feet
   d. 180 cubic feet

2. The area of triangle 3 in *Figure 1* is _____.
   a. 1 in^2
   b. 2 in^2
   c. 3 in^2
   d. 4 in^2

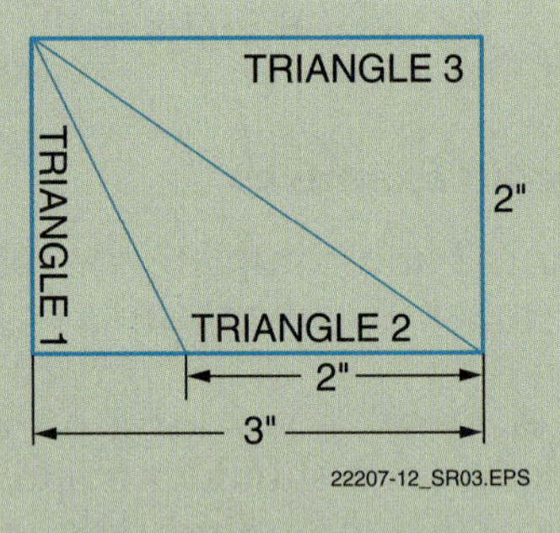

3. A trapezoid can be divided into a _____.
   a. square and a rectangle
   b. rectangle or square and at least one triangle
   c. triangle, a square, and a rectangle
   d. quadrilateral and a square

4. The formula for determining the circumference of a circle is _____.
   a. πr^2
   b. πd
   c. $2\pi r$
   d. πd^2

Section Four

4.0.0 Calculating Volume

Objective 4

Define *volume* and explain the purpose of calculating volume.

a. Calculate the volume of a cube.
b. Calculate the volume of a prism.
c. Calculate the volume of a cylinder.
d. Calculate the volume and weight of simple and complex excavations.

Performance Task 1

Using information provided by the instructor, calculate the volume and weight of a given excavation project.

Volume is the amount of space inside a three-dimensional object. Objects such as boxes, trash cans, coffee cups, and water pipes are three-dimensional objects. Any vessel that can hold a substance is a three-dimensional object, so it can be measured in terms of volume.

Three-dimensional objects have three measurements—length, width, and depth. The length is abbreviated with the letter *l*, width with the letter *w*, and depth with the letter *d*. Volume is measured in cubic units, so an object that is 1 inch in length, width, and depth has a volume of 1 cubic inch (see *Figure 34*), which can be abbreviated $inch^3$ and is calculated as follows:

$$\text{Volume} = l \times w \times d$$
$$\text{Volume} = 1 \text{ inch} \times 1 \text{ inch} \times 1 \text{ inch}$$
$$\text{Volume} = 1 \text{ inch}^3$$

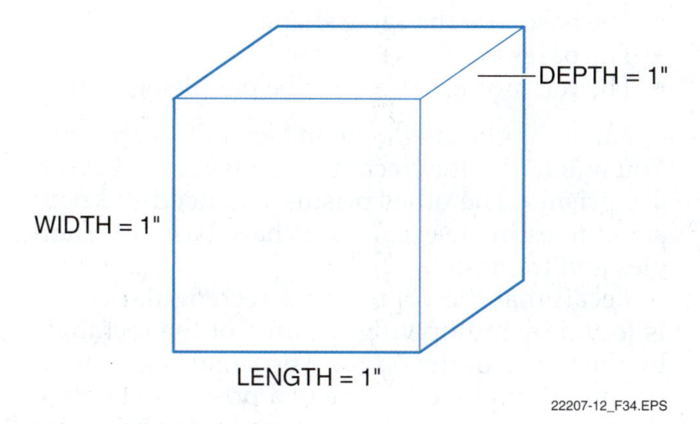

Figure 34 One cubic inch.

When two 1 cubic inch objects are placed together (see *Figure 35*), the volume is 2 cubic inches and is calculated as follows:

$$\text{Volume} = l \times w \times d$$
$$\text{Volume} = 2 \text{ inches} \times 1 \text{ inch} \times 1 \text{ inch}$$
$$\text{Volume} = 2 \text{ inches}^3$$

Just as with area, the same volume measurement can take many shapes. See *Figure 36A*. An object with an area of 2 cubic inches can be divided into two equal parts horizontally (*Figure 36B*), vertically (*Figure 36C*), and diagonally (*Figure 36D*) to form three differently shaped objects each with a volume of 1 cubic inch.

When calculating the volume of an object, all the numbers must be in the same unit of measure. That is, all numbers must be in inches, feet, or yards. It is a good idea to use the smallest measure you can. For example, 1 yard equals 3 feet, and 3 feet equal 36 inches, so the volume of a box that has a length, width, and depth of 1 yard can be calculated as follows:

Volume in yards:

$$\text{Volume} = 1 \text{ yd} \times 1 \text{ yd} \times 1 \text{ yd} = 1 \text{ yd}^3$$

Volume in feet:

$$\text{Volume} = 3 \times 3 \times 3 = 27 \text{ ft}^3$$

Volume in inches:

$$\text{Volume} = 36 \times 36 \times 36 = 46{,}656 \text{ in}^3$$

All of the above represent the same volume, but as you can see, it is much easier to work with the yard measure because it has the smallest numbers.

4.1.0 Cubes and Rectangular Objects

When a square is the base of a three-dimensional object, all the dimensions are equal and the object is called a cube (*Figure 37*). The volume of this object is calculated by multiplying its length by its width by its depth. Since all of the cube's dimensions are equal, each side can be abbreviated with the same letter (*e*), and the formula can be written e^3. The cube shown in *Figure 37* has a length of 2 feet, width of 2 feet, and depth of 2 feet, so its volume is calculated as follows:

$$\text{Volume} = l \times w \times d \text{ or } e^3$$
$$\text{Volume} = 2 \times 2 \times 2 \text{ or } 2^3$$
$$\text{Volume} = 8 \text{ feet}^3$$

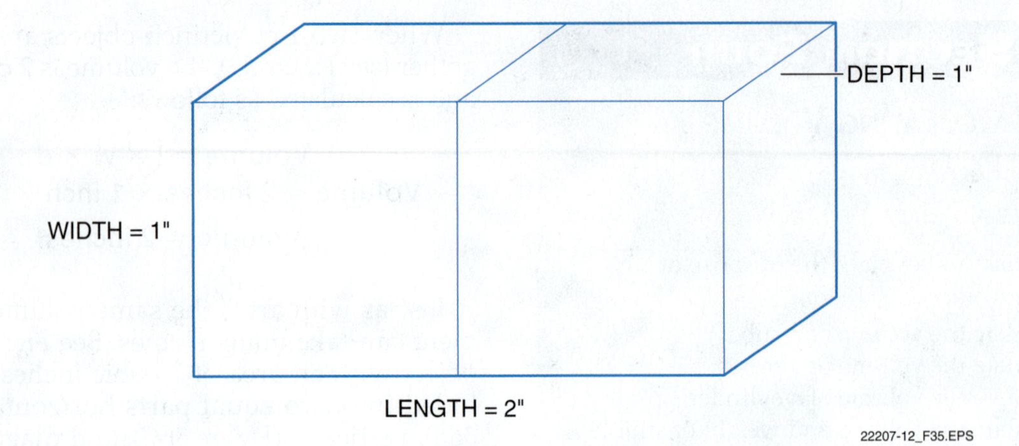

Figure 35 Two cubic inches.

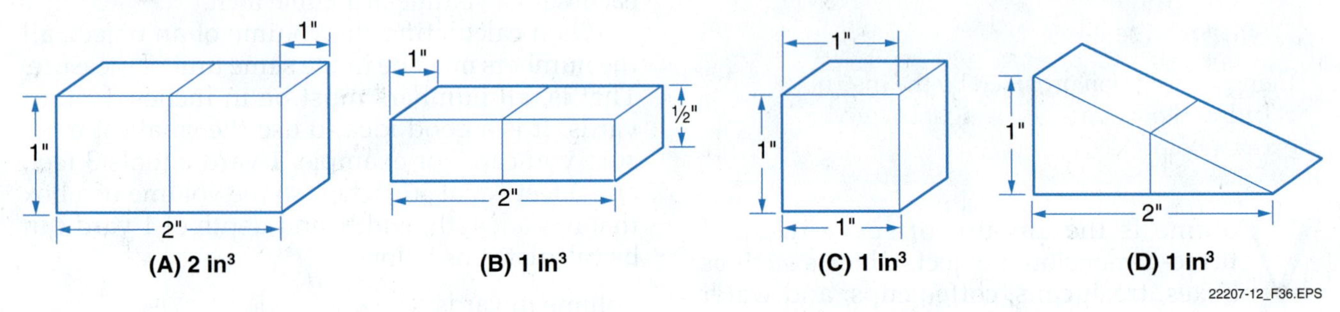

Figure 36 A cubic inch can be different shapes.

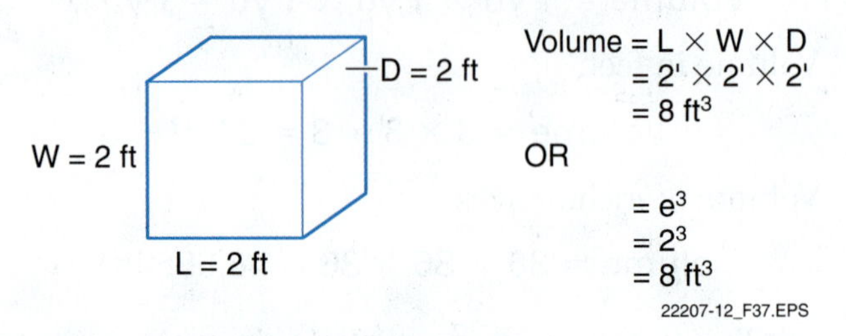

Figure 37 A cube.

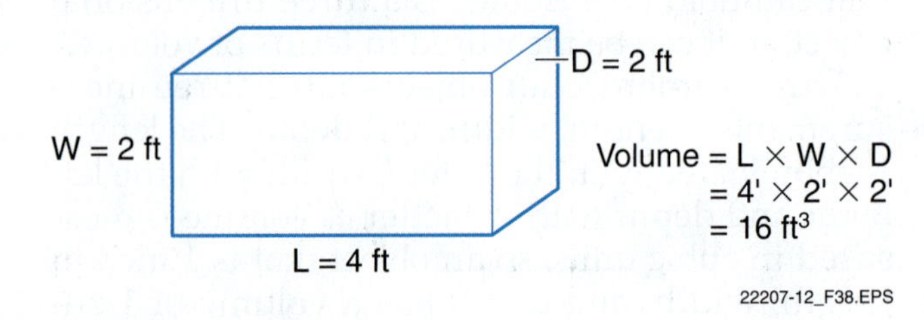

Figure 38 Rectangular object.

When a rectangle is the base of a three-dimensional object, it is called a rectangular object (*Figure 38*). This object's volume is calculated by multiplying its length by its width by its depth, just like a cube. Since each dimension of a rectangle can be different, the length is abbreviated with the letter *l*, the width with the letter *w*, and the depth with the letter *d*. The object shown in *Figure 38* has a length of 4 feet, width of 2 feet, and depth of 2 feet, so its volume is calculated as follows:

$$\text{Volume} = l \times w \times d$$
$$\text{Volume} = 4 \times 2 \times 2$$
$$\text{Volume} = 16 \text{ feet}^3$$

Note that the formula for a cube and a rectangular object contains the formula for area (l × w). To find the volume of a cube, multiply its area by its depth. This will be true for most three-dimensional objects used to estimate excavations.

4.2.0 Prisms

A prism is a multi-sided three-dimensional object that must meet all of the following requirements:

- It has two bases.
- The bases are parallel.
- The bases are the same shape.
- The bases are the same size.
- The remaining sides must be parallelograms.

All of the objects shown in *Figure 39* are prisms. You will notice that rectangular objects and cubes are prisms. The other prisms you need to know about to estimate excavations have bases of triangles and trapezoids.

Recall that the volume of a rectangular object is found by multiplying the area of the rectangle by the depth of the object. The same is true with prisms. To find the volume of a prism, first define the shape of the base, find the area of the base, and then multiply the area by the depth. Use the following procedure to calculate the volume of a prism:

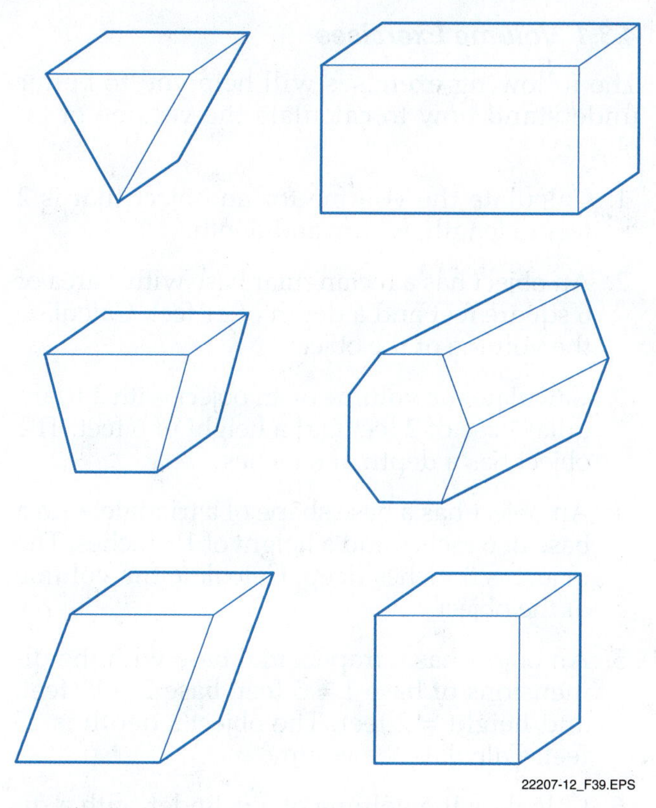

Figure 39 Prisms.

Step 1 Identify the shape of the base.

Step 2 Identify the dimensions of the shape.

Step 3 Calculate the area of the base.

Step 4 Calculate the volume of the prism.

Look carefully at the shape in *Figure 40A* and then perform Steps 1 through 4.

Step 1 Identify the shape of the base.

The base shape is a triangle.

Step 2 Identify the dimensions of the shape.

The dimensions are base = 3 inches, height = 2 inches, and depth = 4 inches.

Step 3 Calculate the area of the base.

$$\text{Area} = \tfrac{1}{2}\text{bh}$$

$$\text{Area} = \tfrac{1}{2}(3 \times 2)$$

$$\text{Area} = \tfrac{1}{2}(6)$$

$$\text{Area} = 3 \text{ inches}^2$$

Step 4 Calculate the volume of the prism.

$$\text{Volume} = \text{area} \times \text{depth}$$

$$\text{Volume} = 3 \text{ inches}^2 \times 4 \text{ inches}$$

$$\text{Volume} = 12 \text{ inches}^3$$

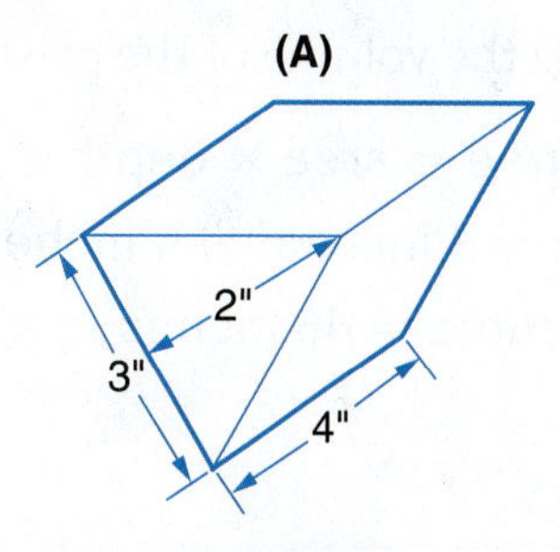

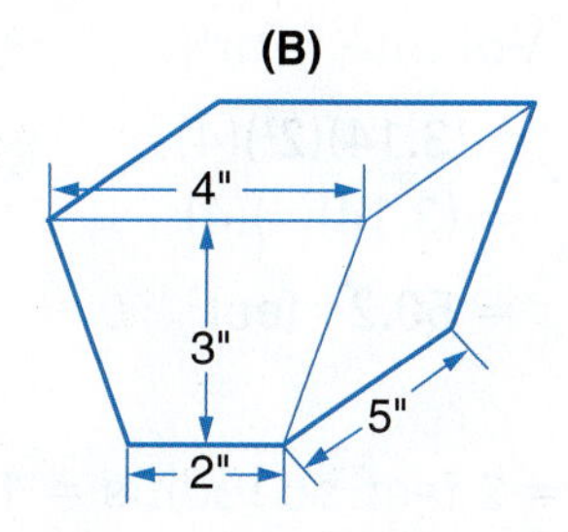

Figure 40 Finding the volume of a prism.

Repeat the procedure for the shape shown in *Figure 40B*.

Step 1 Identify the shape of the base.

The base shape is a trapezoid.

Step 2 Identify the dimensions of the shape.

The dimensions are base 1 = 2 inches, base 2 = 4 inches, height = 3 inches, and depth = 5 inches.

Step 3 Calculate the area of the base.

$$\text{Area} = \tfrac{1}{2}(\text{base 1} + \text{base 2}) \times \text{height}$$

$$\text{Area} = \tfrac{1}{2}(2 + 4) \times 3$$

$$\text{Area} = \tfrac{1}{2}(6) \times 3$$

$$\text{Area} = 9 \text{ inches}^2$$

Step 4 Calculate the volume of the prism.

$$\text{Volume} = \text{area} \times \text{depth}$$
$$\text{Volume} = 9 \text{ inches}^2 \times 5 \text{ inches}$$
$$\text{Volume} = 45 \text{ inches}^3$$

4.3.0 Cylinders

A cylinder is a three-dimensional object with a circle as its base (*Figure 41*). The formula to calculate the area of a circle is πr^2, and the formula to calculate the volume of a cylinder is $\pi r^2 h$.

Cylinder A:

Diameter = 4 feet, so radius = 2 feet
Height = 4 feet

$$\text{Volume} = \pi r^2 h$$
$$= (3.14)(2^2)(4)$$
$$= (3.14)(4)(4)$$
$$= 50.24 \text{ feet}^3$$

Cylinder B:

Diameter = 2 feet, so radius = 1 foot
Height = 2 feet

$$\text{Volume} = \pi r^2 h$$
$$= (3.14)(1^2)(2)$$
$$= (3.14)(1)(2)$$
$$= 6.28 \text{ feet}^3$$

Cylinder C:

Radius = 3 feet
Height = 5 feet

$$\text{Volume} = \pi r^2 h$$
$$= (3.14)(3^2)(5)$$
$$= (3.14)(9)(5)$$
$$= 141.3 \text{ feet}^3$$

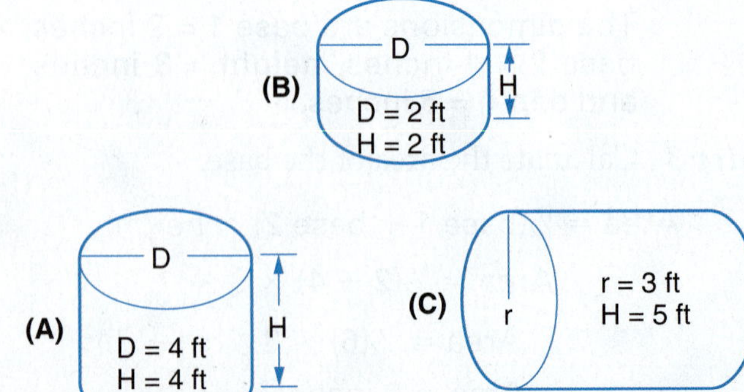

Figure 41 Cylinders.

4.3.1 Volume Exercises

The following exercises will help you to better understand how to calculate the volume of an object.

1. Calculate the volume for an object that is 2 feet in length, width, and depth.
2. An object has a rectangular base with a area of 6 square feet and a depth of 11 feet. Calculate the volume of the object.
3. Calculate the volume of an object with a triangular base of 2 feet and a height of 6 feet. The object has a depth of 6 inches.
4. An object has a base shape of a triangle with a base of 6 inches and a height of 1½ inches. The object is 3 inches deep. Calculate the volume of the object.
5. An object has a trapezoidal base with the dimensions of base 1 = 5 feet, base 2 = 11 feet, and height = 2 feet. The object's depth is 25 feet. Calculate the volume.
6. Calculate the volume of a cylinder with a diameter of 2 feet and a height of 6 feet.
7. Calculate the volume of the shape shown in *Figure 42*.
8. Calculate the volume of a cylinder that has a radius of 2 yards and a height of 3 yards.

4.4.0 Calculating the Volume and Weight of an Excavation

Excavations are measured in cubic yards of soil. You have already learned that 1 cubic yard contains 27 cubic feet and that those 27 cubic feet can take any three-dimensional shape and still have a volume of 1 cubic yard (*Figure 43*). Each of the shapes shown in the figure has a volume of 1 cubic yard or 27 cubic feet. Since volume is the measurement of length, width, and depth, all three dimensions must be considered to calculate volume.

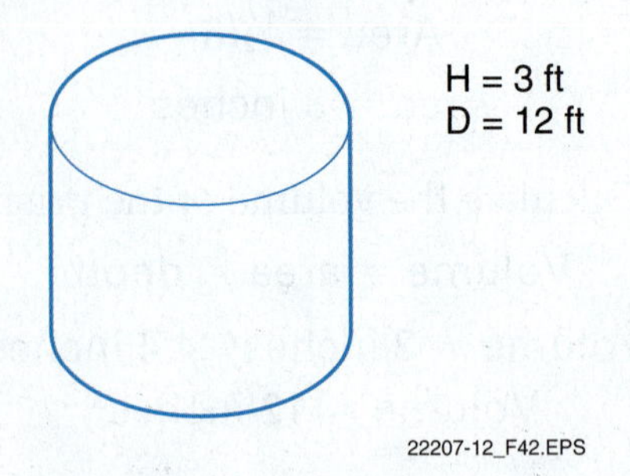

Figure 42 Cylinder Exercise 7.

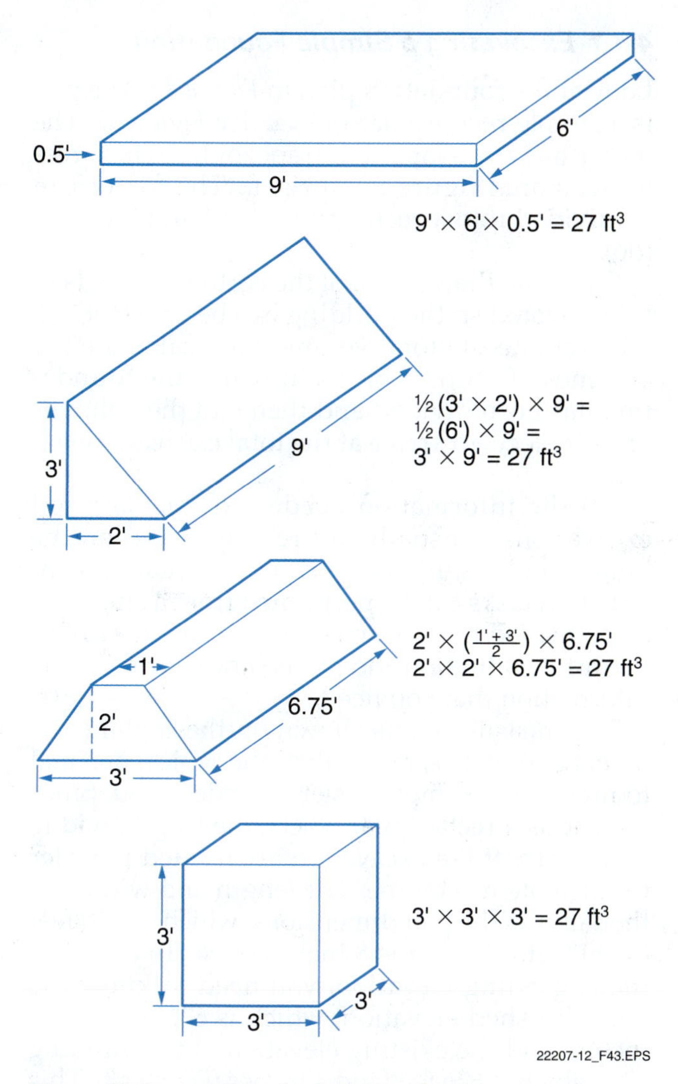

Figure 43 One cubic yard equals 27 cubic feet.

The capacity of heavy equipment is rated by both weight and volume. Operating an overloaded vehicle creates the risk of an accident, so it is important to consider the weight of excavated material as well as the volume when loading the equipment. *Table 1* shows the weights of various materials. To calculate the weight of an excavation, multiply the volume in cubic yards by the material's weight per cubic yard.

Calculating volumes for cuts and fills can be time consuming, but by following a few easy steps, you can soon become skilled at it. First, to estimate excavation volume, assume that the ground is flat. (A way to calculate volume for uneven surfaces is described later in this module.) Second, study the area so it can be divided it into manageable shapes. Third, determine what information you need to make the volume calculations. Finally, gather the needed information. Once you know what shapes to use and have collected the dimensions, you can calculate the volume.

Table 1 Material Weights

Material	Weight in Pounds per Cu Yd (27 Cu Ft)
Clay, dry in lumps	1,701
Clay, compact	2,943
Earth, loamy, dry, loose	2,025
Earth, dry, packed	2,565
Earth, wet	2,970
Gravel, dry, loose	2,970
Gravel, dry, packed	3,051
Gravel, wet, packed	3,240
Limestone, fine	2,700
Limestone, 1½ to 2 inches	2,295
Limestone, above 2 inches	2,160
Sand, dry, loose	2,565
Sand, wet, packed	3,240

The construction plans shown in this section are simplified to make it easier to find the information needed to calculate volumes. Once you gain some experience, you will be able to quickly find the information on actual building plans. Look at the highway plan in *Figure 44*. The cross-section is clearly a trapezoid. (It can also be divided into a rectangle and two triangles, but then three calculations must be performed. There is only one calculation needed if a trapezoid is used.) The three-dimensional object of a trapezoid is a prism. See *Figure 45*.

Since it is assumed that the existing grade is even, the job will involve only fill. To calculate the volume of a prism, measurements are needed for the following:

- Base 1
- Base 2
- Height
- depth

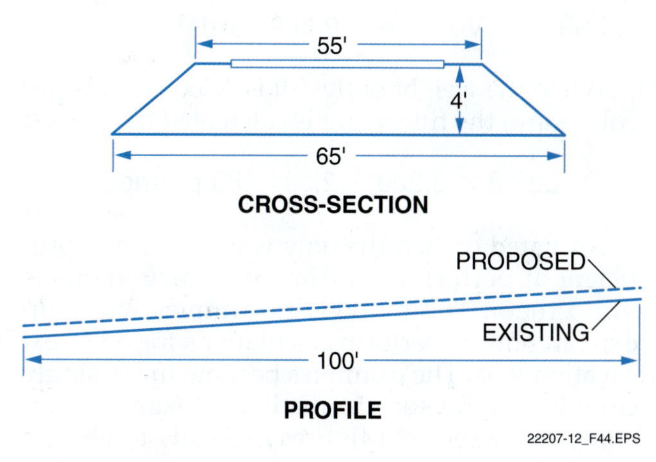

Figure 44 Highway plan.

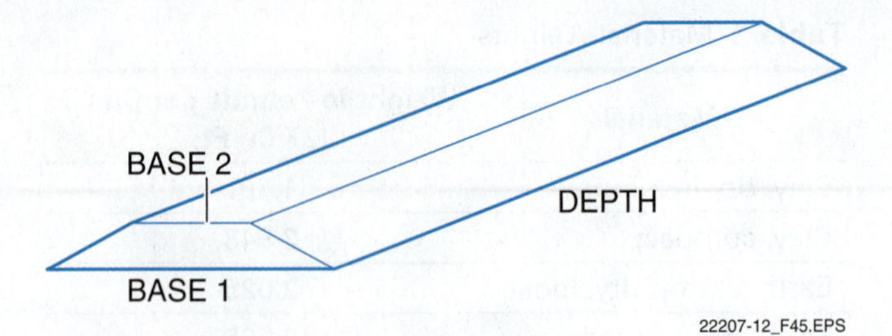

Figure 45 Highway plan as a prism.

Carefully examine the diagrams shown in *Figures 44* and *45*. The cross-section is a trapezoid with the following measurements:

Base 1 = 65 feet

Base 2 = 55 feet

Height = 4 feet

The profile shows that this section of roadway is 100 feet in length, so the trapezoid has a depth of 100 feet.

Example

Use the following dimensions to calculate the amount of fill needed to construct this roadway:

Base 1 = 65 feet

Base 2 = 55 feet

Height = 4 feet

Depth = 100 feet

The volume is calculated as follows:

$$\text{Volume} = \tfrac{1}{2} \times (\text{base 1} + \text{base 2}) \times \text{height} \times \text{depth}$$

$$\text{Volume} = \tfrac{1}{2} \times (65 + 55) \times 4 \times 100$$

$$\text{Volume} = \tfrac{1}{2} \times 120 \times 4 \times 100$$

$$\text{Volume} = 24{,}000 \text{ feet}^3$$

$$\text{Volume} = 24{,}000 \div 27 \text{ cubic feet/cubic yard}$$

$$\text{Volume} = 888.9 \text{ yards}^3$$

When the weight of the fill is 3,200 pounds per cubic yard, the fill weight is calculated as follows:

$$888.9 \times 3{,}200 = 2{,}844{,}480 \text{ pounds}$$

As stated earlier, the only way to become proficient at performing estimating calculations is to do them. The following examples will help explain how to perform calculations for other excavation jobs. The examples become increasingly complex, so be sure to read each example and study the associated figures carefully to be sure you understand the result.

4.4.1 Excavating a Simple Foundation

Look at the foundation plan in *Figure 46*. The pad is a simple rectangular object. See *Figure 47*. The footer has the shape of a trapezoid, so its three-dimensional figure is a prism. (The footer can be divided into a rectangular solid and a prism, too).

Calculate the volume of the earth that needs to be excavated so the building can be constructed. To calculate the total volume, first calculate the volumes of all the objects in which the foundation has been divided and then add the volumes of the objects to arrive at the total excavation volume.

All the information needed to calculate soil excavations is usually not readily found on the building plans. Plans are drawn so that all construction tasks can be performed from a single set of plans. Do not become overwhelmed with all the information on the plans. Look for only the information that you need.

The measurements shown on these plans are in feet and inches, rather than the typical decimal to make the example easier to understand. Since the pad is a rectangular object, the length, width, and depth of the excavation are needed in order to calculate its volume. The length and width are the same as the pad dimensions, which are 30 feet by 30 feet. The pad is 8 inches thick. To calculate the beginning elevation, you need to know the pad's finished elevation, which is 626 feet and 6 inches, and the existing elevation of the building site, which is 626 feet and 4 inches (*Figure 48*). This means that the pad needs to begin at an elevation of 625 feet and 10 inches. This is 6 inches below the existing elevation, so the depth of the excavation is 6 inches.

Unsuitable Soil

Excavation will sometimes uncover soil that is unsuitable for use on the site because it contains too much moisture. This type of soil is easily recognized because the ground will be mushy and will ripple when equipment rolls over it. The initial testing by soil engineers should locate such soil, but patches of it may turn up during excavation. In such cases, the equipment operator should consult a supervisor or engineer before proceeding because it may be necessary to remove the soil and fill the area. In many cases, this work is beyond the scope of the contract, and would require a renegotiation to fund the additional work.

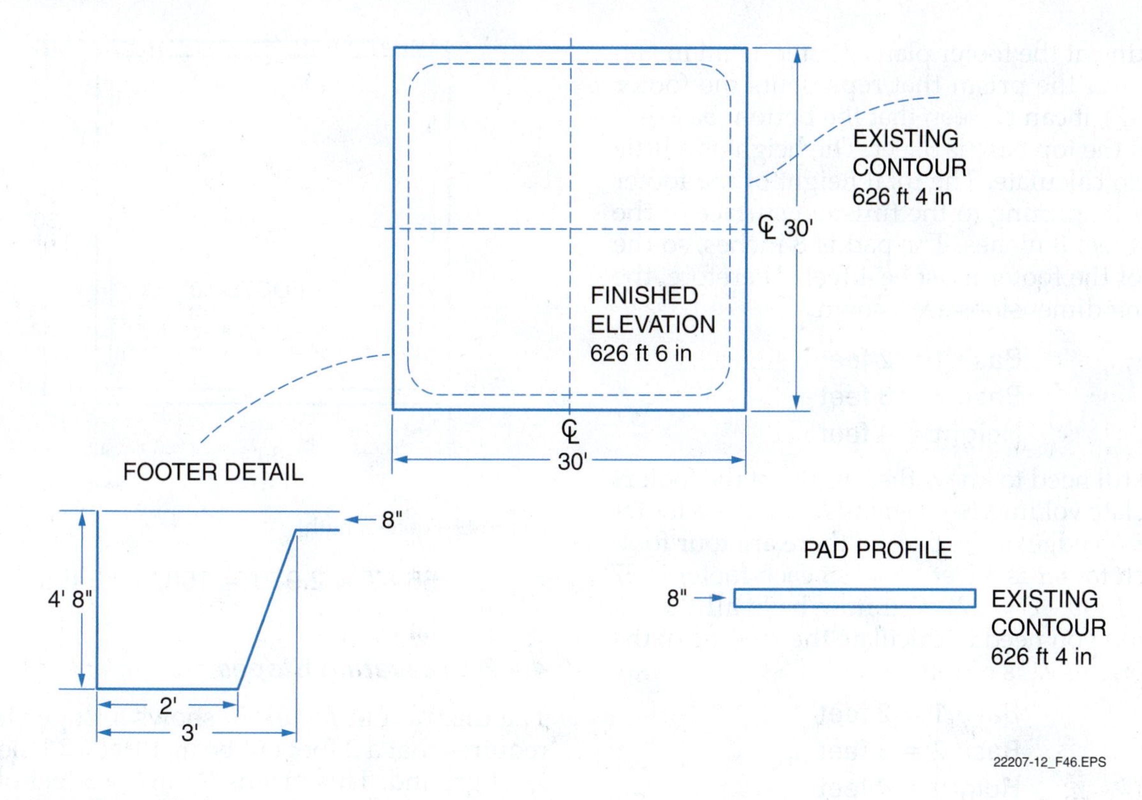

Figure 46 Foundation plan.

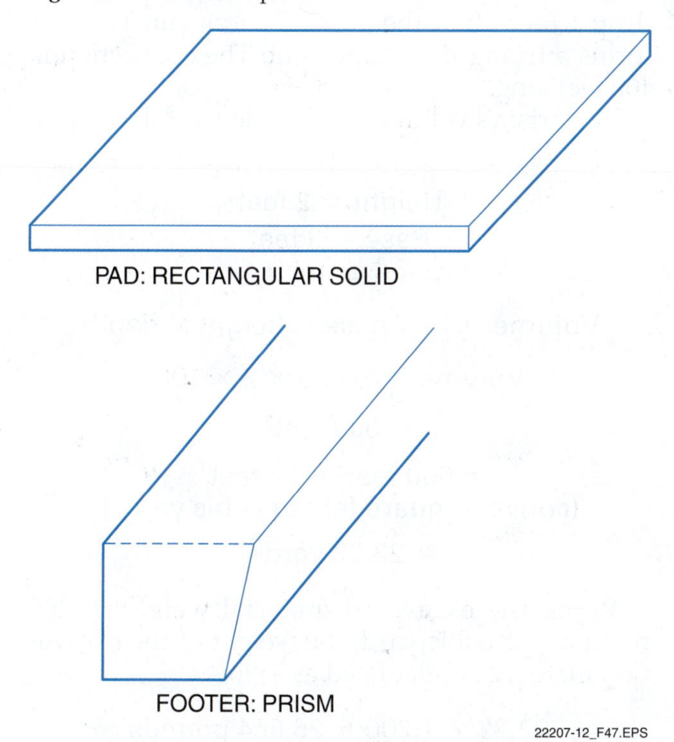

Figure 47 Foundation plan as a rectangular object and a prism.

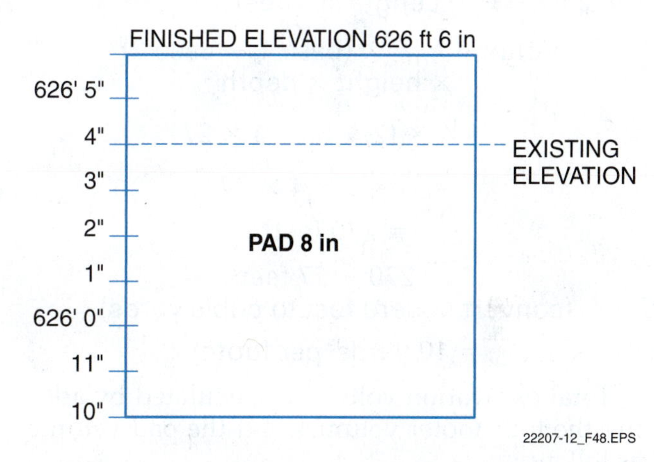

Figure 48 Existing elevation, finished elevation.

You now have all of the information needed to calculate the excavation volume for the slab, which is as follows:

Volume = length × width × depth

Length = 30 feet

Width = 30 feet

Depth = 6 inches = 0.5 feet
(convert inches to feet)

Volume = 30 × 30 × 0.5

Volume = 450 feet^3 ÷ 27 feet^3
(convert cubic feet to cubic yards)

Volume = 16.67 yards^3

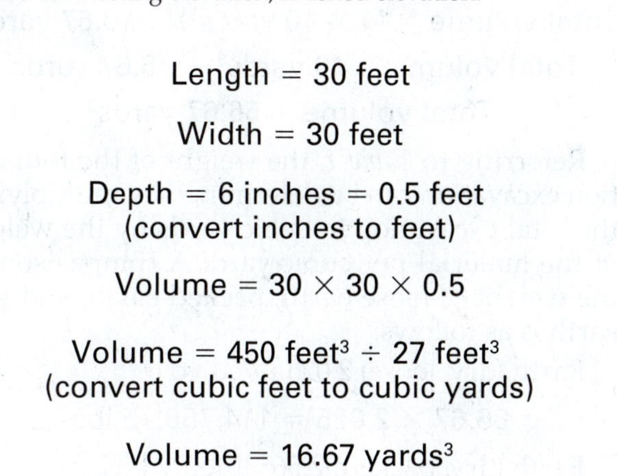

Looking at the footer plans (footer detail in *Figure 46*) and the prism that represents the footer (*Figure 47*), it can be seen that the bottom base is 2 feet and the top base is 3 feet. The height is a little harder to calculate. The total height of the footer from its beginning to the finished surface of the pad is 4 feet 8 inches. The pad is 8 inches, so the height of the footer must be 4 feet. Therefore, the following dimensions are known:

Base 1 = 2 feet
Base 2 = 3 feet
Height = 4 feet

You still need to know the lengths of the footers to calculate volume. See *Figure 49*. There is a footer along each edge of the pad, so there are four footers. Each footer is 3 feet wide, so each footer is 27 feet long (30 – 3 = 27). You now have all the information you need to calculate the volume of the footer prisms. It is as follows:

Base 1 = 2 feet
Base 2 = 3 feet
Height = 4 feet
Length = 7 feet

$$\text{Volume} = \tfrac{1}{2} \times (\text{base 1} + \text{base 2}) \times \text{height} \times \text{depth}$$

$$= \tfrac{1}{2} \times (2 + 3) \times 4 \times 27$$

$$= \tfrac{1}{2} \times 5 \times 4 \times 27$$

$$= 270 \text{ feet}^3$$

$$= 270 \div 27 \text{ feet}^3$$

(convert square feet to cubic yards)

$$= 10 \text{ yards}^3 \text{ per footer}$$

Total excavation volume is calculated by adding the four footer volumes and the pad volume as follows:

$$\text{Total volume} = (4 \times 10 \text{ yards}^3) + 16.67 \text{ yards}^3$$

$$\text{Total volume} = 40 \text{ yards}^3 + 16.67 \text{ yards}^3$$

$$\text{Total volume} = 56.67 \text{ yards}^3$$

Referring to *Table 1*, the weight of the foundation excavation can be determined by multiplying the total excavation in cubic yards by the weight of the material per cubic yard. A comparison of the weight of loose earth, packed earth, and wet earth is as follows:

Earth (dry, loose) 2,025 lb/cu yd

$$56.67 \times 2{,}025 = 114{,}756.75 \text{ lbs}$$

Earth (dry, packed) 2,565 lb/cu yd

$$56.67 \times 2{,}565 = 145{,}358.55 \text{ lbs}$$

Earth (wet) 2,970 lb/cu yd

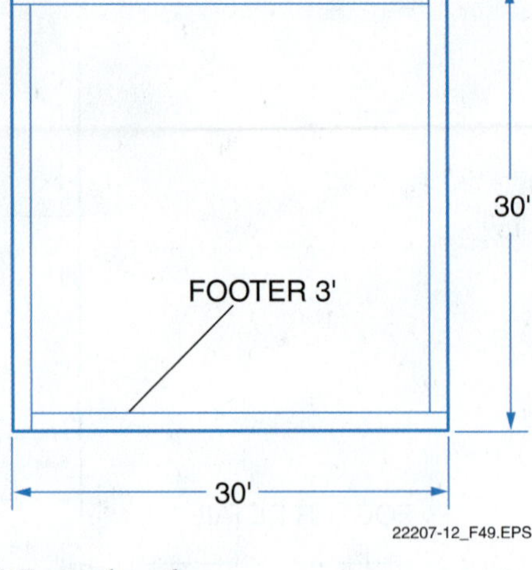

Figure 49 Footer length.

$$56.67 \times 2{,}970 = 168{,}309.9 \text{ lbs}$$

4.4.2 Excavating Slopes

The diagram in *Figure 50* shows a slope stake. It requires that a 2-foot cut be made at a 3:1 slope on level ground. This means for every 3 feet of horizontal travel, the elevation of the ground must drop 1 foot up to the desired 2-foot cut. This cut forms a triangular-shaped cut. The cut section is 100 feet long.

The prism's volume is calculated as follows (see *Figure 51*):

Height = 2 feet
Base = 6 feet
Depth = 100 feet

$$\text{Volume} = \tfrac{1}{2} \times \text{base} \times \text{height} \times \text{depth}$$

$$\text{Volume} = \tfrac{1}{2} \times 6 \times 2 \times 100$$

$$= 600 \text{ feet}^3$$

$$= 600 \text{ feet}^3 \div 27 \text{ feet}^3$$

(convert square feet to cubic yards)

$$= 22.22 \text{ yards}^3$$

When the excavated material weights 1,200 pounds per cubic yard, the weight of the excavation material is calculated as follows:

$$22.22 \times 1{,}200 = 26{,}664 \text{ pounds}$$

4.4.3 Excavating a Complex Foundation

Study the foundation plan in *Figure 52*. This shape can be divided into three to five shapes. With a complex plan such as this one, time can be saved by calculating the area of the entire foundation before calculating the volume.

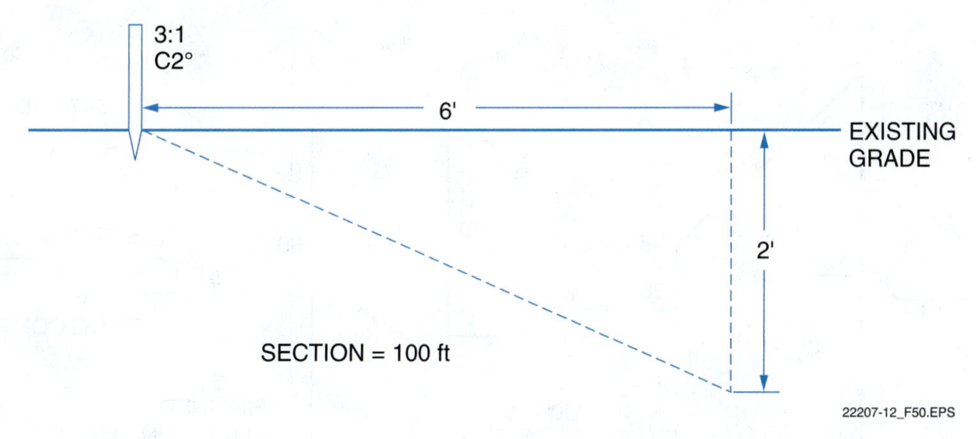

Figure 50 Excavating a slope.

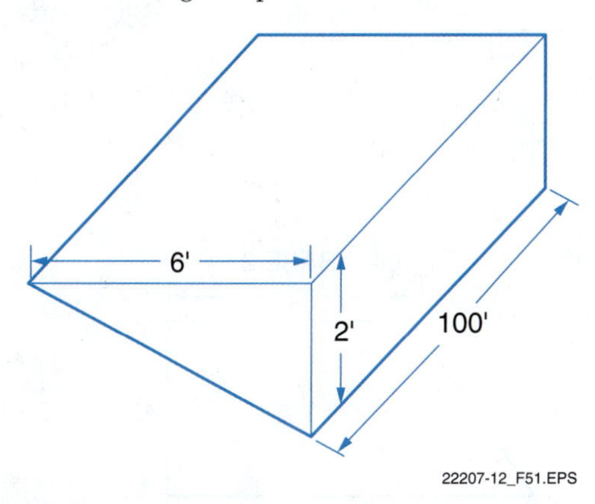

Figure 51 Excavating a slope as a prism.

Figure 53A shows three shapes: two trapezoids and one rectangle. *Figure 53B* shows five shapes: two triangles and three rectangles. It is better to see as many shapes as possible because building plans may not provide enough information to easily gather the measurements needed to calculate volume. In this case, the dimensions shown on *Figure 52* are sufficient to calculate the area of all of the shapes shown in *Figures 53A* and *53B*, so *Figure 53A* can be used because it has the fewest calculations.

When trying to gather information from the building plans, it is easy to become overwhelmed with all the information on the diagram. Concentrate on one shape at a time. For example, Shape 1 in *Figure 54* is a trapezoid; it is the left wing of the building shown in *Figure 52*. To calculate the area of Shape 1, the base 1, base 2, and height measurements are needed. Looking at *Figure 52*, it can be seen that base 1 is 80 feet and height is 40 feet. The base 2 measurement is not obvious so it needs to be calculated. The part of base 2 represented by the solid line is 20 feet and the part of base 2 represented by the dashed line is 30 feet, so the length is 50 feet.

GOING GREEN

Dealing with Contaminated Soil

When sites are being excavated, there is a possibility that the soil being removed is contaminated. The contamination can come from a variety of sources, including leachate from nearby landfills, contaminated surface water, or the prior use of the land as an industrial site. In the past, there were many manufacturing facilities that used toxic chemicals in their processes. These chemicals may have leached into the soil, causing it to become contaminated. Environmental laws regulate disposal of such soil, so it must be decontaminated before it can be moved to another location. Environmental testing is required to ensure that soil on a job site is not contaminated before it is excavated and removed to another location. Treatment approaches can include flushing contaminants out of the soil using water, chemical solvents, or air; destroying the contaminants by incineration; encouraging natural organisms in the soil to break them down; or adding material to the soil to encapsulate the contaminants and prevent them from spreading.

The dimensions for Shape 1 of *Figure 54* are as follows:

$$\text{Base 1} = 80 \text{ feet}$$
$$\text{Base 2} = 50 \text{ feet}$$
$$\text{Height} = 40 \text{ feet}$$

The area for Shape 1 is calculated as follows:

$$\text{Area} = \tfrac{1}{2}(\text{base 1} + \text{base 2}) \times \text{height}$$
$$\text{Area} = \tfrac{1}{2}(80 + 50) \times 40$$
$$\text{Area} = \tfrac{1}{2} \times 130 \times 40$$
$$\text{Area} = 2{,}600 \text{ feet}^2$$

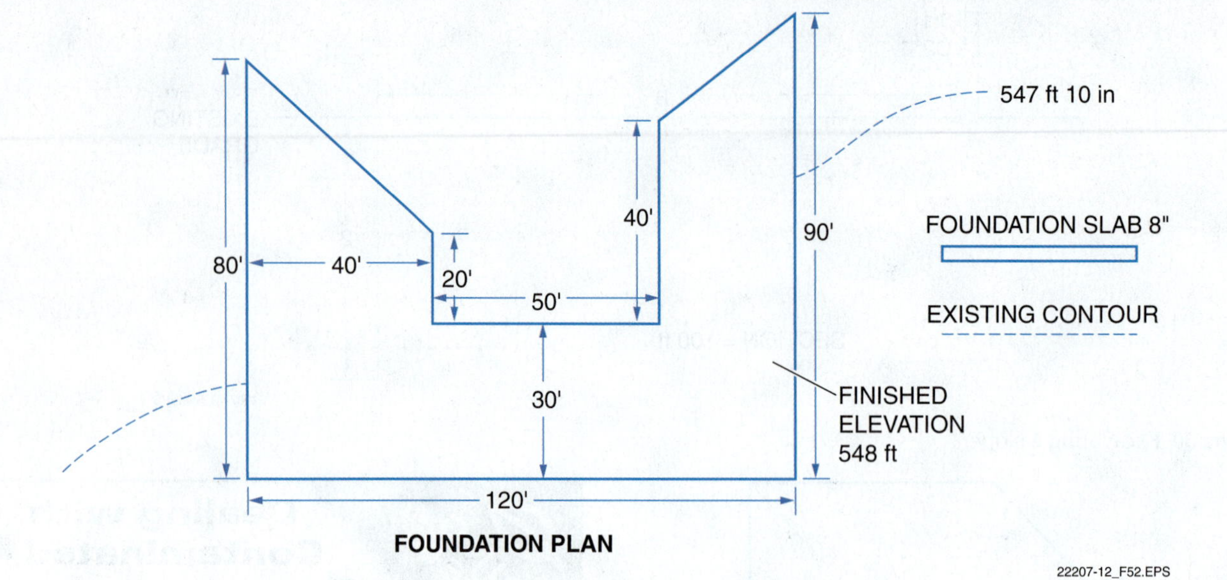

Figure 52 Complex foundation plan.

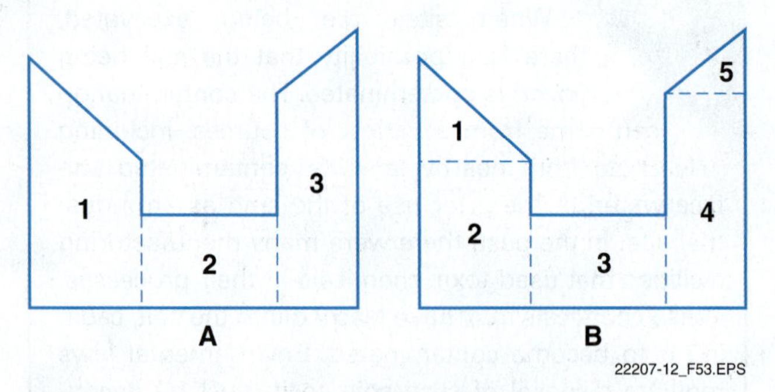

Figure 53 Complex foundation plan shapes.

Shape 2 in *Figure 54* is a little easier. Shape 2 is a rectangle and represents the main part of the building shown in *Figure 53*. You already used the width of the rectangle as part of base 2 in Shape 1, so you know the width is 30 feet. The length of 50 feet is shown at the top of the main building in *Figure 52*, so all the information needed to calculate the area of Shape 2, which is as follows, is available:

$$\text{Length} = 50 \text{ feet}$$
$$\text{Width} = 30 \text{ feet}$$
$$\text{Area} = \text{length} \times \text{width}$$
$$\text{Area} = 50 \times 30$$
$$\text{Area} = 1{,}500 \text{ feet}^2$$

Shape 3 in *Figure 54* is another trapezoid and represents the right wing of the building shown in *Figure 53*. Base 1 is on the far right side of the foundation and is 90 feet. Base 2 is made of the width of Shape 2 (30 feet) and the 40 feet measure on the left side of the wing (*Figure 53*), so base 2 is 70 feet.

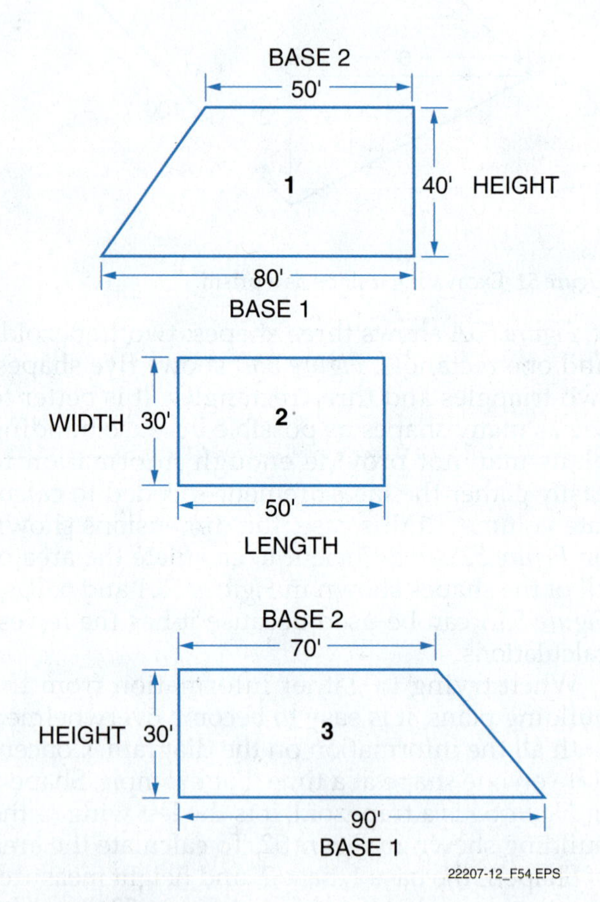

Figure 54 Three shapes.

Now the height of Shape 3 needs to be calculated. You know that the entire length of the face of the building is 120 feet (it is on *Figure 52*) and

the face of the building is made up of the height of shape 1 (40 feet), the length of Shape 2 (50 feet) and the height of Shape 3. Since the dimensions of Shapes 1 and 2 are known the height of Shape 3 can be calculated as follows:

$$\text{Face length} = \text{height Shape 1} + \text{length Shape 2} + \text{height Shape 3}$$

$$120 \text{ feet} = 40 \text{ feet} + 50 \text{ feet} + \text{height Shape 3}$$

$$120 \text{ feet} = 90 \text{ feet} + \text{height Shape 3}$$

$$120 \text{ feet} - 90 \text{ feet} = 90 \text{ feet} + \text{height Shape 3} - 90 \text{ feet}$$

$$30 \text{ feet} = \text{height Shape 3}$$

So the height of Shape 3 is 30 feet. This is all the information you need to calculate the area of Shape 3, as follows:

$$\text{Base 1} = 90 \text{ feet}$$
$$\text{Base 2} = 70 \text{ feet}$$
$$\text{Height} = 30 \text{ feet}$$

$$\text{Area} = \tfrac{1}{2}(\text{base 1} + \text{base 2}) \times \text{height}$$

$$\text{So area} = 2{,}400 \text{ square feet}$$

You still need the depth of each object to calculate its volume. Look on the plans and find the thickness of the slab, the existing contour of the building site, and the finished elevation of the foundation. The slab is 8 inches thick and the finished elevation is 548 feet, so the slab must start at 547 feet 4 inches in elevation. Since the existing elevation is 547 feet 10 inches, 6 inches (or ½ foot) of earth needs to be excavated to achieve the finished elevation after the foundation is poured.

Now that depth has been determined, the volume of each object can be calculated. The answers are added together to get the total excavation volume.

Shape 1:

$$\text{Volume} = 2{,}600 \text{ feet}^2 \times \tfrac{1}{2} \text{ foot}$$

$$\text{Volume} = 1{,}300 \text{ feet}^3$$

Shape 2:

$$\text{Volume} = 1{,}500 \text{ feet}^2 \times \tfrac{1}{2} \text{ foot}$$

$$\text{Volume} = 750 \text{ feet}^3$$

Shape 3:

$$\text{Volume} = 2{,}400 \text{ feet}^2 \times \tfrac{1}{2} \text{ foot}$$

$$\text{Volume} = 1{,}200 \text{ feet}^3$$

Total volume:

$$1{,}300 \text{ feet}^3 + 750 \text{ feet}^3 + 1{,}200 \text{ feet}^3 = 3{,}250 \text{ feet}^3$$

$$3{,}250 \text{ feet}^3 \div 27 \text{ feet}^3 = 120.37 \text{ yards}^3$$

Another way to figure out total volume is to add together the area of all three shapes and then multiply the total area by the depth. Calculate this yourself. You should get a total area of 6,500 square feet.

4.4.4 Complex Calculations

Up until now, the material has focused on how to calculate excavation volumes for uniformly shaped objects, but most building sites are irregularly shaped. There are a number of computer programs that can be used to calculate these volumes quickly and accurately, saving time and money. *Figure 55* shows a screen from one such program.

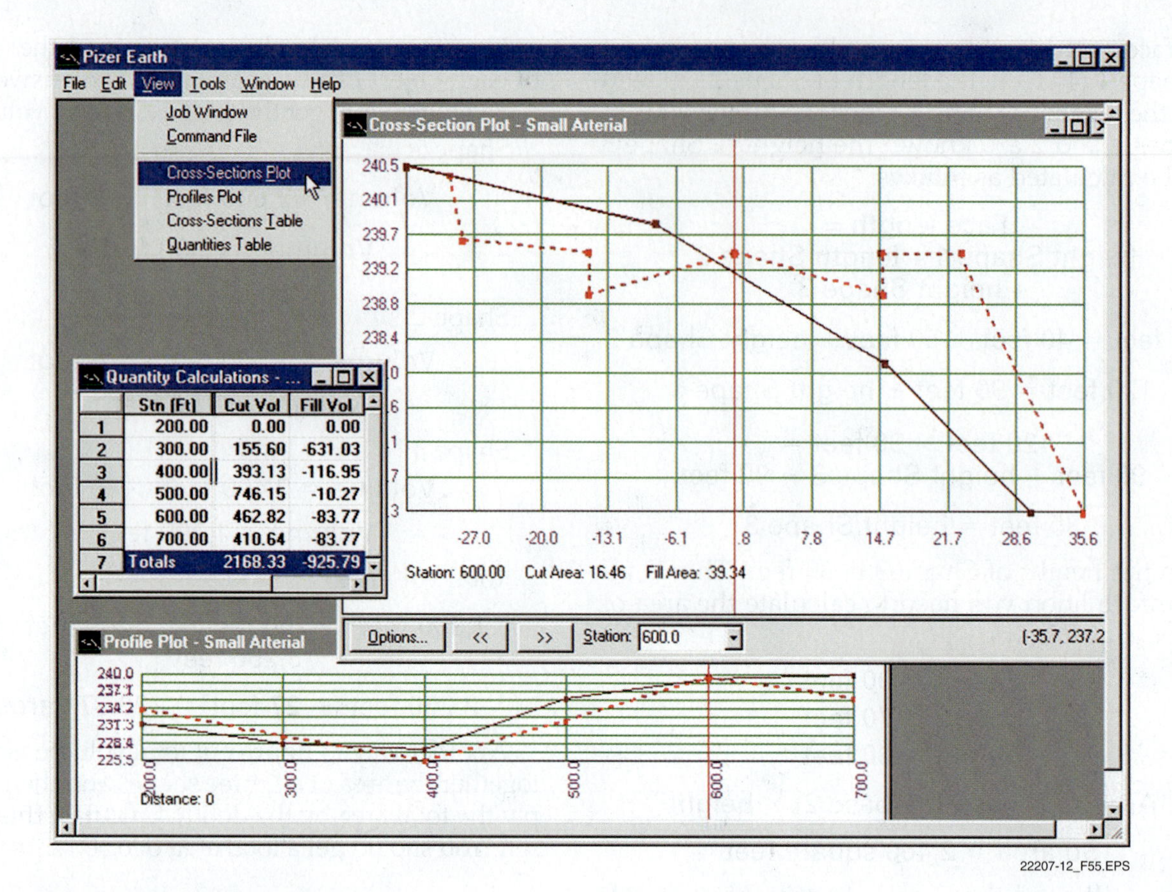

	Stn (Ft)	Cut Vol	Fill Vol
1	200.00	0.00	0.00
2	300.00	155.60	-631.03
3	400.00	393.13	-116.95
4	500.00	746.15	-10.27
5	600.00	462.82	-83.77
6	700.00	410.64	-83.77
7	Totals	2168.33	-925.79

Figure 55 Cut and fill computer program.

You have already learned that you need to calculate the area of an object's base to calculate its volume. Another way to find the area of an irregular shape is to draw a grid over it. *Figure 56* is a benched trench. Each block of the grid in *Figure 57* represents 1 square foot. To estimate the area of the benched trench, you need to count the number of blocks (estimating the size of the partial blocks).

Usually, the ground at the building site is irregular (*Figure 58*), making it difficult to determine the excavation depth precisely. In this case, a more accurate estimation can be made by averaging the elevations of the site. This is done by measuring the elevations at several points and then adding these numbers and dividing by the number of points measured. In *Figure 59*, 14 points are measured; their sum is 93, so the average depth of the excavation is 6.64 inches, which is computed as follows:

$$93 \div 14 = 6.64 \text{ inches}$$

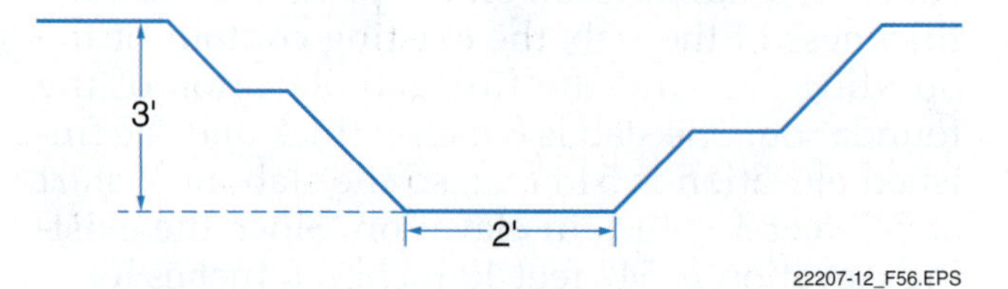

Figure 56 Benched trench.

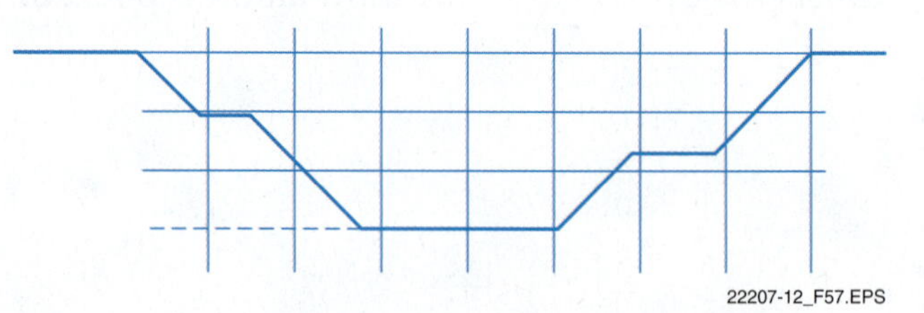

Figure 57 Benched trench with grid.

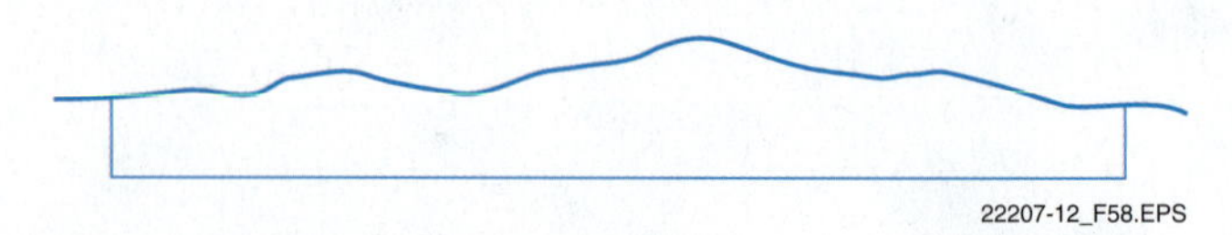

Figure 58 Irregular elevation.

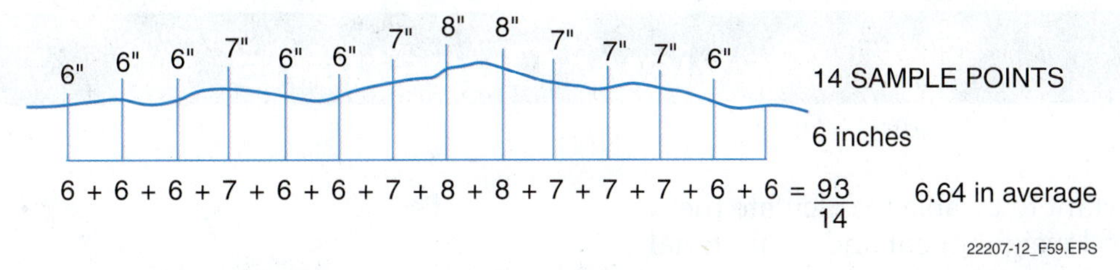

Figure 59 Irregular elevation averaged.

4.0.0 Section Review

1. In order to determine the area of a cube, the product of the length and width are _____.
   a. divided by the depth
   b. multiplied by pi
   c. added to the depth
   d. multiplied by the depth
2. Which of the following is *not* required for a prism?
   a. Two bases
   b. Parallel bases
   c. It must be a cube
   d. Bases are the same size
3. A cylinder is a three-dimensional object with a circle as its base.
   a. True
   b. False
4. Which of the following values is *not* needed in order to calculate the volume of excavation material for a simple foundation?
   a. Depth
   b. Length
   c. Width
   d. Weight

SUMMARY

Calculating the volume and weight of excavations can be a time-consuming task, but it is necessary in order to determine the type of equipment needed on the job. To determine the volume of any excavation, you first calculate the area of a figure and then you use the area to calculate the volume of an object.

Excavations are three-dimensional, but they are usually based on one of several common two-dimensional shapes, such as the square, rectangle, triangle, or circle. Once you have determined the area of the base shape, the volume is calculated by multiplying the base area by the object's depth. Whenever you need to calculate the volume of a complex object, you need to break the object into familiar shapes and then calculate the volume of each shape. Once you have the volume of each shape, it is only a matter of adding them together to calculate the total volume of the object. Cut and fill volumes are measured in cubic yards, so after calculating the excavation volume, it may have to be converted to cubic yards.

The Pythagorean theorem is used to solve for unknown sides of a right triangle. In its basic form, this simple formula is designed to calculate the hypotenuse of the right triangle. It can also be manipulated to solve for any unknown side of the triangle. The 3-4-5 rule, which is derived from the Pythagorean theorem, is used to verify that the corner of a right triangle is perfectly square.

Overloading a vehicle can place the operator at risk for an accident and can also damage the vehicle. Therefore, it is important to use equipment that is rated for the weight of the material. Each type of material has its own weight, so once you have calculated the volume of an excavation, you will need to calculate its weight. This is done by multiplying the total volume in cubic yards by the material weight per cubic yard.

Review Questions

1. It is important to be able to calculate the volume and weight of cut and fill material because _____.
   a. the material cost is established by weight and volume
   b. all heavy equipment is rated by weight and volume
   c. it determines the number of workers needed on a job
   d. trucks use the least amount of fuel when overloaded

2. It is necessary to know the area of a shape in order to calculate its volume.
   a. True
   b. False

3. When an equation has numbers grouped in parentheses, that calculation is performed _____.
   a. first
   b. second
   c. third
   d. last

4. The symbol π in an equation is considered a _____.
   a. variable
   b. root
   c. constant
   d. radius

5. The square root of the number 25 is _____.
   a. 5
   b. 25
   c. 50
   d. 625

6. A right angle contains _____.
   a. 45 degrees
   b. 90 degrees
   c. 180 degrees
   d. 360 degrees

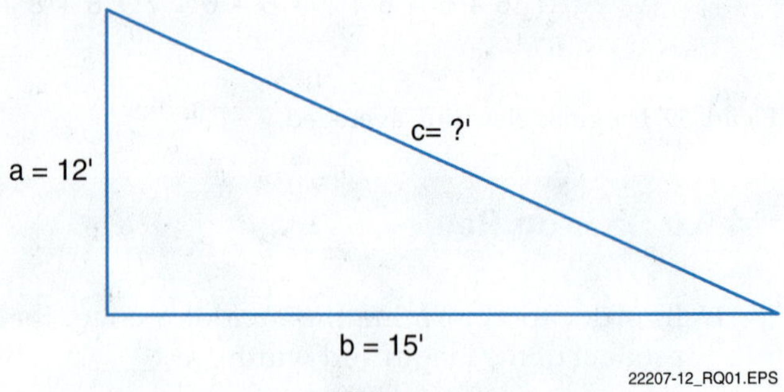

Figure 1

7. The length of the hypotenuse in *Figure 1* is _____.
   a. 27 feet
   b. 19.2 feet
   c. 365 feet
   d. 14 feet

8. Two-dimensional objects can be measured in all of the following values, *except* square _____.
   a. inches
   b. feet
   c. yards
   d. radius

9. One square yard is equal to _____.
   a. 3 square feet
   b. 9 square feet
   c. 27 square feet
   d. 144 square feet

10. One side of a square is 6 inches, so its area is _____.
    a. 3 square inches
    b. 12 square inches
    c. 24 square inches
    d. 36 square inches

11. A rectangle has a length of 6 inches and a width of 2 inches, so its area is _____.
    a. 6 square inches
    b. 12 square inches
    c. 16 square inches
    d. 36 square inches

12. Calculate the area for a triangle with a base of 7 inches and a height of 4 inches.
    a. 12 square inches
    b. 14 square inches
    c. 22 square inches
    d. 28 square inches

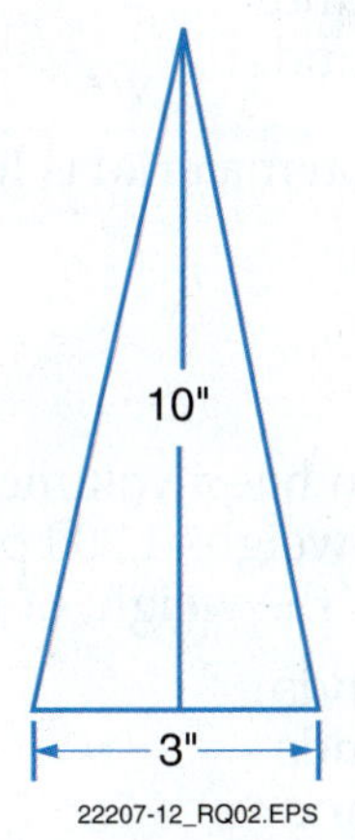

Figure 2

13. The area for the shape shown in *Figure 2* is _____.
    a. 3 square inches
    b. 15 square inches
    c. 30 square inches
    d. 45 square inches

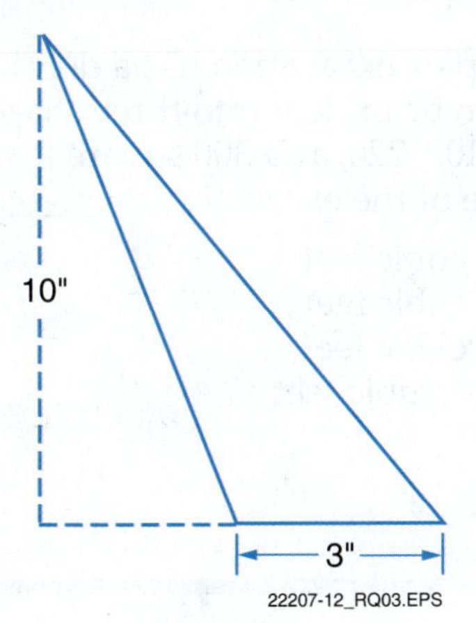

Figure 3

14. The area for the shape shown in *Figure 3* is _____.
    a. 3 square inches
    b. 15 square inches
    c. 30 square inches
    d. 45 square inches

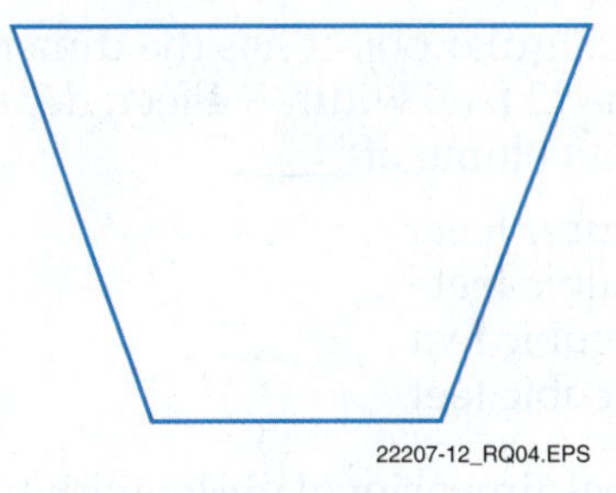

Figure 4

15. The shape shown in *Figure 4* is a _____.
    a. rhombus
    b. parallelogram
    c. trapezoid
    d. rectangle

16. A trapezoid's measurements are base 1 = 2 inches, base 2 = 4 inches, and height = 4 inches. What is its area?
    a. 3 square inches
    b. 12 square inches
    c. 30 square inches
    d. 45 square inches

17. The distance from the center of a circle to the edge of its curved line is called the _____.
    a. angle
    b. circumference
    c. diameter
    d. radius

18. A circle has a diameter of 2 feet. What is its area?
    a. 3.14 square feet
    b. 6.28 square feet
    c. 9.42 square feet
    d. 45.5 square feet

19. A backhoe loader's bucket can hold 7.0 cubic feet or up to 134 pounds, so its volume is _____.
    a. 7.0 cubic feet
    b. 19 pounds per foot
    c. 134 pounds
    d. 938 cubic feet per pound

20. A cube that has a dimension of 1 yard has a volume of _____.
    a. 1 cubic foot
    b. 9 cubic feet
    c. 27 cubic feet
    d. 2 cubic yards

21. A rectangular object has the dimensions of length = 2 feet, width = 4 feet, depth = 3 feet. It has a volume of _____.
    a. 6 cubic feet
    b. 8 cubic feet
    c. 12 cubic feet
    d. 24 cubic feet

22. A three-dimensional object with two identical triangular bases has the following dimensions: base = 9 inches, height = 10 inches, and depth = 2 feet. It has a volume of _____.
    a. 90 cubic inches
    b. 180 cubic inches
    c. 1,080 cubic inches
    d. 2,160 cubic inches

23. A three-dimensional object has a depth of 8 inches and a triangular base that has an area of 12 square inches, so it has a volume of _____.
    a. 8 cubic inches
    b. 20 cubic inches
    c. 48 cubic inches
    d. 96 cubic inches

24. A trapezoidal object has the following dimensions: base 1 = 2 feet, base 2 = 3 feet, height = 2 feet, and depth = 6 feet, so it has a volume of _____.
    a. 5 cubic feet
    b. 25 cubic feet
    c. 30 cubic feet
    d. 72 cubic feet

25. A cylinder has a diameter of 6 feet and a height of 2 feet, so its volume is _____.
    a. 12 cubic feet
    b. 28.3 cubic feet
    c. 56.5 cubic feet
    d. 226 cubic feet

26. A material has a weight of 1,700 pounds per cubic yard, and there are 54 cubic feet to move. What is the total weight of the material?
    a. 3,400 pounds
    b. 10,200 pounds
    c. 30,600 pounds
    d. 91,800 pounds

27. Wet excavation material is lighter than dry material.
    a. True
    b. False

28. An excavation has a volume of 203 cubic feet. The material weighs 1,200 pounds per cubic yard. What is the weight of the excavation?
    a. 2,700 pounds
    b. 9,000 pounds
    c. 81,120 pounds
    d. 243,600 pounds

29. A triangle-shaped slope excavation has a base of 4 feet, height of 6 feet, and length of 100 feet. Its volume is _____.
    a. 800 cubic feet
    b. 1,200 cubic feet
    c. 2,000 cubic feet
    d. 2,400 cubic feet

30. A complex excavation has a depth of 6 inches and can be broken into three shapes with areas of 100, 226, and 300 square feet. The total volume of the excavation is _____.
    a. 104 cubic feet
    b. 313 cubic feet
    c. 626 cubic feet
    d. 3,756 cubic feet

Trade Terms Introduced in This Module

Average: The middle point between two numbers or the mean of two or more numbers. It is calculated by adding all numbers together, and then dividing the sum by the quantity of numbers added. For example, the average (or mean) of 3, 7, 11 is 7 ($3 + 7 + 11 = 21$; $21 \div 3 = 7$).

Constant: A value in an equation that is always the same; for example pi is always 3.14.

Hypotenuse: The long dimension of a right triangle and always the side opposite the right angle.

Parallel: Two lines that are always the same distance apart even if they go on into infinity (forever is called infinity in mathematics).

Parallelogram: A two-dimensional shape that has two sets of parallel lines.

Quadrilateral: A four-sided, closed shape with four angles whose sum is 360 degrees.

Squared: Multiplied by itself.

Variable: A value in an equation that depends on the factors being considered; for example, the lengths of the sides of a triangle may vary from one triangle to another.

Additional Resources

This module presents thorough resources for task training. The following resource material is suggested for further study.

Applied Construction Math: A Novel Approach. Alachua, FL: NCCER.

Figure Credits

Reprinted courtesy of Caterpillar Inc., Module opener

Topaz Publications, Inc., Figure 1

Courtesy of Pizer Inc., Figure 55

Section Review Answers

Answer	Section Reference	Objective
Section One		
1 a	1.0.0	1a
2 d*	1.1.0	1b
3 a	1.3.0	1c
Section Two		
1 b*	2.1.0	2a
2 b*	2.2.0	2b
Section Three		
1 c*	3.1.0	3a
2 c*	3.2.0	3b
3 b	3.3.0	3c
4 b	3.4.0	3d
Section Four		
1 d	4.1.0	4a
2 c	4.2.0	4b
3 a	4.3.0	4c
4 d	4.4.1	4d

Section Review Calculations

1-1. $4 \times 4 \times 4 = 64$

2-1.

$$b = \sqrt{c^2 + a^2}$$

$$b = \sqrt{720^2 - 530^2}$$

$$b = \sqrt{518{,}400 - 280{,}900}$$

$$b = \sqrt{237{,}500}$$

$$b = 487'$$

SR01.EPS

2-2. $24 \div 3 = 8$

$8 \times 4 = 32$

$8 \times 5 = 40$

$24 - 32 = 40$

3-1. $15 \times 12 = 180$

3-2. Triangle 3:

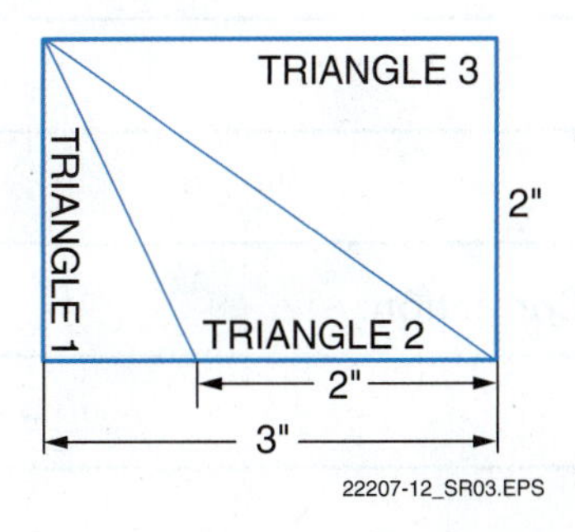

22207-12_SR03.EPS

base = 3 inches

height = 2 inches

Area $= \frac{1}{2} \times (3 \times 2)$

$= \frac{1}{2} \times 6$

= 3 square inches

NCCER CURRICULA — USER UPDATE

NCCER makes every effort to keep its textbooks up-to-date and free of technical errors. We appreciate your help in this process. If you find an error, a typographical mistake, or an inaccuracy in NCCER's curricula, please fill out this form (or a photocopy), or complete the online form at **www.nccer.org/olf**. Be sure to include the exact module ID number, page number, a detailed description, and your recommended correction. Your input will be brought to the attention of the Authoring Team. Thank you for your assistance.

Instructors – If you have an idea for improving this textbook, or have found that additional materials were necessary to teach this module effectively, please let us know so that we may present your suggestions to the Authoring Team.

NCCER Product Development and Revision
13614 Progress Blvd., Alachua, FL 32615
Email: curriculum@nccer.org
Online: www.nccer.org/olf

❑ Trainee Guide ❑ AIG ❑ Exam ❑ PowerPoints Other ______________

Craft / Level: Copyright Date:

Module ID Number / Title:

Section Number(s):

Description:

Recommended Correction:

Your Name:

Address:

Email: Phone:

22209-13

Interpreting Civil Drawings

Overview

Civil drawings define the excavation and grading requirements for roads and building construction sites. The ability to interpret those drawings is important for all personnel involved in site work.

Module Four

Trainees with successful module completions may be eligible for credentialing through NCCER's National Registry. To learn more, go to **www.nccer.org** or contact us at **1.888.622.3720**. Our website has information on the latest product releases and training, as well as online versions of our *Cornerstone* newsletter and Pearson's product catalog.

Your feedback is welcome. You may email your comments to **curriculum@nccer.org**, send general comments and inquiries to **info@nccer.org**, or fill in the User Update form at the back of this module.

This information is general in nature and intended for training purposes only. Actual performance of activities described in this manual requires compliance with all applicable operating, service, maintenance, and safety procedures under the direction of qualified personnel. References in this manual to patented or proprietary devices do not constitute a recommendation of their use.

22209-13
Interpreting Civil Drawings

Objectives

When you have completed this module, you will be able to do the following:

1. Describe the types of drawings usually included in a set of plans and list the information found on each type.
   a. Explain the use of title sheets, title blocks, and revision blocks.
   b. Describe the types of drawings used in highway construction.
   c. Describe the types of drawings used in building site construction.
   d. Describe how as-built drawings are prepared.
2. Read and interpret drawings.
   a. Identify different types of lines and symbols used on drawings.
   b. Define common abbreviations used on drawings.
   c. Interpret building site and highway drawings to determine excavation requirements.
3. Explain specifications and the purpose of specifications.
   a. Identify the types of information contained in specifications.
   b. Explain the common format used in specifications.

Performance Tasks

Under the supervision of your instructor, you should be able to do the following:

1. Determine the scale of different drawings.
2. Interpret a set of drawings to determine the proper type and sequence of excavation and grading operations needed to prepare the site.

Trade Terms

Change order
Contour lines
Easement
Elevation view
Invert
Loadbearing
Monuments
Plan view
Property lines
Request for information
Setback
Uniform Construction Index

Industry Recognized Credentials

If you are training through an NCCER-accredited sponsor, you may be eligible for credentials from NCCER's Registry. The ID number for this module is 22209-13. Note that this module may have been used in other NCCER curricula and may apply to other level completions. Contact NCCER's Registry at 888.622.3720 or go to **www.nccer.org** for more information.

Contents

Topics to be presented in this module include:

Figures

Section One

1.0.0 Types of Civil Drawings

Objective 1

Describe the types of drawings usually included in a set of plans and list the information found on each type.

a. Explain the use of title sheets, title blocks, and revision blocks.
b. Describe the types of drawings used in highway construction.
c. Describe the types of drawings used in building site construction.
d. Describe how as-built drawings are prepared.

Trade Terms

Change order: A formal instruction describing and authorizing a project change.

Contour lines: Imaginary lines on a site/plot plan that connect points of the same elevation. Contour lines never cross each other.

Easement: A legal right-of-way provision on another person's property (for example, the right of a neighbor to build a driveway or a public utility to install water and gas lines on the property). A property owner cannot build on an area where an easement has been identified.

Elevation view: A drawing giving a view from the front or side of a structure.

Loadbearing: A base designed to support the weight of an object of structure.

Monuments: Physical structures that mark the locations of survey points.

Plan view: A drawing that represents a view looking down on an object.

Property lines: The recorded legal boundaries of a piece of property.

Request for information (RFI): A form used to question discrepancies on the drawings or to ask for clarification.

Setback: The distance from a property line in which no structures are permitted.

Every construction project is defined in detail by a set of drawings. In addition, many projects have written specifications that contain further details about the quality of work and the materials to be used. The drawings and specifications are prepared by architects and/or engineers and become part of the construction contract.

During the planning phase of a new construction project, architects and engineers develop detailed plans for the project using data gathered by the survey team. The survey team verifies the site boundaries, which are established by a licensed surveyor, and then set up the construction project's precise position in relation to the property boundaries. In addition, they establish the exact location of road and utility rights-of-way (ROW), **easements**, and other important features of the project. This information is reflected on the project drawings.

Equipment operators must be able to interpret construction drawings and specifications correctly. Failure to do so may result in costly rework and unhappy customers. Depending on the severity of the mistake, it can also expose you and your employer to legal liability.

An equipment operator's duties vary with the type and size of the construction project. Highway construction jobs may require the constant use of heavy equipment for hauling fill, cutting grades, leveling the roadbed, and spreading paving materials. Residential or commercial building projects may require the intense use of heavy equipment at the beginning of the project to prepare the site for construction and at end of the project to complete final grades and landscaping tasks.

Depending on the type of project, preparation of the project drawings may be the responsibility of an architect or an engineering firm. A complete set of project drawings is likely to include three classifications of drawings:

- *Architectural* – The architectural drawings deal primarily with the appearance and finish of the structure. Architectural drawings might include building elevations, floors plans, window and door details, and finish schedules.
- *Structural* – Structural drawings contain information specific to the structure, including foundations, footings, and other **loadbearing** structures.

- *Civil* – The civil drawings are of special interest to the heavy equipment operator because they deal with excavation and grading. These drawings cover site preparation, including clearing and grubbing; cut and fill plans; grading plans; underground utility plans; road profiles and sections; and paving plans.

Most drawings are drawn to scale, which is prominently displayed on the drawing. The scale used depends on the size of the project. A large project often has a small scale, such as 1" = 100', while a small project might have a large scale, such as 1" = 10'. The dimensions shown on drawings can be in engineering scale, which is in feet and tenths of a foot, or architectural scale, which is in feet, inches, and fractions of an inch. Plans used in excavation and grading work are likely to use engineering scale. It is common for a set of drawings to include on the first sheet a map that locates the project (*Figure 1*). In this case, the map also shows the detour that will be required while the work is being done.

The plans for a roadway project differ significantly from those for a building site. Because roadways can stretch across many miles, these plans need to consider the topography and soil conditions over long distances, while building plans need to consider the terrain over a comparatively small area. Equipment operators must understand how to read and interpret both types of plans in order to perform their jobs.

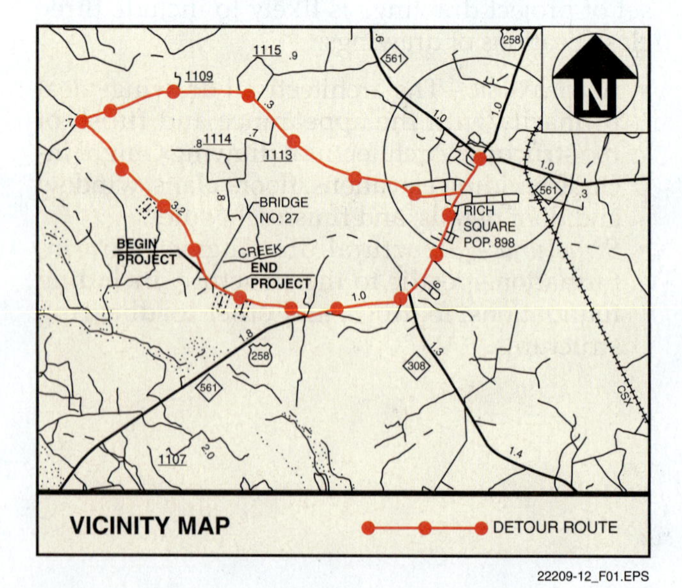

Figure 1 Example of a project location map.

1.1.0 Title Sheets, Title Blocks, and Revision Blocks

A title sheet is normally placed at the beginning of a set of drawings or at the beginning of a major section of drawings. It may provide an index to the other drawings; a list of abbreviations used on the drawings and their meanings; a list of symbols used on the drawings and their meanings. The title sheet often contains other project data, such as the project location, the size of the land parcel, and the building size. It is important to use the title sheet that comes with the drawing in order to understand the specific symbols and abbreviations used on the drawings. These symbols and abbreviations may vary from job to job.

A title block or box (*Figure 2*) is normally placed on each sheet in a set of drawings. It is usually located in the bottom right-hand corner of the sheet, but this location can vary. The drawing set should be folded so that the title block faces up.

The title block serves several purposes in terms of communicating information. It contains the name of the firm that prepared the drawings, the owner's name, and the address and name of the project. It also gives locator information, such as the title of the sheet, the drawing or sheet number, the date the sheet was prepared, the scale, and the initials or names of the people who prepared and checked the drawing.

A revision block is normally shown on each sheet in a set of drawings. Typically, it is located in the upper right-hand corner or bottom right-hand corner of the drawing, near or within the title block. It is used to record any changes (revisions) to the drawing. An entry in the revision block usually contains the revision number or letter, a brief description of the change, the date, and the initials of the person making the revision(s). When using drawings, it is essential to note the revision designation on each drawing and use only the latest issue; otherwise, costly mistakes can result. If there is a conflict between drawings, or if you are in doubt about the revision status of a drawing, check with your supervisor to make sure that you are using the most recent version of the drawing. Also, check to see if any **requests for information (RFI)** or sketches have been included in the latest plan revision.

1.2.0 Highway Plans

Highway construction projects are called horizontal projects because they cover long distances and have almost no height. On highway jobs, the emphasis is on grading the roadbed and nearby areas according to engineering plans, which are

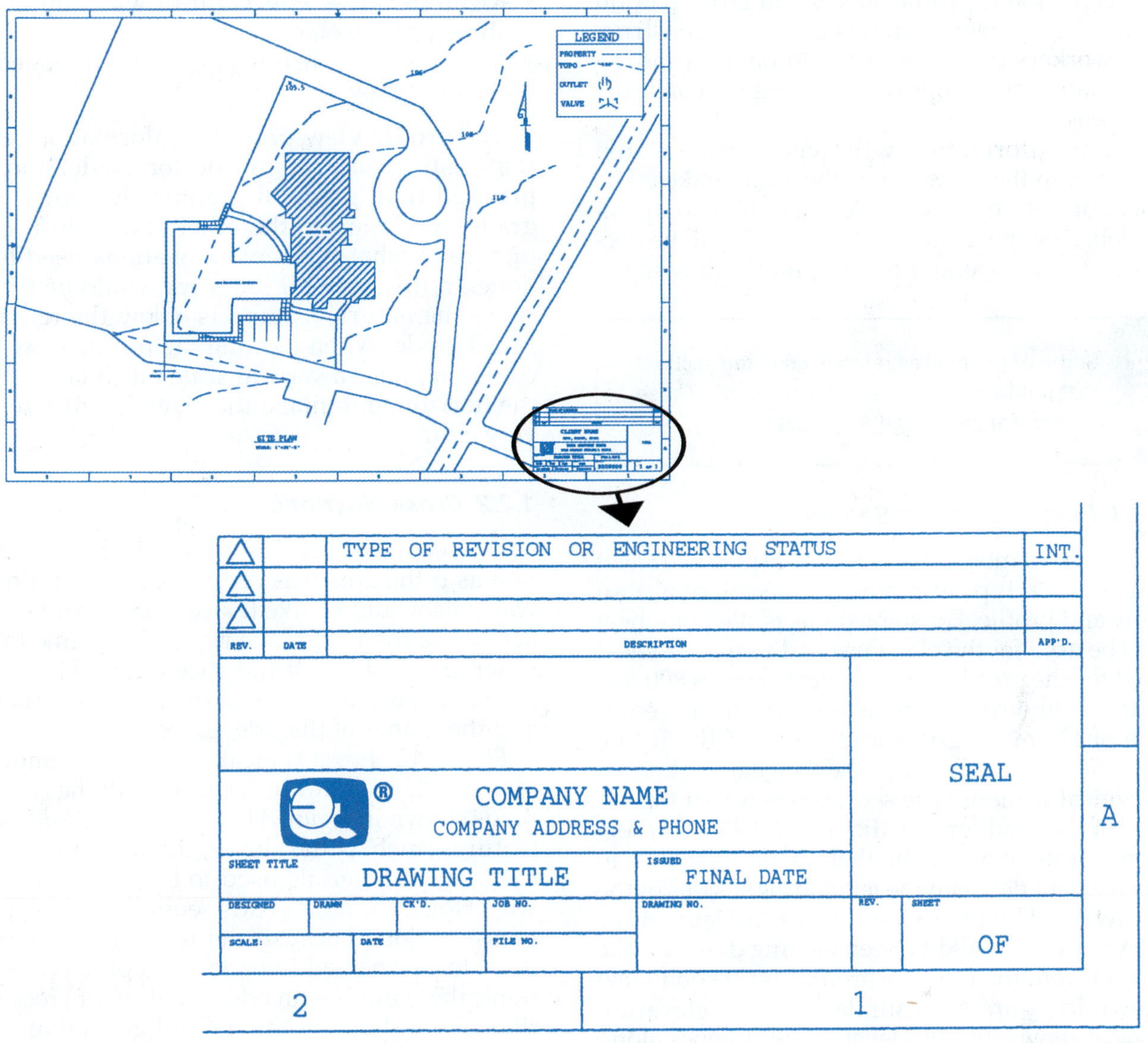

Figure 2 Title and revision blocks.

Technology at Work

Drawings have been used for centuries to define the construction of buildings, roads, bridges, and other structures. Until late in the twentieth century, the drawings were painstakingly done by drafters who worked with mechanical instruments at tall drawing boards. Today, most drawings are done using software in what is known as a computer-aided design, or CAD, system. This software greatly simplifies the process and makes the drawings easy to change. Many CAD systems are capable of producing three-dimensional renderings.

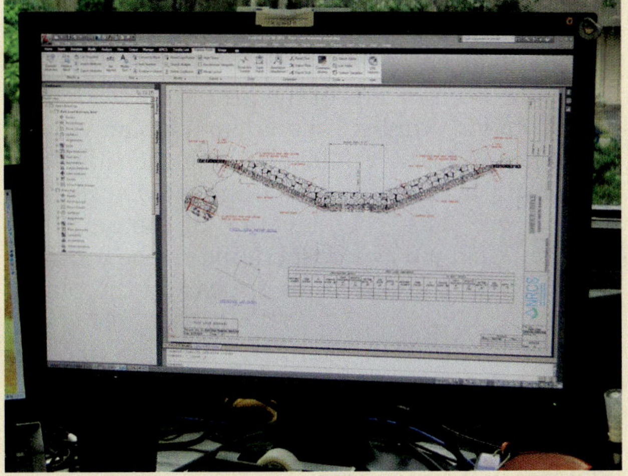

called plan and profile sheets and cross-section sheets. Equipment operators and other construction workers use these plans to obtain grading information at various times during the construction project.

Grade information will generally be transferred from the drawings to the grade stakes. The operator will read the grade stakes to accomplish the job. In some instances, however, it will be necessary for the operator to interpret the drawings.

> **NOTE**
> Some of the figures referenced in the sections that follow are located in the *Appendix*. These figures are numbered A1, A2, etc.

1.2.1 Plan and Profile Sheets

Refer to this module's *Appendix*. *Figure A1* (Sheet C8) shows a typical roadway construction **plan view** and profile. For some projects, a whole sheet may be used for the plan view and a separate sheet used for the profile view. *Figure A1* covers 800 feet of the A-line road as measured from the center line of Highway 270, and 225 feet of the B-line road. *Figure A2* (Sheet C7) shows what are known as typical sections. The six sections reflect the design of the road for specific station-to-station segments of the roads. Note that the designs vary to some extent between the left and right sides of the roadways. The plan view shown in *Figure A1* is the view that would be seen looking down on the project from the top. The profile and section view shown in *Figure A2* is similar to a side **elevation view**. It shows the key elevations and slopes along the center line of the route. The main information shown on these sheets includes the following:

- *Direction* – The directional arrow always points north. Stations on the drawings are numbered from west to east and south to north.
- *Station numbers* – The station numbers listed along the bottom axis of the profile correspond to the station numbers shown on the plan view.
- *Elevations* – These are listed along the sides of the profile according to the designated scale.
- *Natural ground* – The elevation of the natural ground is drawn as a dashed line on the profile sheet, while **contour lines** appear on the plan view.
- *Planned grade* – The planned elevation of the grade is drawn as a continuous line on the profile.
- *Center line* – The center line of the roadway is plotted on the plan view.
- *Right-of-way* – The right-of-way limits are shown on the plan view.
- *Benchmarks* – Benchmarks, if any, are noted on the plan view.

The profile view provides information about the existing natural ground (or grade) and the planned final grade. It graphically shows each grade in relation to the other and gives a good picture of what types of excavations need to be done. In this case, all the work would be fill because the natural ground is below the required finish grade. When plotting the profile view, it is common to use a vertical scale much larger than the horizontal to make the elevation differences very clear.

1.2.2 Cross-Sections

Cross-section sheets are views of the construction as if the area was cut crosswise. For a highway, this would be like taking a knife and slicing across the road from one right-of-way line to the other and looking at the slice taken. The cross-section shows the layers of the road construction and the shapes of the side slopes and ditches.

Figure A2 shows typical sections or templates for a two-lane highway. It covers both the A and B roads shown in *Figure A1*. A typical section shows features, such as the slopes, ditches, and ramps, along with materials used to build up the roadway. However, most roads require many typical cross-sections because the terrain varies from point to point. In addition to the typical cross-sections, there may be an additional set of cross-section sheets showing the natural ground and the shape of the finish grade every 50 feet or other established distance. These sheets must be checked frequently to get grade information about the section of road that is being worked on because grade details can change as the terrain varies.

1.3.0 Building Construction Site Drawings

Every project requires a site plan to show the locations of buildings and other structures on the site. The site plan is often divided into additional plans that cover excavation, grading, utilities, and drainage. The site plan and its subordinate plans are the civil drawings of primary interest to heavy equipment operators. In addition to the site plans, there are some sheets of the architectural and structural drawings that may be of use during excavation and grading. These include the building elevations and the foundation plan.

Plan view drawings are drawings that show the site looking down from above. The object is projected from a horizontal plane. Typically, plan view drawings are made to show the overall construction (site plan), the structure's foundation (foundation plan), and the structure's floor plans.

1.3.1 Site Plans

Man-made and topographical (natural) features and other relevant project information, including the information needed to correctly locate structures on the site, are shown on a site plan. The site plan is sometimes called a plot plan. Man-made features include roads, sidewalks, utilities, and buildings. Topographical features include trees, streams, springs, and existing contours. Project information includes the building outline, general utility information, proposed sidewalks, parking areas, roads, landscape information, proposed contours, and any other information that conveys what is to be constructed or changed on the site. A prominently displayed north direction arrow is included for orientation purposes on site plans. Sometimes a site plan contains a large-scale map of the overall area that indicates where the project is located on the site. Examples of two different site plans are shown in *Figures 3* and *A3*. The plan in *Figure 3* is for a small site and shows topographical features in addition to the locations of structures. The plan in *Figure A3* is for a much larger site.

Typically, site plans show the following types of detailed information:

- Coordinates of control points or property corners
- Direction and length of **property lines** or control lines
- Description, or reference to a description, for all control and property **monuments**
- Location, dimensions, and elevation of the structure on site
- Finish and existing grade contour lines
- Finish elevations of building floors
- Location of utilities
- Location of existing elements such as trees and other structures
- Locations and dimensions of roadways, driveways, and sidewalks
- Names of all roads shown on the plan
- Locations and dimensions of any easements

Like other drawings, site plans are usually drawn to scale. The scale is prominently displayed on the drawing. The scale used depends on the size of the project. A project covering a large area typically has a small scale, such as 1" = 100', while a project on a small site might have a large scale, such as 1" = 10'.

NOTE

Not all drawings are made to scale. Those that are not scaled should be marked "NOT TO SCALE" or "NTS."

Normally, the dimensions shown on site plans are stated in feet and tenths of a foot (engineer's scale). However, some site plans state the dimensions in feet, inches, and fractions of an inch (architect's scale). Dimensions measured between the property lines and the structures are shown to verify that the locations of structures meet building code requirements. Building codes typically establish minimum **setbacks** that are measured from a property line. Front setbacks may be measured from the center line of the road rather than the property line. A front setback specifies the minimum distance that must be maintained between the property line (or the center line of the road) and the front of a structure (building line). Side and rear setbacks are also established by code. Building lines reflecting these setbacks are often included on the site plan (see *Figure 3*). Normally, side yard setbacks are specified to allow for access to rear yards and to reduce the possibility of fire spreading to adjacent buildings.

Site plans and survey maps often show areas of easement on the property. Easements are legal rights of persons other than the owner to use the property. The most common reason for an easement is to provide access to utility lines such as sewer, water, and electricity. Easements are also granted to municipalities for the purpose of maintaining drainage swales. The property owner

Blueprints

Many people still refer to construction drawings as blueprints, even though today's drawings are usually black ink on white paper. The term *blueprint* derives from the ammonia-based process once used to copy drawings, a process that turned the paper blue. Although it has been many years since this process was in common use, the term *blueprint* is still used by many people in the construction industry.

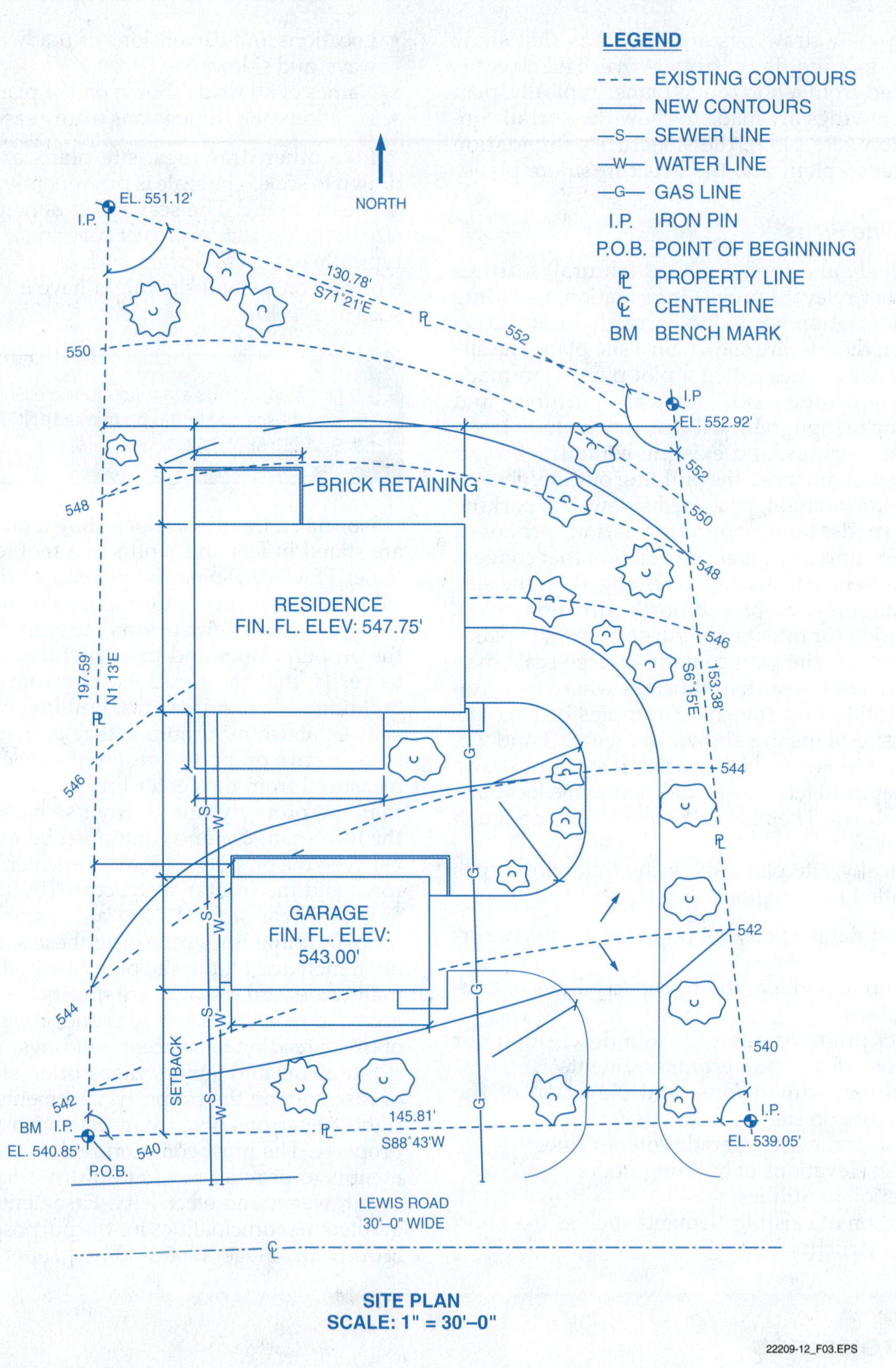

Figure 3 Example of a simple site plan showing topographical features.

is prohibited from building any structure on an easement or otherwise obstructing access to it.

Site plans show finish grades (also called elevations) for the site, based on data provided by a surveyor or engineer. It is necessary to know these elevations for grading the lot and for construction of the structure. Finish grades are typically shown for all four corners of the lot, as well as other points within the lot. Finish grades or elevations are also shown for the corners of the structure and relevant points within the building.

Heavy equipment operators need to pay particular attention to existing and proposed contour lines on these plans. These lines define how deep the cuts will be or how much fill is required to bring the site to the correct elevation.

It is important to study the site plans because they show details such as locations of property lines, survey markers, and utilities that help equipment operators avoid problems. Follow these rules while you are working:

- Be aware of the locations of property boundaries. Do not cross property boundaries unless you know that the owner has given the managers of the project an easement. Heavy equipment can damage terrain and underground structures such as drainage pipes, culverts, and septic tanks.
- Know the locations of surveyors marks. Do not operate heavy equipment in the location of a benchmark, monument, or control points until you are sure of its location. Damaging these references can cause costly delays.
- Know the locations of buried utilities. Do not operate heavy equipment near any buried utilities unless you are certain of their locations. Not only can hitting underground gas lines and power cables cause delay, it can be fatal. If the property you are working on was previously developed, the chances are it has buried utilities. These utilities must be located and marked before any excavation can begin.

Take a few moments to study the site plan shown in *Figure 3*. Find the North marker and review the legend to become familiar with the symbols used on the drawing. Look for property boundaries, utilities, and existing and proposed grades (contours).

All the finish grade references shown are keyed to a reference point, called a benchmark or job datum. This is a reference point established by the surveyor on or close to the property, usually at one corner of the lot. At the site, this point may be marked by a plugged pipe driven into the ground, a brass marker, or a wood stake. The location of the benchmark is shown on the plot plan with a grade figure next to it. This grade figure may indicate the actual elevation relative to sea level, or it may be an arbitrary elevation, such as 100.00' or 500.00'. All other grade points shown on the site plan, therefore, are relative to the benchmark. In *Figure 3*, this point is labeled P.O.B. for point of beginning and is located at the southwest corner of the property.

Scaling Drawings

Measuring the length of a line on a drawing, then converting that measurement to an actual length is known as scaling. Scaling can be done using an engineer's scale, but the task can be simplified by using an electronic plan wheel scaler like the one shown. The device is first set to match the drawing scale. Then, as the scaler is rolled along a line on the drawing, its digital readout gives a direct reading of the length of the line.

22209-12_SA02.EPS

Enforcing Setbacks

Municipal inspectors can be very strict in enforcing setback requirements. If an addition to a building penetrates a setback, for example, the inspector may refuse to issue a certificate of occupancy until the problem is corrected. As a result, the property owner would be forced to change or remove the structure or obtain a variance by appealing to a special review board. Both methods can be costly and time consuming.

A site plan usually shows the finish floor elevation of the building. This is the level of the first floor of the building relative to the job-site benchmark. For example, if the benchmark is labeled 100.00' and the finish floor elevation indicated on the plan is marked 105.00', the finish floor elevation is 5' above the benchmark. During construction, many important measurements are taken from the finish floor elevation point.

On *Figure 3*, the benchmark is located at the southwest property corner and is 540.85' (P.O.B.). Since the finish floor elevation of the residence is 547.75', the finish floor elevation is 6.9' above the benchmark.

Depending on the size and complexity of the site, and sometimes on the requirements of the local municipality, the site plan may be subdivided to include one or more additional plans showing specific details of site preparation. These additional plans might include the following:

- Excavation plan
- Utility plan
- Grading plan
- Drainage plan

Excavation plan – Heavy equipment operators play a key role in the cut and fill work that needs to be done in preparing a job site for construction. The plans that guide cut and fill operations are part of the drawing set that defines a project. *Figure 4* shows cross-sections of a construction site. The red lines on the charts represent locations of rock; the blue lines are finish grade; and the green lines are existing grade. *Figure 5* is a cut and fill plan for the same site. It shows in detail how much cut and fill work is needed to complete the project. "F" represents fill, and each F bubble indicates how much fill is needed. For example, F1+72 means fill 1 foot, 7 tenths, and 2 hundredths. The "C" (cut) bubbles are interpreted the same way.

Utility plan – Most, if not all, sites have buried utilities such as sanitary sewer lines; stormwater drain lines; fresh water piping; natural gas lines; and any electrical and communications cabling. A utility plan is prepared to show the locations of these utilities, as well as the depth at which they are buried. If the property does not have access to some municipal services, the locations of on-site services such as wells, septic systems, and gas tanks are included on the plan. *Figure A4* is an example of a utility plan showing a water line. Additional sheets are used to show other utilities, such as sewer, electrical, and gas lines.

Grading and drainage plans – A grading plan is used to show how the surface of the construction site will be altered in order to accommodate the construction. One of the main functions of the grading plan is to ensure proper drainage of the site to the established stormwater removal system. This plan also helps to verify that drainage from the site under construction will not adversely affect adjoining properties. Municipal engineers and planning boards, who review and approve the site plans, will want to see that required swales, culverts, and drains have been accounted for. *Figure A5* shows an example of a grading plan. The contour lines on the plan represent the surface features of the site. The solid contour lines show the current topographical features, while the dashed lines show the planned configuration. Site layout crews will used this plan to place cut and fill stakes.

In some instances, a separate drainage plan is required to show how rainwater runoff will be contained on the property under construction. *Figure 6* is an example of a drainage plan. Project engineers will analyze the soil on the property under construction to determine its ability to absorb water. They will also take into account the amount of impervious area that will be created by the construction. The construction of buildings, roads, and parking lots, as well as the removal of trees, reduces the capacity of a site to absorb water. In such cases, it may be necessary to create holding areas for runoff water. Such areas are known as detention or retention ponds. A detention pond is an excavation intended to hold overflow water until it can be absorbed naturally or evaporate. A detention pond is dry much of the time. A retention pond is an excavation that holds water continuously. Retention ponds are often installed where there is a natural source of water, such as a spring. Retention ponds such as the one shown in *Figure 7* are often used as landscape features. Another method of controlling runoff water is to install drains that are connected through buried pipes to the stormwater system (*Figure 8*).

NOTE

On smaller sites, the grading, drainage, and utility plans may be included on the same drawing.

1.3.2 Foundation Plans

Foundation plans, such as the one shown in *Figure 9*, give information about the location and dimensions of footings, grade beams, foundation walls, stem walls, piers, equipment footings, and windows and doors. The specific information shown on the plan is determined by the type of construction involved, such as full-basement foundation,

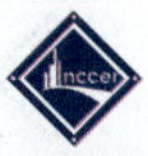

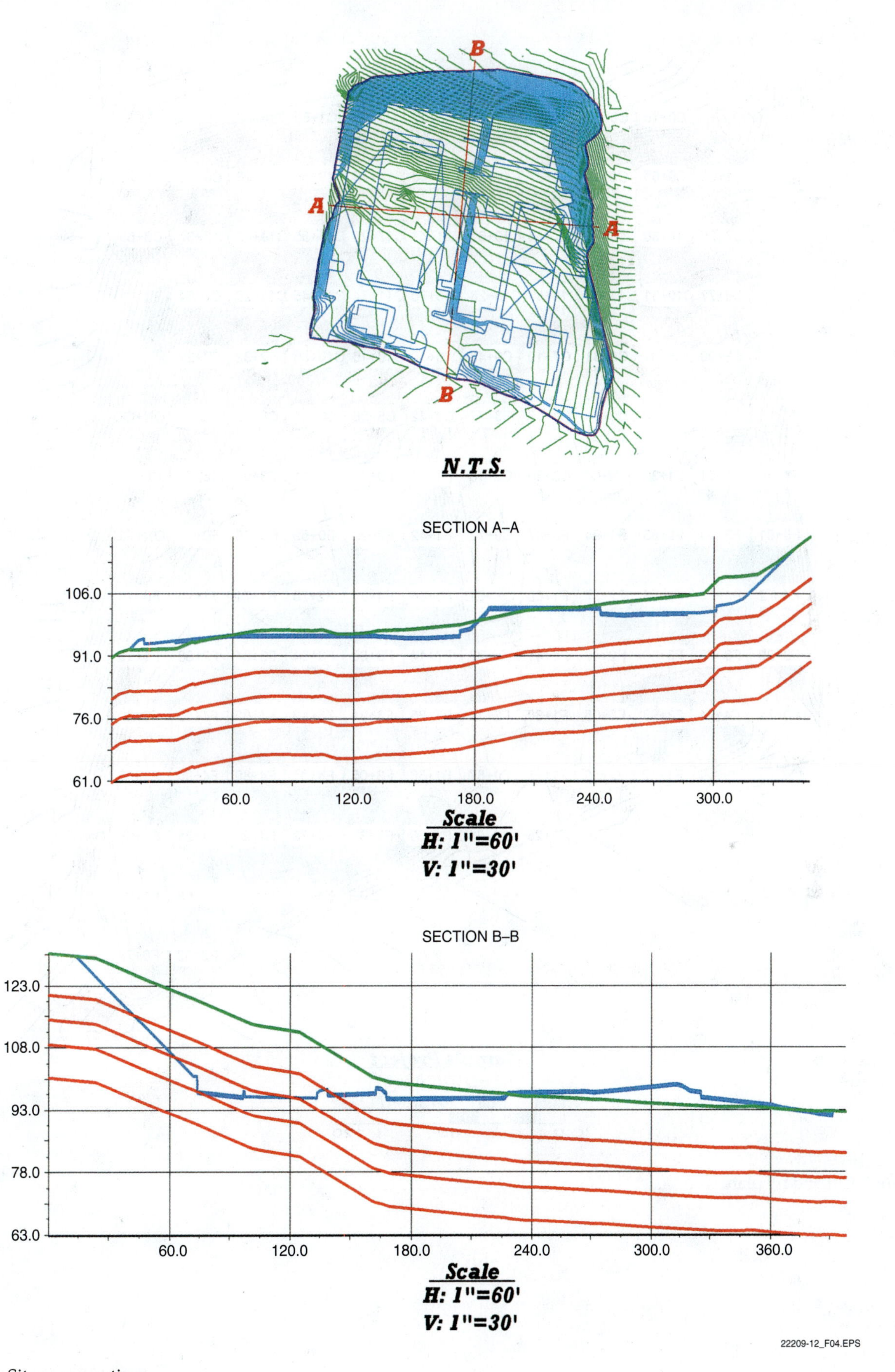

Figure 4 Site cross-sections.

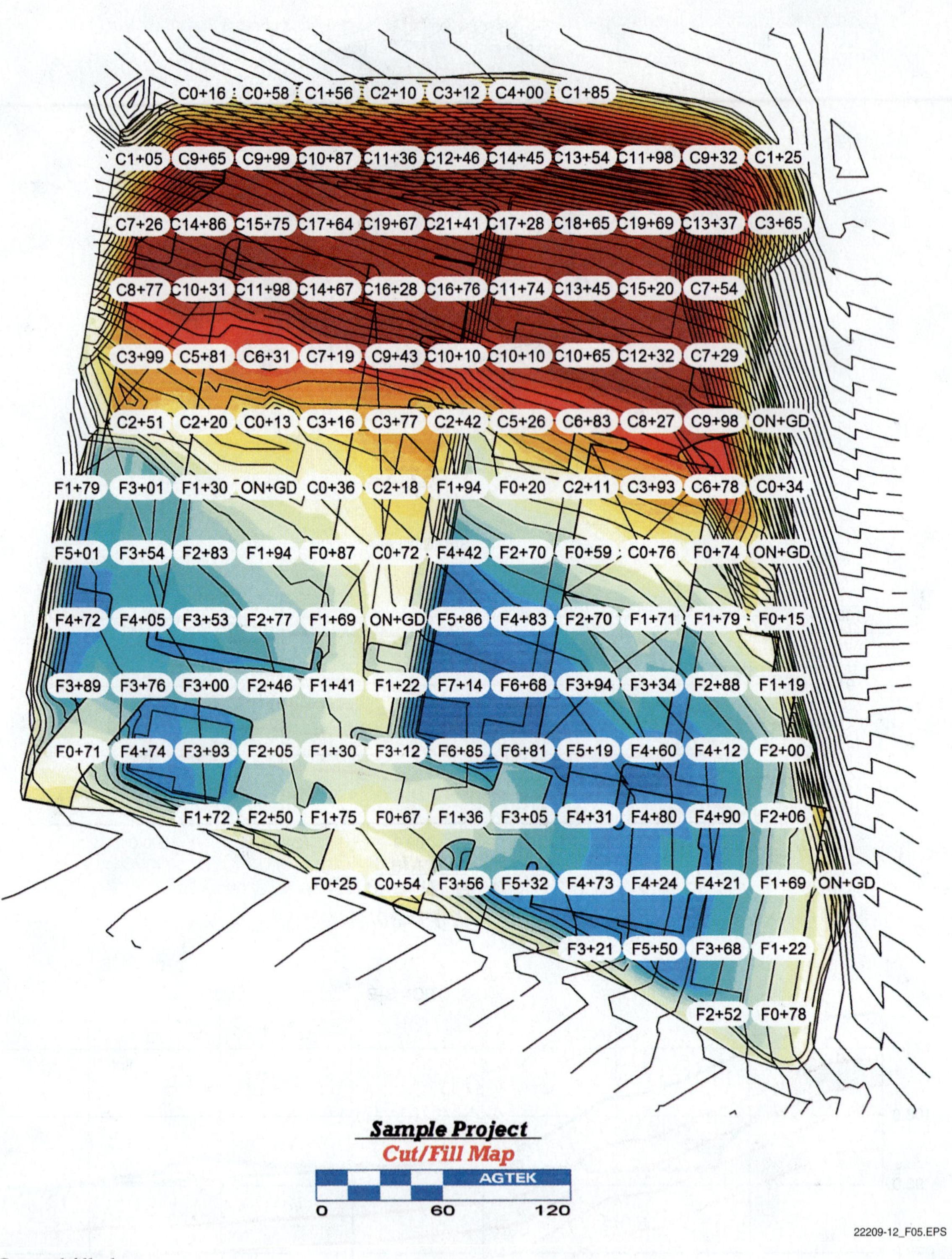

Figure 5 Cut and fill plan.

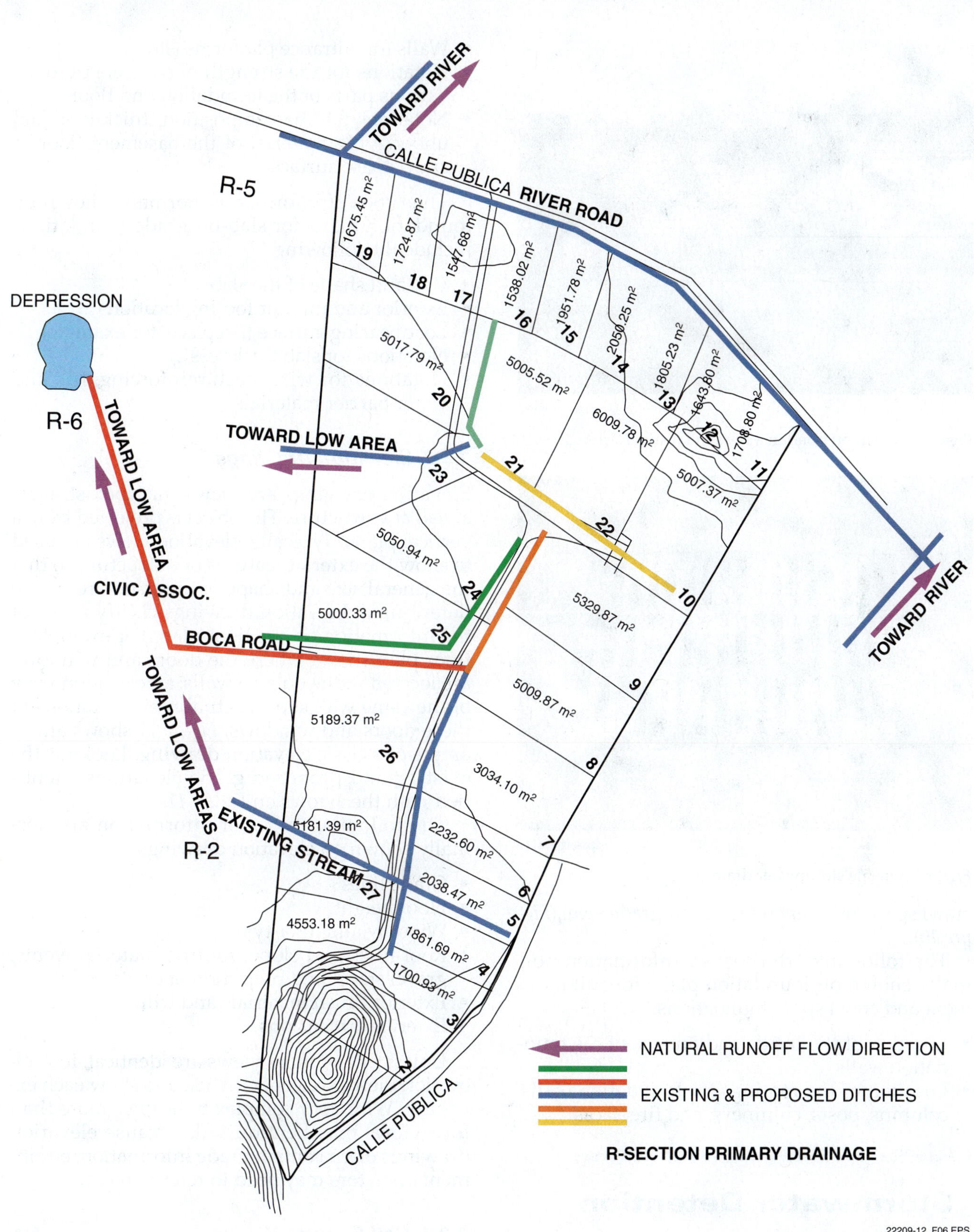

Figure 6 Example of a drainage plan.

22209-12_F07.EPS

Figure 7 Retention pond in a residential subdivision.

22209-12_F08.EPS

Figure 8 On-site stormwater drain.

crawl space, or a concrete slab-on-grade level (*Figure 10*).

The following are types of information normally shown on foundation plans for full-basement and crawl space foundations:

- Location of the inside and outside of the foundation walls
- Location of the footings for foundation walls, columns, posts, chimneys, and fireplaces

Stormwater Detention

Sandy soil does not absorb water very well, so heavy rain tends to run off. Construction sites often require acres of detention ponds to compensate for the impervious area created by the construction of buildings, roads, and parking lots.

- Walls for entrance platforms (stoops)
- Notations for the strength of concrete used for various parts of the foundation and floor
- Notations for the composition, thickness, and underlaying material of the basement floor or crawl space surface

The types of information normally shown on foundation plans for slab-on-grade foundations include the following:

- Size and shape of the slab
- Exterior and interior footing locations
- Loadbearing surface (fireplace, for example)
- Notations for slab thickness
- Notations for wire mesh reinforcing, fill, and vapor barrier materials

1.3.3 Elevation Drawings

Elevation drawings are views that look straight ahead at a structure. The object is projected from a vertical plane. Typically, elevation views are used to show the exterior features of a structure so that the general size and shape of the structure can be determined. Elevation drawings clarify much of the information on the floor plan. For example, a floor plan shows where the doors and windows are located in the outside walls; an elevation view of the same wall shows actual representations of these doors and windows. *Figure 11* shows an example of a basic elevation drawing. Look for the existing and proposed grade elevations, identified with the arrows on *Figure 11*.

The following types of information are normally shown on elevation drawings:

- Grade lines
- Floor height
- Window and door types
- Roof lines and slope, roofing material, vents, gravel stops, and projection of eaves
- Exterior finish materials and trim
- Exterior dimensions

Unless one or more views are identical, four elevation views are generally used to show each exposure. With very complex buildings, more than four views may be required. Because elevation drawings often contain grade information, equipment operators may need to refer to them.

1.3.4 Soil Reports

Soil conditions are among the factors that determine the type of foundation best suited for a structure. This information can be vital in determining how you do your job. A structure that is built on soil that lacks consistent quality and compaction

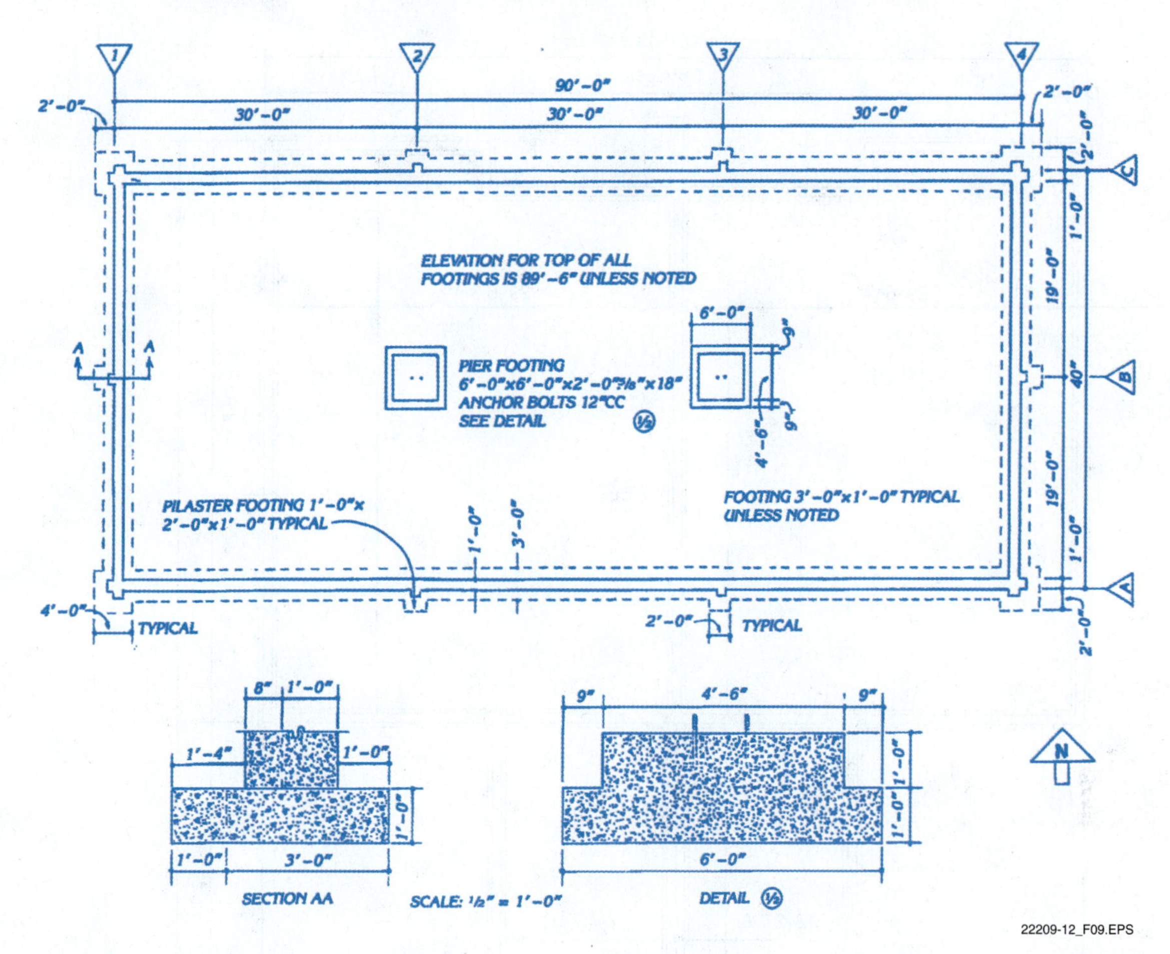

Figure 9 Foundation plan.

will settle unevenly. This can result in cracks in the foundation and structural damage to the rest of the building. Therefore, in designing the foundation for a structure, the architect must consider the soil conditions on the building site. Typically, the architect consults a soil engineer, who makes test bores of the soil on the building site and analyzes the samples. The results of the soil analysis are summarized in a soil report issued by the engineer. This report is often included as part of the drawing set. When using this information, consider all aspects of the soil report, including elevation of the water table. The type of soil on the job site will determine the types of equipment needed to do the job. For example, a backhoe can easily excavate sandy soil, but hard-packed clay may require some other equipment to break it up before it can be excavated with a backhoe.

1.4.0 As-Built Drawings

As-built drawings are formally incorporated into the drawing set to record changes made during construction. These drawings are marked up on the job by the various trades to show any differences between what was originally shown on a plan by the architect or engineer and what was actually built. Such changes result from the need to relocate equipment to avoid obstructions; relocate utilities; or because the architect has changed a certain detail in the site design in response to customer preferences. On many jobs, any such changes to the design can only be made after a **change order** has been generated and approved by the project engineer or other designated person. Depending on the complexity of the change, changes to the drawings are typically outlined with a unique design such as a cloud symbol. Changes should be made in red ink to make sure they stand out. Changes must be dated and initialed by the responsible party.

A supervisor or the project engineer will determine if it is necessary to deviate from the design plans for some reason, but it is part of the operator's job to make sure that the changes get marked on the as-built drawings. One of the most important entries on these plans is any deviation in the placement of underground utilities.

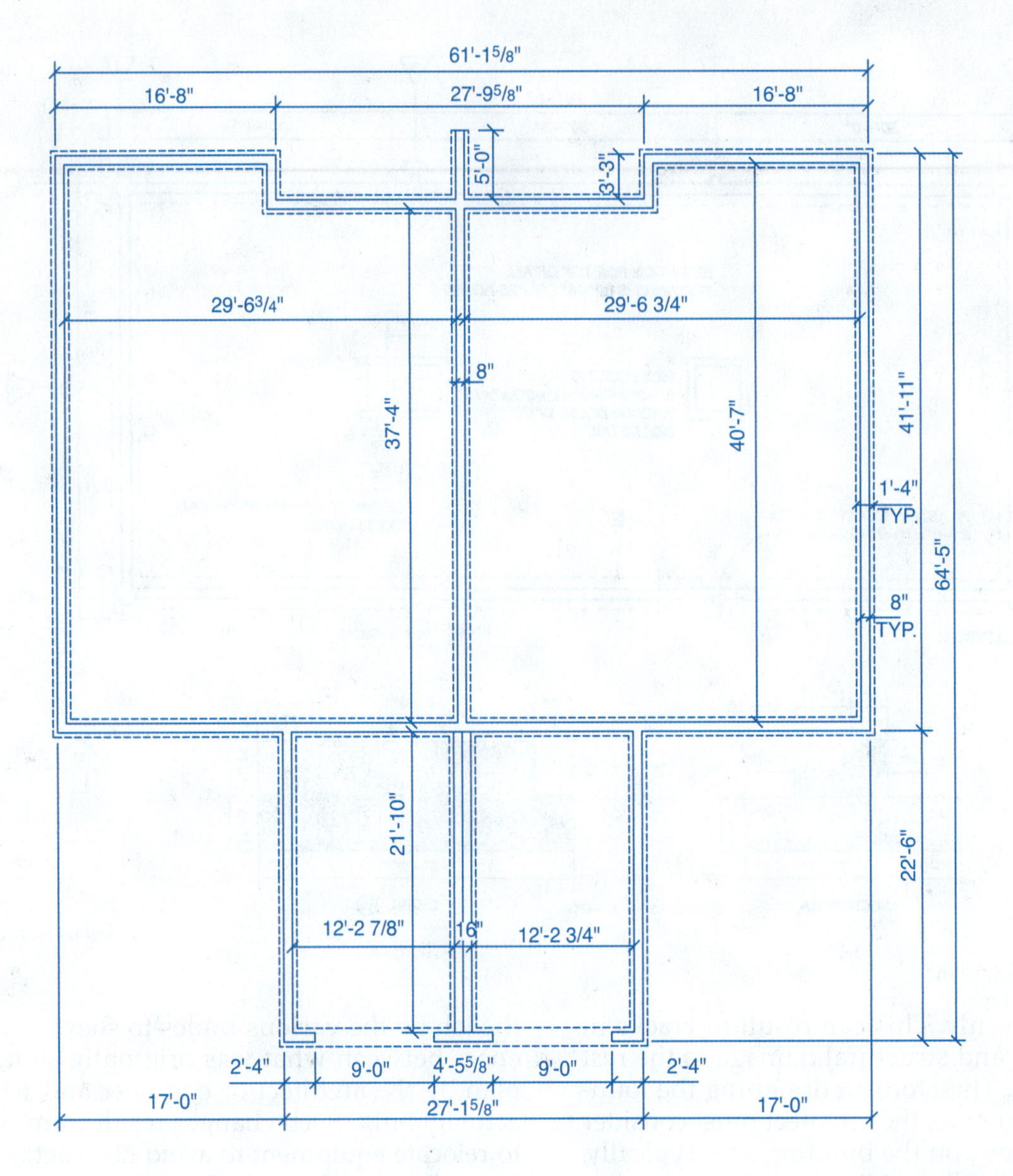

Figure 10 Slab-on-grade plan.

As-Built Drawings

The job specifications generally contain a section defining how as-built drawings are to be prepared. Changes made by various contractors are usually marked on a master set of drawings that have been set aside for that purpose. On some jobs, it may be necessary to obtain an approved change order before making any change to the drawings.

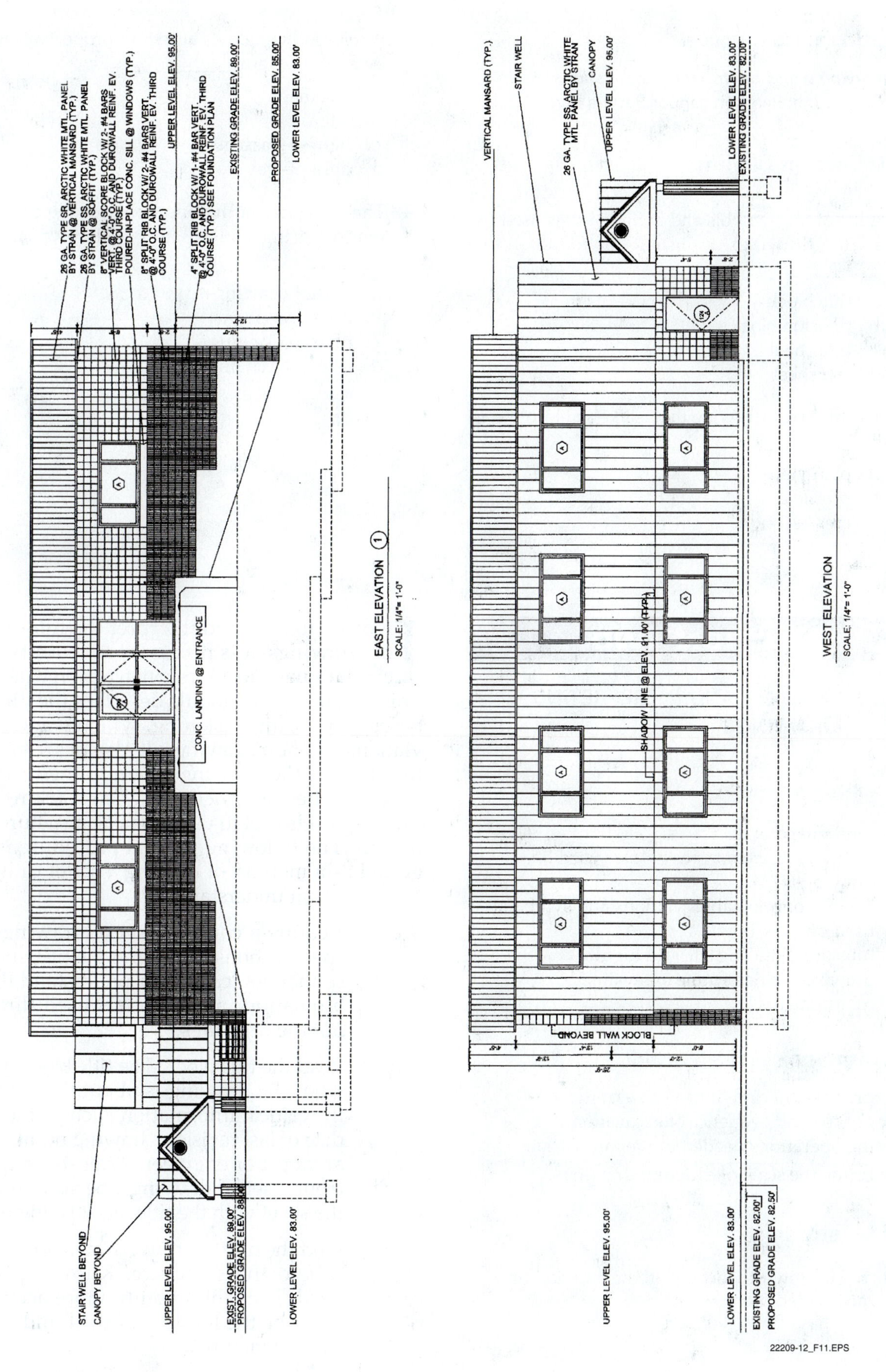

Figure 11 Elevation drawing.

Additional Resources

Surveying with Construction Applications, Barry F. Kavanaugh; Pearson, Upper Saddle River, NJ.

1.0.0 Section Review

1. A list of the symbols and abbreviations used on a set of drawings can usually be found on the _________.
   a. title block
   b. revision block
   c. title sheet
   d. last sheet

2. The highway drawing that is like an elevation is the _________.
   a. plan view
   b. profile
   c. cross-section
   d. grading plan

3. A release that allows access to property owned by another is a(n) _________.
   a. easement
   b. setback
   c. right-of-passage
   d. right-of-way

4. The drawing set that incorporates the changes made during construction is known as the _________.
   a. master drawings
   b. final drawings
   c. revised drawings
   d. as-built drawings

Section Two

2.0.0 Reading and Interpreting Drawings

Objective 2

Read and interpret drawings.
a. Identify different types of lines used on drawings.
b. Define common abbreviations and symbols used on drawings.
c. Interpret building site and highway drawings to determine excavation requirements.

Performance Tasks 1 and 2

Interpret a set of drawings to determine the proper type and sequence of excavation and grading operations needed to prepare the site.
Determine the scale of different drawings.

Trade Terms

Invert: The lowest portion of the interior of a pipe, also called the flow line.

In order to read and interpret the information on drawings, it is necessary to learn the special language used in construction drawings. This section of the module describes the different types of lines, dimensioning, symbols, and abbreviations used on drawings. It also describes how to interpret the drawings used in highway and building site construction. When working with drawings, it is best to use a logical, structured approach. The following general procedure is suggested as a method of reading a set of drawings for maximum understanding:

Step 1 Acquire a complete set of drawings and specifications, including the title sheet(s), so that you can better understand the abbreviations and symbols used throughout the drawings.

Step 2 Read the title block. The title block defines what the drawing is about. Take note of the critical information such as the scale, date of last revision, drawing number, and architect or engineer. After using a sheet from a set of drawings, be sure to refold the sheet with the title block facing up.

Step 3 Find the north arrow. Always orient yourself to the structure. Knowing where north is enables you to more accurately describe the location of the building and other structures.

Step 4 Always be aware that the drawings work together as a group. The reason the architect or engineer draws plans, elevations, and sections is that drawings require more than one type of view to communicate the whole project. Learn how to use more than one drawing when necessary to find the information you need.

Step 5 Check the list of drawings in the set. Note the sequence of the various plans. Some drawings have an index on the front cover. Notice that the prints in the set are of several categories:

- Architectural
- Structural
- Civil
- Mechanical
- Electrical
- Plumbing
- Landscape

Step 6 Study the site plan to determine property boundaries and carefully note the location of any benchmarks. Further, determine the location of the building to be constructed, as well as the various utilities, roadways, and any easements. Note the various elevations and the existing and proposed contours.

Step 7 Check the floor plan for the orientation of the building. Observe the locations and features of entries, corridors, offsets, and any special features to get an idea of the finished construction.

Step 8 Check the foundation plan for the sizes and types of footings, reinforcing steel, and loadbearing substructures.

Step 9 Check the floor construction and other details relating to excavations.

Step 10 Study the utility drawings and structural plans for features that affect earthwork.

Step 11 Check the notes on various pages, and compare the specifications against the construction details.

Step 12 Browse through the sheets of drawings to become familiar with all the plans and details.

Step 13 Recognize applicable symbols and their relative locations in the plans. Note any special excavation details.

Step 14 After you are acquainted with the plans, walk the site so that you can relate the plans to the site.

2.1.0 Lines and Symbols

Many different types of symbols and lines are used in the development of a set of plans. Lines are drawn wide, narrow, dark, light, broken, and unbroken, with each type of line conveying a specific meaning. *Figure A6* shows the most common lines and symbols used on site drawings. The following describes the types of lines commonly found on drawings:

- *Object lines* – Heavier-weight lines used to show the main outline of the structure, including exterior walls, interior partitions, porches, patios, sidewalks, parking lots, and driveways.
- *Dimension and extension lines* – Provide the dimensions of an object. An extension line is drawn out from an object at both ends of the part to be measured to indicate the part being measured. Extension lines are not supposed to touch the object lines. This is so they cannot be confused with the object lines. A dimension line is drawn at right angles between the extension lines and a number placed above, below, or to the side of it to indicate the length of the dimension line. Sometimes a gap is made in the dimension line and the number is written in the gap.
- *Leader line* – Connects a note or dimension to a related part of the drawing. Leader lines are usually curved or at an angle from the feature being distinguished to avoid confusion with dimension and other lines.
- *Center line* – Designates the center of an area or object and provides a reference point for dimensioning. Center lines are typically used to indicate the centers of roadways and the center of objects such as columns, posts, footings, and door openings. On roadways, the center line is a common reference point.

Walking the Site

The plans provide a two-dimensional perspective, but walking the site is the best way to visualize the project. Most racecar drivers walk the track before a race. They can look at a map of the track and identify locations of the entry, apex, and exit points of each turn, but walking the course helps them visualize and memorize the track.

- *Cutting plane (section line)* – Indicates an area that has been cut away and shown in a section view so that the interior features can be seen. The arrows at the ends of the cutting plane indicate the direction in which the section is viewed. Letters identify the cross-sectional view of that specific part of the structure. More elaborate methods of labeling section reference lines are used in larger, more complicated sets of plans (*Figure 12*). The sectional drawing may be on the same page as the reference line or on another page.
- *Break line* – Shows that an object or area has not been drawn in its entirety.
- *Hidden line* – Indicates an outline that is invisible to an observer because it is covered by another surface or object that is closer to the observer.
- *Phantom line* – Indicates alternative positions of moving parts, such as a damper's swing, or adjacent positions of related parts. It may also be used to represent repeated details.
- *Contour line* – Contour lines are used to show changes in the elevation and contour of the land. The lines may be dashed or solid. Generally, dashed lines are used to show the natural or existing grade, and solid lines show the finish grade to be achieved during construction.

Symbols are used on drawings to show different kinds of materials, objects, fixtures, and structural members. The meanings of symbols and the types used are not standardized and can vary from location to location. A set of drawings generally includes a sheet that identifies the specific symbols used and their meanings (see *Figure A6*). When using any drawing set, always refer to this sheet of symbols to avoid making mistakes when reading the drawings .

Some symbols give a good idea of what the object it represents looks like, but others are used to show the position of the object. Examples of this type of symbol include door and window designators that refer to door and window schedules where the different types are described. Still other symbols are used to show the orientation of the object, showing the direction or side (north, south, front, back, and so on).

2.2.0 Abbreviations Used on Drawings

Many written instructions are needed to complete a set of construction drawings. It is impossible to print out all such references, so a system of abbreviations is used. By using standard abbreviations, such as BRK for brick or CONC for concrete, the architect ensures that the drawings will be accurately interpreted. *Figure A7* contains a list of abbreviations commonly used on site drawings. Note that some architects and engineers may use different abbreviations for the same terms. Normally, the title sheet in a drawing set contains a list of abbreviations used in the drawings. For this reason, it is important to get a complete set of drawings and specifications, including the title sheet(s), in order to better understand the exact abbreviations used. Some practices for using abbreviations on drawings are as follows:

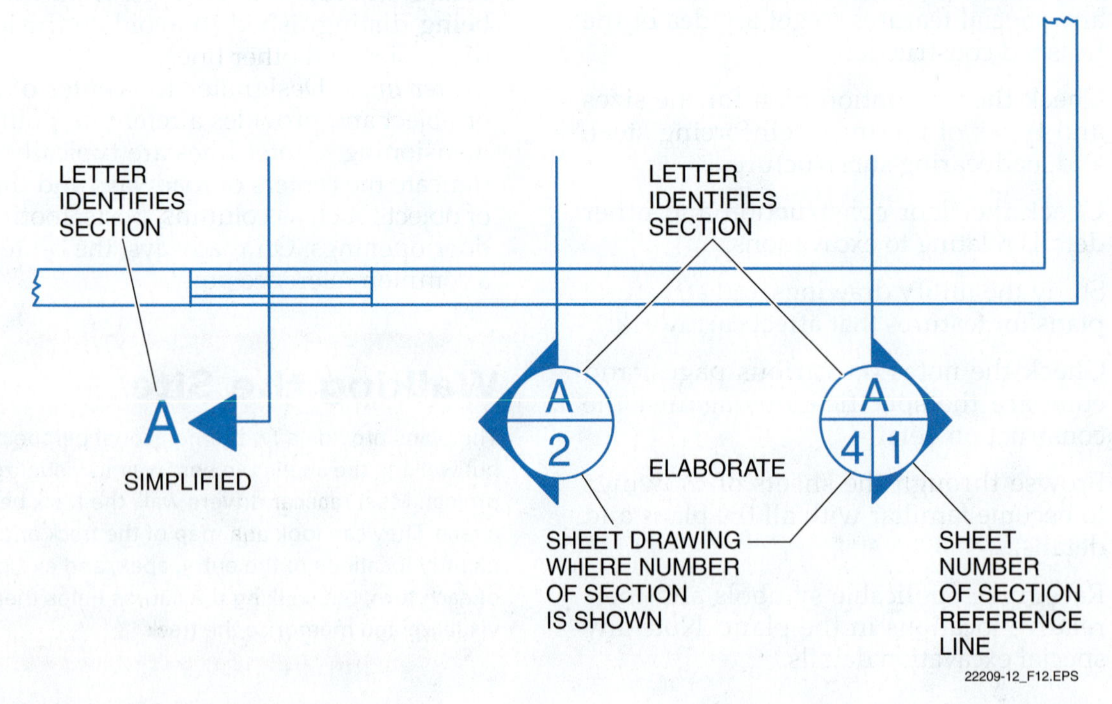

Figure 12 Methods of labeling section reference lines.

- Most abbreviations are capitalized.
- Periods are used when abbreviations might look like a whole word.
- Abbreviations are the same whether they are singular or plural.
- Several terms have the same abbreviations and can only be identified from the context in which they are found.

2.3.0 Interpreting Civil Drawings

The ability to read drawings is an important skill. A person in the construction industry who lacks this ability is limiting their growth potential and relies on others when it is necessary to use the project drawings. Learning to read drawings takes patience and practice, but it is a skill that can help improve productivity and reduce errors.

2.3.1 Site Plans

Site plans show the positions and sizes of all relevant structures on the site, as well as the features of the terrain. It would be difficult to show the amount of information on these drawings without using symbols and abbreviations. *Figure 13* shows symbols commonly used on site plans. It is important to take time to become familiar with these symbols so that you can quickly identify them on plans.

Each contour line across the plot of land represents a line of constant elevation relative to some point such as sea level or a local feature. *Figure 14* shows an example of a contour map for a hill. As shown, contour lines are drawn in uniform elevation intervals called contour intervals. Commonly used intervals are 1', 2', 5', and 10'.

On some plans and surveys, every fifth contour line is drawn using a heavier-weight line and is labeled with its elevation to help the user more easily determine the contour. This method of drawing contour lines is called indexing contours. The elevation is marked above the contour line, or the line is interrupted for it.

As shown in *Figure 14*, contour lines can form a closed loop within the map. These lines represent an elevation (hill) or depression. If you start at any point on these lines and follow their path, you eventually return to the starting point. A contour may close on a site plan or map, or it may be discontinued at any two points at the borders of the plan or map. Examples of this can be seen on the site plan shown in *Figure 3*. Such points mark the ends of the contour on the map, but the contour does not end at these points. The contour is continued on a plan or map of the adjacent land.

Some rules for interpreting contours include the following:

- Contour lines do not cross.
- Contour lines crossing a stream point upstream.
- The horizontal distance between contour lines represents the degree of slope. Closely spaced contour lines represent steep ground and widely spaced contour lines represent nearly level ground with a gradual slope. Uniform spacing indicates a uniform slope.
- Contour lines are at right angles to the slope. Therefore, water flow is perpendicular to contour lines.
- Straight contour lines parallel to each other represent man-made features such as terracing.

The existing contour is shown as a dashed line; the new or proposed contour is shown as a solid line (*Figure 15*). Both are labeled with the elevation of the contour in the form of a whole number. The spacing between the contour lines is at a constant vertical increment, or interval. The typical interval is five feet, but intervals of one foot are not uncommon for site plans requiring greater detail, or where the change in elevation is gradual.

A known elevation on the site that is used as a reference point during construction is called a benchmark. The benchmark is established in reference to the datum and is commonly noted on the site print with a physical description and its elevation relative to the datum. For example, "Northeast corner of catch basin rim—Elev. 102.34'" might be a typical benchmark found on a site plan. When individual elevations, or grades, are required for other site features, they are noted with a + and the grade. Grades vary from contours in that a grade has accuracy to two decimal places, whereas a contour is expressed as a whole number.

Always review materials symbols, dimensioning and scaling, and fundamental construction techniques before attempting to understand a full set

Drawing Revisions

Always be sure to use the latest set of drawings. When a set of construction drawings has been revised and reissued, the superseded set, or the sheets that were replaced, should be discarded. If they are kept for record purposes, they should be marked with a notation such as "Obsolete Drawing – Do Not Use." This will prevent them from being used in error.

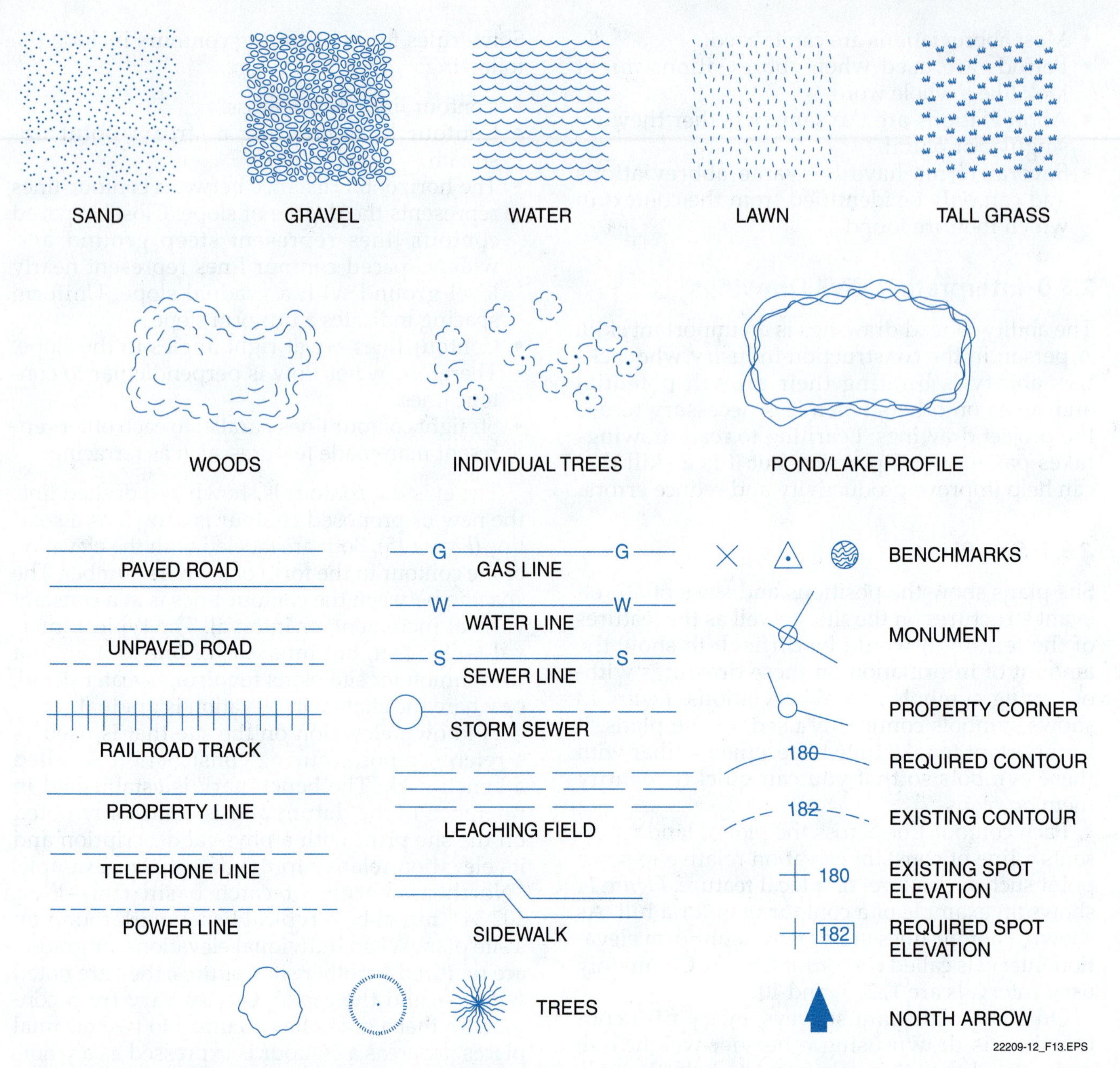

Figure 13 Common site/plot plan symbols.

of plans. Carefully review the site plan and get an overall concept of the work required. Look at the contours to determine where excavations are required. For example, if a point on *Figure 15* is at an existing elevation of about 68 feet and a finish elevation of 70 feet is required, 2 feet of fill is required at this location. It is often helpful to divide the site plan into sections or by grid lines to fully understand the amount of work required on the site.

Site plans are drawn using any convenient scale. This may be ⅛ inch to 1 foot, or it may be an engineering scale, such as 1 inch to 20 feet. Larger projects have several site plans showing different scopes of related or similar work. One such plan is the drainage and utility plan. Utility drawings show locations of the water, gas, sanitary sewer, and electric utilities that will service the building. Drainage plans detail how surface water will be collected, channeled, and dispersed on-site or off-site. Drainage and utility plans illustrate in plan view the size and type of pipes, their length, and the special connections or terminations of the various piping. The elevation of a particular pipe below the surface is given with respect to its **invert**. The invert is the lowest point on the inside of the pipe. The invert is also referred to as the flow line. It is typically noted with the abbreviation for invert and an elevation, for example: INV. 543.15. An example of this can be seen in *Figure A4*. The inverts are shown at the intersections of pipes or other changes in the continuous run of piping, such as a manhole, catch basin, sewer manholes,

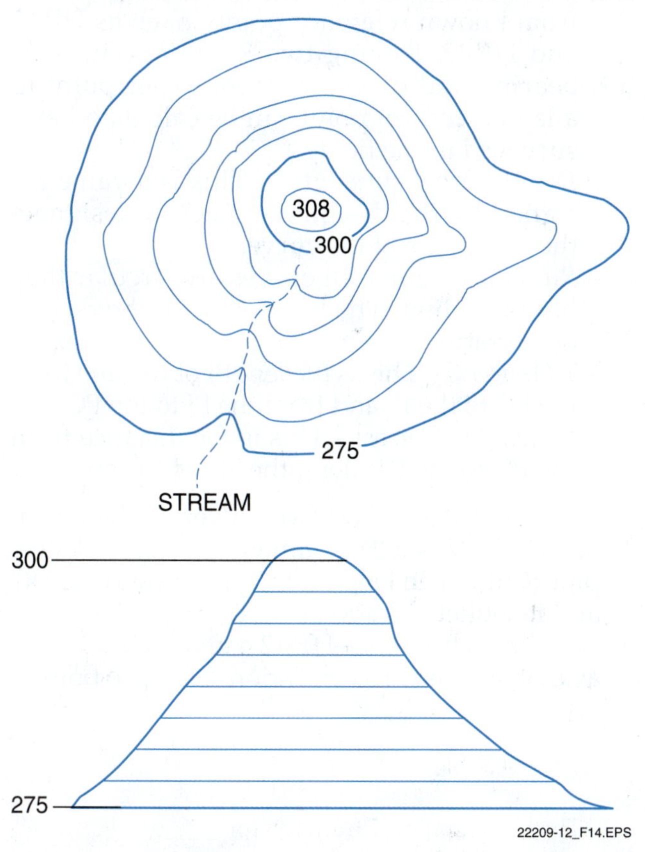

Figure 14 Contour map of a hill.

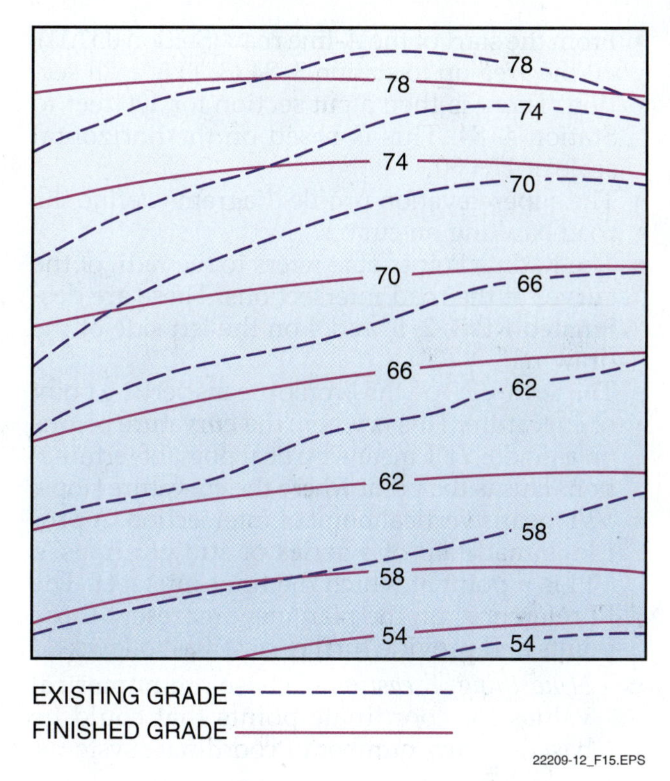

Figure 15 Contour lines used for grading.

and so on. Inverts are usually given for piping that has a gravity flow or pitch. Using benchmarks, contours, or spot elevations, the distance of the piping below the surface and the direction of the flow can be quickly calculated.

For projects of a more complex nature, separate drawings showing various site improvements may be needed for clarification. Site improvements may include such items as curbing, walks, retaining walls, paving, fences, steps, benches, and flagpoles. *Figure A8* is one example of such a drawing showing sign placement and installation requirements.

2.3.2 Reading Highway Drawings

Highway construction and improvement projects are a source of employment for many heavy equipment operators. The drawings used for these projects are much different than those used for building sites. This section contains instructions for interpreting highway drawings in *Figures A1* and *A2*. These drawings are located in the *Appendix*.

Figure A2 is an example of a highway cross-section drawing. The sections show the finished road design for several segments of the road, including the pavement buildup and finish elevation. If the grade is the same on both sides of the road, it is only necessary to show one side on the drawing. In this case, both sides are shown because the grades vary and require different levels of cut and fill. The section table at the left of *Figure A2* shows the station segments to which the sections apply. For example, the design shown in Section C applies to four different segments on the left side of the A-line road.

> **NOTE**
>
> *Figure A1* only goes to Station 8+00, so some of the entries in the Section Table on *Figure A2* refer to another sheet of the drawing package (C6), which is not included here.

The distances shown on the *Figure A2* are measured from the center line of the road to the various outer points. Note that the A-line road is wider than the B-line road by 2.5 feet, as measured from the center line to the hinge point (also called the shoulder break). The Slope Rounding Chart is used to establish the rounding at the top and bottom of a slope in order to provide a finished appearance.

Referring to the profile and plan view on *Figure A1*, there are a few key points to be brought out:

- Station numbers appear along the upper edge of the plan view and the bottom of the profile.
- The starting point for construction is 11.11' from the center line of Highway 270.

- From the start of the A-line road (Station 0.11.11), all the area up to station 3+84 (384') is a fill section. There is then a cut section for 100 feet to Station 4+84. This is based on the horizontal scale of 1" = 80'.
- The superelevation profile diagrams define the road banking on curves.
- The Radius Point Table refers to the radii of the curves at the road intersections. These are designated RP#1, 2, 3, and 4 on the left side of the drawing.
- The term *VPC* on the profile means vertical point of curvature. This is where the curvature begins on a grade. VPT means vertical point of termination. This is the point where the curvature stops.
- VPI means vertical point of intersection. A profile is made up of a series of straight lines; a VPI is a point at which the lines intersect. The PI references on the plan view represent those points and provide further detail as follows:
  - *N northing, E easting* – These are numerical values for coordinate points that could be based on any number of coordinate systems. They are used for layout of control points from known reference points, such as HP-12 and HP-13. Having two known points, the bearing and distances from either point to a layout control point can be calculated and surveyed in the field.
  - *DC (degree of curvature)* – This is a value assigned by the design engineer to designate the sharpness of the curve.
  - *R (radius)* – Since all curves are circular they have a radius, which varies with the degree of curvature.
  - *T (Tangent)* – This is the length of the line from the PC to the PI and from the PI to the PC.
  - *L (length of curve)* – This is the distance from the PC to the PT along the arc of the curve.
- The plan view shows two culverts. The one at Station 3+27 is a 24" diameter corrugated steel pipe (CSP) 71' in length. Its inlet elevation is 140' and its outlet elevation is 134'.
- Two benchmarks—HP-12 and HP-13—are available to serve as elevation reference points.

Additional Resources

Surveying with Construction Applications, Barry F. Kavanaugh; Pearson, Upper Saddle River, NJ.

Guideline for Preparation and Design of Construction Drawings, Reference Manual 10A;

US Department of the Interior National Park Service; *http://cadd.den.nps.gov/downloads/Support/RefMan10A.pdf*

2.0.0 Section Review

22209-12_SR01.EPS

1. The symbol shown in *Figure 1* represents a _____.
   a. soil boring
   b. property corner
   c. stone curb
   d. monument

2. The abbreviation CB stands for _______.
   a. curbstop valve
   b. cubic yards
   c. catch basin
   d. combined sewer

3. The horizontal distance between contour lines on a site plan represents the _______.
   a. degree of slope
   b. elevation
   c. amount of fill needed
   d. amount of cut needed

4. In order to determine the finish elevation of a road at a given station, it would be necessary to refer to the _______.
   a. cross-section
   b. profile
   c. plan view
   d. superelevation

SECTION THREE

3.0.0 SPECIFICATIONS

Objective 3

Explain specifications and the purpose of specifications.

a. Identify the types of information found on specifications.
b. Explain the common format used in specifications.

Trade Term

Uniform Construction Index: The construction specification format adopted by the Construction Specification Institute (CSI). Known as the CSI format.

Specifications, commonly called specs, are written instructions developed by architectural and engineering firms for use by the contractors and subcontractors involved in the construction. Specifications are just as important as the drawings in a set of plans. They furnish what the drawings cannot in that they define the quality of work to be done and the materials to be used. Specifications serve several important purposes:

- Clarify information that cannot be shown on the drawings
- Identify work standards, types of materials to be used, and the responsibility of various parties to the contract
- Provide information on details of construction
- Serve as a guide for contractors bidding on the construction job
- Serve as a standard of quality for materials and workmanship
- Serve as a guide for compliance with building codes and zoning ordinances
- Provide the basis of agreement between the owner, architect, and contractors in settling any disputes

Equipment operators use the specification to determine the quality of fill, the depth of top soil, special finish grading requirements, and landscaping requirements. The plans are often more specific to the job than the specifications. Therefore, notes on the plans may be considered by the architect/owner to be the true intent. Equipment operators must be very careful to watch for discrepancies between the plans and specifications and report them to a supervisor immediately.

3.1.0 Organization of Specifications

Specifications consist of various elements that may differ somewhat for particular construction jobs. Basically, two types of information are contained in a set of specifications: special and general conditions, and technical aspects of construction.

3.1.1 Special and General Conditions

Special and general conditions cover the nontechnical aspects of the contractual agreements. Special conditions cover topics such as safety and temporary construction. General conditions cover the following points of information:

- Contract terms
- Responsibilities for examining the construction site
- Types and limits of insurance
- Permits and payments of fees
- Use and installation of utilities
- Supervision of construction
- Other pertinent items

The general conditions section is the area of the construction contract where misunderstandings often occur. Therefore, these conditions are usually much more explicit on large, complicated construction projects. Part of a typical residential material specification is shown in *Figure 16.*

The earthwork sections of this specification are shown as Item 1 (Excavation) and under the headings *Other Onsite Improvements and Landscaping, Planting, and Finish Grading.*

NOTE

Residential specifications often do not spell out general conditions and are basically material specifications only.

3.1.2 Technical Aspects

The technical aspects section includes information on materials that are specified by standard numbers and by standard testing organizations such as the American Society for Testing and Materials (ASTM) International. The technical data section of specifications can be any of three types:

- *Outline specifications* – These specifications list the materials to be used in order of the basic parts of the job, such as foundation, floors, and walls.
- *Fill-in specifications* – This is a standard form filled in with pertinent information. It is typically used on smaller jobs.

Form RD 1924-2
(Rev. 7-99)

UNITED STATES DEPARTMENT OF AGRICULTURE
U.S. DEPARTMENT OF HOUSING AND URBAN DEVELOPMENT-FEDERAL HOUSING ADMINISTRATION
U.S. DEPARTMENT OF VETERANS AFFAIRS

FORM APPROVED
OMB NO. 0575-0042

☐ **Proposed Construction**

DESCRIPTION OF MATERIALS

No. ______ (To be inserted by Agency)

☐ **Under Construction**

Property address ______ *City* ______ *State* Oklahoma

Mortgagor or Sponsor ______ (Name) ______ (Address)

Contractor or Builder ______ (Name) ______ (Address)

INSTRUCTIONS

1. For additional information on how this form is to be submitted, number of copies, etc., see the instructions applicable to the FHA Application for Mortgage Insurance, VA Request for Determination of Reasonable Value or other, as the case may be.

2. Describe all materials and equipment to be used, whether or not shown on the drawings, by marking an X in each appropriate check-box and entering the information called for in each space. If space is inadequate, enter"See misc," and describe under item 27 or on an attached sheet: THE USE OF PAINT CONTAINING MORE THAN THE PERCENT OF LEAD BY WEIGHT PERMITTED BY LAW IS PROHIBITED.

3. Work not specifically described or shown will not be considered unless required, then the minimum acceptable will be assumed. Work exceeding minimum requirements cannot be considered unless specifically described.

4. Include no alternates,"or equal" phrases, or contradictory items. (Consideration of a request for acceptance of substitute materials or equipment is not thereby precluded.)

5. Include signatures required at the end of this form.

6. The construction shall be completed in compliance with the related drawings and specifications, as amended during processing. The specifications include this Description of Materials and the applicable building code.

1. **EXCAVATION:**
Bearing soil, type Firm clay; Note: Where fill is in excess of 18", concrete piers to be installed at 8' O.C. and the cost will be added to the contract.

2. **FOUNDATIONS:**
Footings: concrete mix transite 14"x18" ftg ; strength psi 2500 PSI Reinforcing (4) 5/8" steel rebar
Foundation wall: material 2500 PSI concrete Reinforcing
Interior foundation wall: material 2500 PSI concrete Party foundation wall 2500 concrete
Columns: material and sizes ______ Piers: material and reinforcing ______
Girders: material and sizes ______ Sills: material W.Coast Utility Douglas Fir w/sill sealer
Basement entrance areaway ______ Window areaways ______
Waterproofing waterproof mix in concrete Footing drains open mortar joints
Termite protection Pretreat soil... Chlordane and issue 5 year warranty.
Basementless space... ; foundation vents
Special f... ...turbed soil.

...tion construction: 6"x8" poured monolithic with the sl...
...floor: 4" concrete slab with 6x6-W1.4xW1.4 WWF smooth trowel finis...

TERRACES:
stoops: 4" concrete slab- smooth trowel finish.
patio: 4" concrete slab- smooth trowel finish. - see plans for size.

GARAGES: automatic garage door opener
foundation: 14"x 18" concrete footing with (4) 5/8" rebars; 6" concrete stem wall
floor: 4" concrete with 6x6-W1.4xW1.4 WWF; smooth trowel finish floors.
interior: 3/8" prefinished Sheetrock on walls; texture and paint 1/2" Sheetrock on ceiling

WALKS AND DRIVEWAYS:
Driveway: width see plans ; base material tamped earth; thickness 4 "; surfacing material concrete ; thickness 4
Front walk: width 36" ; material concrete ; thickness 4 ". Service walk: width ______ ; material ______ ; thickness ______
Steps: material ______ ; treads ______ "; risers ______ ". Check walls ______

OTHER ONSITE IMPROVEMENTS:
(Specify all exterior onsite improvements not described elsewhere, including items such as unusual grading, drainage structures, retaining walls, fence, railings, and accessor structures.)
NOTE: All dimensions to be rechecked on site prior to beginning construction by builder and builder shall be responsible for the same.

LANDSCAPING, PLANTING, AND FINISH GRADING:
Topsoil ______ " thick: ☐ front yard: ☐ side yards; ☐ rear yard to ______ feet behind main building.
Lawns *(seeded, sodded, or sprigged)*:. ☐ front yard ______ ; ☐ side yards ______ ; ☐ rear yard ______
Planting: ☐ as specified and shown on drawings; ☐ as follows:
______ Shade trees, decidous. ______ " caliper. ______ Evergreen trees ______ ' to ______ ', B & B.
______ Low flowering trees, decidous. ______ ' to ______ ' ______ Evergreen shrubs ______ ' to ______ ', B & B.
______ High-growing shrubs, decidous. ______ ' to ______ ' ______ Vines, 2-years
______ Medium-growing shrubs, decidous, ______ ' to ______ '
______ Low-growing shrubs, decidous. ______ ' to ______ '

IDENTIFICATION. This exhibit shall be identified by the signature of the builder, or sponsor, and/or the propsed mortgagor if the latter is known at the time of application.

Date ______ Signature ______

Signature ______

HUD-FHA 2005
VA Form 26-1852

4

22209-12_F16.EPS

Figure 16 Parts of a typical materials specification.

- *Complete specifications* – For ease of use, most specifications written for large construction jobs are organized in the Construction Specification Institute format called the **Uniform Construction Index**. This is known as the CSI format and is explained in the next section.

3.2.0 Format of Specifications

The most commonly used specification-writing format in North America is the *MasterFormat*™. This standard was developed jointly by the Construction Specifications Institute (CSI) and Construction Specifications Canada (CSC). In this format, the specifications are divided into a series of sections dealing with the construction requirements, products, and activities. Using this format makes it easy to write and use the specification, and it is easily understandable by the different trades.

For many years prior to 2004, the organization of construction specifications and suppliers catalogs was based on a standard with 16 sections, otherwise known as divisions, where the divisions and their subsections were individually identified by a five-digit numbering system. The first two digits represented the division number and the next three individual numbers represented successively lower levels of breakdown. For example, the number 13213 represents division 13, subsection 2, sub-subsection 1 and sub-sub-subsection 3. In this older version of the standard, electrical systems, including any electronic or special electrical systems, were lumped together under Division 16 – *Electrical*. Today, specifications conforming to the 16-division format may still be in use, so it is a good idea to be familiar with both formats.

In 2004, the *MasterFormat*™ standard underwent a major change. What had been 16 divisions was expanded to four major groupings and 49 divisions with some divisions reserved for future expansion. The first 14 divisions are essentially the same as the old format. Subjects under the old Division 15 – *Mechanical* have been relocated to new divisions 22 and 23. The basic subjects under old Division 16 – *Electrical* have been relocated to new divisions 26 and 27. In addition, the numbering system was changed to 6 digits to allow for more subsections in each division, which allows for finer definition. In the new numbering system, the first two digits represent the division number. The next two digits represent subsections of the division and the two remaining digits represent the third level sub-subsection numbers. The fourth level, if required, is a decimal and number added to the end of the last two digits. This allows tasks to be divided into finer definitions. For example, Division 31 is entitled *Earthwork* (see *Figure 17*), so much of your work will fall under this division. Use *Figure 17* to look up code 312219.13 and you will find it relates specifically to spreading and grading topsoil.

31 00 00 Earthwork

31 01 00 Maintenance of Earthwork

31 01 10 Maintenance of Clearing
31 01 20 Maintenance of Earth Moving
31 01 40 Maintenance of Shoring and Underpinning
31 01 50 Maintenance of Excavation Support and Protection
31 01 60 Maintenance of Special Foundations and Load Bearing Elements
31 01 62 Maintenance of Driven Piles
31 01 62.61 Driven Pile Repairs
31 01 63 Maintenance of Bored and Augered Piles
31 01 63.61 Bored and Augered Pile Repairs
31 01 70 Maintenance of Tunneling and Mining
31 01 70.61 Tunnel Leak Repairs

31 05 00 Common Work Results for Earthwork

31 05 13 Soils for Earthwork
31 05 16 Aggregates for Earthwork
31 05 19 Geosynthetics for Earthwork
31 05 19.13 Geotextiles for Earthwork
31 05 19.16 Geomembranes for Earthwork
31 05 19.19 Geogrids for Earthwork
31 05 19.23 Geosynthetic Clay Liners
31 05 19.26 Geocomposites
31 05 19.29 Geonets
31 05 23 Cement and Concrete for Earthwork

31 06 00 Schedules for Earthwork

31 06 10 Schedules for Clearing
31 06 20 Schedules for Earth Moving
31 06 20.13 Trench Dimension Schedule
31 06 20.16 Backfill Material Schedule
31 06 40 Schedules for Shoring and Underpinning
31 06 50 Schedules for Excavation Support and Protection
31 06 60 Schedules for Special Foundations and Load Bearing Elements
31 06 60.13 Driven Pile Schedule
31 06 60.16 Caisson Schedule
31 06 70 Schedules for Tunneling and Mining

31 08 00 Commissioning of Earthwork

31 08 13 Pile Load Testing
31 08 13.13 Dynamic Pile Load Testing
31 08 13.16 Static Pile Load Testing

31 09 00 Geotechnical Instrumentation and Monitoring of Earthwork

31 09 13 Geotechnical Instrumentation and Monitoring
31 09 13.13 Groundwater Monitoring During Construction
31 09 16 Foundation Performance Instrumentation
31 09 16.26 Bored and Augered Pile Load Tests

31 10 00 Site Clearing

31 11 00 Clearing and Grubbing

31 12 00 Selective Clearing

111

22209-12_F17A.EPS

Figure 17 CSI Division 31 – *Earthwork* (1 of 6).

31 13 00		**Selective Tree and Shrub Removal and Trimming**
	31 13 13	Selective Tree and Shrub Removal
	31 13 16	Selective Tree and Shrub Trimming
31 14 00		**Earth Stripping and Stockpiling**
	31 14 13	Soil Stripping and Stockpiling
	31 14 13.13	Soil Stripping
	31 14 13.16	Soil Stockpiling
	31 14 13.23	Topsoil Stripping and Stockpiling
	31 14 16	Sod Stripping and Stockpiling
	31 14 16.13	Sod Stripping
	31 14 16.16	Sod Stockpiling
31 20 00		**Earth Moving**
31 21 00		**Off-Gassing Mitigation**
	31 21 13	Radon Mitigation
	31 21 13.13	Radon Venting
	31 21 16	Methane Mitigation
	31 21 16.13	Methane Venting
31 22 00		**Grading**
	31 22 13	Rough Grading
	31 22 16	Fine Grading
	31 22 16.13	Roadway Subgrade Reshaping
	31 22 19	Finish Grading
	31 22 19.13	Spreading and Grading Topsoil
31 23 00		**Excavation and Fill**
	31 23 13	Subgrade Preparation
	31 23 16	Excavation
	31 23 16.13	Trenching
	31 23 16.16	Structural Excavation for Minor Structures
	31 23 16.26	Rock Removal
	31 23 19	Dewatering
	31 23 23	Fill
	31 23 23.13	Backfill
	31 23 23.23	Compaction
	31 23 23.33	Flowable Fill
	31 23 23.43	Geofoam
	31 23 33	Trenching and Backfilling
31 24 00		**Embankments**
	31 24 13	Roadway Embankments
	31 24 16	Railway Embankments
31 25 00		**Erosion and Sedimentation Controls**
	31 25 14	Stabilization Measures for Erosion and Sedimentation Control
	31 25 14.13	Hydraulically-Applied Erosion Control
	31 25 14.16	Rolled Erosion Control Mats and Blankets
	31 25 24	Structural Measures for Erosion and Sedimentation Control
	31 25 24.13	Rock Barriers
	31 25 34	Retention Measures for Erosion and Sedimentation Controls
	31 25 34.13	Rock Basins
31 30 00		**Earthwork Methods**
31 31 00		**Soil Treatment**

112

22209-12_F17B.EPS

Figure 17 CSI Division 31 – *Earthwork* (2 of 6).

MasterFormat® 2012 – Numbers and Titles	April 2012
31 31 13	Rodent Control
31 31 13.16	Rodent Control Bait Systems
31 31 13.19	Rodent Control Traps
31 31 13.23	Rodent Control Electronic Systems
31 31 13.26	Rodent Control Repellants
31 31 16	Termite Control
31 31 16.13	Chemical Termite Control
31 31 16.16	Termite Control Bait Systems
31 31 16.19	Termite Control Barriers
31 31 19	Vegetation Control
31 31 19.13	Chemical Vegetation Control
31 32 00	**Soil Stabilization**
31 32 13	Soil Mixing Stabilization
31 32 13.13	Asphalt Soil Stabilization
31 32 13.16	Cement Soil Stabilization
31 32 13.19	Lime Soil Stabilization
31 32 13.23	Fly-Ash Soil Stabilization
31 32 13.26	Lime-Fly-Ash Soil Stabilization
31 32 16	Chemical Treatment Soil Stabilization
31 32 16.13	Polymer Emulsion Soil Stabilization
31 32 17	Water Injection Soil Stabilization
31 32 19	Geosynthetic Soil Stabilization and Layer Separation
31 32 19.13	Geogrid Soil Stabilization
31 32 19.16	Geotextile Soil Stabilization
31 32 19.19	Geogrid Layer Separation
31 32 19.23	Geotextile Layer Separation
31 32 23	Pressure Grouting Soil Stabilization
31 32 23.13	Cementitious Pressure Grouting Soil Stabilization
31 32 23.16	Chemical Pressure Grouting Soil Stabilization
31 32 33	Shotcrete Soil Slope Stabilization
31 32 36	Soil Nailing
31 32 36.13	Driven Soil Nailing
31 32 36.16	Grouted Soil Nailing
31 32 36.19	Corrosion-Protected Soil Nailing
31 32 36.23	Jet-Grouted Soil Nailing
31 32 36.26	Launched Soil Nailing
31 33 00	**Rock Stabilization**
31 33 13	Rock Bolting and Grouting
31 33 23	Rock Slope Netting
31 33 26	Rock Slope Wire Mesh
31 33 33	Shotcrete Rock Slope Stabilization
31 33 43	Vegetated Rock Slope Stabilization
31 34 00	Soil Reinforcement
31 34 19	Geosynthetic Soil Reinforcement
31 34 19.13	Geogrid Soil Reinforcement
31 34 19.16	Geotextile Soil Reinforcement
31 34 23	Fiber Soil Reinforcement
31 34 23.13	Geosynthetic Fiber Soil Reinforcement
31 35 00	**Slope Protection**
31 35 19	Geosynthetic Slope Protection
31 35 19.13	Geogrid Slope Protection
31 35 19.16	Geotextile Slope Protection

113

22209-12_F17C.EPS

Figure 17 CSI Division 31 – *Earthwork* (3 of 6).

MasterFormat® 2012 – Numbers and Titles April 2012

31 35 19.19	Slope Protection with Mulch Control Netting
31 35 23	Slope Protection with Slope Paving
31 35 23.13	Cast-In-Place Concrete Slope Paving
31 35 23.16	Precast Concrete Slope Paving
31 35 23.19	Concrete Unit Masonry Slope Paving
31 35 26	Containment Barriers
31 35 26.13	Clay Containment Barriers
31 35 26.16	Geomembrane Containment Barriers
31 35 26.23	Bentonite Slurry Trench
31 36 00	**Gabions**
31 36 13	Gabion Boxes
31 36 19	Gabion Mattresses
31 36 19.13	Vegetated Gabion Mattresses
31 37 00	**Riprap**
31 37 13	Machined Riprap
31 37 16	Non-Machined Riprap
31 37 16.13	Rubble-Stone Riprap
31 37 16.16	Concrete Unit Masonry Riprap
31 37 16.19	Sacked Sand-Cement Riprap
31 40 00	**Shoring and Underpinning**
31 41 00	**Shoring**
31 41 13	Timber Shoring
31 41 16	Sheet Piling
31 41 16.13	Steel Sheet Piling
31 41 16.16	Plastic Sheet Piling
31 41 19	Metal Hydraulic Shoring
31 41 19.13	Aluminum Hydraulic Shoring
31 41 23	Pneumatic Shoring
31 41 33	Trench Shielding
31 43 00	**Concrete Raising**
31 43 13	Pressure Grouting
31 43 13.13	Concrete Pressure Grouting
31 43 13.16	Polyurethane Pressure Grouting
31 43 16	Compaction Grouting
31 43 19	Mechanical Jacking
31 45 00	**Vibroflotation and Densification**
31 45 13	Vibroflotation
31 45 16	Densification
31 46 00	**Needle Beams**
31 46 13	Cantilever Needle Beams
31 48 00	**Underpinning**
31 48 13	Underpinning Piers
31 48 19	Bracket Piers
31 48 23	Jacked Piers
31 48 33	Micropile Underpinning
31 50 00	**Excavation Support and Protection**
31 51 00	**Anchor Tiebacks**
31 51 13	Excavation Soil Anchors
31 51 16	Excavation Rock Anchors

114

22209-12_F17D.EPS

Figure 17 CSI Division 31 – *Earthwork* (4 of 6).

31 52 00	**Cofferdams**
31 52 13	Sheet Piling Cofferdams
31 52 16	Timber Cofferdams
31 52 19	Precast Concrete Cofferdams
31 53 00	**Cribbing and Walers**
31 53 13	Timber Cribwork
31 54 00	**Ground Freezing**
31 56 00	**Slurry Walls**
31 56 13	Bentonite Slurry Walls
31 56 13.13	Soil-Bentonite Slurry Walls
31 56 13.16	Cement-Bentonite Slurry Walls
31 56 13.19	Slag-Cement-Bentonite Slurry Walls
31 56 13.23	Slag-Cement-Bentonite Slurry Walls
31 56 13.26	Pozzolan-Bentonite Slurry Walls
31 56 13.29	Organically-Modified Bentonite Slurry Walls
31 56 16	Attipulgite Slurry Walls
31 56 16.13	Soil-Attipulgite Slurry Walls
31 56 19	Slurry-Geomembrane Composite Slurry Walls
31 56 23	Lean Concrete Slurry Walls
31 56 26	Bio-Polymer Trench Drain
31 60 00	**Special Foundations and Load-Bearing Elements**
31 62 00	**Driven Piles**
31 62 13	Concrete Piles
31 62 13.13	Cast-in-Place Concrete Piles
31 62 13.16	Concrete Displacement Piles
31 62 13.19	Precast Concrete Piles
31 62 13.23	Prestressed Concrete Piles
31 62 13.26	Pressure-Injected Footings
31 62 16	Steel Piles
31 62 16.13	Sheet Steel Piles
31 62 16.16	Steel H Piles
31 62 16.19	Unfilled Tubular Steel Piles
31 62 19	Timber Piles
31 62 23	Composite Piles
31 62 23.13	Concrete-Filled Steel Piles
31 62 23.16	Wood and Cast-In-Place Concrete Piles
31 63 00	**Bored Piles**
31 63 13	Bored and Augered Test Piles
31 63 16	Auger Cast Grout Piles
31 63 19	Bored and Socketed Piles
31 63 19.13	Rock Sockets for Piles
31 63 23	Bored Concrete Piles
31 63 23.13	Bored and Belled Concrete Piles
31 63 23.16	Bored Friction Concrete Piles
31 63 26	Drilled Caissons
31 63 26.13	Fixed End Caisson Piles
31 63 26.16	Concrete Caissons for Marine Construction
31 63 29	Drilled Concrete Piers and Shafts
31 63 29.13	Uncased Drilled Concrete Piers

115

22209-12_F17E.EPS

Figure 17 CSI Division 31 – *Earthwork* (5 of 6).

31 63 29.16	Cased Drilled Concrete Piers
31 63 33	Drilled Micropiles
31 64 00	**Caissons**
31 64 13	Box Caissons
31 64 16	Excavated Caissons
31 64 19	Floating Caissons
31 64 23	Open Caissons
31 64 26	Pneumatic Caissons
31 64 29	Sheeted Caissons
31 66 00	**Special Foundations**
31 66 13	Special Piles
31 66 13.13	Rammed Aggregate Piles
31 66 15	Helical Foundation Piles
31 66 16	Special Foundation Walls
31 66 16.13	Anchored Foundation Walls
31 66 16.23	Concrete Cribbing Foundation Walls
31 66 16.26	Metal Cribbing Foundation Walls
31 66 16.33	Manufactured Modular Foundation Walls
31 66 16.43	Mechanically Stabilized Earth Foundation Walls
31 66 16.46	Slurry Diaphragm Foundation Walls
31 66 16.53	Soldier-Beam Foundation Walls
31 66 16.56	Permanently-Anchored Soldier-Beam Foundation Walls
31 66 19	Refrigerated Foundations
31 68 00	**Foundation Anchors**
31 68 13	Rock Foundation Anchors
31 68 16	Helical Foundation Anchors
31 70 00	**Tunneling and Mining**
31 71 00	**Tunnel Excavation**
31 71 13	Shield Driving Tunnel Excavation
31 71 16	Tunnel Excavation by Drilling and Blasting
31 71 19	Tunnel Excavation by Tunnel Boring Machine
31 71 23	Tunneling by Cut and Cover
31 72 00	**Tunnel Support Systems**
31 72 13	Rock Reinforcement and Initial Support
31 72 16	Steel Ribs and Lagging
31 73 00	**Tunnel Grouting**
31 73 13	Cement Tunnel Grouting
31 73 16	Chemical Tunnel Grouting
31 74 00	**Tunnel Construction**
31 74 13	Cast-in-Place Concrete Tunnel Lining
31 74 16	Precast Concrete Tunnel Lining
31 74 19	Shotcrete Tunnel Lining
31 75 00	**Shaft Construction**
31 75 13	Cast-in-Place Concrete Shaft Lining
31 75 16	Precast Concrete Shaft Lining
31 77 00	**Submersible Tube Tunnels**
31 77 13	Trench Excavation for Submerged Tunnels
31 77 16	Tube Construction (Outfitting Tunnel Tubes)
31 77 19	Floating and Laying Submerged Tunnels

116

22209-12_F17F.EPS

Figure 17 CSI Division 31 – *Earthwork* (6 of 6).

3.0.0 Section Review

1. In a specification, safety is covered under _____.
   a. special conditions
   b. general conditions
   c. technical details
   d. its own section

2. In a specification, information on earthmoving requirements are likely to be found in _____.
   a. the general conditions
   b. Division 16
   c. Division 26
   d. Division 31

SUMMARY

Construction plans show where a project will be located and how it will be built. All members of the construction team use these drawings, so they are often crowded with information and can be confusing to read. Regardless of that, part of the operator's job is to be able to find the information needed to complete the excavation and grading work. The main concern will be to interpret existing and proposed elevation readings in order to ensure that the site is prepared for construction according to the specification. Proper leveling and grading are vital to the successful completion and durability of any construction project. Highways and buildings are only as stable as the ground they are built on, so the grading, cut and fill, and compacting tasks are completed as called for on the plans.

In addition to the drawings, most jobs have specifications that provide detailed information not included on the drawings. Specifications define and clarify the scope of a job and identify specific materials and components to be used in the construction.

Review Questions

1. The revision status block of a drawing appears only on the first sheet of the drawing set.
   a. True
   b. False

2. The elevation of the existing ground is shown as a _____.
   a. solid line on the cross-section drawing
   b. dashed line on the plan view
   c. solid line on the profile drawing
   d. dashed line on the profile drawing

3. An operator has been assigned to a new job site and needs to become acquainted with the project. What is the best plan to study to identify the location of existing roads, easements, and utility information, as well as proposed construction?
   a. Structural drawings
   b. Floor plan
   c. Foundation drawings
   d. Site plan

4. Elevation drawings show _____.
   a. a straight ahead view of a building
   b. the natural grade of a project
   c. the proposed grade of a project
   d. the height of the roof line

5. A line on a drawing representing an object or area that is *not* shown in its entirety is known as a _____.
   a. phantom line
   b. cutting plane
   c. leader line
   d. break line

6. Contour lines show elevation changes on diagrams. Existing contours are usually shown as _____ lines and proposed finish grade contours are shown as _____ lines.
   a. heavy; light
   b. dashed; solid
   c. solid; dashed
   d. light; heavy

7. Contour lines that make closed loops on a plan represent _____.
   a. property boundaries
   b. depressions or hills
   c. building foundations
   d. roadway direction

8. Widely spaced contour lines represent _____.
   a. steep terrain
   b. hilly terrain
   c. rocky terrain
   d. level terrain

9. What drawing symbol represents a property line?
   a. + + + + + PPPPPPP++++++++
   b. — —
   c. p
   d. _ _ _ _ _ _ _ b _ _ _ _ _ _ _ _ _ _

10. Refer to *Figure 1*. Which of these symbols is used to represent a benchmark?

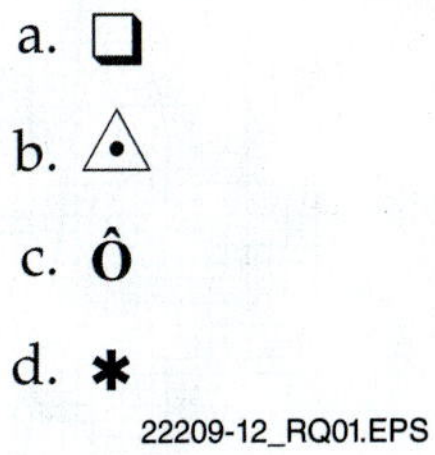

a. ❑

b. ⚠ (triangle with dot)

c. Ô

d. ✱

22209-12_RQ01.EPS

Figure 1

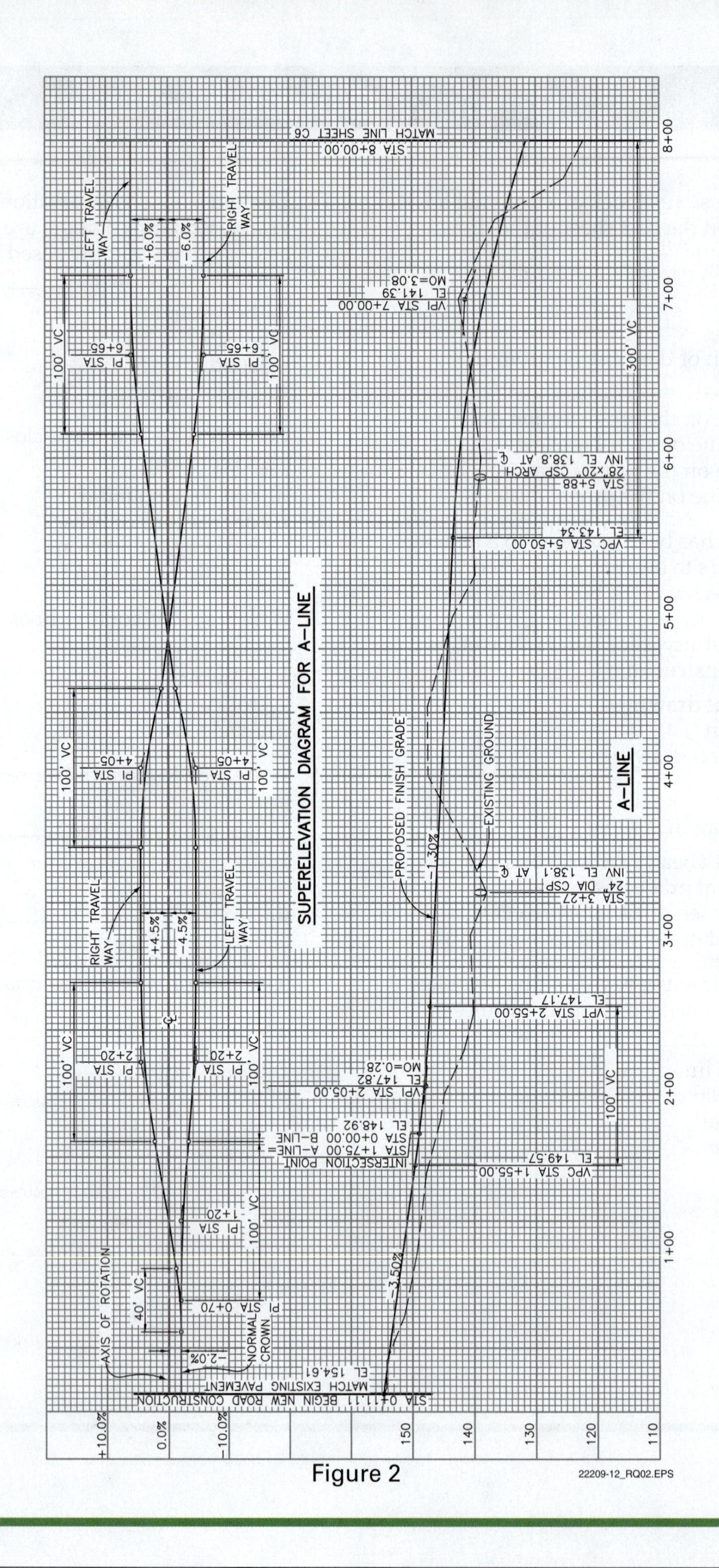

Figure 2

22209-12_RQ02.EPS

11. On *Figure 2*, the elevation of the finished roadway at Station 7+00 is approximately _____ .

    a. 130'
    b. 135'
    c. 138'
    d. 141'

12. On *Figure 2*, the total change in elevation from the beginning of the A-line road to its end is approximately _____.

    a. 45'
    b. 33'
    c. 28'
    d. 23'

13. Specifications are used to expand on the drawing set. When there is a difference between a drawing and a specification, you should _____.

    a. notify your supervisor
    b. follow the drawing set
    c. follow the specification
    d. notify the owner

14. When you are new to a job and need to review the plans, you need to look at only the site plan, since all of the grading information can be found on it.

    a. True
    b. False

15. The CSI *MasterFormat*™ of 2004 identifies how many divisions?

    a. 16
    b. 25
    c. 32
    d. 49

Trade Terms Introduced in This Module

Change order: A formal instruction describing and authorizing a project change.

Contour lines: Imaginary lines on a site/plot plan that connect points of the same elevation. Contour lines never cross each other.

Easement: A legal right-of-way provision on another person's property (for example, the right of a neighbor to build a driveway or a public utility to install water and gas lines on the property). A property owner cannot build on an area where an easement has been identified.

Elevation view: A drawing giving a view from the front or side of a structure.

Invert: The lowest portion of the interior of a pipe, also called the flow line.

Loadbearing: A base designed to support the weight of an object of structure.

Monuments: Physical structures that mark the locations of survey points.

Plan view: A drawing that represents a view looking down on an object.

Property lines: The recorded legal boundaries of a piece of property.

Request for information (RFI): A form used to question discrepancies on the drawings or to ask for clarification.

Setback: The distance from a property line in which no structures are permitted.

Uniform Construction Index: The construction specification format adopted by the Construction Specification Institute (CSI). Known as the CSI format.

Appendix

Examples Of Civil Drawings

This appendix contains drawings that are referenced in the text. The drawings are inserted in 11 × 17 form to make them more readable. This appendix contains the following drawings:

- Figure A1 – Sample Roadway Plan and Profile (C8) – 499/41001
- Figure A2 – Sample Roadway Cross-Sections (C7) – 499/41001
- Figure A3 – Sample Project Overview Site Plan (G1) – 499/41001A
- Figure A4 – Sample Water Line Plan and Profile (C9) – 499/41001
- Figure A5 – Sample Parking Area Grading Plan (C5) – 499/41001
- Figure A6 (1 of 2) – Sample Symbol Sheet (C2) – 499/41001
- Figure A6 (2 of 2) – Sample Mapping Symbols (C3) – 499/41001
- Figure A7 – Sample Abbreviation Sheet (C1) – 499/41001
- Figure A8 – Sample Plan and Details for Roadway Signs and Pavement Markings (C12) – 499/41001

NOTE

These drawings are taken from a book of sample drawings produced by the National Park Service. They are not necessarily related to the same project.

Additional Resources

This module presents thorough resources for task training. The following resource material is suggested for further study.

Surveying with Construction Applications, Barry F. Kavanaugh; Pearson, Upper Saddle River, NJ.

Guideline for Preparation and Design of Construction Drawings, Reference Manual 10A;

US Department of the Interior National Park Service; **http://cadd.den.nps.gov/downloads/Support/RefMan10A.pdf**

Figure Credits

Reprinted courtesy of Caterpillar Inc., Module opener

Courtesy of Carolina Bridge Company, Inc., Figure 1

Topaz Publications, Inc., SA01, SA02, Figures 7 and 8

John Hoerlein, Figure 3

Courtesy of Mark Jones, Figures 4 and 5

Courtesy of Michael Reif, **nosarasprings.com**, Figure 6

The Construction Specifications Institute (CSI), Figure 17

MasterFormat® Numbers and Titles used in this book are from *MasterFormat®*, published by The Construction Specifications Institute (CSI) and Construction Specifications Canada (CSC), and are used with permission from CSI. For those interested in a more in-depth explanation of *MasterFormat®* and its use in the construction industry visit **www.masterformat.com** or contact:

The Construction Specifications Institute

110 South Union Street, Suite 100

Alexandria, VA 22314

800-689-200; 703-684-0300; **www.csinet.org**

Courtesy of National Park Service Denver Service Center, Appendix

Section Review Answers

Answer	Section Reference	Objective
Section One		
1 c	1.1.0	1a
2 b	1.2.1	1b
3 a	1.3.1	1c
4 d	1.4.0	1d
Section Two		
1 b	2.1.0, Figure A6	2a
2 c	2.2.1, Figure A7	2b
3 a	2.3.1	2c
4 b	2.3.2	2c
Section Three		
1 a	3.1.1	3a
2 d	3.2.0	3b

NCCER CURRICULA — USER UPDATE

NCCER makes every effort to keep its textbooks up-to-date and free of technical errors. We appreciate your help in this process. If you find an error, a typographical mistake, or an inaccuracy in NCCER's curricula, please fill out this form (or a photocopy), or complete the online form at **www.nccer.org/olf**. Be sure to include the exact module ID number, page number, a detailed description, and your recommended correction. Your input will be brought to the attention of the Authoring Team. Thank you for your assistance.

Instructors – If you have an idea for improving this textbook, or have found that additional materials were necessary to teach this module effectively, please let us know so that we may present your suggestions to the Authoring Team.

NCCER Product Development and Revision

13614 Progress Blvd., Alachua, FL 32615

Email: curriculum@nccer.org

Online: www.nccer.org/olf

❑ Trainee Guide ❑ AIG ❑ Exam ❑ PowerPoints Other ______________________

Craft / Level: Copyright Date:

Module ID Number / Title:

Section Number(s):

Description:

Recommended Correction:

Your Name:

Address:

Email: Phone:

22210-13

Site Work

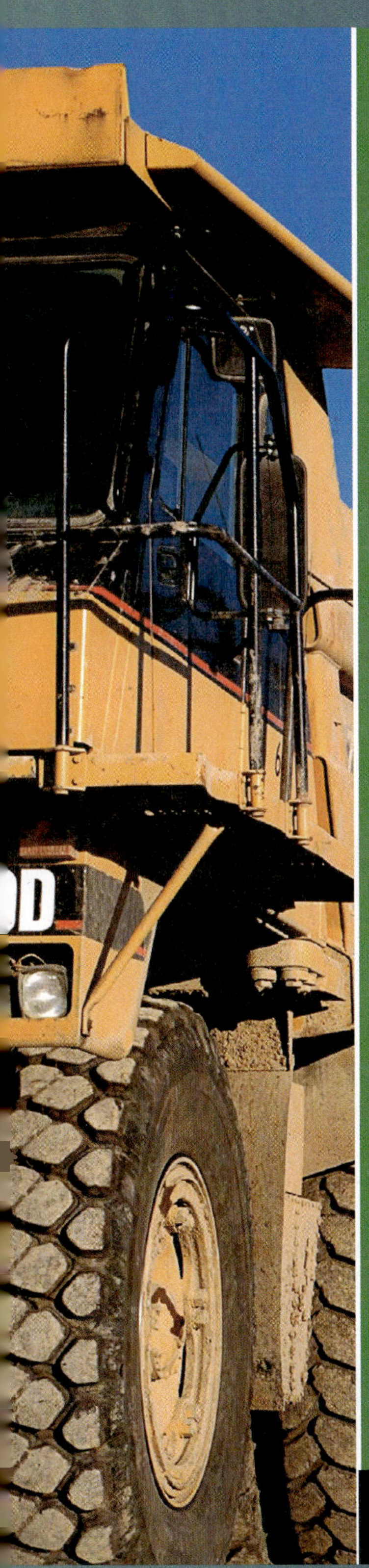

Overview

Understanding how to operate various pieces of heavy equipment is only the beginning for a heavy equipment operator. In addition, an operator must understand safety issues, equipment transportation, groundwater control, operational costs, and advanced grading methods.

Module Five

Trainees with successful module completions may be eligible for credentialing through NCCER's National Registry. To learn more, go to **www.nccer.org** or contact us at **1.888.622.3720**. Our website has information on the latest product releases and training, as well as online versions of our *Cornerstone* newsletter and Pearson's product catalog.

Your feedback is welcome. You may email your comments to **curriculum@nccer.org**, send general comments and inquiries to **info@nccer.org**, or fill in the User Update form at the back of this module.

This information is general in nature and intended for training purposes only. Actual performance of activities described in this manual requires compliance with all applicable operating, service, maintenance, and safety procedures under the direction of qualified personnel. References in this manual to patented or proprietary devices do not constitute a recommendation of their use.

22210-13
SITE WORK

Objectives

When you have completed this module, you will be able to do the following:

1. Describe the safety practices associated with site grading work.
   a. Explain the purpose of a site safety program.
   b. Describe why safety inspections and investigations are important.
   c. Explain how hazardous materials are controlled on a job site.
   d. Describe safety practices associated with trenching and excavations.
   e. Describe how to prepare heavy equipment for transporting.
2. Describe the methods used to control water on job sites.
   a. Explain the importance of maintaining proper drainage on a job site.
   b. Describe the methods used to control ground water and surface water.
   c. Describe the safety practices and construction methods used when working around bodies of water.
3. Explain how grades are established on a job site.
   a. Describe how to set grades from a benchmark.
   b. Describe how grades are set for highway construction.
   c. Describe how grades are set for building construction.
   d. Explain how grading operations are performed.
   e. Describe the use of stakeless and stringless grading systems.
4. Describe grading and installation practices for pipe-laying operations.
   a. Explain how grades are established for pipe-laying operations.
   b. Describe the equipment and methods used to lay pipe.

Performance Tasks

Under the supervision of the instructor, you should be able to do the following:

1. Interpret layout and marking methods to determine grading requirements and operation.
2. Set up a level and determine the elevations at three different points, as directed by the instructor.

Trade Terms

Aquifer
Balance point
Bedding material
Berm
Boot
Competent person
Cross braces
Dewater
Erosion
Groundwater
Sedimentation
Shielding
Stations
Stormwater
String line
Subsidence
Sump
Swale
Topographic survey
Uprights
Walers

Industry Recognized Credentials

If you are training through an NCCER-accredited sponsor, you may be eligible for credentials from NCCER's Registry. The ID number for this module is 22210-13. Note that this module may have been used in other NCCER curricula and may apply to other level completions. Contact NCCER's Registry at 888.622.3720 or go to **www.nccer.org** for more information.

Contents

Topics to be presented in this module include:

Figures

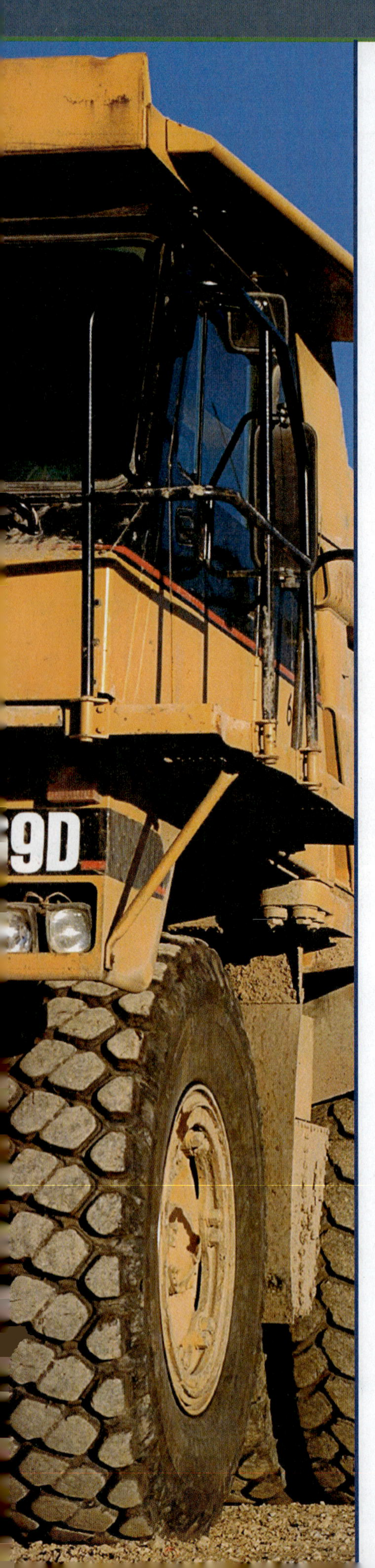

Figures

SECTION ONE

1.0.0 JOB SITE SAFETY

Objective 1

Describe the safety practices associated with site grading work.

a. Explain how to a site safety program is set up.
b. Describe safety inspection and investigation practices.
c. Explain how hazardous materials are controlled on a job site.
d. Describe safety practices associated with trenching and excavations.
e. Describe how to prepare heavy equipment for transporting.

Trade Terms

Competent person: A person who is capable of identifying existing and predictable hazards in the area or working conditions that are unsanitary, hazardous, or dangerous to employees, and who has the authority to take prompt corrective measures to fix the problem.

Cross braces: The horizontal members of a shoring system installed perpendicular to the sides of the excavation, the ends of which bear against either uprights or walers.

Shielding: A structure that is able to withstand the forces imposed on it by a cave-in and thereby protect employees within the structure.

Subsidence: Pressure created by the weight of the soil pushing on the walls of the excavation. It stresses the excavation walls and can cause them to bulge.

Uprights: The vertical members of a trench shoring system placed in contact with the earth and usually positioned so that individual members do not contact each other. Uprights placed so that individual members are closely spaced, in contact with, or interconnected to each other, are often called sheeting.

Walers: Horizontal members of a shoring system or coffer dam placed parallel to the excavation face whose sides bear against the vertical members of the shoring system or the earth. Also, supports for piles in a coffer dam.

Safety is everyone's responsibility. You and your co-workers have an obligation to your employer to operate your equipment safely. You are also obligated to make sure that anyone you supervise works safely. Your employer has an obligation to you to provide you with safe machinery and equipment, and to maintain a safe workplace for all employees.

The Occupational Health and Safety Administration (OSHA) sets forth a number of safety regulations that you, your co-workers, and your employer must follow. The *Code of Federal Regulation (CFR) 29* has to do with labor standards, while *CFR 29, Part 1926* has to do specifically with standards for the construction industry. Your employer's safety program is based on these standards.

Safety begins with the proper training and orientation of employees. New operators should not begin work until they have read and demonstrated an understanding of the company's safety program. In addition, operators should not perform work in hazardous areas or operate new types of equipment until they have received a safety orientation and checkout from their supervisor or safety officer.

It is important to take the time to learn as much about safe operating practices as possible. Do not be put off by safety requirements or regulations that seem unnecessary. They are established for a reason. Follow them for your own protection, as well as for the protection of workers around you.

The basic rule to follow every working day is to report everything that you consider unsafe. These situations cannot be corrected if they are not reported. If you are a lead operator or a first-line supervisor, it is your responsibility to follow up on these reports and see that the situations are corrected.

The person who has the greatest effect on safety is the worker. Without safety-conscious employees, an employer cannot have a good safety record. Part of a worker's responsibility for safety is to follow good safety practices while working, and to report any unsafe conditions to the employer. Workers have an obligation to themselves, their families, and their co-workers to report any unsafe conditions. Safe workers take steps to remain alert while working. This can be difficult when performing repetitive tasks. The repetition can be monotonous and a person can become distracted. To combat this, it is important to get adequate rest. Caffeine and other stimulants are no substitute for sleep. In addition, eat on a regular schedule so that your body has the fuel it needs to work and think. Avoid any prescription or over-the-counter medications that can cause drowsiness. Recreational drug use and excessive alcohol use must be avoided during off-duty hours. Never use drugs or alcohol on the job.

Some hazards have an immediate effect on health. For example, exposure to excessive levels of carbon monoxide or an oxygen-deficient environment can be deadly in a matter of minutes. Other hazards may not have any noticeable effect until years after the exposure. For example, repeated exposure to noise has a cumulative effect rather than an immediate effect.

In addition to protecting yourself for your family's sake, you need to protect your family's health. Dust particles that cling to your clothing can contain materials that are hazardous to the health of people—especially children—with whom you are in close contact. If you are working with any potentially hazardous materials, you can help protect your family by showering immediately after work. If workplace showers are available, take advantage of them. Change into clean clothing before socializing with your family or others. If possible, immediately launder your work clothing after work. If this is not possible, keep work clothing separate from all other clothing and launder it separately.

1.1.0 Site Safety Program

Although there is no specific requirement for a company to have a safety officer, it makes sense to designate someone who can oversee the safety program and act as a point of contact for all safety-related matters. The safety officer also coordinates safety programs and activities. Most companies have an appointed person responsible for dealing with safety issues. Larger job sites often have a site safety officer. Larger companies generally have a full-time safety officer or engineer. The safety officer is responsible for training employees in safety and for the development of the company's safety program. This safety officer is an important resource for safety concerns.

The general activities of a safety officer are as follows:

- Ensuring that all safety rules are communicated and followed. The safety rules are enforced by making periodic inspections of all job sites and offices. If the rules are not enforced consistently and uniformly, their importance will be diminished.
- Initiating training programs to ensure that every employee is aware of any safety and health hazards associated with their assigned tasks or areas.
- Investigating accidents. When an accident occurs, the safety officer should review the accident report to help supervisory personnel determine the cause of the accident and recommend methods to prevent a recurrence.
- Conducting safety meetings. The safety officer is often the person best suited to conduct safety meetings to ensure that employees receive the most from the meeting and get involved in the safety program.
- Performing inspections of offices and job sites upon request to detect unsafe conditions or safety rule violations.
- Analyzing work practices to detect unsafe conditions, and to develop procedures to eliminate accidents or injuries.

Depending upon the size and the type of the company, the safety officer may have several other related responsibilities in the areas of project planning, equipment maintenance, and operator certification.

1.1.1 Safety Meetings

Safety meetings are used to keep employees informed of safe operating practices, new techniques and equipment, and workplace hazards. These meetings should be an everyday work activity. Regularly scheduled safety meetings are held to reinforce safety policy and procedures and to inform workers of any new industry standards or safety practices the company is implementing.

Daily meetings at the job site are short and topical and are intended to keep attention focused on task-related safety. Toolbox meetings are usually held first thing in the morning when everyone is alert and before people have a chance to spread out. Topics can include accident prevention, safe working practices, emergency procedures, tool and equipment operation, and personal protective equipment (PPE). The meeting discussions and demonstrations may be led by a safety officer or supervisor who is qualified to present safety material or information. There are many resources for prepared typical toolbox meetings. A good online resource can be found at **www.toolboxtopics.com**. The meeting subject and names of those in attendance are usually documented on a form (*Figure 1*).

In addition to toolbox meetings, companies often have formal training programs that operators are required to attend periodically. Topics of these meetings include first aid, emergency procedures, OSHA and other government safety requirements

JOBSITE SAFETY MEETING REPORT

DATE: ________ TIME: ________ SUPERVISOR: ________________ PROJECT # ________

STANDARD TOPICS:

- ☐ EMERGENCY RESPONSE PROCEDURES
- ☐ HARD HATS / VESTS / CLOTHING
- ☐ EYE / EAR / HAND PROTECTION
- ☐ PROPER LIFTING TECHNIQUES
- ☐ EQUIPMENT WALK AROUNDS
- ○ FIRE EXTINGUISHER
- ○ BACK-UP ALARMS

JOBSITE TOPICS DISCUSSED: ________________

MAIN POINTS EMPHASIZED: ________________

REVIEWED "MSDS" YES / NO (circle one) SUBJECTS:

☞ ________

☞ ________

SUGGESTIONS MADE BY CREWS: ________________

CORRECTIVE ACTION TAKEN: ________________

OTHER COMMENTS: ________________

ATTENDANCE: ________________

SUPERVISOR'S SIGNATURE: ________________

22210-12_F01.EPS

Figure 1 Safety meeting attendance form.

and regulations, hazardous materials handling, and licensing and certification requirements. It is your responsibility to attend these meetings and pay attention to what is being presented.

Many people do not like to go to meetings or training sessions because they think the meetings are a waste of time. Others are always eager to attend because they think they are getting out of work, and relaxing at the company's expense. Safety training sessions should be treated like any other part of the job. Keep focused and learn everything you can about the subject being discussed. In the case of safety information, it could save your life.

1.1.2 Safety Committees

Some employers have safety committees. If you work for such a company, you have an obligation to that committee to help maintain a safe working environment. This means following the safety committee's rules for proper working procedures and practices and reporting any unsafe equipment or conditions to the safety committee or the appropriate supervisor.

Creating a committee is a good way to get more people involved in safety activities. You may be asked to participate on such a committee—either to help out with particular project, or to represent a specific part of your organization. If you are asked to be on a safety committee, you should feel honored that your organization thinks of you as someone who is safety-conscious. Involvement with this effort gives you an opportunity to make an impact on your company's safety performance, and you will gain the satisfaction of having diversity in your job and seeing your ideas put into practice. Safety committee activities may include the following:

- Reviewing company safety policies and procedures.
- Conducting training in certain areas of expertise.
- Promoting safety awareness and holding safety campaigns.
- Assisting in the investigation of incidents and accidents.
- Recommending new safety training and information programs.
- Assisting in the recognition and/or correction of hazards.

1.2.0 Safety Investigations and Inspections

In spite of good education and careful work practices, accidents can happen. The term *accident* applies to those occurrences resulting in property damage, as well as those involving physical injury. Incidents, or near misses, are defined as any event that could have resulted in damage or physical injury, but did not. All accidents must be reported to your employer, who should ensure that your supervisor perform an accident investigation. Not all incidents are reported; those that are reported should be investigated just as if an accident had occurred.

Investigation of an incident or accident is similar to the investigation of a crime; the object is to solve the mystery, and then take a course of corrective action to prevent it from happening again. It is essential to find out what happened (investigate) and then change or isolate the conditions that allowed the accident to happen in the first place.

Productive accident investigation is not driven by a desire to hand out punishment. When human error is identified as the cause, it should be confronted in a problem-solving, non-threatening way. What is important is not fault finding, but that facts are discovered which result in corrective action.

Investigations of accidents and incidents may result in an OSHA inquiry, depending on the type of occurrence. Regardless of OSHA requirements, the company should carry out its own investigation to determine what went wrong, and why.

1.2.1 Accident and Incident Reports

Investigating an accident requires interviewing the people involved and reviewing the accident scene. This should be done promptly, with care and attention to detail, so that all the facts are uncovered. An accident report form can be extremely useful in guiding investigators through an investigation and providing information necessary for later analysis. *Figure 2* shows an example of an accident investigation report. The accident report form used in your internal investigations should contain enough information to help determine what happened and what should be done to prevent recurrence.

An accident report is similar to an incident report, but an accident report must include the name of the employee involved in the accident and the extent of his or her injuries, as well as the names of witnesses. There is no standard format for these forms, although much of the basic information, such as names, dates, locations, and accident type, is standard. Accident report forms vary from company to company. Some forms rely on a check-off system to categorize many of the items of information such as operation, accident type, location, and severity. By using this approach, data can be summarized quickly, and more direct comparisons can be made without having to interpret a narrative description of the accident.

The following types of questions should be asked to obtain information for the accident report:

- Who was involved in the accident? What do they do?

SUPERVISOR'S ACCIDENT INVESTIGATION REPORT

Type of Accident: ☐ Motor Vehicle ☐ Property Damage ☐ Equipment Damage

Personal Injury: ☐ Employee ☐ Non-Employee

Project Name: ____________

Project No. ____________

Equipment No. ____________

Name of Employee ____________ How long employed on crew ____________

Job Assignment ____________ Length of known prior experience ____________

Date of Accident ____________ Time ____________ First reported to me on ____________

Exact Place of Accident ____________

Where was the supervisor at the time of the accident? ____________

What happened? ____________

Diagram on back ☐

Describe injury/damage ____________

Was treatment provided: First Aid ☐ Doctor ☐ Admitted to Hospital ☐

What caused the accident to happen? ____________

Corrective action taken to prevent this happening again ____________

Recommendations to avoid similar accidents ____________

What safety equipment was being used? ____________

Witnesses (please print) ____________

Number of accidents on my crew this year ____________ For This employee ____________

Supervisor's signature ____________ Date ____________

Supervisor's printed name ____________

Project Manager's signature ____________ Date ____________

Manager's printed name ____________

Use back of form for additional detail.

22210-12_F02.EPS

Figure 2 Accident report form.

- How much experience do they have at this particular job?
- Who witnessed the accident?
- When did the accident occur?
- Where was the accident?
- What was the extent of injury or damage?
- What actually happened? What exactly was the person (or persons) doing at the time of the accident. What task or step of the job cycle?
- What was the direct cause of the accident? Were there contributing factors?

- What corrective action (in the investigator's opinion) should be taken to prevent this accident from recurring?

Some suggestions for completing an accident report form are:

- *Identify the direct cause or causes* – The direct causes can include unsafe actions, unsafe conditions, or, most likely, a combination of both. Try to state the cause or causes in simple, direct terms. Avoid vague and general statements; the clearer your statements are, the easier it will be to determine the appropriate corrective action. Be sure that the information is accurate.
- *Identify all contributing factors* – The factors that contributed to the accident might include horseplay, fatigue, lack of attention, inexperience, shortcutting recommended procedures, improper use of tools, poor supervision, and so on. Again, try to state the factors in simple, clear terms and as accurately as possible.
- *Document the incident* – Documentation may be done in writing and with photos or video.
- *Recommend corrective action* – Corrective action must be taken to prevent a recurrence of this type of accident. Further training or retraining, changes in job procedures, environmental changes, equipment modifications, or improvements in housekeeping are examples of possible corrective actions.

Investigative efforts for an accident depend on several factors, including the magnitude, severity, and complexity of the occurrence. Some accidents are simple to solve, because a witness observed the procedure or act that caused it. Many of these cases can be associated with human error while performing a task or operating a piece of machinery. Other accidents are caused by the gradual failure of some component or material until finally the accident occurs; these accidents may be much more difficult to investigate. Any job-related injuries or illness must be logged and reported to OSHA by the safety director or manager.

Incidents and accidents should be investigated by the supervisor responsible for the person(s) involved and the area where the accident occurred. This is important because the immediate supervisor is the one who should implement a plan of corrective action. The supervisor would also be the person most familiar with the operation and the work environment where the accident occurred. The company safety officer should review the accident investigation report to ensure that a complete and proper investigation has taken place. Because safety officers are trained in accident investigation and analysis, their expertise can be a valuable tool in preventing a recurrence.

There is no such thing as a minor accident. Any accident is symptomatic of a larger problem that should be identified and corrected.

Incident reports are similar to accident reports. While an accident report documents events that caused property damage or bodily injury, incident reports document near misses or events that could have resulted in property damage or bodily injury. The incident report is not difficult to complete, and simply requires basic information about what happened. Since incidents are not accidents, it may not be necessary to report them to OSHA. Incident reports are very valuable, however, because they provide the safety officer with data that can be used to identify and analyze patterns of unsafe work activities before an accident occurs.

1.2.2 OSHA Inspections

To enforce the safety standards defined in the Occupational Safety and Health Act of 1970 (and later revisions), OSHA is authorized to conduct workplace inspections. Every employer covered by the Act is subject to inspection (states with approved safety programs are also authorized to conduct inspections). There are a number of situations in which OSHA will conduct a job-site inspection. For example, an inspection will be conducted any time there is an on-the-job fatality or when three or more workers are hospitalized.

OSHA is responsible for developing and implementing legally enforceable standards based on requirements in the law. States with OSHA-approved occupational safety and health programs must set standards that are at least as effective as the federal programs. It is the responsibility of employers to become familiar with standards applicable to their line of business, and to ensure that employees have (and use) appropriate personal protective equipment when required for safety. Employees must comply with all rules and regulations that affect their own actions and conduct.

Prior to any inspection, the OSHA Compliance Safety and Health Officer (CSHO) becomes familiar with as many relevant facts as possible about the workplace, taking into account such things as the history of the establishment, the nature of the business, and the particular safety standards that may apply. The inspection process is detailed in the OSHA standards and in numerous other OSHA publications.

A typical inspection takes place as follows:

- The CSHO shows their credentials. All CSHOs carry US Department of Labor credentials that can be verified by contacting the nearest OSHA office.
- Company and employee representatives are designated to accompany the CSHO on the inspection.
- The route and duration of the inspection is determined by the CSHO, who may stop and confer with employees in private if necessary. The CSHO observes general conditions, consults with employees, takes pictures and instrument readings, and examines certain records, such as the OSHA 300 log.
- At the closing conference, the CSHO discusses all unsafe and unhealthful conditions observed, and indicates the apparent violations for which the company may be cited.

After the inspection and the report of findings by the CSHO, OSHA's area director determines what citations (if any) will be issued, and what penalties will be imposed. Citations inform the employer and employees of the regulations and standards alleged to have been violated and the length of time proposed for their correction. The employer receives the citations and notices of proposed penalties by certified mail. The employer must post a copy of each citation at or near the place where the violation occurred and leave it there for three days or until the violation is corrected, whichever is longer.

During an OSHA inspection, employees should be polite and respectful to the CSHO. OSHA inspections are not the time to get back at your employer for some injustice, nor is it the time to cover up unsafe conditions. Always answer questions directly and honestly. Lying to the CSHO is just like lying to a police officer.

1.3.0 Hazardous Materials

OSHA Standard Part 1910.1200 is titled *Hazard Communication*, and it specifically requires that a material safety data sheet (MSDS) be maintained for hazardous chemicals in the workplace and that all MSDSs be available for everyone's use (*Figure 3*). Employers must establish a written, comprehensive hazardous communications (HAZCOM) program that includes provisions for container labeling, MSDSs, and employee safety training activities. OSHA even has documents to help your employer develop a HAZCOM program.

Figure 3 HAZCOM.

The program must include the following:

- A list of hazardous chemicals in each work area.
- The means the employer uses to inform employees of the hazards of nonroutine tasks.
- The procedure the employer uses to inform other companies of the hazards to which their employees may be exposed.

Employers must establish a training and information program for employees at the time of their initial assignment if they will be exposed to hazardous chemicals, and again whenever a new hazard is introduced into their work area. The discussion must include at least the following topics:

- The existence of the HAZCOM program and the requirements of the standard.
- The components of the HAZCOM program in the employees' workplaces.
- Operations in the work areas where hazardous chemicals are present.
- Location where the employer keeps the written hazard evaluation procedures, communications program, list of hazardous chemicals, and the required MSDS forms.

The employee training plan must consist of the following elements:

- A description of how the HAZCOM program is implemented in their workplace; how to read and interpret information on the labels and the MSDSs; and how employees can obtain and use the available hazard information.
- An explanation of the hazards of the chemicals in a specific work area. The hazards may be discussed by individual chemical, or by hazard categories such as flammability.

- A discussion of safety measures that employees must follow to protect themselves from these hazards.
- Specific procedures put into effect by the employer to provide protection such as engineering controls, work practices, and the use of personal protective equipment.
- Methods of observation (such as visual appearance or smell) that workers can use to detect the presence of a hazardous chemical.

Anyone who works on a construction site is likely to work around hazardous materials. Fluids such as gasoline, diesel fuel, and solvents used in the workshop are all considered hazardous materials under the OSHA HAZCOM Standard. This section has general information concerning how to act when using or working around hazardous material, and how to respond to accidents or spills.

Employers rely on the MSDS provided by the manufacturers of any hazardous materials used at the workplace, including materials used by subcontractors. The form and style of the MSDS may vary, but each sheet includes sections that detail information as required by the OSHA standard. These sheets contain information about the hazards of a product. They help determine safe handling practices and procedures, emergency response in the event of an accident, and waste disposal requirements. Material safety data sheets are normally kept with the HAZCOM program documentation and placed in work areas for easy reference. This should be the first stop before working with a new chemical.

An MSDS provides information the manufacturer considers necessary to determine what chemicals are in the product and the actions required to protect persons using the product. The sheet is divided into the following sections:

- Basic information, such as the manufacturer, the name of the hazardous material or substance, and an emergency telephone number.
- Hazardous ingredients in the substance, including information on exposure limits. Such limits may be expressed as Permissible Exposure Limit (PEL), Threshold Limit Value (TLV), or in other ways.
- Physical data describing how to identify the material or substance by observation or odor.
- Fire and exposure data describing, in technical terms, the degree of hazard and how to extinguish fires should they occur.
- Reactivity data indicating chemical stability of the substance and situations where a hazardous reaction can occur.
- Health hazard data describing, in technical terms, the degree of hazard, as well as emergency and first-aid procedures.
- Special protection information describing how to protect against hazardous exposure, what to do if a spill or leak occurs, and how to properly dispose of waste materials. Precautions describing any special requirements for handling or storing material and information not provided elsewhere.

Another source of hazardous chemical information is the log of hazardous materials and substances. It serves as a summary list of the individual MSDSs. The list includes the common or trade name, the manufacturer's name, and the chemical name of materials and substances onsite. The location and quantity of certain large volumes of highly toxic substances may also be recorded. The log is usually kept with the HAZCOM program documentation.

Employees should obtain an MSDS for any hazardous material they use. Anyone purchasing or using hazardous materials should verify that all containers are clearly labeled as to contents, and that the label contains the appropriate hazard warning and the name and address of the manufacturer.

WARNING!

No container of hazardous material should be accepted unless it is properly labeled by the dispensing person or agency.

Employers should provide training for all employees who work with or who will potentially be exposed to hazardous materials. Working safely with chemicals is a two-way street. The employer will provide access to needed information and proper protective gear, but it is up to the worker to handle the chemicals safely and use proper safety equipment and safe work procedures when working around chemicals. Always follow these rules when using hazardous materials:

- Know what chemicals you are working with and how strong (concentrated) they are. Use appropriate personal protective equipment as required.
- In case of skin or eye contact, flush with cool water for at least 15 minutes, but do not rub the skin or eyes. This can break the skin or delicate eye tissue and give the substance a path into the body.

- Keep different types of chemicals stored in different areas or in cabinets like the one shown in *Figure 4*.
- Clean up spills promptly. If you cannot identify a chemical or do not know how to handle it, check with your supervisor or safety officer before taking any action.
- Storage areas should be kept free of unnecessary items such as chemicals that present fire hazards. Paints, thinners, pesticides, solvents, and other chemicals should be kept in a well-ventilated area under lock and key.
- Any unknown chemical or unlabeled containers should be treated as extremely hazardous until it is identified.

> **NOTE**
>
> Beginning on June 1, 2015, OSHA will require new labeling methods for hazardous chemicals. The new labels are intended to make it easier for people to recognize and identify the chemical and its associated hazards. The new labels will include a product identifier; a pictogram representing the nature of the hazard; a signal word such as Danger; precautionary statements; and identification of the supplier.

22210-12_F04.EPS

Figure 4 Safety cabinet.

1.4.0 Excavation Safety

Safety is crucial during any excavation job. Trenches are a common form of excavation. A trench is a narrow excavation made below the surface of the ground in which the depth is greater than the width and the width does not exceed 15 feet, as measured at the bottom of the trench. Working around trenches is one of the most dangerous situations for a construction worker. When earth is removed from the ground, the excavation walls are not supported, so any pressure on the ground around the trench can cause the walls to collapse. To prevent this, the walls need to be properly secured by shoring, sloping, or **shielding**.

One cubic yard of earth weighs about 3,000 pounds. That is the weight of a small car and is more than enough weight to seriously injure or kill a worker. In fact, each year in the United States, more than 100 people are killed and many more are seriously injured in cave-in accidents.

When operating machinery around excavations, it is important to stay alert for workers on foot. Always keep co-workers in sight. If you lose track of a worker, stop the equipment until you are certain he or she is safe. The following guidelines must be enforced to ensure everyone's safety:

- Be alert. Watch and listen for possible dangers.
- Never enter an excavation without the approval of the **competent person** on site.
- Never operate your machinery above workers in an excavation.
- Ensure that the OSHA-approved competent person inspects the excavation daily for changes in the excavation environment, such as rain, frost, or severe vibration from nearby heavy equipment.
- Stay alert for other machinery and stay clear of any vehicle that is being loaded.
- Keep the excavated soil (spoil) at least two feet from the edge of the excavation.
- Stop work immediately if there is any potential for a cave-in. Make sure any problems are corrected before starting work again.
- Use shoring, trench boxes, benching, or sloping for excavations and trenches over five feet deep.

- Keep equipment back from the edge of the excavation to prevent cave-ins.

> **WARNING!**
>
> Keep in mind that gases such as carbon monoxide can displace oxygen in a trench. The gas sinks to the bottom of the trench and workers can be overcome. Trenches should be tested in the same way as confined spaces and respirators may have to be worn by those working in the trench. Some flammable gases, such as acetylene gas can form a layer in a trench, creating an explosion hazard. In general, trenches that exceed four feet in depth should be considered confined spaces until confined-space hazards have been ruled out or dealt with by a competent person.

> **NOTE**
>
> For any excavation deeper than 20 feet, the shoring design must be developed and stamped by a professional engineer.

1.4.1 Indications of an Unstable Trench

A number of stresses and weaknesses can occur in an open trench or excavation. For example, increases or decreases in moisture content can affect the stability of a trench or excavation. The following sections discuss some of the more frequently identified causes of trench failure. These conditions are shown in *Figure 5*.

Tension cracks usually form a quarter to half way down from the top of a trench. Sliding or slipping may occur as a result of tension cracks. In addition to sliding, tension cracks can cause toppling. Toppling occurs when the trench's vertical face shears along the tension crack line and topples into the excavation. **Subsidence** is when pressure on the surface of the ground around the excavation stresses the excavation walls and can cause the wall to bulge. If uncorrected, this condition can cause wall failure and trap workers in the trench or topple equipment near the excavation. Bottom heaving is caused by downward pressure created by the weight of adjoining soil. This pressure causes a bulge in the bottom of the cut. These conditions can occur even when shoring and shielding are properly installed.

Another indication of an unstable trench is boiling. Boiling is when water flows upward into the bottom of the cut. A high water table is one of the causes of boiling. Boiling can happen quickly and can occur even when shoring or trench boxes are used. If boiling starts, stop what you are doing and leave the excavation area immediately.

1.4.2 Making the Excavation Safe

There are several ways to make an excavation site a safer place to work. Heavy equipment operators are called to install shoring or trenching systems, so it is important to be familiar with these procedures. Sometimes, plans call for the excavation walls to be sloped away from the excavation floor to relieve pressure and avoid cave-ins. Shoring, shielding, and sloping are different methods used to protect workers and equipment. It is important to recognize the differences between them.

Shoring in a trench is placed against the excavation walls to support them and prevent their movement and collapse. Shoring does more than provide a safe environment for workers in a trench. Because it restrains the movement of trench walls, shoring also stops the shifting of surrounding soil, which may contain buried utilities or on which sidewalks, streets, building foundations, or other structures are built.

Trench shields, also called trench boxes, are placed in unshored excavations to protect personnel from excavation wall collapse. They provide no support to trench walls or surrounding soil. But for specific depths and soil conditions, trench shields withstand the side weight of a collapsing trench wall to protect workers in the event of a cave-in.

1.4.3 Shoring and Shielding Systems

Shoring systems are metal, hydraulic, mechanical, or timber structures that provide a framework to support excavation walls. Shoring uses **uprights**, **walers**, and **cross braces** to support walls. Because of their great weight, these structures need to be lifted into the excavation with heavy equipment. *Figure 6* shows a shoring system in place.

When excavating near existing structures or performing short-term excavations, vertical sheeting may be used and supported with hydraulic walers. The walers support the vertical sheeting and are held in place against the trench walls by hydraulic spreaders (*Figure 7*). Other types of spreaders, including screw jacks and trench jacks, are also available.

Interlocking steel sheeting may be specified under certain conditions such as deep excavations and excavations near buildings or building foundations. Interlocking steel sheeting is commonly used on Department of Transportation (DOT) right-of-ways. It prevents damage to subbase pavement caused by vibration from vehicle traffic. Interlocking steel sheeting is required when working in waterways. Steel sheeting con-

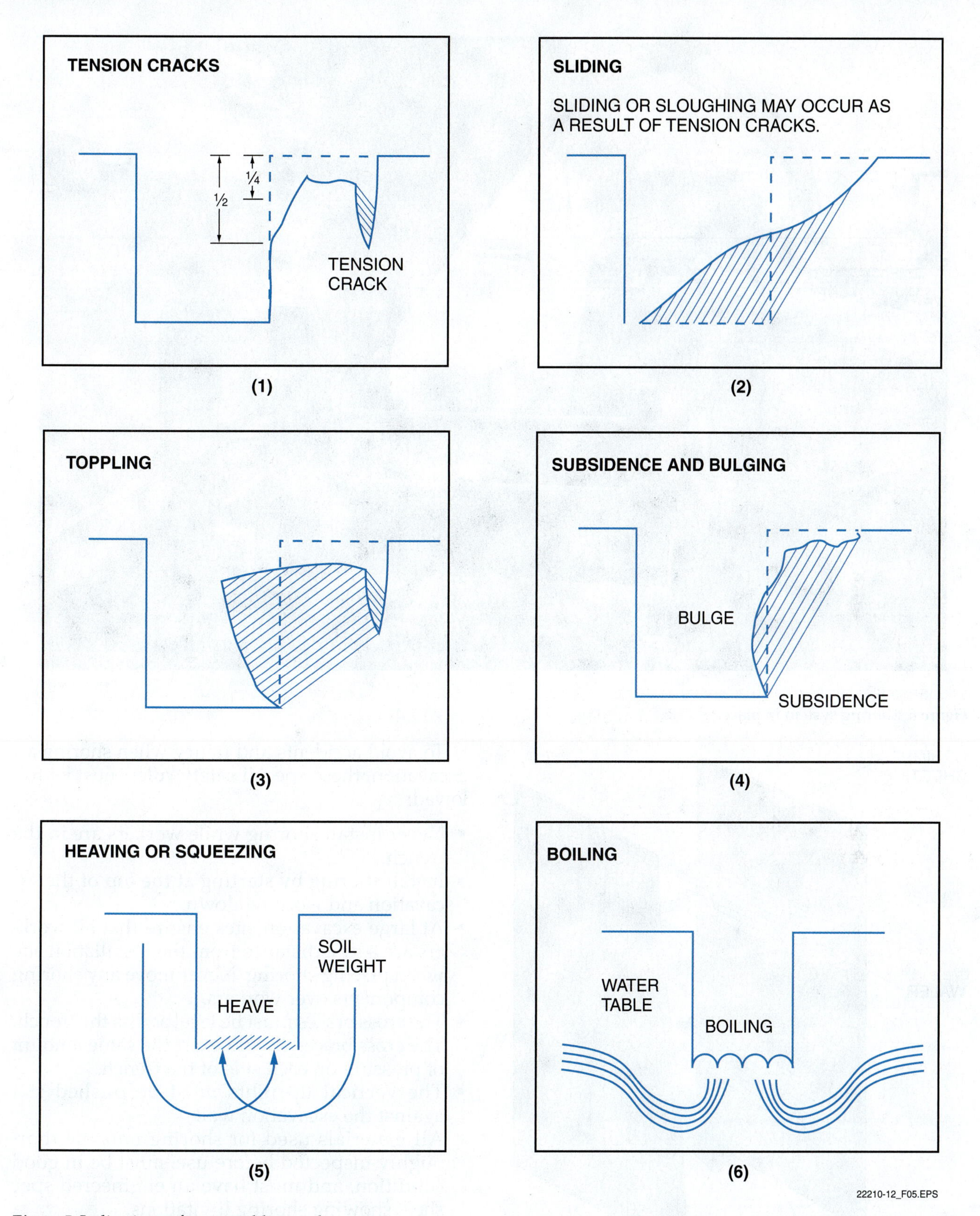

Figure 5 Indications of an unstable trench.

sists of interlocking panels of steel reinforced with cross members. Steel sheeting is engineered for a particular application. It must be installed precisely in accordance with the engineer's specifications. Steel sheeting is commonly installed by driving it into the ground using a vibrating hydraulic hammer. It can also be driven with a drop hammer.

Figure 6 Shoring system in place.

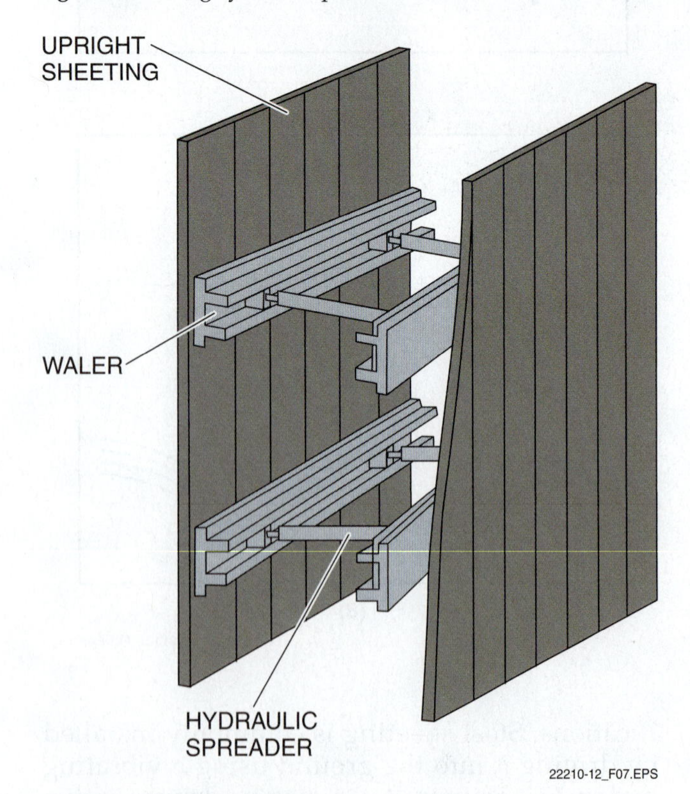

Figure 7 Vertical sheeting system.

To avoid accidents and injury when shoring an excavation, these special safety rules must be followed:

- Never install shoring while workers are in the trench.
- Install shoring by starting at the top of the excavation and working down.
- At large excavation sites ensure that all workers are a safe distance from the installation site when placing shoring. Never move any shoring components over workers.
- The cross braces must be level across the trench. The cross braces should exert the same amount of pressure on each side of the trench.
- The vertical uprights must be pushed flat against the excavation wall.
- All materials used for shoring must be thoroughly inspected before use, must be in good condition, and must have an engineered spec sheet showing shoring limitations.
- Shoring is removed by starting at the bottom of the excavation and going up.
- The vertical supports are pulled out of the trench from above.
- Every excavation must be backfilled immediately after the support system is removed.

A shielding system is a structure that is able to withstand the forces imposed on it by a cave-in, and to protect employees within the excavation. Shields can be permanent structures or can be designed to be portable and moved along as work progresses. The shielding system is also known as a trench box or trench shield. *Figure 8* shows workers protected by a trench box. If the trench will not stand long enough to excavate for the shield, the shield can be placed high and pushed down as material is excavated.

The excavated area between the outside of the trench box and the face of the trench should be as small as possible. The space between the trench box and the excavation side must be backfilled to prevent side-to-side movement of the trench box. Remember that job-site and soil conditions, as well as trench depths and widths, determine what type of trench protection system should be used. A single project can include several depth or width requirements and varying soil conditions. This may require that several different protective systems be used for the same site. A registered engineer must certify shields for the existing soil conditions. The certification must be available at the job site during assembly of the trench box. After assembly, certifications must be made available upon request.

If used correctly, trench boxes protect workers from the dangers of a cave-in. All safety guidelines for excavations also apply to trench boxes. Follow these safety guidelines when using a trench box:

- Be sure that the vertical walls of a trench box extend at least 18 inches above the lowest point where the excavation walls begin to slope. *Figure 9* shows proper trench box placement.
- Never enter the trench box during installation or removal.
- Never enter an excavation behind a trench box (between the excavation wall and the trench box).
- Never enter an excavation after the trench box has been moved.
- Backfill the excavation as soon as the trench box has been removed.
- If a trench box is to be used in a pit or a hole, all four sides of the trench box must be shielded.

22210-12_F08.EPS

Figure 8 Workers in a trench box.

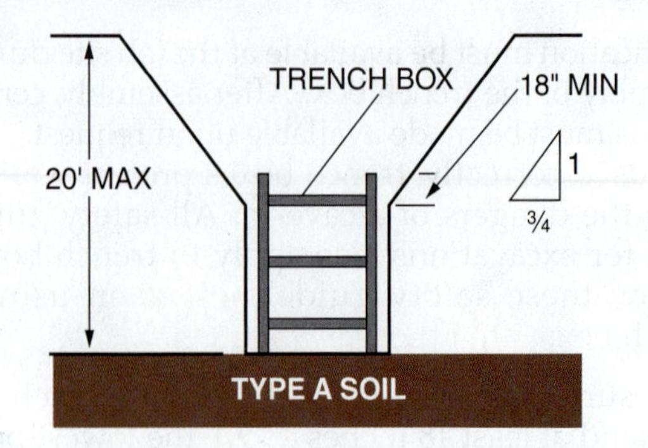

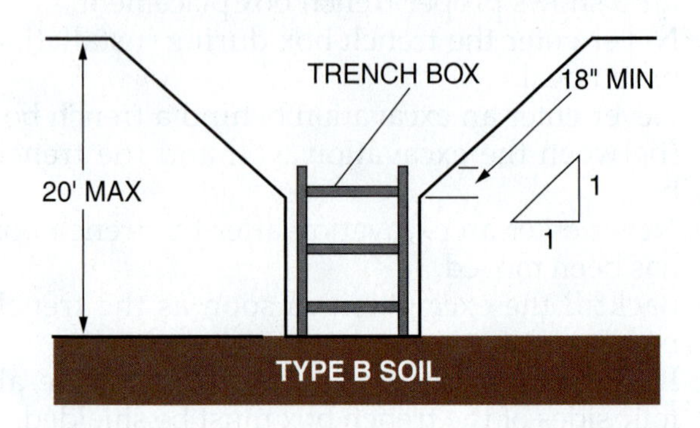

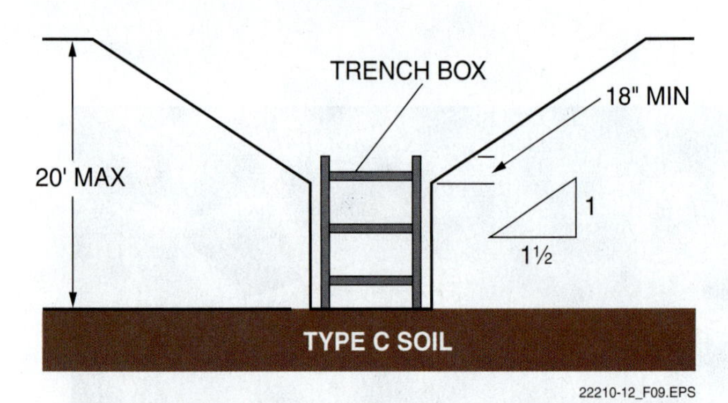

Figure 9 Proper trench box placement.

Remember that trench boxes are designed to protect workers in the excavation. They do not stabilize the walls of the trench, so be alert for changes in the ground surrounding the trench—they could indicate that a cave-in is imminent.

1.5.0 Transporting Equipment Safely

Heavy equipment is usually transported to job sites on flatbed trailers known as low-boys. The driver of the transport vehicle must have a CDL. Loading and securing heavy equipment is a critical operation that calls for knowledge, training, skill, and experience. Never attempt to load or unload equipment without the proper training. The loading procedure for all heavy equipment is not the same, so specific training is needed to safely load and unload each piece of equipment.

Always consult the manufacturer's instructions in the operator's manual for the equipment being transported before you transport any equipment. Follow these general guidelines when loading heavy equipment for transportation to another location:

- Ensure that the transport equipment is in good working order and that tires are inflated according to the manufacturer's specifications. Test lights and turn signals before starting to load the equipment.
- Use loading ramps and other equipment that are adequately rated for the weight of the equipment.
- Use ramps with a surface that gives the equipment being loaded good traction.
- Ensure that the transport vehicle is specified for the load.
- Chock the wheels of the transport vehicle before loading the equipment.

Soil Freezing

As technology becomes available, new methods and practices may be used to help ensure excavation safety. Soil freezing is one such technology. Although this exciting technology has been used with success on some jobs, its use is not widespread. It involves laying chiller pipes at the perimeter of the excavation site, and then a freezing solution is circulated through the pipes, freezing the surrounding soil. The system is left in place for several weeks until the desired level of freezing is reached. Then normal excavation work is performed inside the frozen area. The chiller system is left in place until excavation work is completed. When the pipes are removed, the soil thaws over the course of several weeks until the normal temperature is reached. This method has the added advantage of controlling groundwater flow.

- Inspect tie-down material before and after its use. Discard any worn or damaged material.
- Ensure that the equipment is securely fastened to the transport vehicle in all directions to prevent any movement (*Figure 10*). Use chains, wire rope, or other device as specified by the transporter manufacturer's instruction manual.
- Be certain to use protection on sharp edges to avoid damaging the tie-down material and the equipment.
- Use proper warning signs and flags.
- Secure all attachments and loose gear to prevent shifting of the load during transportation.
- If possible, have another competent person inspect the vehicle after it is loaded to be certain the payload is secure.

Often equipment is delivered at night or early in the morning so the customer can get maximum use during business hours. Keep in mind that loading and unloading equipment is a noisy job. Some areas have noise ordinances that forbid construction noise during the night. Be certain of local laws before attempting to load or unload after normal business hours.

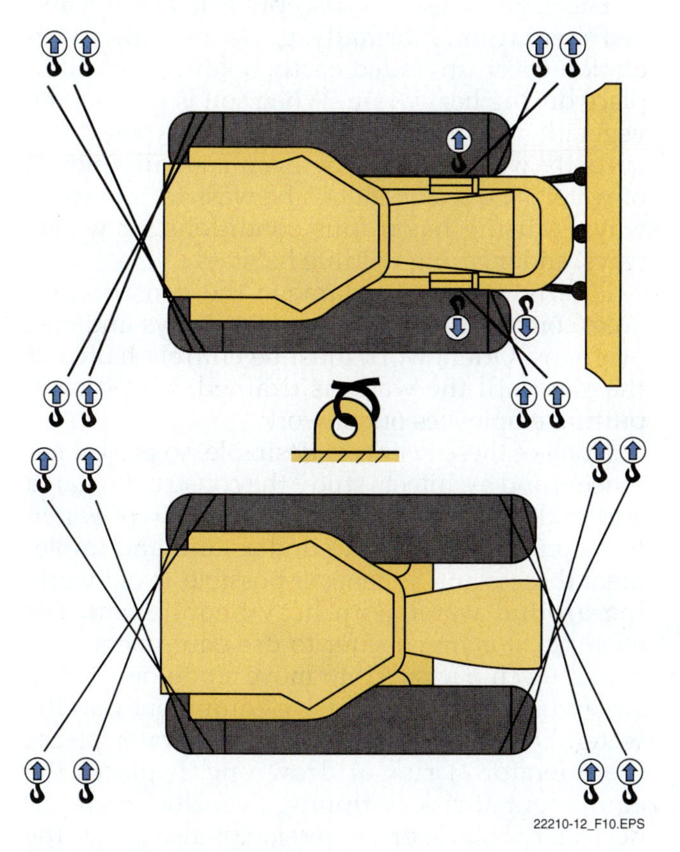

Figure 10 Machine transport.

Additional Resources

OSHA Trench Safety Quick Card: **www.osha.gov/Publications/quickcard/trenching_en.pdf**

1.0.0 Section Review

1. Which of the following is one of the responsibilities of a company safety officer?
   a. Approves the construction plans and drawings
   b. Analyzes work practices to improve work conditions
   c. Requests that the local OSHA office visit the site
   d. Checks out all heavy equipment before it is used

2. There must be an accident investigation whenever _____.
   a. a near miss occurs
   b. there is an employee complaint
   c. the local OSHA inspector requests one
   d. there is injury or property damage reported

3. An MSDS must be available for _____.
   a. all chemicals ever used by the company
   b. all chemicals in the OSHA Standards
   c. only those chemicals that are combustible
   d. all hazardous chemicals in the work area

4. One cubic yard of earth weighs approximately _____.
   a. 1,000 pounds
   b. 2,000 pounds
   c. 3,000 pounds
   d. 4,000 pounds

5. When loading equipment for transport, ensure that the equipment is secured _____.
   a. side to side
   b. front to back
   c. in all directions
   d. with the parking brake

Section Two

2.0.0 Controlling Water at a Job Site

Objective 2

Describe the methods used to control water on job sites.

a. Explain the importance of maintaining proper drainage on a job site.
b. Describe the methods used to control ground water and surface water.
c. Describe the safety practices and construction methods used when working around bodies of water.

Trade Terms

Aquifer: An underground layer of water-bearing permeable rock or unconsolidated materials through which water can easily move.

Berm: A raised bank of earth.

Dewater: To remove water from a site.

Erosion: The removal of soil from an area by water or wind.

Groundwater: Water beneath the surface of the ground.

Sedimentation: Soil particles that are removed from their original location by water, wind, or mechanical means.

Stormwater: Water from rain or melting snow.

Sump: A small excavation dug below grade for the purpose of draining or retaining subsurface water. The water is then usually pumped out of the sump by mechanical means.

Swale: A shallow trench used to direct the flow of water.

Unchecked water flow can damage excavations, stop work, make work more difficult, and cause accidents. For these reasons, a great deal of time and effort goes into the project's **dewatering** plans. Federal and local agencies regulate all aspects of water control, so it is very important that you know and follow the plans developed by dewatering experts.

A construction site may contain surface water, **groundwater**, or both. Surface water comes from rain and snow, while groundwater exists below the surface of the ground. Surface water slowly seeps into the groundwater, thus recharging groundwater supplies. Some groundwater exists in underground **aquifers.** Groundwater sometimes breaks through the surface of the earth as a freshwater stream.

A great deal of planning is needed to complete a job safely in a wet environment. Construction workers face a number of hazards when working around water. First, water from either the ground or the surface can weaken or flood excavations, endangering workers. Trenches must be inspected for safety and integrity daily and after any event that could weaken them, such as rain.

Second, drinking water comes from groundwater aquifers, so it is important to protect groundwater from contamination. When surface water washes across construction debris, such as oil, gasoline, chemicals, and other contaminates, those contaminants can make their way into groundwater supplies. Workers can help to protect surface water and groundwater by performing maintenance in the designated area, avoiding chemical spills, and placing all trash in its proper receptacle.

Third, graded soil is susceptible to **erosion** and **sedimentation.** Normally, grass and other vegetation cover ungraded earth, holding the soil in place during heavy rain. When soil is graded, the vegetation is stripped away and the exposed soil is easily washed away by even a small amount of water. The sediment can be washed into roadways, causing hazardous conditions, or waterways, endangering wildlife habitats.

Finally, water can damage the construction area, contributing to scheduling delays and cost overruns. Often, work must be entirely halted at the site until the water is drained, temporarily putting employees out of work.

None of these events is desirable, so project engineers and architects study the construction area and make plans to minimize the flow of water. It is up to the workers to understand and implement these plans. Whenever possible, avoid working around water with heavy equipment. For example, it is much safer to use equipment with a long reach (*Figure 11*) to move a boulder from a streambed than it is to drive equipment into the water. Driving equipment into the water places the operator at risk of drowning. It places the equipment at risk of tipping over due to an underwater obstacle or depression. It also places the environment at risk because any soil or chemicals clinging to the equipment can be washed off into the water.

22210-12_F11.EPS

Figure 11 Long-reach excavator.

2.1.0 Establishing Proper Drainage

Proper drainage is a very important part of the construction process. It helps to maintain the stability of highway roadbeds and building foundations. Drainage is also necessary to reduce or eliminate erosion on embankments and slopes. If the job site has a **stormwater** control plan, it is your responsibility to ensure that you are familiar with its contents.

Drainage work required during construction falls into three broad categories: natural and constructed drainage; control ditches; and drains and collection systems. Natural drainage and constructed drains are placed to make sure water is drained away from the structure. Control ditches are constructed to keep water from entering the construction area and causing damage to work. Drains and collection systems collect unwanted water within the construction site and provide a method of removal, allowing the soil or other material to dry out and not be saturated.

2.1.1 Natural Drainage and Drains

Roads are always crowned and sloped so that water drains from the road and minimizes water ponding on the surface. Shoulders should be graded to slope as much or more than the road in order to keep water flowing to the ditches. For example, a paved roadway with an 11-foot lane and 4-foot shoulder should have a total crown (from center line to outside edge of shoulder) of not less than 3.5 inches. Roads with steeper grades may require higher crowns, because the water tends to flow down the road rather than across the crown.

Ditches and channels must be dug and maintained to avoid damage to the roadway. Their primary function is to carry water away from the roadway for absorption, or to another area, such as a detention basin. Ditches must be properly shaped for safety and ease of maintenance, as well as proper water-flow and erosion control. A ditch should be at least one foot below the bottom of the roadway base to properly drain the pavement.

It is very important that water flows through the ditches and does not pond. Ponding water may saturate the subsurface material beneath a roadway, preventing it from draining during a storm. Ditches with at least a 1-percent grade are required to ensure proper flow.

Ponding water on a roadway surface is a result of poor grading and compaction and the lack of drainage ditches. Ponding water in ditches is the result of insufficient grading or poor definition of the grade line.

Springs or seepage areas under the road require special treatment. Rock-filled trenches known as French drains, or perforated pipes, are used to drain subsurface water into ditches or streams. Pipe culverts should be opened as soon as they are finished to help control water flow.

Some soils used in embankments are subject to high capillary action, which means that water easily creeps up towards the surface; this creeping action is also called wicking. To reduce this process, a granular blanket of sand or gravel is placed over these soils. This blanket, between the embankment soils and the subgrade, help protect the subgrade material from the water wicking upward. Requirements for a granular blanket are usually specified in the plans.

2.1.2 Control Ditches

Control ditches (*Figure 12*), also called intercepting ditches, need to be placed wherever there is a possibility of water intrusion from a source outside the construction area. This is a precautionary action based on knowledge of the terrain as well as possible rain and flooding conditions. These ditches drain water away from the construction site and deposit it in a nearby drainage ditch.

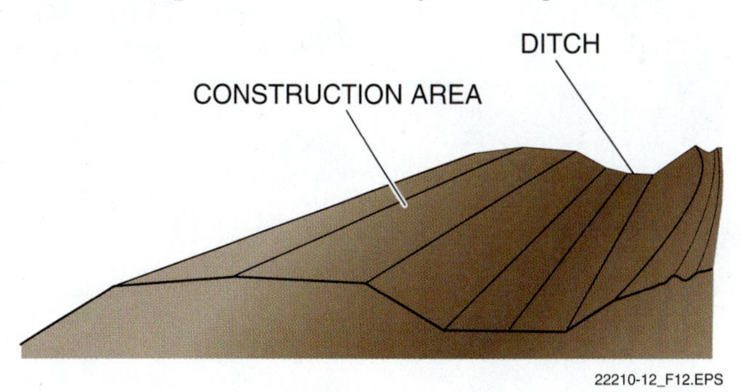

22210-12_F12.EPS

Figure 12 Control ditch.

Intercepting ditches for highway construction should be placed where the original ground outside of the finished cut sections slopes toward the center line. Seeping water in cut sections may indicate the need for a control ditch. The project engineer should be informed of these conditions, so intercepting ditches or another corrective action can be started.

2.1.3 Drains and Collection Systems

Excavation of pits and trenches can run into trouble because of unexpected pockets of water or a high groundwater table. When excavating below the water table, there is normally some intrusion into the excavated area. Because this water is below the natural ground line, it cannot be channeled out by digging a control ditch or another drain.

Channels must be dug in the bottom of the excavation to collect the water and direct it into a **sump**, where it can be pumped out. The sump should be placed as close as possible to the greatest source of the flow. The channel depth below the excavation floor should be about four feet, but depends on the pumping equipment being used. When a large amount of water flow exists, you should increase the number of sumps, not their size. *Figure 13* provides two examples of different types of drainage in an excavation.

The first example (*Figure 13A*) shows water coming from only one face of an excavation. A drain along this wall channels the water into sumps at either end. Pump-out is done from the sumps. The second example (*Figure 13B*) is similar, except that the water comes off two faces of the excavation. In this case, the drains are designed to channel water into a sump in the corner between the two faces. Pump-out is done from this one point.

When water is coming in from the bottom of the excavation, the bottom must be graded to drain to each side. Then a layer of stone and damp-proof paper is placed to provide a stable work platform. Channels are dug on the sides of the excavation to collect the water and the water is pumped out by the sumps. The sumps always need to be placed as close to the major source of water flow, using multiple sumps if necessary.

Soil conditions dictate the method of constructing drains. Drainage layouts are governed not only by soil conditions, but also by the location, direction, and quantity of flow. Drains should be limited in width and depth to minimum requirements. They should intercept the ground water as close to its source as possible.

2.2.0 Controlling Ground Water and Surface Water

The US Environmental Protection Agency (EPA) requires that contractors disturbing one acre or more of land obtain a National Pollutant Discharge Elimination System (NPDES) stormwater permit before work begins on the site. The purpose of the permit is to ensure that the contractor has a workable plan to prevent pollution and to control stormwater runoff from the site, as well

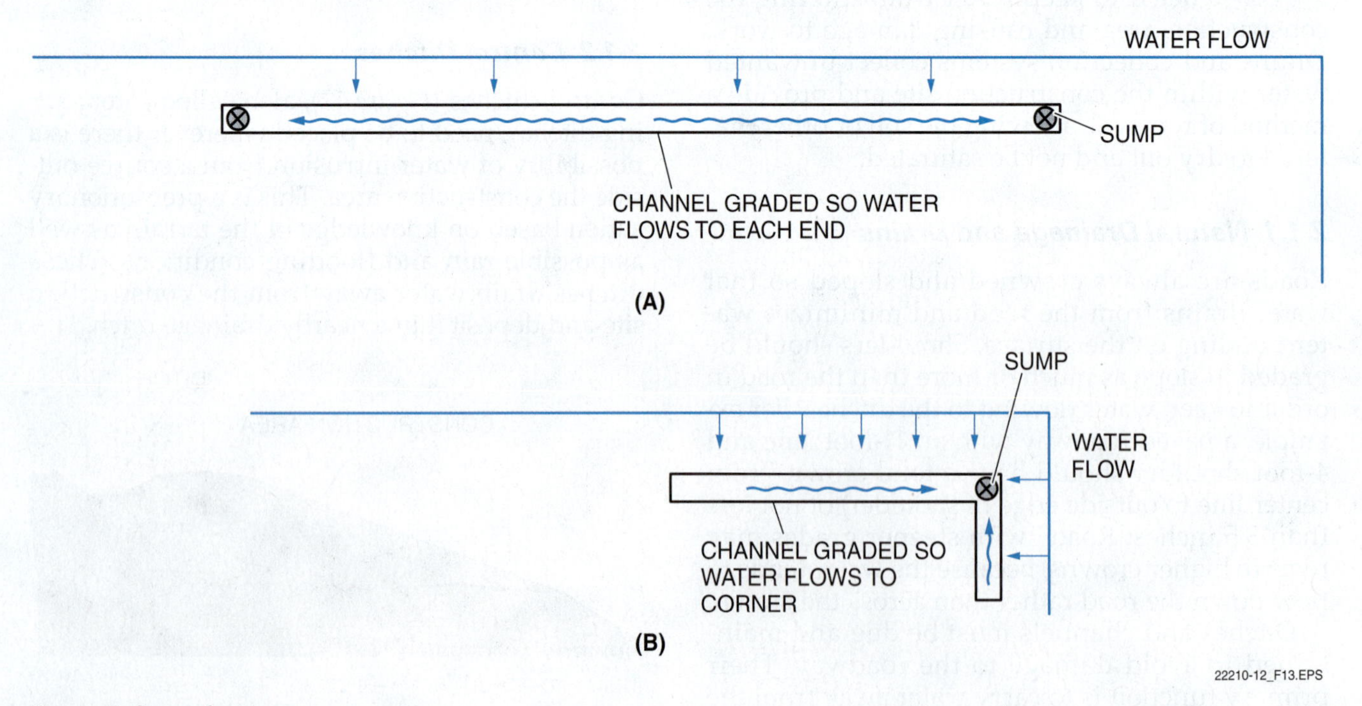

Figure 13 Sump drainage.

as to prevent erosion. Heavy equipment operators need to know and understand the site's erosion, sediment, and runoff-prevention plans.

> **NOTE**
> Construction sites that are smaller than one acre, but are part of a larger development, need to have an NPDES permit too.

Contractors need to use adequate measures to prevent erosion and sedimentation caused by stormwater runoff. Erosion is the eating away of soil by water or wind, while sedimentation refers to soil that has been moved from its original place by wind, water, or other means. Stormwater runoff is water from rain and snow that is shed from the ground rather than being absorbed. Runoff increases as an area is developed with roads and buildings because there is less available ground surface to absorb water.

Sediment is soil that has been moved from its original place, so the best way to avoid sediment is to prevent erosion. In earthmoving, some erosion is unavoidable, so most sites use various devices, such as installing silt fences (*Figure 14*) and blocking stormwater drains (*Figure 15*) to trap displaced soil. Other methods, such as **berms** and **swales** are used to guide and trap runoff. Workers can help prevent sedimentation by avoiding walking or driving across disturbed soil, washing soil off truck tires before leaving the site, and limiting vehicle operation to approved haul roads.

Water that flows through the ground is called groundwater. Any excavation, even one of only a few feet, has the potential of being affected by and affecting groundwater. Before a large project is approved, an environmental survey may be required. The survey includes a map of known groundwater hazards, such as buried chemical storage tanks, or landfills containing possible pollutants. In many jurisdictions, there are legal requirements for plans to remedy such hazards, and these plans must be supplied to and approved by the governing agencies before digging begins. Failing to follow the approved plan can make your employer liable for legal action and fines.

It is normal to encounter groundwater in deep trenches and excavations, especially where the groundwater is close to the surface. This is known as a high water table. The depth of the water table varies widely and depends on the amount of annual precipitation, the terrain, and the amount of water pumped from it for drinking, irrigation, and other uses. When groundwater begins to enter the trench or excavation, it causes several problems.

22210-12_F14.EPS

Figure 14 Silt fence.

22210-12_F15.EPS

Figure 15 Stormwater drain barrier.

First, the soil of the trench walls becomes more likely to collapse or to flow into the bottom of the trench. Second, the water makes it very difficult to keep digging. Further, when digging through water, some amount will fill the bucket, increasing the weight and forcing the equipment to use more fuel.

Soil helps protect groundwater by filtering chemicals and debris before it reaches the aquifer. Since excavations remove soil, chemicals that normally would not reach the underground level can contaminate the water. On many construction jobs, monitoring wells are dug to check the quality of groundwater, especially where blasting or toxic chemicals may be involved.

The easiest and best way to control water damage at an excavation is to prevent the water from entering the site. If this is not possible, control measures need to be as close to the source of water as possible. Various types and sizes of ditches are used to collect and channel the flow of water away from excavation areas. Some of these ditches are permanent and some temporary. All of them, however, are designed and specified according to the permeability of the soil, conditions at the site, and on the ability to handle the predicted quantity of water. Some common ditch configurations are shown in *Figure 16A*.

More active measures are sometimes needed to evacuate water from an area. In this case, the plans may call for sumps. Sumps are open pits or holes constructed to collect water (see *Figure 16B*). *Figure 17* shows examples of methods used to control ground water in road construction.

NOTE

A stilling basin is designed to dissipate the energy of fast-moving water in order to prevent erosion.

Pumps are placed in the sump to remove the water mechanically. The requirements for sumps and pumps vary with the depth and the flow rate of the water. A variety of pumps are described in the following section.

2.2.1 Equipment Used to Control Water

Portable pumps are widely used in construction for various applications, including draining water from excavations. The three most common types of portable pumps are trash pumps, mud pumps, and submersible pumps. These pumps can be pneumatic, electric, or gasoline engine-driven pumps.

Trash pumps – Trash pumps are centrifugal pumps specifically designed for use with water containing solids such as sticks, stones, sand, gravel, and other foreign materials that would clog a standard centrifugal pump. Trash pumps are adequate for pumping water from a maximum depth of 20 feet. To remove water from a depth greater than 20 feet, a submersible pump should be used. *Figure 18* shows a portable trash pump.

CAUTION

Any time that pumps are to be left outside during cold weather, remove all water from the pump to prevent the water from freezing inside and cracking the pump housing.

Trash pumps are continuous flow pumps that use centrifugal force generated by rotation. The liquid enters the impeller at the center, or the eye, and the rotation of the impeller blades causes a rotary motion of the liquid. Centrifugal force moves the liquid away from the center. As this happens, the liquid's velocity increases until it is finally released through the discharge outlet.

Mud pumps – Most mud pumps (*Figure 19*) are diaphragm pumps designed for use where low, continuous flow is required for highly viscous fluids, such as thick mud, air mixed with water, and water containing a considerable amount of solids or abrasive materials.

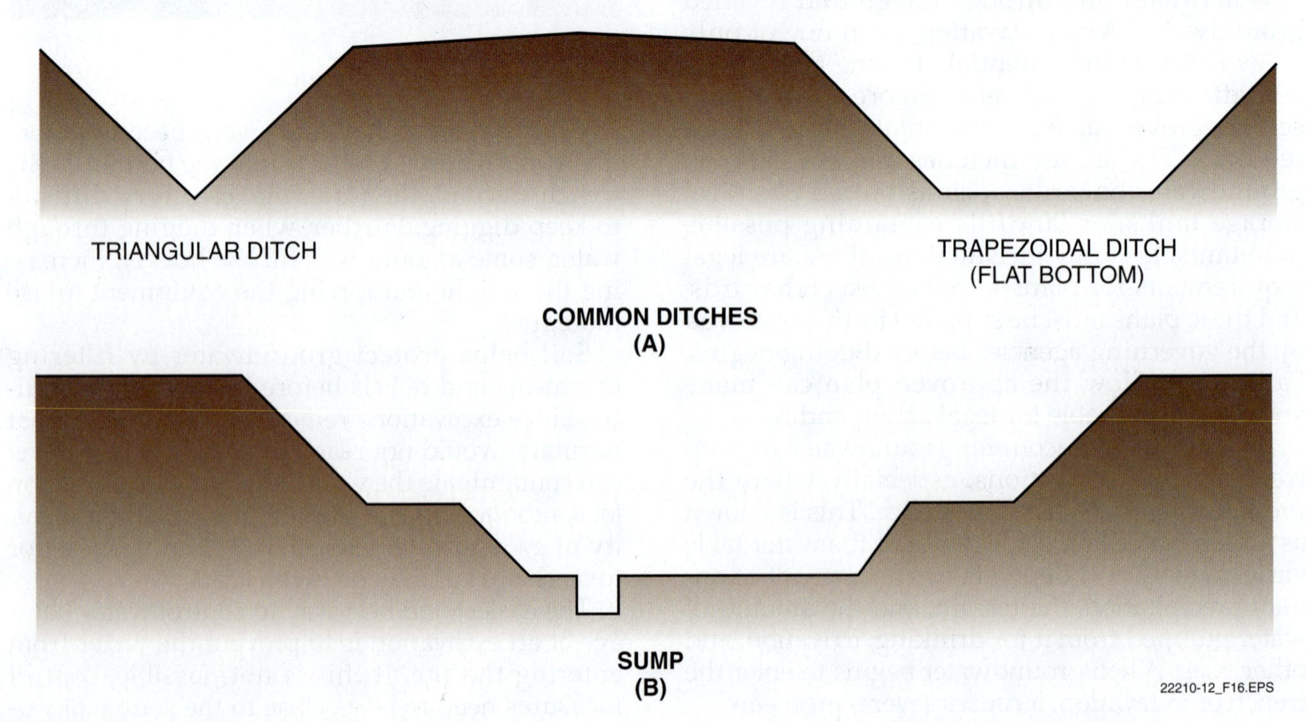

Figure 16 Groundwater control.

(A) STILLING BASIN

(B) INLET SEDIMENT TRAP

(C) TEMPORARY SLOPE DRAIN PIPE

(D) SLOPE DRAIN PIPE AND SILT CHECK DAM

22210-12_F17.EPS

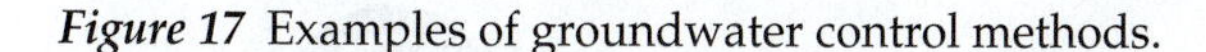

Figure 17 Examples of groundwater control methods.

Most diaphragm pumps use hydraulic pressure delivered by a piston to actuate the flexible diaphragm. The side of the diaphragm that produces the pumping action is called the power side, and the side doing the pumping is called the fluid side. A suction check valve and a discharge check valve open and close to move fluids through the pump. *Figure 20* shows the operation of a diaphragm pump.

Submersible pumps – A submersible pump (*Figure 21*) is encased with its motor in a protective housing that enables the entire unit to operate underwater. Submersible pumps are used in many residential, commercial, and industrial wells. Submersible pumps can be controlled by an optional float switch located inside the well. The float switch controls the pump by manual or automatic control switches outside the well. The float switch is a mercury-to-mercury switch that is controlled by the position of an attached float. As long as the water level is at a height adequate enough to remove water from the well, the pump remains on. When the water level falls too low for pumping, the float drops and turns the pump off. As the water level increases, the float rises and turns the pump on again. *Figure 22* shows a submersible pump controlled by a float switch.

2.2.3 Digging in Wet Soil

Water always seeks the lowest level, so it is not unusual for the bottom of an excavation to be covered with water from either ground or surface water sources. A small amount of water can actually be helpful to ensure a level bottom grade. However, too much water saturates the soil and makes work hazardous.

If a large amount of water is found when digging out the soil, the job can take much longer than expected and the equipment is forced to

22210-12_F18.EPS

Figure 18 Portable trash pump.

22210-12_F19.EPS

Figure 19 Mud pump.

work harder, thus increasing cost. The water can run back into the excavation, setting up a never-ending cycle of digging. Soil mixed with water is more liquid than dry soil so spoil piles may slide back into the excavation. It may be necessary to move them further from the open excavation than planned, thus increasing cycle time.

Under some conditions, the best solution would be to pump the water out of the excavation before digging. Pumping is an extra step that delays excavations and increases costs, so there are a number of other actions that may get the job done. These include compartment digging and using a spoil barrier.

Compartment digging – On a large excavation where a small amount of water is present, the water can be confined in a compartment, and thus kept out of the bucket. A hole must be dug in the exca-

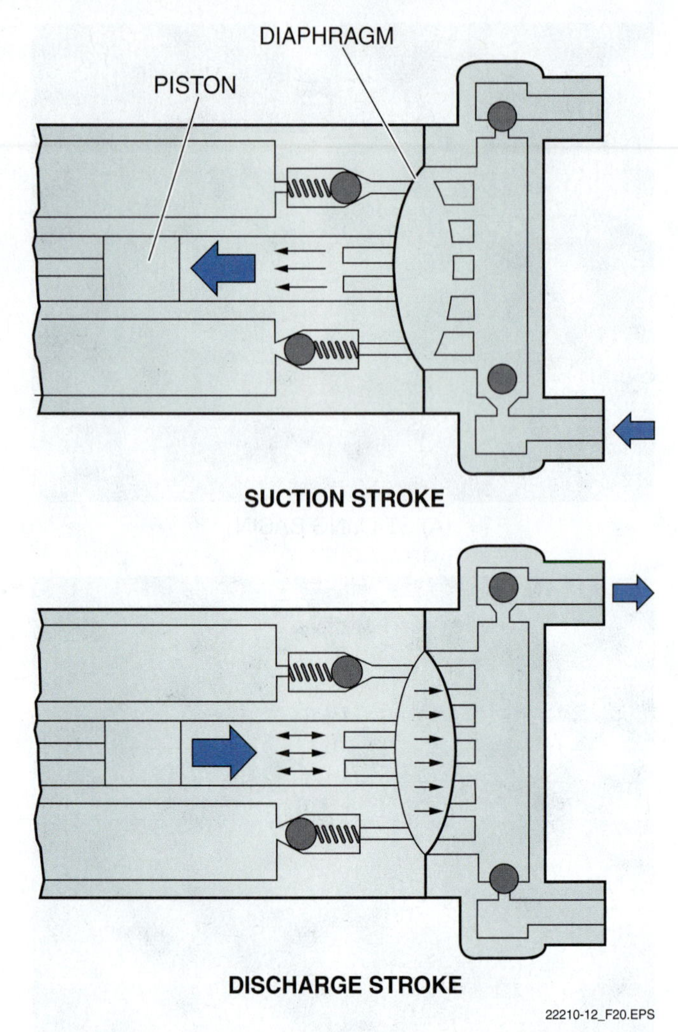

22210-12_F20.EPS

Figure 20 Operation of diaphragm pump.

22210-12_F21.EPS

Figure 21 Submersible pump.

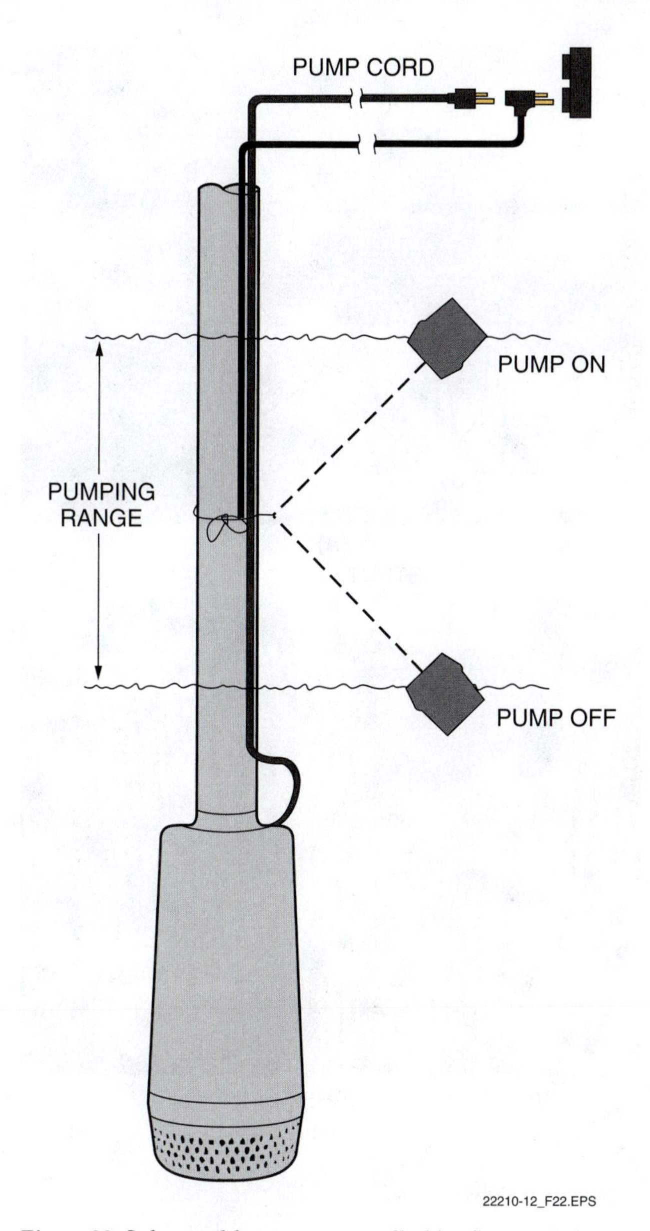

Figure 22 Submersible pump controlled by float switch.

vation large enough to hold the water. The floor of the excavation is sloped so the water runs into the hole. When the water has drained into the hole, the remainder of the excavation can be dug to grade, leaving a wall of soil between the water compartment and the area where digging is being done. Once at grade, break through the wall and allow the water to flow into the deep side of the excavation, then dig the compartment floor to grade. This process may need to be repeated several times, alternating sides to achieve the correct grade.

Spoil barrier – Muddy spoil is a problem. It can easily flow back into the excavation, increasing the amount of work that needs to be done. The best way to handle muddy spoil is to move the spoil pile far from the excavation. When space is limited, this is not a possibility. When dry soil is available, it is possible to build a dike with dry soil that will act as a barricade between the excavation and the spoil. In this case, the dry soil can be piled a safe distance from the excavation, and the muddy spoil can be placed behind it. Keep in mind that this method will increase the cycle time (and thus costs), because the equipment needs to travel further from the excavation to dump the spoil.

NOTE

When muddy spoil is loaded onto a truck for immediate transport from the site, ensure that the floor of the bed is covered with dry soil so the load does not stick when it is dumped.

2.3.0 Working Around Water

Some construction projects are performed on or near water. Anyone working in such an environment needs to follow some basic safety guidelines. Sometimes work needs to be done on building or bridge foundations that are underwater. In this case, there must be some means of draining and controlling water so that workers can accomplish their work safely. Two structures that are used under these conditions are coffer dams and caissons.

2.3.1 Safety Guidelines

The first rule to working safely around water is to stay as far away from the water as possible. Heavy equipment operations are risky on level dry ground, but operations can become perilous near the water. When working around water, follow these guidelines:

- Understand that some municipalities have laws and regulations about heavy equipment entering water, such as streams, lakes, and rivers. Ask your supervisor for instructions before entering water.

NOTE

Special permits may be required when working in or around water. Check federal, state, and local regulations.

- Select the correct equipment for the job. If equipment is available that permits the work to be done from dry ground, use it, but never overextend the reach of any equipment.
- Before approaching any water body with heavy equipment, carefully examine the terrain for obstacles, depressions, and soft spots. Walk the area if possible. Observe the water current before driving the equipment into the water.

- Before approaching the water, inspect the equipment to ensure that it is in good working order and clean of mud or any chemical residues such as oil that can be washed off into the water. Make a plan about what to do in case of mechanical problems.
- Before entering the water, assess the task. Think about what could go wrong and be prepared.
- While in the water with the equipment, use the buddy system. Have a responsible person nearby who can help in case there is a problem.
- When moving equipment through water, move cautiously. Water masks obstacles and hazards.
- Once in the water, perform the task as quickly and as safely possible to minimize the time spent in the water.
- Special precautions are required when working around open water. These precautions include wearing a flotation device, having a throw ring available, and having a boat of some type to perform a rescue in case of a man-overboard situation.

2.3.2 Coffer Dams

A coffer dam is a temporary structure used to keep water out of a construction site so that work can be done in a dry area. Although most coffer dams are built to create a dry work space in a waterway, some are used to prevent ground water from entering work sites. This is common in the construction of roadways, as well as buildings that need deep foundations. *Figure 23A* shows a coffer dam under construction. *Figure 23B* shows the same coffer dam completed and in the process of being drained.

Coffer dams are generally built in place, but may be prefabricated and dropped into place using cranes. That type is known as a *gravity dam*. Once coffer dam construction is complete, water is pumped out of it. *Figure 24* shows workers inside the drained coffer dam previously seen in *Figure 23*. Coffer dams used in the construction of piers, docks, bridges and the like are usually made from sheet piles that fit together tightly enough to keep water infiltration at a minimum. If it is necessary to make the work area completely dry, there are various methods available to seal the interlocking joints in the sheet piles.

Sheet pile systems are popular because the piles are easy to install and remove and can be reused. A braced coffer dam like the one shown in *Figure 23* is made of a single row of sheet piles. Because the coffer dam must be able to withstand the pressure of the surrounding water, it is braced with walers

Figure 23 A coffer dam.

Figure 24 Completed coffer dam.

and struts. Braced coffer dams are commonly used for bridge piers and other structures in shallow water. The coffer dam in *Figures 23* and *24* is being used for construction of a boat-launching ramp.

A double-walled coffer dam (*Figure 25*) has two parallel rows of sheet piles. The sheet piles are driven into the bottom and connected with heavy-duty tie rods. The space between rows is often filled with earth removed from the excavation. The walls of double-wall coffer dams are often spaced widely apart so that the space between them can be used as a roadway for construction vehicles. Double-walled coffer dams are used in locations where a large area is to be excavated.

When metal sheeting is used to build a coffer dam, the sheets are usually lifted into place with a crane and then driven into the floor of the waterway with a vibratory hammer pile driver (*Figure 26*). The sheets must be driven in to a point that is solid enough to prevent seepage, and then supported with bracing to counteract the force of the water. Once the area is enclosed, the water is pumped out so that work can begin. An excavator with the bucket removed is sometimes fitted with a special attachment and used to drive sheet piles. This approach is common in situations where there is not enough overhead space to accommodate a crane, such as the underside of an elevated roadway.

Wooden piles can be used instead of metal piles. Construction is done in a similar fashion. Earth and rock coffer dams are also used. An earthen dam is built by dumping earth fill into the water and shaping it to surround the construction site. This is slow work and requires a lot of fill, so it is not used often. Earthen coffer dams are usually restricted to shallow areas in waterways that have a very low current, because the earth tends to be washed away in strong currents.

2.3.3 Caissons

Caisson is French for box. Caissons are self-contained boxes or chambers that are used so construction work can take place underwater. Caissons are lifted into the water with a crane or other suitable equipment, floated on the water to the place they are needed, then sunk, and pumped dry. Caissons that are used in shallow water are usually open at the top and bottom, but caissons that are used in deep water can be quite sophisticated—these caissons are completely enclosed and pressurized with air to keep water out. *Figure 27* shows an example of a caisson being used for construction of a bridge abutment.

22210-12_F25.EPS

Figure 25 Double-wall coffer dam.

22210-12_F26.EPS

Figure 26 A crane driving sheet piles for a coffer dam.

WARNING!

Coffer dams and caissons should be treated as confined spaces, in that they must be monitored for proper oxygen levels and the presence of toxic gases or other airborne contaminants. It is possible that toxic gases such as methane can bubble up from the bottom of the excavation, for example. If any airborne contaminants exist in the excavation, personnel must be equipped with the proper respiratory equipment. Multigas monitors should be used to monitor for the presence of toxic gases.

22210-12_F27.EPS

Figure 27 Example of a caisson.

Additional Resources

Soil properties that affect groundwater: **www.co.portage.wi.us/groundwater/undrstnd/soil.htm**

2.0.0 Section Review

1. In what situation would a road be built without a crown?
   a. In a subdivision
   b. In the desert
   c. On steep hills
   d. Never

2. Trash pumps are adequate for pumping water from a maximum depth of _____.
   a. 5 feet
   b. 10 feet
   c. 20 feet
   d. 50 feet

3. Which of the following is used to keep water away from a bridge foundation while it is being repaired?
   a. Trench box
   b. Shielding system
   c. Detention system
   d. Coffer dam

Section Three

3.0.0 Establishing Grade on a Job Site

Objective 3

Explain how grades are established on a job site.

a. Describe how to set grades from a benchmark.
b. Describe how grades are set for highway construction.
c. Describe how grades are set for building construction.
d. Explain how grading operations are performed.
e. Describe the use of stakeless and stringless grading systems.

Performance Tasks 1 and 2

Interpret layout and marking methods to determine grading requirements and operation.

Set up a level and determine the elevations at three different points, as directed by the instructor.

Trade Terms

Balance point: The location on the ground that marks the change from a cut to a fill. On large excavation projects there may be several balance points.

Boot: A special name for laths that are placed by a grade setter to help control the grading operation. The boot can also be the mark on the lath, usually 3, 4, or 5 feet above the finish grade elevation, which can be easily sighted. This allows the grade setter to check the grade alone instead of having to use another person to hold a level rod on the top of the grade stake.

Stations: Designated points along a line or a network of points used to survey and lay out construction work. The distance between two stations is normally 100 feet or 100 meters, depending on the measurement system used.

String line: A tough cord or small diameter wire stretched between posts or pins to designate the line and elevation of a grade. String lines take the place of hubs and stakes for some operations.

Topographic survey: The process of surveying a geographic area to collect data indicating the shape of the terrain and the location of natural and man-made objects.

Finish grading is normally done with graders. Dozers and scrapers are also used in grading operations. The site plan for a project shows the topographical features of a site, as well as project information such as the building outline, existing and proposed contours, and other information that is relevant to the project. Based on this information, survey teams set grade stakes. Effective grading operations depend on proper staking of the project and the ability of the equipment operators to follow the grading instructions on the stakes. The grading work begins after a survey crew places the initial grade stakes. Each phase of construction may require more grading work, so the survey team sets new stakes at each phase.

Before construction starts, a **topographic survey** and property survey of the site are performed. This information is used to develop the project plans. The plans include information about elevation requirements at the construction site. Based on the plan, stakes are set by the project survey team to inform workers of the earthwork needed to bring the building site to the specified elevations. Methods for staking grades vary greatly with the type, location, and size of project. Methods used by different engineering and construction organizations also vary; some companies use electronic equipment, while others use manual instruments. Regardless of how the stakes are set, unless the project is very small, they be set in reference to benchmarks.

3.1.0 Using Benchmarks to Establish Grade

Benchmarks are used to show a precise elevation in relation to sea level. On building projects, benchmarks are used as reference points from which grade stakes are set. Benchmarks may be permanent or temporary and can be set by federal or local governments or by the project survey team. Benchmarks must not be disturbed. It is very important to know the location of any benchmarks in the area so that you can avoid them with your equipment.

Control Points

Good practice dictates that a control point should always be placed so that three other control points are visible from it at all times. This is done in case the control point becomes covered up or destroyed. If this happens, two tapes can be stretched from the other points in order to relocate the control point.

3.1.1 State Plane Coordinates

Benchmarks on the state plane coordinate system are permanent reference points set by the federal government. They are permanent markers placed in specially designed locations and have specific reference information engraved on a bronze cap (*Figure 28*). These markers are part of the National Geodetic Survey, which is part of the National Oceanic and Atmospheric Administration (NOAA). These markers are frequently accompanied by a witness post to make them easier to find.

The markers provide an accurate reference point for other surveying activities within that general area. Many such markers are located throughout the United States. They are usually set in concrete on the ground to help preserve them. It is important that the markers not be moved or otherwise disturbed. If a permanent marker is damaged it may not be used as a benchmark. It is important to report it to a supervisor or the project engineer. The supervisor or engineer will notify the proper agency so that it may be reset.

States may set similar markers in accordance with their own specifications. State markers are also intended as permanent markers, so if there is one in the work area that is not recorded on the project plan, do not disturb it. Put some type of stake or marker next to it so that you are able to find it again, and notify the project engineer. The engineer will notify the proper authorities to see if the monument needs to be relocated by a survey crew before any further work is done in that area.

3.1.2 Permanent Project Benchmarks

Sometimes a project is developed over a long period. If additional construction is scheduled in the future, there needs to be a control point established close to the construction site so that new construction work can be easily referenced to the previous work. If no other benchmarks are close by, then a permanent project benchmark must be established. This benchmark may be a special cap of some type set in concrete on the ground or on the side of a structure, such as a building or bridge abutment.

3.1.3 Temporary Project Benchmarks

Temporary project benchmarks can be placed in the ground, on a structure, or on a large sturdy tree. They are used only for the specific project. The location and elevation of these benchmarks is usually derived from permanent benchmarks on the state plane coordinate system described earlier. *Figure 29* shows an example of how temporary project benchmarks are placed.

3.2.0 Highway and Other Horizontal Construction

For rough grading on highway and other horizontal construction, there are several methods to set grade control depending on the type of construction and the surrounding terrain. The types of stakes used in highway construction depends on the number of lanes and the existence of curbs and medians.

22210-12_F28.EPS

Figure 28 Benchmark of the National Geodetic Survey.

22210-12_F29.EPS

Figure 29 Temporary project benchmark.

3.2.1 Setting Highway Grade Control

For road construction, reference stakes are located outside the construction limits, and are the first stakes set by the project survey team. They are used by the contractor as a reference to perform rough grading. These stakes are located outside the work area, so it is unlikely that they will be damaged when the area is being prepared for construction.

The center line stakes are set by the survey crew using benchmarks and control points. Because the location and elevation of each control point is known, all center line stakes can be referenced directly or by referencing one stake to the stake behind it until the center line is tied into another control point further up the line. This tie-in to a second control point provides a double-check on the center line staking process, as well as any other stakes that must be precisely set. *Figure 30* shows an example of a highway plan and profile sheet that includes references to a benchmark used on this job. The benchmark is located at station 109+16.00.

All members of the construction team use the drawing set, so sometimes the plan and profile drawings look confusing (refer to *Figure 30*). The profile portion of the road project is shown at the bottom of the drawing. On the right side of the sheet, the drawing shows the center line profile of the road. The drawing ends at station 115+00 with an elevation of about 2,842 feet (read from the elevation scale located on the right side of the drawing). The center drawing also shows the center line profile of the road. It starts at station 115+00 at an elevation of about 2,842 feet. The center line profile ends at station 110+20.89, which is written on the drawing, at an elevation of about 2,847 feet. The left side of the drawing does not show the roadway center line. The two profile curves shown on this portion of the drawing represent the sides of the road. The bottom drawing is the right side of the road and the top one is the left side of the road.

If reference stakes are used for rough grading, they can be placed relative to the center line stakes or referenced directly from the benchmarks. They are placed at the right-of-way limits at the distance from the center line shown in the plans. Because the stakes are placed at the right-of-way limit, there is little chance that they will be disturbed during earthmoving operations. These stakes are usually placed at each **station** (every 100 feet) but may be placed at closer intervals on corners and alignment changes.

The cut or fill amount is indicated on each stake. This is the vertical distance between the profile grade stake and a line on the stake. *Figure 31* shows a typical cross-section of a cut. The stake shown in the figure identifies the station, amount of cut, and the point from which the cut or fill is measured. Most slope stakes would be marked in a similar manner.

Slope stakes, which can be used as cut and fill stakes, are set by the survey team at points where the cut slopes and fill slopes intersect a hinge point. For rough grading, it is unlikely that the survey team will set cut and fill stakes, but the team will probably use some other type of marker to guide equipment operators in the initial cut and fill operation. Such temporary markers have no information written on them, but should be familiar to the equipment operator. If you are confused about the stakes, talk to our supervisor before you begin grading work. Finish grade stakes show information about the elevation and offset distance.

To visualize the grade for a center line, look at the profile sheet in the plans. *Figure 32* shows an example of the profile sheet and the cut and fill areas of the roadway. From Station 10 to Station 40+31.05, the proposed grade is higher than the existing grade, so the area needs to be filled. At Station 40+31.05 the existing grade crosses the proposed grade and both are maintained until Station 50+10.00. This area is called the **balance point**—it needs neither cut nor fill. At Station 50+10.00, the existing elevation is greater than the proposed elevation, which will require a cut to achieve the desired elevation.

The slopes of the proposed grade are also shown, along with the reference axis. The vertical axis on the profile sheet shows the elevation. Usually, the elevation is shown at a much greater scale than the horizontal axis. The horizontal axis shows the stations along the center line. Therefore, you can identify the points on the natural ground where the elevation of the proposed grade is the same as the natural ground.

3.2.2 Finish Grade Reference Points

When the roadbed is close to the proposed grade, the survey crew places finish grade stakes to guide the final grading operation. If any type of stabilization of the subgrade is required, then the subgrade has to be mixed, regraded, and compacted to the proper density. After compaction, the subgrade must again be shaped to conform to the finish grades and cross-sections shown in the plans. This finish work is often performed by motor graders.

By setting a **string line** at a convenient height above each grade stake, measurements can easily

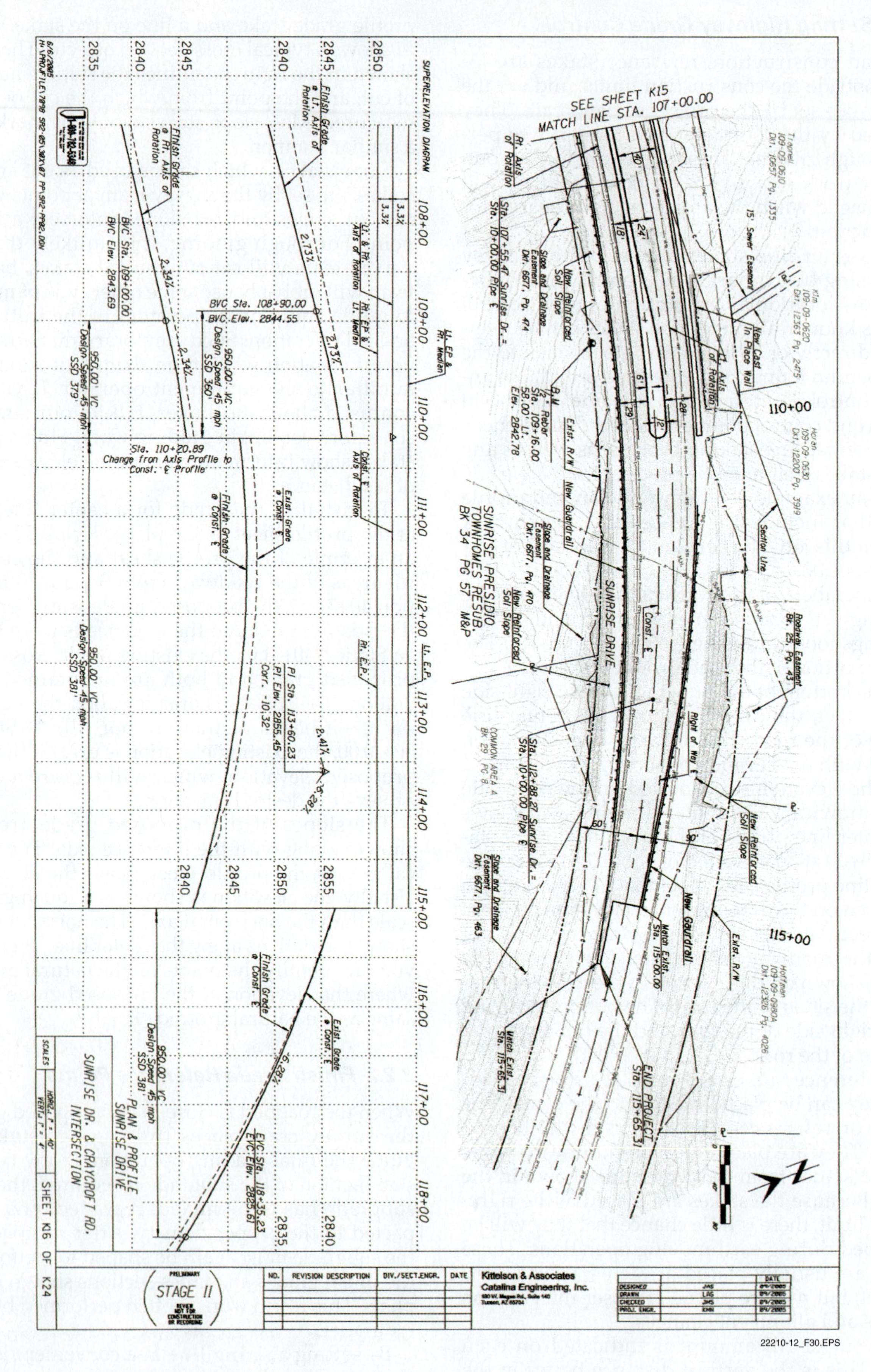

Figure 30 Highway plan sheet showing benchmark reference.

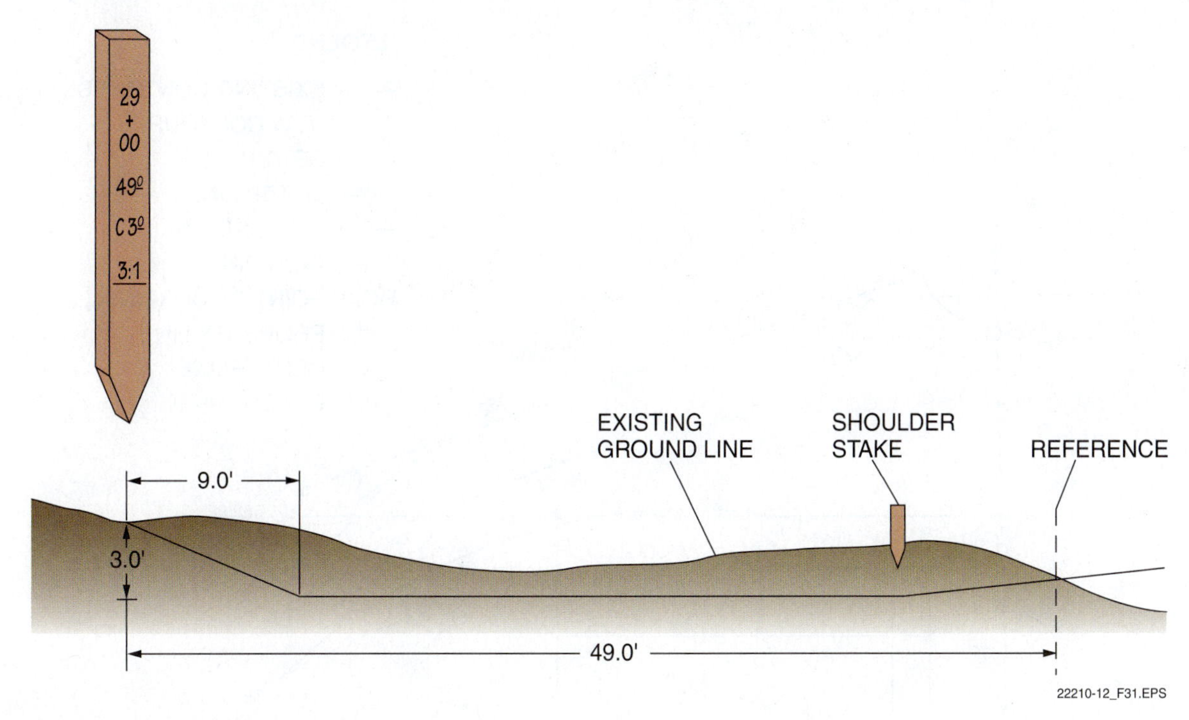

Figure 31 Cross-section of a cut.

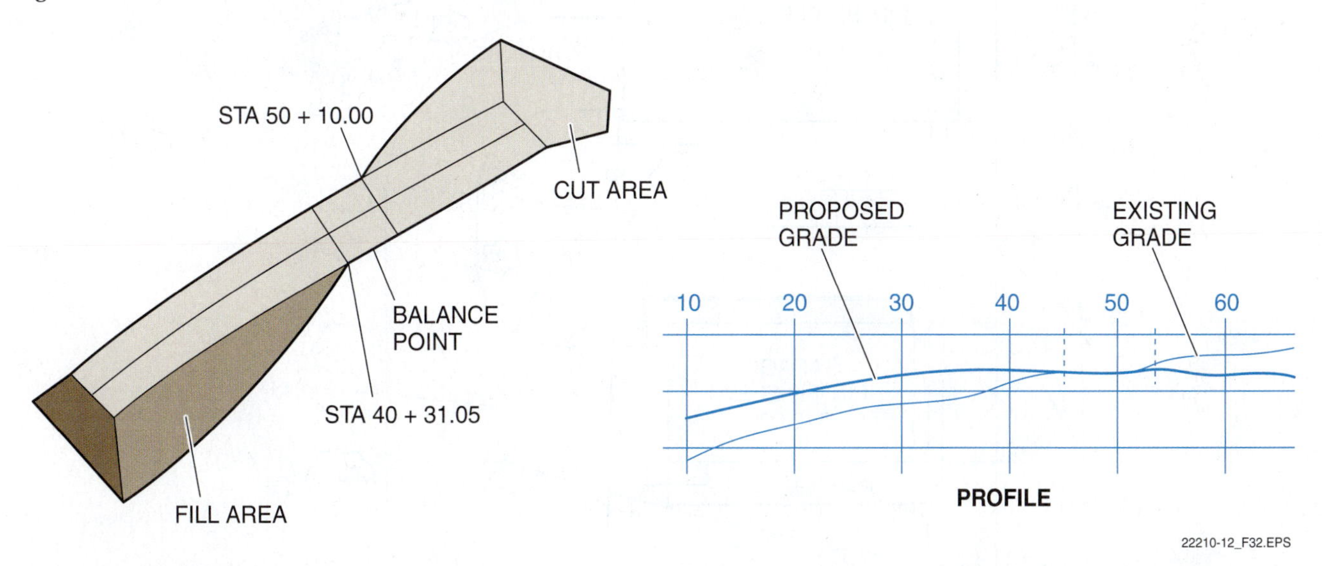

Figure 32 Example of cut and fill sections.

be made down to the finished subgrade surface. To check the lines of the completed subgrade, little flags or other markers can be attached to the string line at the edge of the pavement, shoulder, or other appropriate points.

3.3.0 Building Foundations and Pads

Site plans vary for buildings and other structures. A large building or development will usually have a detailed plan that shows existing elevations and the finished elevation of structure's foundation. All site plans should show the general outline of the structure and the property boundaries (*Figure 33*). On small jobs, a site plan may not have detailed elevations, so you may need to ask the site supervisor for verbal instructions before you begin earthwork operations.

Study the site plan before starting work in order to verify the locations of the property boundaries and to avoid crossing them. Do not cross a property line without permission. The site plan may also show the existing elevation of the property, so the operator or the site supervisor can determine the amount of excavation needed to ensure that the finished elevation of the building foundation is correct.

For structures with foundations, temporary stakes or markers are usually set at the corners as a rough guide for starting the excavation. These

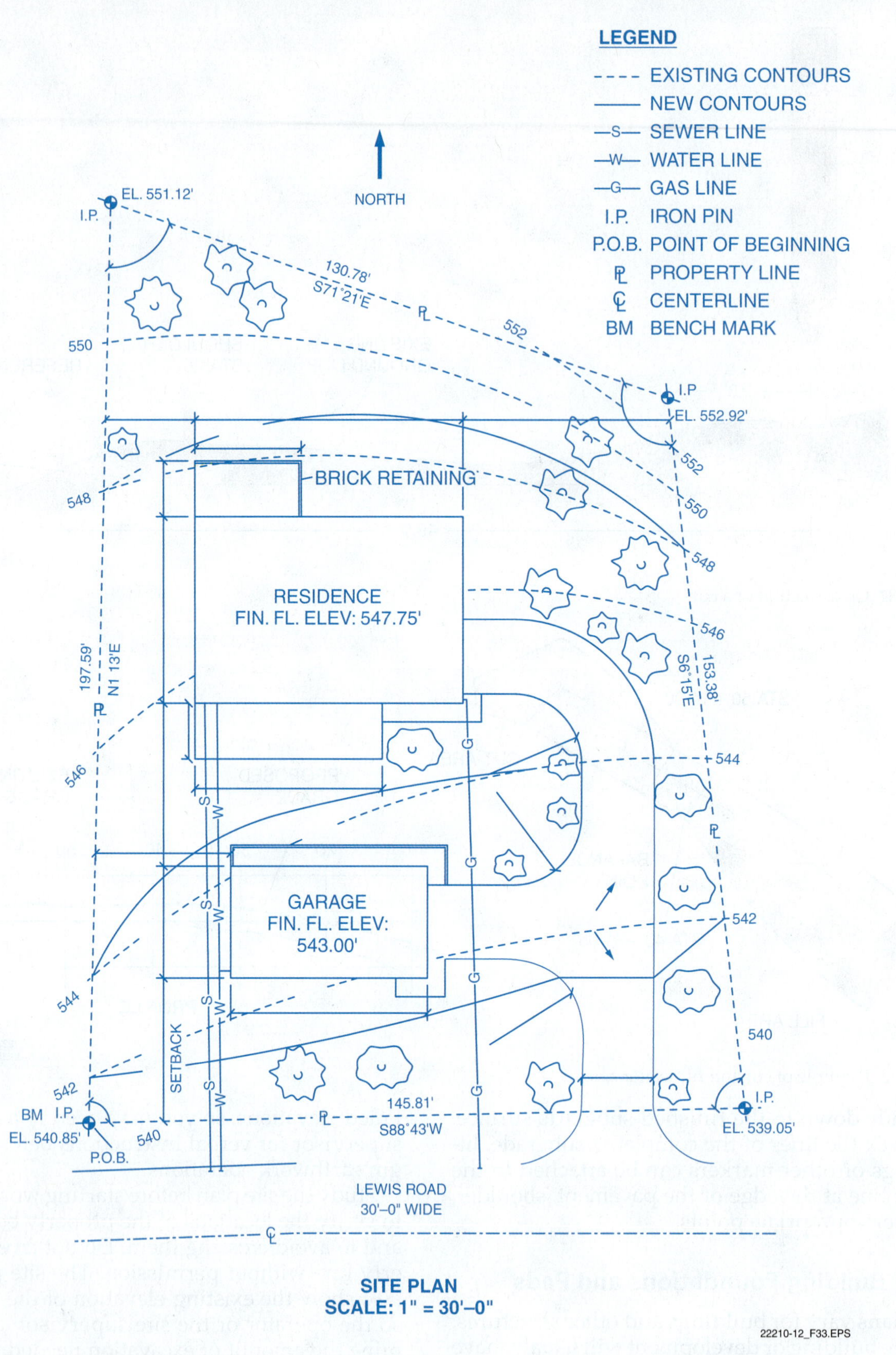

Figure 33 Site plan showing topographical features.

markers are set outside the construction limits or area of disturbance, but close enough to be convenient. Permanent control points are set to allow for measurements during the construction process.

Stakes or other markers are set on important lines to mark clearly the limits of the work. For small building foundations and trenches, grade information is usually provided using batter boards.

3.3.1 Establishing the Proper Grade

To establish the proper grade for an excavation or fill, the grade setter must locate a benchmark or hub set by the surveyor. (Typically, a surveyor will set hubs for a project.) The grade setter will then begin setting **boots**. A boot is a mark on a stake or lath at a convenient height that can be used as a reference elevation relative to the finished subgrade, pavement, or structure foundation. *Figure 34* shows a typical boot placed by the grade setter. The boot will be 3, 4, or 5 feet above the finished grade. It provides a convenient height for the grade setter to view with a hand level.

For subdivision, office park, and commercial sites, the grade setter sets boots based on the finish grade, then adds the curb or road subgrade information when checking grade from these boots. After placing each grade mark, the cut needed to reach subgrade will be written on all the cut stakes.

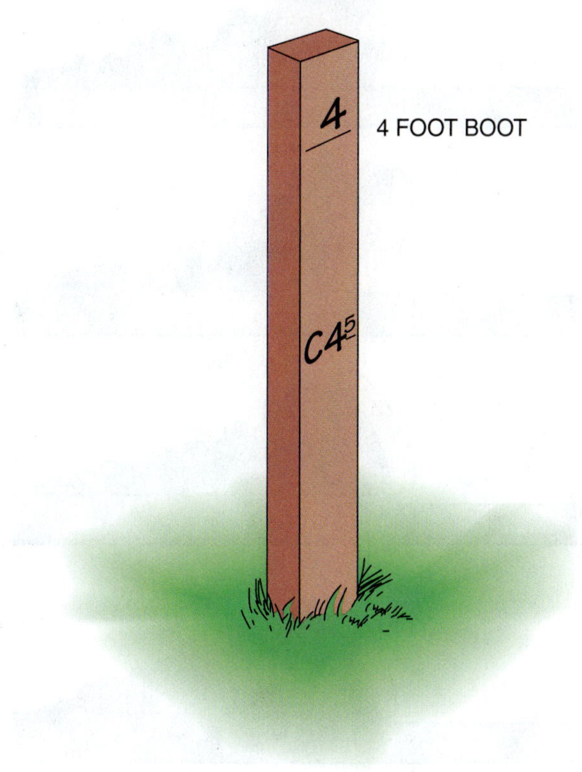

Figure 34 Grade setter boot.

This way the equipment operator can see what cut is needed directly without having to add the curb and road section height to the surveyor's cut marks.

3.4.0 Grading Operations

Some grading operations are automated. A computer-controlled grader is driven over the work site, and the computer adjusts the level and angle of the blade based on a topographical survey and Global Positioning System (GPS) coordinates. For grading operations that are not automated, the successful and efficient completion of the job relies on the skills of the equipment operators. The information presented in this section can provide you with knowledge that can help you develop grading skills, but the only way that you can actually develop grading skills is through experience. Keep the following points in mind while you are learning to cut and fill grades:

- One of the most common problems new operators have is that they are reluctant to make deep cuts. While you are learning this is wise, but skilled operators use their equipment to the fullest. Experienced operators are able to cut to grade with a minimum of passes by keeping the blade or bucket full at most times.
- After each pass, evaluate your work. Did you leave too many uncut areas that require a second or third pass? If so, adjust the blade or bucket attachment.
- When cutting close to grade, it may be wise to make smaller cuts and more passes, rather than risk cutting too much. It is better to make a second pass with a blade than it is to cut too much material, which will require that fill and a compactor be brought in to fix the cut.
- Always follow the site plans. Many sites have a stormwater control plan to help prevent erosion and sedimentation. Be sure to know and follow the plans. When you are not sure, ask your supervisor.
- Even if the site does not have a stormwater control plan, avoid driving the equipment across graded areas to avoid increased sedimentation.
- Never cut in an area that does not need grading. This can increase sedimentation and erosion.
- When performing rough grading or filling, be sure to leave enough material to trim during finish grading.
- When filling areas, thin layers of fill are easier to spread, have fewer air holes, and compress more easily than thick layers of fill.

Motor graders are specially designed to be used in many grading operations. All graders have blades that are used to cut and level grading material. The angle and height of the blade can be adjusted from the cab and the frame is designed to keep the blade stable even when driving over rough terrain. The grader's front wheels can be tilted to the right or left, which helps to increase the steering ability of the grader while cutting.

3.4.1 Grader Blade

The grader blade (*Figure 35*) is made up of a moldboard, endbits, and a cutting edge. The moldboard is shaped so that grading material roll and mixes as it is cut. The end bits and cutting edge protect the moldboard edges from the abrasive action of the grading material. Inspect the blade daily and replace the endbits and/or cutting edge as necessary.

The grader blade may be positioned in the center of the grader frame or off to one side (*Figure 36*). The blade may be set perpendicular to the frame or it may be angled. When the blade is angled, the front edge is called the toe and the rear edge is the heel. Grading material spills off the heel of the blade.

> **CAUTION**
>
> When the blade is set at a sharp angle, it may hit a tire when the grader is turned, causing tire damage. Use caution when operating a grader.

The pitch of the blade can be adjusted for the desired results. For most grading operations, the blade will be upright (*Figure 37A*), but for more cutting, the blade may be pitched back slightly (*Figure 37B*). To increase the mixing of the graded material, the blade is pitched slightly forward (*Figure 37C*). Pitching the blade forward sharply (*Figure 37D*) will help to compact the graded material and ensure that low spots are filled.

22210-12_F35.EPS

Figure 35 Grader blade.

3.4.2 Grader Wheels

A grader's front wheels are positioned to help stabilize the grader while cutting, and to make it easier to steer while grading. The wheels are usually tilted in the direction of the heel of the blade. In

22210-12_F36.EPS

Figure 36 Grader with blade positioned slightly to the side.

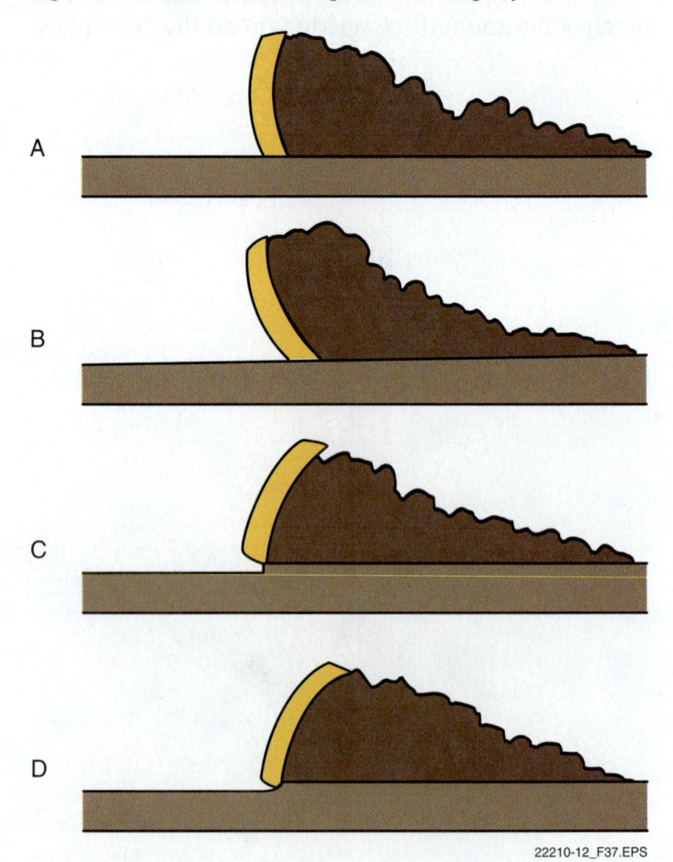

22210-12_F37.EPS

Figure 37 Blade pitch.

Figure 38, the grader is forming the ditch slope. The blade is pushing dirt on the right side. The force of the dirt on the blade would normally tend to pivot the grader into the ditch. By tilting the wheels away from the ditch, the loader stays on a straight course.

3.4.3 Cutting a Ditch

Most grading operations require several passes to complete. As discussed previously, the most efficiency is achieved when the operator is skillful enough to keep the blade full and to complete the operation with a minimum number of passes. Experienced operators can often cut a ditch with two or three passes. These operators are familiar with the equipment and have developed a sense for the soil conditions. Initially, it is best to remove the smallest amount of material on your first pass and then gradually increase the depth of your cuts until you are comfortable with your skills.

When cutting a ditch, the first pass is called the marking cut (*Figure 39A*). It is a 3- to 4-inch deep cut made with the toe of the blade and it is used as a guide to help you to cut a straight ditch. In the second cut, the blade is angled more and positioned over the marking cut (*Figure 39B*). When the second cut is made, more material is deposited onto the road from the heel of the blade. At some point, the cut graded material needs to be spread toward the center of the road—this is called shoulder pickup (*Figure 39C*). These steps are repeated until the ditch is cut to the desired grade.

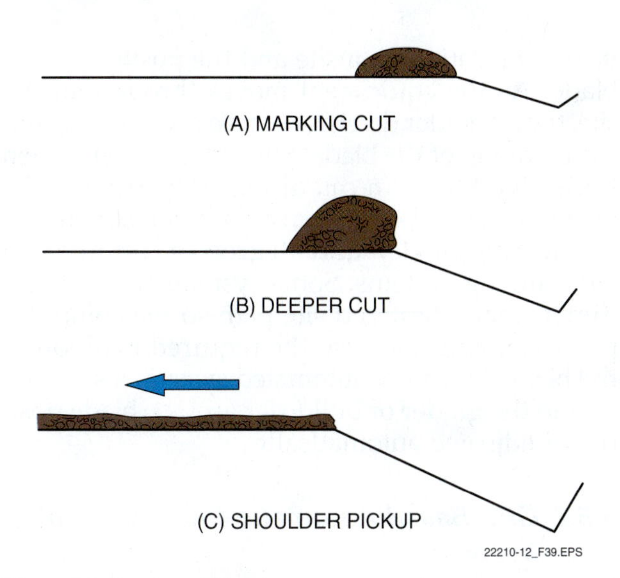

Figure 39 Cutting a ditch.

3.5.0 Stakeless and Stringless Grading Systems

At one time, site layout and grading were done by manually setting stakes and strings. This method is being rapidly replaced with methods that rely on laser and satellite technology that make it possible for grading operations to be partially or fully automated. Using computer technology, entire projects can be completed without setting a single grade stake. Automatic grade control systems can use lasers, the GPS, or both to get a job done faster and more accurately than conventional staking, thus saving time and money.

3.5.1 Laser-Based Automatic Grade Control Systems

Laser-based automatic grade control systems use complicated technology, but their operation is very simple. *Figure 40* shows a typical automatic grade control system. Signals from an off-board laser transmitter are used as the reference. The grader or bulldozer is outfitted with electronic components that permit a computer processor to determine

22210-12_F38.EPS

Figure 38 Tires tilted to cut a ditch.

22210-12_F40.EPS

Figure 40 Caterpillar Accugrade laser grade control system.

its exact location on a site and the position of the blade. As the equipment moves throughout the site, the computer determines its exact position and the elevation of the blade. This information is then transmitted to an in-cab display (*Figure 41*) so the operator can make necessary blade adjustments.

The complexity and degree of automation vary among systems. Some systems use a three-dimensional digitized site plan so the computer processor can calculate the required position of the blade. In a fully automated system, a signal is sent to the grader or bulldozer and the blade position is adjusted automatically.

3.5.2 GPS-Based Automatic Grade Control Systems

GPS-based automatic grade control systems operate in a similar fashion to the laser-based systems, except the reference signal is from the GPS. The GPS was developed by the US military to provide precise position information for vehicle, troop, and equipment movement. It is made up of a series of satellites that circle Earth at various orbits. These satellites broadcast radio signals that contain information about satellite locations for 24 hours a day.

GPS-based systems are especially helpful on jobs that require a great deal of contouring. These systems compare the blade position to a three-dimensional digitized site plan to determine whether the blade position requires any adjustments. *Figure 42* shows a typical GPS-based automatic grade control system. The information can be displayed on an in-cab display (*Figure 43*). The operator can make the required blade adjustments or a signal can be sent to the grader or bulldozer control system to automatically adjust the blade.

Figure 41 Caterpillar Accugrade laser grade control system display and control.

22210-12_F42.EPS

Figure 42 Caterpillar Accugrade GPS grade control system.

22210-12_F43.EPS

Figure 43 Caterpillar Accugrade GPS grade control system display and control.

Additional Resources

Excavators Handbook Advanced Techniques for Operators. Reinar Christian. Addison, IL: The Aberdeen Group, A division of Hanley-Wood, Inc.

GPS Systems for the Earthmoving Contractor: **www.constructionequipment.com/gps-systems-earthmoving-contractor**

3.0.0 Section Review

1. A permanent benchmark shows a point's _____.
   a. exact elevation
   b. precise location
   c. exact azimuth
   d. precise value

2. A balance point is _____.
   a. the point of an excavation that changes from a cut to a fill
   b. the edge of a batter board where a level can be balanced
   c. when a job has used all of the cut material for fill
   d. the center of a benchmark that is marked by a tack

3. When you are working on a grading job for a new building, it is important that you know where the property boundaries are so that you can _____.
   a. park your equipment over the line
   b. be sure to grade right up to the line
   c. stay 10 feet from the boundary line
   d. avoid crossing over the property line

4. On a grader blade, which of the following is adjustable?
   a. Toe
   b. Heel
   c. Blade pitch
   d. Cutting edge

5. GPS-based systems are especially helpful on jobs that _____.
   a. are close to water
   b. require a great deal of contouring
   c. require three or more graders
   d. do not have a power supply

SECTION FOUR

4.0.0 PIPE-LAYING OPERATIONS

Objective 4

Describe grading and installation practices for pipe-laying operations.

a. Explain how grades are established for pipe-laying operations.
b. Describe the equipment and methods used to lay pipe.

Trade Term

Bedding material: Select material that is used on the floor of a trench to support the weight of pipe. Bedding material serves as a base for the pipe.

Setting the proper grade for pipe placement in trenches is very important because the drainage through the pipe is controlled by gravity. After the pipe is laid and the trench is backfilled, it is expensive to re-excavate if it will not drain properly. In addition, an uneven grade will put extra pressure on a pipe, causing it to break or leak at the joints.

WARNING!

Make sure the trench is barricaded if required and that proper protection is provided for anyone working in the trench.

4.1.0 Setting the Trench Grade

Setting the proper grade for pipe or other conduits can be accomplished in different ways. The most popular method involves using a laser level. Other methods include the use of a string line. Regardless of the grading method, a survey crew establishes hubs that can be used as control points for setting conduit or pipe. Before these hubs are set, a decision is made about how far the offset should be and on which side of the trench it should be placed. The width of the trench and the direction the spoil will be thrown usually determines these two factors.

Often, some type of **bedding material** is used to help set the pipe to the proper grade. This material can be any material that does not readily compress. However, it is important that the mate-

rial uniformly support the pipe to avoid damaging it. Many workers find that it is helpful to use a shovel or long pole to position the pipe once it is in the trench. The pipe should be placed so that it contacts the bedding all at once.

Once the pipe is positioned, the trench can be backfilled. Filling the trench with thin layers of fill rather than large loads helps to ensure that the pipe is not pushed off grade. It also helps to reduce air pockets and decrease the need for compaction.

4.1.1 Setting Grade for Pipe Using a Laser

The center line for a pipe trench is located on the ground with stakes or other marks at 50-foot intervals where the grade is uniform. In rough terrain, stakes are set at a much closer interval. Hubs are set parallel to the center line but far enough away so they are not disturbed by the excavation process. These stakes are placed using a manual transit or electronic survey device.

When the trench has been excavated, a laser pipe level (*Figure 44*) can be used to provide a grade line at some elevation above the bottom of the trench. The pipe laser receiver is set up at one end of the segment, and the transmitter is set at the other end of the segment to check the grade. The laser level can be used first to check the rough grading of the trench, and then to recheck it after the pipe has been placed in the trench, and before the trench is backfilled.

CAUTION

A laser level may not be a good choice if the work site is dusty or foggy. The laser beam is bent by dust and water particles and may read the incorrect grade. In addition, some laser levels do not operate well when part or all of the trench or pipe is very hot. Check the manufacturer's instructions for more information.

4.1.2 Setting Grade Using a String Line

String lines can be used to set grades if no laser level or other surveying instrument is available. String lines are more effective in adverse environmental conditions such as heavy fog or dust that can hinder the operation of a laser level.

The initial setup for the excavation is similar to the procedure described earlier. Once the offset has been determined, the survey crew places the required number of stakes. This information can then be transferred to the string line setup, as shown in *Figure 45*. It is necessary to decide how high above the trench the string line should be placed over the top of the trench and how far apart the batter boards need to be spaced. A recommended maximum space for the batter boards is 25 feet. The string line sags if support is spaced any further apart. The height of the string line depends on the terrain and the amount of slope required at the bottom of the trench. The height of the string line is set by measuring a given elevation above the required grade of the trench plus the offset distance for the line.

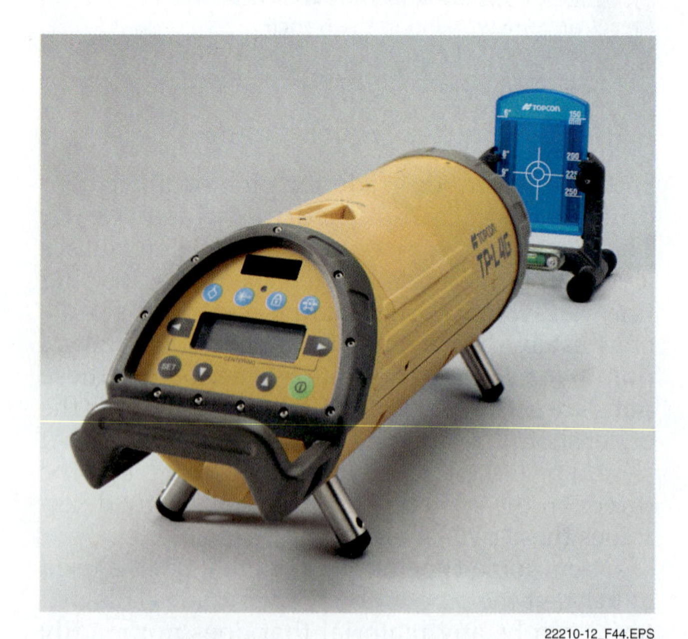

22210-12_F44.EPS

Figure 44 A laser pipe level.

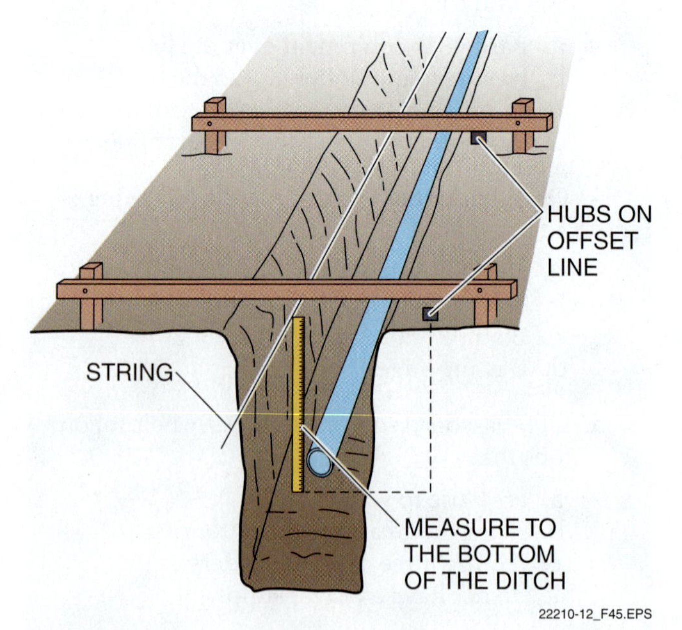

22210-12_F45.EPS

Figure 45 Using a string line to check grades.

To determine the exact elevation of the pipe, measure the distance from the string to the bottom of the trench, and subtract the height of the batter boards from the hub. Then subtract the result from the hub elevation. Repeat this process for each station.

When setting any string line, keep three stations set up along the line so you can sight down the line to correct errors or movement that might occur. As you sight down the string line, each station should blend with the next, with no sudden rise or dip from station to station. If you spot a sudden grade change, check all measurements. If you cannot find any error in your work, check the rate of slope shown on the plans. If the plans do not show a sudden grade change, it is possible the surveyors made an error that has been carried through to the grade line.

4.2.0 Laying Pipe

On large building sites, cranes are usually available to move and place lengths of pipe. But on most jobs, the pipeline is away from other construction, and it may not be practical to keep a crane around to lay pipe. On these sites, a backhoe, excavator, or other equipment may be used to lift and place pipe. There are a number of attachments available to make pipe laying easy. But even if your equipment does not have an attachment, you can still lay pipe (*Figure 46*).

22210-12_F46.EPS

Figure 46 Excavator laying pipe.

Use approved lifting devices to lift the pipe. Be certain to select equipment that is strong enough to lift the pipe safely. You also need several shackles (*Figure 47*) and rigging hooks (*Figure 48*).

> **WARNING!**
> Always make sure the rigging is rated for the load, rated for overhead lifting, and properly tagged. The working load limit (WLL) should be stamped on all items of rigging equipment. WLL was formerly called SWL (safe working load), which may appear on some rigging equipment.

Before starting, look over the job site for potential obstructions. Remember that once the pipe is lifted, it oscillates, creating a dangerous pendulum effect. Minimize the number of workers in the area. You must always know the location of any workers in the area. Ideally, set the pipe pieces as close to the trench as possible to minimize the length of time the pipe is in the air.

Because the pipe may not have attachment points, a choker hitch may be used to attach the sling to the pipe. Chances are good that a rigger will be attaching the pipe to the lifting device, but it is necessary to know how it is done so you can recognize improper rigging. Never lift pipe if you believe it is improperly rigged.

> **WARNING!**
> Remember that the choker hitch decreases the capacity of the sling by a minimum of 25 percent. Keep this in mind when selecting equipment.

The bucket of the backhoe or excavator has an eye or a pin to which the cable or chain can be attached using a shackle. Short lengths of pipe can be transported with a single cable or choker hitch (*Figure 49*), but any pipes over 12 feet in length should use a double choker hitch (*Figure 50*). Be certain that all hitch hardware is oriented properly. Also make sure that the load is balanced to minimize oscillations and to make the pipe easier to place.

Pipe should be lifted slowly and smoothly. Do not raise the pipe any higher than necessary. To help ensure stability, make sure a tag line is used so workers do not need to be close to the pipe. As the pipe is being lowered into the trench, the ground worker can control the alignment of the pipe with the tag line. The pipe must be held close to the grade line so that the entire length of the pipe contacts the ground simultaneously. If all of the weight rests on one end of the pipe, it will

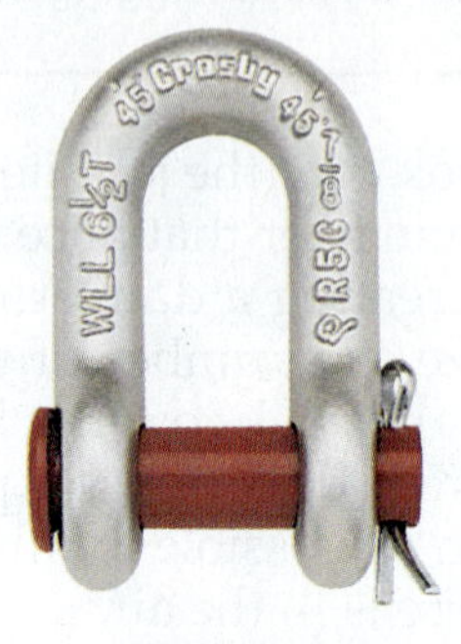

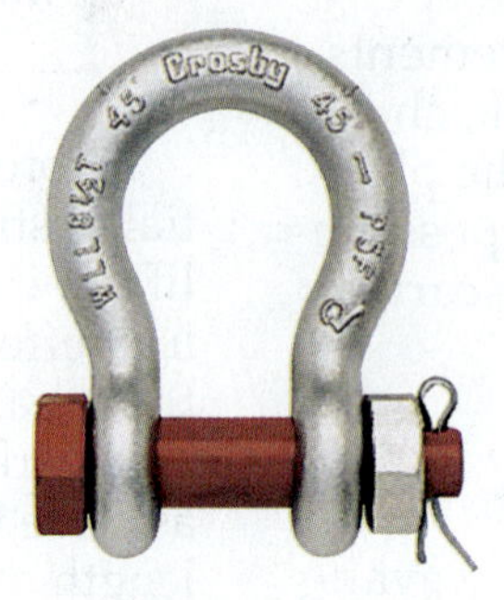

Figure 47 Shackles.

compact the soil in the trench. This can change the grade of the trench floor.

After each section of pipe is in place, a pipe level or builder's level is used to check the grade. At the end of each workday, the last joint of pipe is placed into position and the end is closed with a cap or a blind flange to prevent water, animals, or contaminants from entering the pipe. When a run of pipe is finished, a shutoff valve or blowoff assembly is attached. The run is flushed to remove any foreign material.

4.2.1 Using a Dozer Side Boom

Pipe booms or side booms are used extensively for heavy hoisting and carrying, particularly in pipeline work. The boom is attached to one side of the bulldozer with the hoisting mechanism and counterweights attached on the opposite side. A power takeoff drives two drums that are controlled through separate clutches and brakes. One controls the boom height while the other controls the load line that is attached to a hoist block. The counterweight is hinged so that it can be brought in close to the dozer for traveling and handling light loads. It can be extended away from the dozer by hydraulic pistons to counterbalance heavy loads on the boom.

There are two approaches to determining the safe operation of a dozer with a side boom. Both have to do with the load rating of the boom.

Where stability governs lifting performance, load rating at the minimum load overhang may be based on structural competence rather than stability. The margin of stability for the determination of load ratings, with booms of specific lengths at a given load for the various types of tractor mountings, is established by taking a percentage of the loads that will produce a condition of tipping.

Where structural competence governs lifting performance, load ratings are governed by the stability of the side boom tractor. This is the load required to tip the side boom tractor at a given load overhang.

The following procedure summarizes the steps for handling a load with a side boom:

- Make sure the weight of the load does not exceed the maximum load capacity.
- Ensure that the hoist rope is not kinked or twisted on itself.
- Ensure that the load is secured and balanced in the sling or lifting device before it is lifted more than a few inches off the ground.

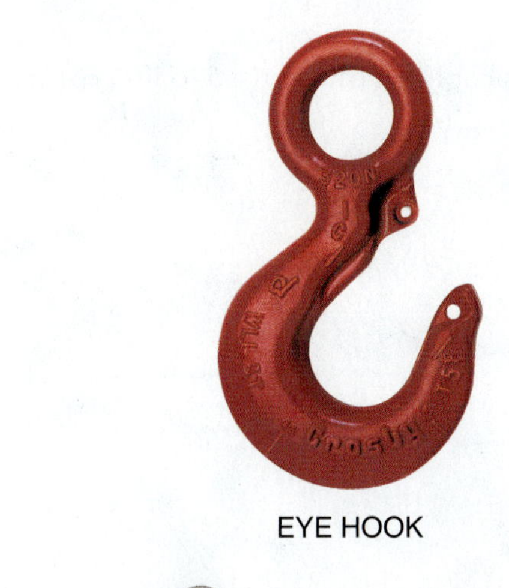

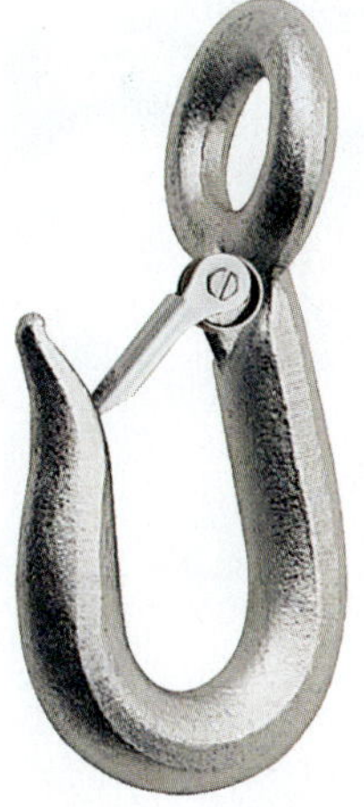

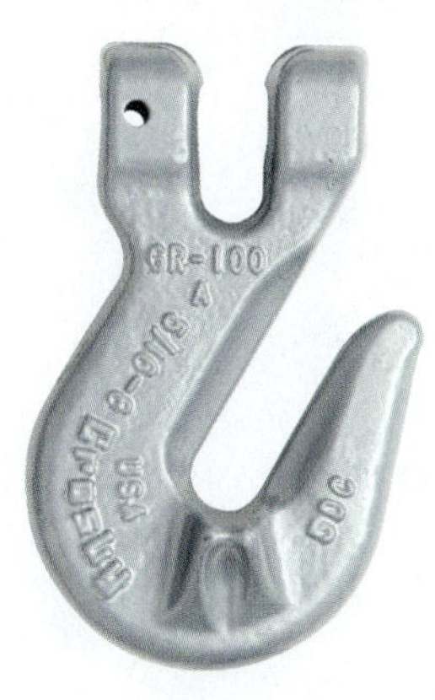

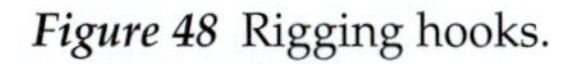

Figure 48 Rigging hooks.

- When two or more side boom tractors are used to lift one load, one designated person should be responsible for the operation. This person should analyze the operation and instruct all personnel involved in the proper positioning, rigging of the load, and the movements to be made.

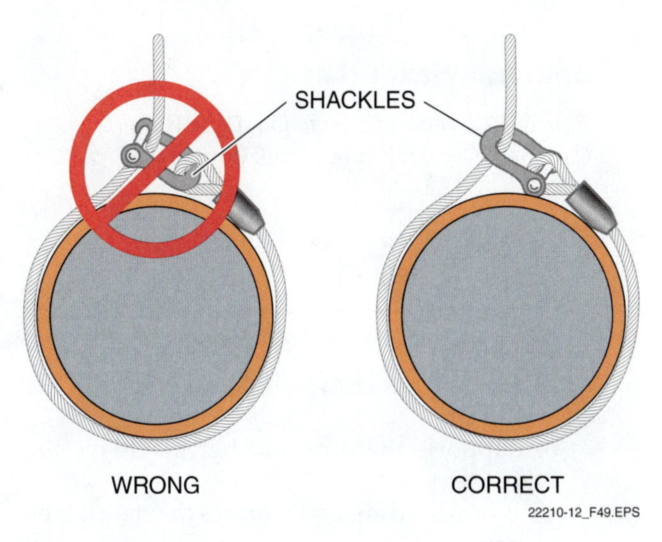

Figure 49 Choker hitch.

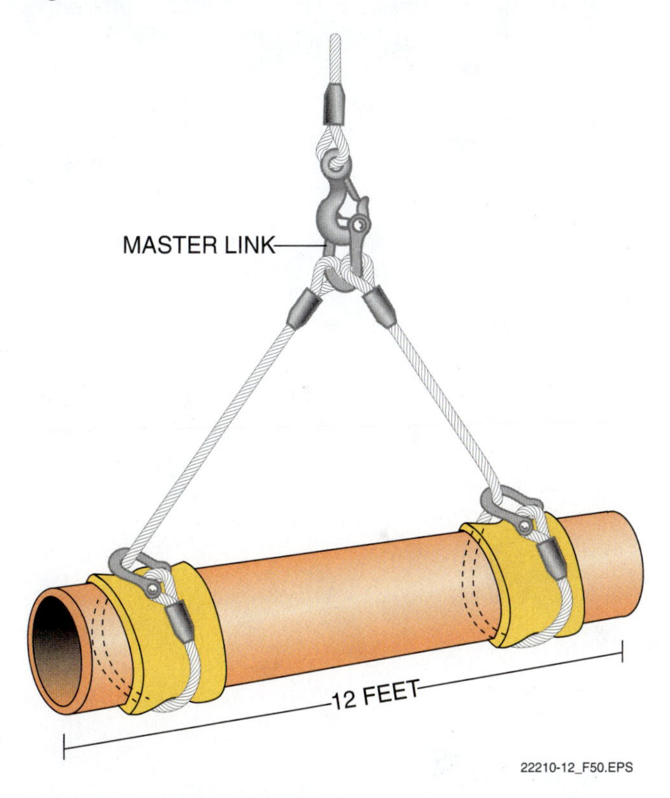

Figure 50 Double choker hitch.

- When operating at a fixed boom overhang, the bottom hoist pawl or other positive locking device should be engaged.

> **WARNING!**
> Be especially cautious when operating the equipment near electric power lines. The boom could come in contact with the lines and cause electrocution to the operator. Maintain the required minimum distance.

Additional Resources

Pipe & Excavation Contracting. Dave Roberts. Carlsbad, CA: Craftsman Book Company.

4.0.0 Section Review

1. Placement of bedding material in a trench _____.
   a. makes it easier to work in the bottom of the trench
   b. supports the pipe and helps achieve the proper grade
   c. is purely cosmetic—it makes the trench look better
   d. makes it easier for the inspector to move the pipe

2. A single choker hitch can be used to lift pipe no longer than _____.
   a. 6 feet
   b. 12 feet
   c. 18 feet
   d. 24 feet

SUMMARY

The most important part of any safety program is the individual worker. A company cannot have a good safety record without safety conscious employees. An employer is required to ensure that employees have safe equipment and a safe environment in which to work, but safety is ultimately up to the employee. Safe workers know, understand, and use safety procedures and devices that their employer has made available. Employees need to incorporate safe habits into their work on a daily basis, including attending required safety training sessions, participating in accident investigations, and learning how to work safely around hazardous materials.

During excavation work, it is extremely important to be aware of any potential hazards in the area. This means constantly monitoring the area for signs of instability and potential failures, as well as keeping co-workers in sight when they need to work near your equipment. It is essential to follow construction specifications whenever it is necessary to install excavation protective systems, such as shoring, trench boxes, or sloping, which are designed to ensure the safety of those working in an excavation.

Poor drainage on a construction site can cause major problems. If surface water is a problem, temporary drainage ditches or channels can be constructed to drain the water to retaining ponds. If groundwater problems are encountered in a cut area, control (intercept) ditches can be built to collect the outflow and reduce the possibility of damage to the construction site. In excavated pits or trenches, drains can be installed to channel water to sumps where it can be held until removed by pumping.

Equipment operators sometimes are required to read and use project plans in their work. At the beginning of a project, the grading supervisor or project engineer will usually discuss the grading and excavation operation with the equipment operators to familiarize them with the area. Operators must know the basic steps to go from natural ground to the finish grade, be able to follow the operation on the plans, and understand the workflow at the site. You will need to work with the plan and profile sheet, the cross-section plans, and the grading plans to get information about grading requirements on the job. Each of these sheets gives specific information about the grading requirements and the elevations of the grade, so you must be able to read and understand each part in order to do your job.

Most projects require the project survey team to set reference points, center lines, slope stakes, and grading stakes for equipment operators and other workers to follow. After these controls are set, the contractor may use his or her own personnel to set the detailed stakes that direct the grading operations. Equipment operators must be aware of the survey requirements and various survey activities going on throughout the duration of the project. They must understand the information on the stakes and be able to visualize the completed grading project based on the many stakes that are placed. In some cases, an equipment operator may be asked to assist a grade setter or surveyor in staking activities. Although operators need to understand basic surveying functions, it is not their responsibility to set stakes independently.

Initial surveying requirements involve setting grades and control points from benchmarks or known reference points. The benchmarks are usually permanent markers installed by federal or state governments. If one of these markers is found on the construction site, it should not be disturbed until the proper authorities are notified and a decision is made about relocation.

Heavy equipment, such as excavators and backhoes, is commonly used to place pipe in trenches once the trenches have been dug. Establishing and maintaining the correct grading of the trench is necessary to ensure that the liquid in the pipe will flow naturally.

On today's construction sites, grading is often controlled by lasers or GPS. Operators must know how to interact with these tools in order to work effectively.

Review Questions

1. Who has the most effect on worker safety?
   a. Supervisor
   b. Employer
   c. OSHA
   d. Worker

2. Toolbox safety training sessions should be chaired by a safety officer or _____.
   a. any safety committee member at the site
   b. supervisor qualified to discuss the subject
   c. the local OSHA agent
   d. a professional presenter

3. One function of the safety committee is _____.
   a. reviewing all accident reports
   b. selecting the company safety officer
   c. performing pop inspections
   d. promoting safety awareness

4. Last week five workers were injured in an on-the-job accident. All are still hospitalized. This is cause for a(n) _____.
   a. OSHA inspection
   b. plant shutdown
   c. increase in health insurance premiums
   d. investigation by the local newspaper

5. If you cannot identify a chemical in an unlabeled container, you can assume it is safe.
   a. True
   b. False

6. Material safety data sheets must be available for use by _____.
   a. everyone on the work site
   b. the work site supervisor
   c. OSHA inspectors only
   d. the site safety officer

7. If a worker finds a partially full, unlabeled container on a job site, what should he or she do?
   a. Do nothing. The container is unlabeled, so it is not hazardous.
   b. Since you do not know what is in the container, pour it down the drain.
   c. Treat the container as if it contains an extremely hazardous chemical.
   d. Dial 9-1-1 and call in the local fire department's HAZMAT team.

8. If a person's eyes come in contact with a hazardous chemical, the correct course of action is _____.
   a. rub it out with a rag
   b. wait until medical aid is available
   c. apply compresses and seek medical aid
   d. flush the eyes with cool water for 15 minutes

9. A trench is defined as an excavation in which the depth is no greater than the width and the width is no greater than _____.
   a. 5 feet
   b. 10 feet
   c. 15 feet
   d. 20 feet

10. One cubic yard of earth weighs approximately the same amount as a _____.
    a. gallon of water
    b. backhoe
    c. small car
    d. bag of cement

11. The foreman sent you to work with your backhoe near an excavation site. You notice a worker in the excavation below you, so you _____.
    a. keep the worker in sight as you complete your work
    b. ignore him because the foreman must know the worker is there
    c. stop work until the worker leaves the excavation
    d. attract the worker's attention so he knows to be careful

12. To relieve pressure on the walls and prevent material from falling into the excavation, the excavation walls are _____.
    a. sloped
    b. moistened
    c. shielded
    d. vibrated

13. A device laid up against the excavation wall to support the wall is called _____.
    a. shoring
    b. a waler
    c. shielding
    d. sloping

14. To protect workers from a cave-in in an excavation, use a _____.
    a. slide
    b. waler
    c. trench box
    d. ladder

15. When a trench box is removed from an excavation, it is important to _____.
    a. inspect the trench box for damage
    b. immediately backfill the trench
    c. clean the dirt from the trench box
    d. check the stability of the trench

16. In a braced coffer dam, the walers are the _____.
    a. walls of the coffer dam
    b. braces that go across the dam
    c. connections between the sheet piles
    d. horizontal support members around the walls

17. Caissons are used _____ .
    a. so that work can take place underwater
    b. to shore up the sides of a narrow trench
    c. when coffer dams are not practical
    d. as a shield to prevent water pollution

18. Grade stakes are set by _____.
    a. the project survey team to inform equipment operators of grading needs
    b. government workers to inform the survey team of a site's exact elevation
    c. equipment operators to inform the project survey team of existing elevations
    d. government workers to inform equipment operators of grading needs

19. The posts the National Oceanic and Atmospheric Administration (NOAA) sets near its benchmarks to make them easier to find are called _____.
    a. witness posts
    b. goal post
    c. fence posts
    d. trial stakes

20. Whenever you find a permanent benchmark that is damaged, you _____ .
    a. need to stop work and repair it immediately
    b. must report it to the site supervisor or engineer
    c. can ignore it because there are plenty of other markers
    d. stop work since the grade will be incorrect

21. A temporary benchmark is often placed _____ .
    a. on a structure
    b. in a foundation of concrete
    c. on the site's office trailer
    d. in the general work area

22. To check the grade of an excavation, the grade setter can set grade boots that he or she can _____ .
    a. sight with a hand level
    b. use with a pipe level
    c. use to check subgrades only
    d. look at through a laser level

23. The motor grader blade is made up of a moldboard, end pieces, and a cutting edge. Every day the end pieces and cutting edge should be _____ .
    a. replaced
    b. used
    c. inspected
    d. polished

24. The first pass when cutting a ditch is used _____ .
    a. to make a marking cut
    b. for shoulder pick-up
    c. to level the ditch area
    d. to spread loose material

25. A double choker hitch is used to lift pipe longer than _____ .
    a. 6 feet
    b. 12 feet
    c. 18 feet
    d. 24 feet

Trade Terms Introduced in This Module

Aquifer: An underground layer of water-bearing permeable rock or unconsolidated materials through which water can easily move.

Balance point: The location on the ground that marks the change from a cut to a fill. On large excavation projects there may be several balance points.

Bedding material: Select material that is used on the floor of a trench to support the weight of pipe. Bedding material serves as a base for the pipe.

Berm: A raised bank of earth.

Boot: A special name for laths that are placed by a grade setter to help control the grading operation. The boot can also be the mark on the lath, usually 3, 4, or 5 feet above the finish grade elevation, which can be easily sighted. This allows the grade setter to check the grade alone instead of having to use another person to hold a level rod on the top of the grade stake.

Competent person: A person who is capable of identifying existing and predictable hazards in the area or working conditions that are unsanitary, hazardous, or dangerous to employees, and who has the authority to take prompt corrective measures to fix the problem.

Cross braces: The horizontal members of a shoring system installed perpendicular to the sides of the excavation, the ends of which bear against either uprights or walers.

Dewater: To remove water from a site.

Erosion: The removal of soil from an area by water or wind.

Groundwater: Water beneath the surface of the ground.

Sedimentation: Soil particles that are removed from their original location by water, wind, or mechanical means.

Shielding: A structure that is able to withstand the forces imposed on it by a cave-in and thereby protect employees within the structure.

Stations: Designated points along a line or a network of points used to survey and lay out construction work. The distance between two stations is normally 100 feet or 100 meters, depending on the measurement system used.

Stormwater: Water from rain or melting snow.

String line: A tough cord or small diameter wire stretched between posts or pins to designate the line and elevation of a grade. String lines take the place of hubs and stakes for some operations.

Subsidence: Pressure created by the weight of the soil pushing on the walls of the excavation. It stresses the excavation walls and can cause them to bulge.

Sump: A small excavation dug below grade for the purpose of draining or retaining subsurface water. The water is then usually pumped out of the sump by mechanical means.

Swale: A shallow trench used to direct the flow of water.

Topographic survey: The process of surveying a geographic area to collect data indicating the shape of the terrain and the location of natural and man-made objects.

Uprights: The vertical members of a trench shoring system placed in contact with the earth and usually positioned so that individual members do not contact each other. Uprights placed so that individual members are closely spaced, in contact with, or interconnected to each other, are often called sheeting.

Walers: Horizontal members of a shoring system or coffer dam placed parallel to the excavation face whose sides bear against the vertical members of the shoring system or the earth. Also, supports for piles in a coffer dam.

Additional Resources

This module presents thorough resources for task training. The following resource material is suggested for further study.

OSHA Trench Safety Quick Card: **www.osha.gov/Publications/quickcard/trenching_en.pdf**

Soil properties that affect groundwater: **www.co.portage.wi.us/groundwater/undrstnd/soil.htm**

Excavators Handbook Advanced Techniques for Operators. Reinar Christian. Addison, IL: The Aberdeen Group, A division of Hanley-Wood, Inc.

GPS Systems for the Earthmoving Contractor: **www.constructionequipment.com/gps-systems-earthmoving-contractor**

Pipe and Excavation Contracting. Dave Roberts. Carlsbad, CA: Craftsman Book Company.

Figure Credits

Reprinted courtesy of Caterpillar Inc., Module opener, Figures 11, 35, 36, 38, 40–43, 46

Courtesy of Datum Filing Systems at **www.datumfiling.com**, Figure 3

Courtesy of Justrite Mfg. Co. LLC, Figure 4

Courtesy of Efficiency Production, Inc., Figures 6 and 8

Topaz Publications, Inc., Figures 14, 15, 29

Courtesy of Carolina Bridge Company, Inc., Figure 17A–D

Courtesy of Wacker Neuson, Figures 18, 19, 21

Courtesy of Chesterfield Associates Inc., Figures 23, 24, 26

US Army Corps of Engineers, Figure 25

Courtesy of Mike Pearson Construction, Inc., Figure 27

National Oceanic and Atmospheric Admin./Department of Commerce, Figure 28

Courtesy of Kittelson & Associates, Figure 30

John Hoerlein, Figure 33

Courtesy of Topcon Positioning Systems, Inc., Figure 44

The Crosby Group LLC, Figures 47 and 48

Section Review Answers

Answer	Section Reference	Objective
Section One		
1-1. b	1.1.0	1a
1-2. d	1.2.0	1b
1-3. d	1.3.0	1c
1-4. c	1.4.0	1d
1-5. c	1.5.0	1e
Section Two		
2-1 d	2.1.1	2a
2-2 c	2.2.1	2b
2-3 d	2.3.2	2c
Section Three		
3-1 a	3.1.0	3a
3-2 a	3.2.1	3b
3-3 d	3.3.0	3c
3-4 c	3.4.1	3d
3-5 b	3.5.2	3e
Section Four		
4-1 b	4.1.0	4a
4-2 b	4.2.0	4b

NCCER CURRICULA — USER UPDATE

NCCER makes every effort to keep its textbooks up-to-date and free of technical errors. We appreciate your help in this process. If you find an error, a typographical mistake, or an inaccuracy in NCCER's curricula, please fill out this form (or a photocopy), or complete the online form at **www.nccer.org/olf**. Be sure to include the exact module ID number, page number, a detailed description, and your recommended correction. Your input will be brought to the attention of the Authoring Team. Thank you for your assistance.

Instructors – If you have an idea for improving this textbook, or have found that additional materials were necessary to teach this module effectively, please let us know so that we may present your suggestions to the Authoring Team.

NCCER Product Development and Revision

13614 Progress Blvd., Alachua, FL 32615

Email: curriculum@nccer.org
Online: www.nccer.org/olf

❑ Trainee Guide ❑ AIG ❑ Exam ❑ PowerPoints Other ____________________

Craft / Level: Copyright Date:

Module ID Number / Title:

Section Number(s):

Description:

Recommended Correction:

Your Name:

Address:

Email: Phone:

22308-13

Soils

OVERVIEW

Soil conditions vary widely from site to site. In order to effectively dig, place, and finish earthwork, a heavy equipment operator must be able to identify soil types and understand how these soils will react to various conditions.

Module Six

Trainees with successful module completions may be eligible for credentialing through NCCER's National Registry. To learn more, go to **www.nccer.org** or contact us at **1.888.622.3720**. Our website has information on the latest product releases and training, as well as online versions of our *Cornerstone* newsletter and Pearson's product catalog.

Your feedback is welcome. You may email your comments to **curriculum@nccer.org,** send general comments and inquiries to **info@nccer.org**, or fill in the User Update form at the back of this module.

This information is general in nature and intended for training purposes only. Actual performance of activities described in this manual requires compliance with all applicable operating, service, maintenance, and safety procedures under the direction of qualified personnel. References in this manual to patented or proprietary devices do not constitute a recommendation of their use.

22308-12
SOILS

Objectives

When you have completed this module, you will be able to do the following:

1. Describe the different types and characteristics of soils.
   a. Identify the types of soils.
   b. Describe the properties of soils.
   c. Explain how soil density is determined.
   d. Explain how moisture affects soil.
2. Describe the factors that affect soil excavation.
   a. Explain what the swell factor is and how to calculate the swell factor of soils.
   b. Explain what the shrink factor is and how to calculate the shrink factor of soils.
   c. Describe how swell and shrink factors affect cycle times and equipment selection.
3. Describe working in various soil conditions.
   a. Describe the weight-bearing and flotation properties of different soils.
   b. Explain how soil characteristics affect machine performance.
   c. Describe how soil conditions can affect trenching safety.

Performance Tasks

Under the supervision of the instructor, you should be able to do the following:

1. Identify five basic types of soils, and summarize their characteristics.
2. Read results from a field density test and explain what additional compaction effort is needed.
3. Compute shrinkage and relative compaction for two different types of soils.

Trade Terms

American Association of State Highway and Transportation Officials (AASHTO)
American Society of Testing Materials (ASTM)
Banked
Bedrock
Capillary action
Cohesive
Consolidation
Density
Elasticity
Expansive soil
Fines
Friable
Horizon
Humus
In situ
Inorganic
Loading
Liquid limit
Optimum moisture
Organic
Peat
Plastic limit
Plasticity
Settlement
Shrinkage
Swell
Swell factor
Voids
Water table
Well-graded

Industry Recognized Credentials

If you are training through an NCCER-accredited sponsor, you may be eligible for credentials from NCCER's Registry. The ID number for this module is 22308-13. Note that this module may have been used in other NCCER curricula and may apply to other level completions. Contact NCCER's Registry at 888.622.3720 or go to **www.nccer.org** for more information.

Contents

Topics to be presented in this module include:

Figures and Tables

SECTION ONE

1.0.0 TYPES AND CHARACTERISTICS OF SOIL

Objective 1

Describe the different types and characteristics of soils.

a. Identify the types of soils.
b. Describe the properties of soils.
c. Explain how soil density is determined.
d. Explain how moisture affects soil.

Performance Task 1

Identify five basic types of soils, and summarize their characteristics.

Trade Terms

American Association of State Highway and Transportation Officials (AASHTO): An organization representing the interest of all state government highway and transportation agencies throughout the United States. This organization establishes design standards, materials-testing requirements, and other technical specifications concerning highway planning, design, construction, and maintenance.

American Society of Testing Materials (ASTM): A national organization that establishes standards for testing and evaluation of manufactured and raw materials.

Bedrock: The solid layer of rock under Earth's surface. Its solid-rock state distinguishes it from boulders.

Capillary action: The tendency of water to move into free space or between soil particles, regardless of gravity.

Cohesive: The ability to bond together in a permanent or semipermanent state. To stick together.

Density: Ratio of the weight of material to its volume.

Elasticity: The property of a soil that allows it to return to its original shape after a force is removed.

Expansive soil: A soil that expands and shrinks with moisture. Clay is an expansive soil.

Fines: Very small particles of soil. Usually particles that pass the No. 200 sieve.

Friable: Crumbles easily.

Horizon: Layers of soil that develop over time.

Humus: Dark swamp soil or decaying organic matter. Also called peat.

In situ: In the natural or original place on site.

Inorganic: Derived from other than living organisms, such as rock.

Liquid limit: The amount of moisture that causes a soil to become a fluid.

Loading: Applying a force to soil. A building can be a permanent load at a site, and a truck can be a passing load on a roadway.

Optimum moisture: The percent of moisture at which the greatest density of a particular soil can be obtained through compaction.

Organic: Derived from living organisms, such as plants and animals.

Peat: Dark swamp soil or decaying organic matter. Also called humus.

Plastic limit: The amount of water that causes a soil to become plastic (easily shaped without crumbling).

Plasticity: The range of water content in which a soil remains plastic or is easily shaped without crumbling.

Voids: Open space between soil or aggregate particles. A reference to voids usually means that there are air pockets or open spaces between particles.

Water table: The depth below the ground's surface at which the soil is saturated with water.

Well-graded: Soil that contains enough small particles to fill the voids between larger ones.

The stability of any building relies on the stability of the soil on which it is built. There are various types of soils. Each type has its own characteristics and requires different preparations for construction. The most important property for any soil for construction purposes is its strength, which is its ability to support a building or road without compressing or otherwise deforming.

Soil strength is measured by **density**, or how tightly the soil is packed. Soil that is densely packed and has few air pockets, called **voids**, weigh more per cubic unit than lightly packed soil, which has many air pockets. See *Figure 1*. Density is measured in weight per volume and can be expressed as pounds of wet or dry soil per cubic foot. The greater the weight per unit, the more tightly the soil is packed and the greater its density and strength. Densely compacted soil is usually made of a mixture of different sizes and types of soil particles.

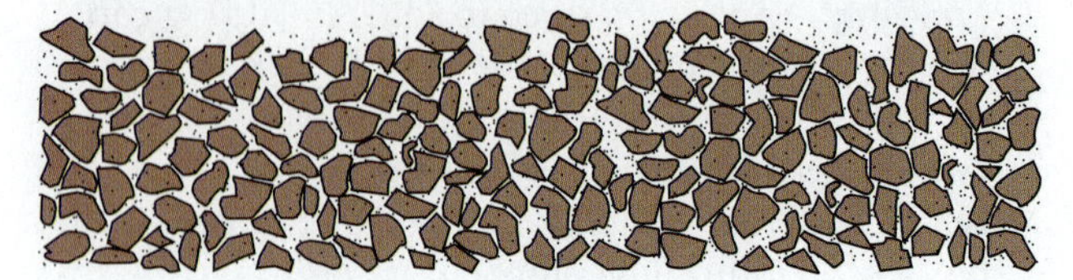

Figure 1 Soil density.

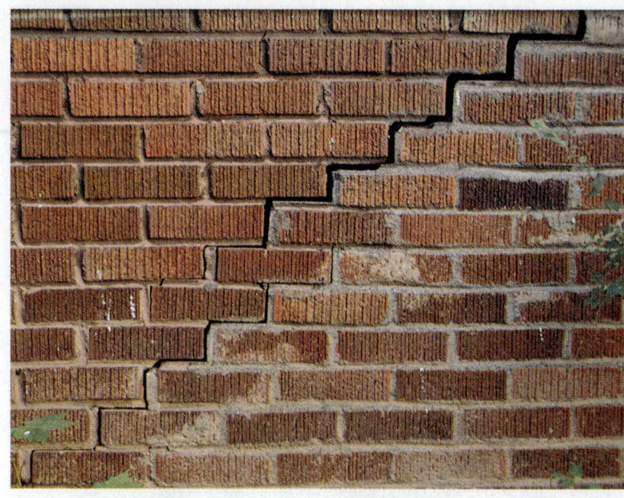

Figure 2 Cracks caused by foundation settling.

In construction, soil is often compacted to improve its density and strength. Compaction is the deliberate application of pressure or vibration to eliminate air pockets and increase the density of soil. It is one of the most important tasks in construction, and it helps ensure the durability and safety of a building or roadway. Poorly compacted soil naturally settles over time. The resulting movement of the ground can cause a cracked foundation (see *Figure 2*) or road surface failure.

Soils that have poor structural strength, even with compaction, can sometimes be improved by using additional materials or chemicals that stabilize the soil. Different geographic regions have predominant soil types, but most areas have a mixture of soils. It is possible to find many different kinds of soils on the same project.

1.1.0 Types of Soils

Soil is the loose material on the surface of Earth that is laid on the area's **bedrock**. Soil develops over time. When a hole is dug, definite layers of soil are noticeable. These layers, called **horizons**, represent the various times in which the soil developed. *Figure 3* shows the soil layers from an excavated lime rock bed in Florida.

There are several soil classification systems used, but the most common are the Unified Soil Classification System, which is also known as the **American Society for Testing and Materials (ASTM)**

The Importance of Recognizing Soil Types

Engineers, architects, and specially trained workers identify soils on a site and make technical decisions about how to prepare the site for construction. A heavy equipment operator must implement those decisions in order to safely operate the equipment in various soils and environmental conditions. To do this, the operator must be able to identify and understand the general behavioral characteristics of the different types of soils, as well as the various soil factors that affect equipment operation.

Figure 3 Soil layers.

Unified Soil Classification System, and the **American Association of State Highway and Transportation Officials (AASHTO)** Soil Classification System. Other classification systems are used in agriculture and geology; however, these names occasionally show up on construction documents.

For construction purposes, soil is classified into two types: granular or fine-grained. Fine-grained soil is often referred to as **fines.** Soils with particles that can be seen with the naked eye are granular. They include sand, gravel, and rock. Soils with particles that cannot be seen with the naked eye are fines. They include clays and silts.

Table 1 shows soil characteristics for the Unified Soil Classification System. This system classifies soils as coarse-grained or fine-grained, and gives each subcategory a two-letter designation. For example, gravel is designated with the letter G and clay with the letter C, so gravel and clay mixtures are designated as GC. The table shows that silts and clays are not measured by particle size, but by **liquid limit.** Liquid limit refers to the percent of moisture content in the soil at the point where it turns into a liquid. Silts and clays have very small soil particles and are referred to as fine-grained soils. These soils have almost no strength, but small amounts of these soils mixed with coarser soils give the mixture other desirable characteristics for building.

NOTE

Soils are usually classified by particle size, but keep in mind that the various soil classification organizations may use the same name for a different particle size. The important thing to remember is that the particle size greatly influences soil behavior. The tables shown in this section are for reference. It is unlikely that you will need this kind of information often. If you do need the information, don't rely on memory; look it up!

Table 1 Unified Soil Classification and Symbol Chart

Unified Soil Classification and Symbol Chart		
Coarse-grained soils (more than 50% of material is larger than No. 200 sieve size.)		
Gravel More than 50% of the coarse material is larger than No. 4 sieve size.		Clean Gravels (less than 5% fines):
	GW	Well-graded gravels, gravel-sand mixtures, little or no fines
	GP	Poorly graded sands, gravelly sands, little or no fines
		Gravels With Fines (more than 12% fines):
	GM	Silty gravels, gravel-sand-silt mixtures
	GC	Clayey gravels, gravels-sand-clay mixtures
Sands More than 50% of the coarse material is smaller No. 4 sieve size.		Clean Sands (less than 5% fines):
	SW	Well-graded sands, gravelly sands, little or no fines
	SP	Poorly graded sand, gravelly sands, little or no fines
		Sands With Fines (more than 12% fines):
	SM	Silty sands, sand-silt mixtures
	SC	Clayey sands, sand-clay mixtures
Fine-grained soils (50% or more of the material is smaller than No. 200 sieve size.)		
Silts and Clays Liquid limit less than 50%.	ML	Inorganic silts and very fine sands, rock flour, silty or clayey fine sands or clayey silts with slight plasticity.
	CL	Inorganic clays of low to medium plasticity, gravelly clays, sandy clays, silty clays, lean clays
	OL	Organic silts and organic silty clays of low plasticity.
Silts and Clays Liquid limit 50% more.	MH	Inorganic silts, micaceous or diatomaceous fine sandy or silty soils, elastic silts.
	CH	Inorganic clays of high plasticity, fat clays
	OH	Organic clays of medium to high plasticity, organic silt
Highly Organic Soils	PT	Peat and other highly organic soils.

22308-12_T01.EPS

Table 2 shows soil characteristics for the AASHTO Soil Classification System. This system divides soils into granular and silt-clay materials, and then further by the sizes of the soil particles in a mixture. Granular material includes gravel and sand. The sieve number designates the size of the particles that pass the sieve. No. 10 passes big particles and No. 200 passes very tiny particles. At the bottom of the table, it can be seen that well-mixed material that is mostly granular gets the best rating for construction. Sand and clay are often blended in specified proportions to increase the strength of the soil at a site.

The Unified Soil Classification System and the AASHTO Soil Classification System define the main classes of soils as gravels, sands, silt and clay, and organics. In general, soils are divided as follows for construction purposes:

- *Gravel* – Gravel is any rock-like material above 0.125 inches (⅛ inch) in diameter. Larger particles are called cobbles or stones. Particles larger than 10 inches are called boulders. Gravel occurs naturally or it can be made by crushing rock. Natural gravel is usually rounded from the effects of water, while crushed rock is usually angular (*Figure 4*). Gravel is very strong.
- *Sand (coarse and fine)* – Sand is made of mineral grains measuring 0.002 to 0.125 inches. Sand comes from grinding or decaying rock. It usually contains a high amount of quartz, which is a very hard mineral. It is called granular material because it separates easily, giving it almost no cohesive strength. Coarse sand is frequently rounded like gravel. It is often found mixed with gravel, but fine sand is usually more angular. See *Figure 5*.
- *Inorganic silt* – **Inorganic** silt is very fine sand with particles that are 0.002 inches or less. Silt is sand that has been ground very fine. Silt is often called rock flour or rock dust because it has a dusty appearance and powdery texture when dry. When wet, it sticks together, but silt has almost no cohesive strength. Dried lumps are easily crushed. Silt has a tendency to absorb moisture by **capillary action**, which means that moisture wicks up through the soil (see *Figure 6*), making silts problematic in areas where the **water table** is shallow.

22308-12_F04.EPS

Figure 4 Gravel.

Table 2 AASHTO Soil Classification System

General Classification	Granular Material (35% or less passing No. 200 Sieve size)							Silt-Clay Materials (More than 35% passing No. 200 sieve size)			
Group Classification	A-1		A-3	A-2				A-4	A-5	A-6	A-7
	A-1-a	A-1-b		A-2-4	A-2-5	A-2-6	A-2-7				A-7-5 A-7-6
Sieve Analysis (percent passing)											
No. 10	50 max	–	–	–	–	–	–	–	–	–	–
No. 40	30 max	51 max	–	–	–	–	–	–	–	–	–
No. 200	15 max	25 max	10 max	35 max	35 max	35 max	35 max	36 min	36 min	36 min	36 min
Characteristics of fraction passing No. 40											
Liquid Limit	–		–	40 max	41 min	40 max	41 min	40 max	41 min	40 max	41 min
Plasticity Index	6 max		NP	10 max	10 max	11 min	11 min	10 max	10 max	11 min	11 min
Usual types of significant constituent materials	Stone fragments gravel and sand		Fine Sand	Silty or clayey gravel and sand				Silty soils		Clayey soils	
General rating as subgrade	Excellent to good					Excellent to good					

22308-12_T02.EPS

Figure 5 Fine sand.

- *Clay* – Clay is the finest size of soil particles. Clay is very cohesive. When wet, clayey soils feel like putty and can be easily molded and made into long ribbons (see *Figure 7*). When dry, clay is very strong and clumps are difficult to crush. Clay is an **expansive soil**. It swells and shrinks with moisture changes, so pure clay is not suitable for building. However, a small amount mixed uniformly with granular material is desirable.
- *Organic matter (top soil) and colloid clays* – **Organic** matter is partly decomposed vegetable and animal material. Organic matter is sometimes called **humus** or **peat** and is usually soft, fibrous, and may have an offensive odor when warm. This material is not suitable for building or as fill because as it decays it loses volume, which may cause air pockets that make the ground unstable. Colloidal clays are very fine clay particles that remain suspended in water for long periods of time and don't settle easily under the force of

Figure 7 Clay ribbon.

CAPILLARY ACTION
SILT
WATER TABLE
MOISTURE TRAVELS UP FROM THE WATER TABLE TO THE SURFACE OF THE GROUND.
22308-12_F06.EPS

Figure 6 Capillary action.

gravity. Individual particles cannot be seen with the naked eye. Colloidal clays are very susceptible to swelling and shrinking, so they are not suitable for construction.

1.2.0 Properties of Soils

There are many different types of soils and each type reacts differently to environmental conditions such as moisture and temperature. This reaction helps determine the strength and suitability of a soil for a construction project. The most useful characteristics in predicting the behavior of a soil are grain size, shape, surface texture, and soil composition. *Table 3* summarizes the properties of various soils that may be encountered on a job site.

1.2.1 Grain Size

Sieves are used to measure the grain size of soil particles (*Figure 8*). The number of the sieve tells you how many holes there are per linear inch of screen. A No. 10 sieve has 10 holes per inch, so particles smaller than a tenth of an inch can pass. A No. 200 sieve has 200 holes per inch, so it passes only very small particles. The size and distribution of soil particles throughout the soil mixture are important factors that determine a soil's behavior under load. *Table 4* shows the range in particle sizes of different soil materials based on the AASHTO classification.

Soil-Related Safety

The US Occupational Health and Safety Administration (OSHA) views soils from a safety perspective. OSHA issued its first Excavation and Trenching Standard in 1971 for the purpose of protecting workers from excavation hazards. OSHA developed the standard in order to prevent cave-ins that were injuring and killing workers. The purpose of the standard is to specify sloping, benching, and mechanical protection measures for specific types of soils. The standard requires that excavations be inspected by a competent person, which is defined as "one who is capable of identifying existing and predictable hazards in the surroundings or working conditions which are unsanitary, hazardous, or dangerous to employees, and who has authorization to take prompt corrective measures to eliminate them."

Notice in *Table 4* that silts and clays are not measured by sieve. These soils are measured using a hydrometer, which measures liquid content. For most construction projects, having soil that consists of predominately coarse-grain soil, such as sand, is preferable to fine-grain soil, such as clay. Sands are very permeable, so they quickly drain water. Clay particles attract water, so they swell and shrink with changes in moisture, causing movement in the ground. Permanent structures should never be built on purely clay soils. The best soil mixture for construction is one that contains enough small particles to fill the voids between large particles. This is called **well-graded** soil.

1.2.2 Grain Shape

Soil mixtures that contain a lot of gravel and sand are good for highway and building projects. Crushed rock and gravel have angular surfaces that resist sliding. This increases the stability of the soil under load. Small soil particles, such as fine sand, can easily slide into the voids between the larger and more angular gravel, increasing the density of the soil mass (*Figure 9*). Building specifications often call for a soil mixture that contains certain percentages of crushed rock particles or gravel and fine sands. This mixture of angular particles with rough edges is desirable because it can be compacted to a high density.

1.2.3 Surface Texture

The strength of a soil mixture is partly determined by the surface texture of soil particles. Friction is the resistance two materials have to sliding over each other. Soil particles with a smooth, slick surface (such as quartz sand that has been polished by water action) slip easily over one another; they have a low amount of friction. Soil mixtures that are high in quartz content have little resistance to deformation when under pressure. However, a small amount of this type of material can help the soil mixture compress more easily, because the smooth, slick particles easily slip into air pockets between larger, more angular particles.

Gravels and sands that have rough angular surfaces have a high amount of surface friction and a grainy appearance. They are strong and resist deforming under load. Clays have a smooth texture; clay particles tend to slide over one another easily and stick together. Clays are easily deformed.

Table 3 Soil Properties

Soil Texture	Visual Detection of Particle Size and General Appearance of Soil	Squeezed in Hand	Soil Ribboned Between Thumb and Finger
Sand	Soil has a granular appearance in which the individual grain sizes can be detected. It is free-flowing when in a dry condition.	*When air dry*: Will not form a cast and will fall apart when pressure is released. *When moist*: Forms a cast which will crumble when lightly touched.	Cannot be ribboned.
Sandy Loam	Essentially a granular soil with sufficient silt and clay to make it somewhat coherent. Sand characteristics dominate.	*When air dry*: Forms a cast which readily falls apart when lightlytouched. *When moist*: Forms a cast which will bear careful handling without breaking.	Cannot be ribboned.
Loam	A uniform mixture of sand, silt, and clay. Grading of sand fraction quite uniform from coarse to fine. It is mellow, has a somewhat gritty feel, yet is fairly smooth and slightly plastic.	*When air dry*: Forms a cast which will bear careful handling without breaking. *When moist*: Forms a cast which can be handled freely without breaking.	Cannot be ribboned.
Silt Loam	Contains a moderate amount of the finer grades of sand and only a small amount of clay. Over half of the particles are silt. When dry it may appear quite cloddy, which can be readily broken and pulverized to a powder.	*When air dry*: Forms a cast which can be freely handled. Pulverized, it has a soft, flour-like feel. *When moist*: Forms a cast which can be freely handled. When wet, soil runs together and puddles.	It will not ribbon, but has a broken appearance, feels smooth, and may be slightly plastic.
Silt	Contains over 80 percent silt particles, with very little fine sand and and clay. When dry, it may be cloddy, and readily pulverizes to powder with a soft flour-like feel.	*When air dry*: Forms a cast which can be handled without breaking. *When moist*: Forms a cast which can be freely handled. When wet, it readily puddles.	It has a tendency to ribbon with a broken appearance, feels smooth.
Clay Loam	Fine textured soil breaks into very hard lumps when dry. Contains more clay than silt loam. Resembles clay in a dry condition; identification is made on the physical behavior of moist soil.	*When air dry*: Forms a cast which can be handled freely without breaking. *When moist*: Forms a cast which can be freely handled without breaking. It can be worked into a dense mass.	Forms a thin ribbon which readily breaks, barely sustaining its own weight.
Clay	Fine textured soil breaks into very hard lumps when dry. Difficult to pulverize into a soft flour-like powder when dry. Identification is based on cohesive properties of the moist soil.	*When air dry*: Forms a cast which can be freely handled without breaking. *When moist*: Forms a cast which can be freely handled without breaking.	Forms long, thin, flexible ribbons. Can be worked into a dense, compact mass. Considerable plasticity.
Organic Soils	Identification based on the high organic content. Muck consists of thoroughly decomposed organic material with considerable amount of mineral soil finely divided with some fibrous remains. When considerable fibrous material is present, it may be classified as peat. The plant remains or sometimes the woody structure can easily be recognized. Soil color ranges from brown to black. They occur in lowlands, swamps, or swales. They have high shrinkage upon drying.		

22308-12_F08.EPS

Figure 8 Soil sieves.

1.2.4 Soil Composition

What is normally considered to be soil is usually a mixture of two or more soil types, organic matter, and multiple chemicals. No single soil type is acceptable for construction. Gravel and sand have great strength but no significant **elasticity**, **plasticity**, or cohesiveness. Once gravel or sand is deformed under load, it stays deformed. Silt and clay are very plastic and **cohesive**, but have almost no strength. These soils easily compress and break apart under load. Soil that is easy to crush, crumble, or break apart is said to be **friable**.

The best soil for construction is a mixture that contains a high amount of large-particle soils, such as gravel and sand, and a small amount of small particle soil, such as silt and clay. Soils that contain moderate to large amounts of organic matter are not suitable for construction because as the matter decays, it will decrease in mass and cause settling to occur in the soil. This may damage building foundations and road surfaces. Looking back at *Table 1* and *Table 2*, it can be seen that soils with excellent to good subgrade ratings contain a high amount of gravel and sand, and a small amount of silt or clay.

Although it is difficult to determine the chemical components of a soil, they can affect the performance of the soil under load. In particular, chemicals may affect the amount of water the soil can absorb, which changes the strength of the soil.

1.2.5 Engineering Properties of Soil

For the soil at a site to be strong enough for construction, it must have characteristics that allow it to be stable under a variety of environmental conditions. These characteristics are called the engineering properties of soil and include the following:

- Permeability
- Elasticity
- Plasticity
- Cohesion
- Shearing strength
- Shrinkage and swelling
- Frost susceptibility

Permeability is the ability of a soil to allow water to flow through it. Soils that hold water swell and cause building foundations and road surfaces to move. Coarse-grained soil, such as sand, is more permeable than fine-grained soil, such as clay, so

22308-12_F09.EPS

Figure 9 Different sized particles make soil easy to compact.

Table 4 Soil Particle Size Using US Sieve Sizes

Soil Type	Particle Diameter		Passing	Retained
	millimeters	inches		
Gravel	76.2 to 2.0	3.0 to 0.08	3-inches	No. 10
Coarse Sand	2.0 to 0.42	0.08 to 0.017	No. 10	No. 40
Fine Sand	0.42 to 0.074	0.017 to 0.003	No. 40	No. 200
Silt	0.074 to 0.005	0.003 to 0.0002	No. 200	—
Clay	0.005 to 0.001	Less than 0.0002		
Colloidal Clay	Less than 0.001	Less than 0.00004		

22308-12_T04.EPS

untreated clayey sites are poor choices for construction. *Table 5* lists the coefficients of permeability for various soil mixtures. The coefficient is measured in the amount of water that can drain from the soil in a period of time. In this case, it is the number of feet of water that drain in a minute.

Elasticity is the ability of the soil to return to its original shape after having been deformed by some temporary force, such as vehicular traffic. In most soils, this deformation is small, but it is important in highway engineering. Coarse-grained soils have almost no elasticity, so mixing clay with sand gives the resulting mixture some elasticity.

Plasticity is the ability of a soil to be molded into a different shape without breaking apart. Plasticity is used to classify fine-grained soils and is directly related to their moisture content. As moisture increases from 0 percent, plastic soils go from solids to liquids (*Figure 10*). The moisture content at which the soil goes from semisolid to plastic is called the **plastic limit**, and the moisture content at which the soil goes from plastic to liquid is called the liquid limit. These are referred to as the Atterberg limits, named after the Swedish scientist who defined them Albert Atterberg.

Clay is a plastic soil. When clay is saturated with water, it is a very thick liquid, but it dries into solid clumps. At some point between the liquid and solid states, the clay can be molded and is plastic.

The shear strength of a soil is its ability to withstand pressure without breaking apart. A soil's shear strength is determined by its cohesiveness and friction. Cohesiveness is the ability of the soil particles to stick together, while friction is the resistance the soil particles have to sliding over one another. Clay is a cohesive soil, while sand has a high amount of friction. Cohesion and friction, and thus shear strength, vary with moisture, **loading**, and other factors. Clay is extremely cohesive when dry, but less cohesive when wet.

Table 5 Coefficient of Permeability for Various Soil Mixtures

Type of Sand (Unified Soil Classification System)	Coefficient of Permeablility ft/min
Sandy Silt	0.001 to 0.004
Silty Sand	0.004 to 0.01
Very Fine Sand	0.01 to 0.04
Fine Sand	0.04 to 0.1
Fine to Medium Sand	0.1 to 0.2
Medium Sand	0.2 to 0.3
Medium to Coarse Sand	0.3 to 0.4
Coarse Sand and Gravel	0.4 to 10

22308-12_T05.EPS

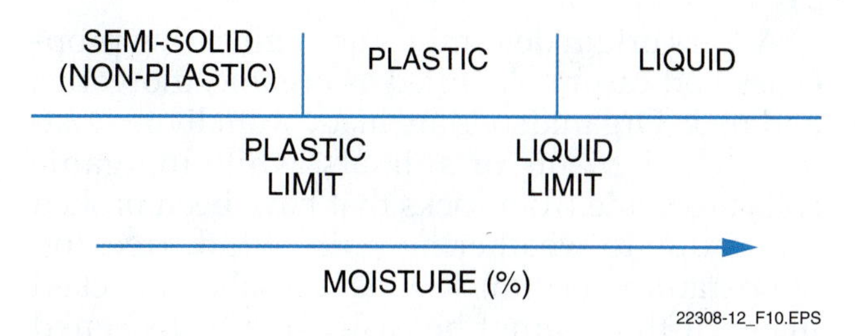

Figure 10 Atterberg limits.

Shrinkage and swelling of soils due to moisture changes are problematic in construction because the resulting movement of the ground may damage building foundations, underground utilities, and road surfaces (*Figure 11*). Clay soils are particularly prone to shrinkage and swelling with moisture changes.

Frost is troublesome for soil in cold climates. In freezing weather, moisture in the soil freezes and causes the soil to expand; the movement of the ground may damage unprotected building foundations and road surfaces. The moisture can come from above the ground, such as melting snow, or below the ground, such as a shallow water table. Where freezing temperatures are prolonged and the frost line penetrates deep into the soil, ice layers cause the soil to heave at the surface.

1.3.0 Soil Density

The most important characteristic to a building site is its **in situ** soil density. Soil density is a ratio of a soil's weight to its volume (usually per cubic foot). Each site will have a natural density that depends on the engineering properties of the soil, as well as the stressors to which the site is subjected. Keep in mind that no single soil type is ideal for construction, but rather it is a mixture of types that make a good building site.

22308-12_F11.EPS

Figure 11 Cracked road surface.

A soil's origin determines its engineering properties and can be classified as organic, inorganic, and rock. Organic soils are made from living matter such as plants or animals, while inorganic soils are made from rocks that have been broken down due to weathering (rain, wind, freezing temperatures, or sun). Rock is any solid or packed material that cannot be easily dug or loosened by machinery. The soils that you work with will most likely be a mixture of these three categories.

Stressors that affect a site's soil density can be natural or man-made. Weather is the primary natural stressor; soil that is subject to heavy rains and freeze/thaw cycles have a greater density than the same soil that is not exposed to such conditions. Man-made stressors are usually determined by land use. Soils at a site subjected to a great deal of traffic or vibration from nearby roadways have a higher density than if the site were located far away from heavy traffic.

Engineers measure the soil density to determine whether the site needs additional compacting or stabilization. It is doubtful that an equipment operator will be required to do soil density tests, so they are discussed very briefly in the following paragraphs to familiarize you with the technology. There are three types of tests primarily in use today: sand cone testing, penetrometer testing, and nuclear density testing.

1.3.1 Sand Cone Test

Many engineers consider the sand cone test to be the most accurate testing method. To perform this test, dig a round hole with a volume of one-tenth of a cubic foot. Weigh the soil taken out of the hole. Send the soil removed from the hole to a laboratory. At the laboratory, the soil is dried and weighed again to determine how much of the total weight is water. An additional sample is taken and analyzed the same way so the lab can plot a moisture density curve of the site.

The specific volume of the hole is determined by using a calibrated jar of dry sand. The sand is calibrated so that the volume can be easily determined from the weight. Weigh the jar full of dry sand. Fill the hole with sand using the jar and cone device shown in *Figure 12*. Weigh the container and remaining sand again. Subtract that from the original weight to determine the weight of sand in the hole. Use the conversion factor to convert the weight of the sand into the volume of the hole.

The dry weight of the soil removed from the hole is divided by the volume of sand needed to fill the hole. This gives the density of the compacted soil in pounds per cubic foot. This result is compared to the theoretical maximum density, which gives the relative density of the soil that was just compacted.

Soil Density

In the 1930s, experiments conducted by American architect R.R. Proctor established the relationship between soil density and moisture content. Proctor determined that varying the water content of soil had a direct bearing on the amount of compacting required. Soil with no moisture (such as sand) is impossible to compact. As water is added, the soil becomes easier to compact. The conclusion to be drawn from Proctor's studies is that in order to be compacted, soil must contain the correct amount of water.

1.3.2 Penetrometer Testing

The cone penetrometer is used to determine the bearing capacity of soil. *Figure 13* shows a portable penetrometer. There are also truck-mounted penetromers that are used for extensive soil testing. The penetrometer density test is also called a cone test because of the conical attachment that is screwed onto the end of the probe. The cone and the friction sensor behind it contain instrumentation. The cone reacts to the pressure of the soil, while the friction sleeve detects the resistance (friction) of the soil.

1.3.3 Soil Density Testing

Soil density testing involves either placing the testing machine on the ground to get a reading or inserting a probe attached to the machine into a small hole drilled into the soil. Either method sends impulses into the soil that are reflected back to the device and recorded. Denser soils absorb more impulses. The more a soil is compacted, the fewer impulses are returned. The lab technician then creates a moisture density curve of the site in the same way as the sand cone test. The test can be performed with nuclear instruments such as the one shown in *Figure 14A*, or non-nuclear instruments.

WARNING!

Nuclear equipment should only be used by trained personnel. Exposure to nuclear materials may cause serious illness.

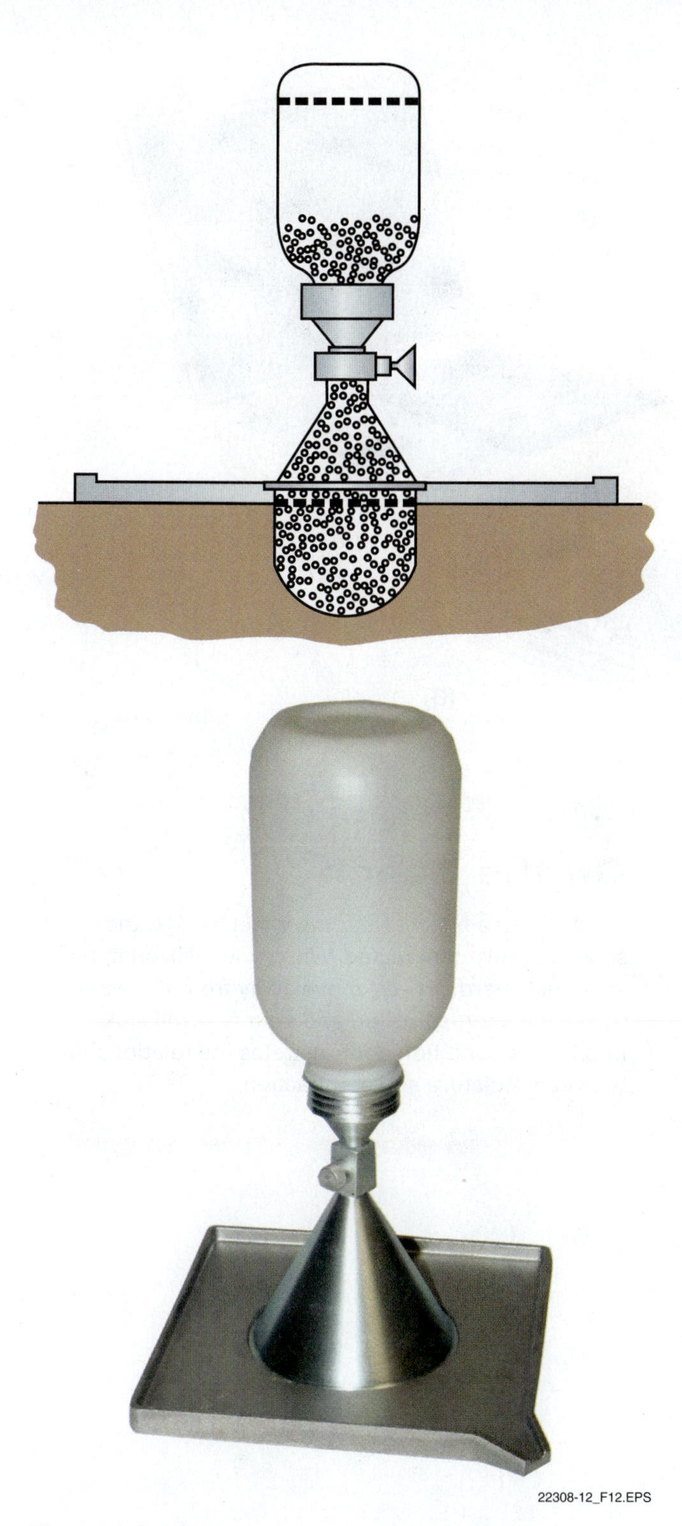

22308-12_F12.EPS

Figure 12 Sand cone test.

The electrical density gauge (EDG) shown in *Figure 14B* is a non-nuclear instrument that measures the physical properties of compacted soils used in roadbeds and foundations. This device is battery-operated and can determine the wet and dry density, gravimetric moisture content, and percent of compaction. The kit includes a console, four electrodes, a hammer, soil sensor and cables, template, temperature probe, and battery charger.

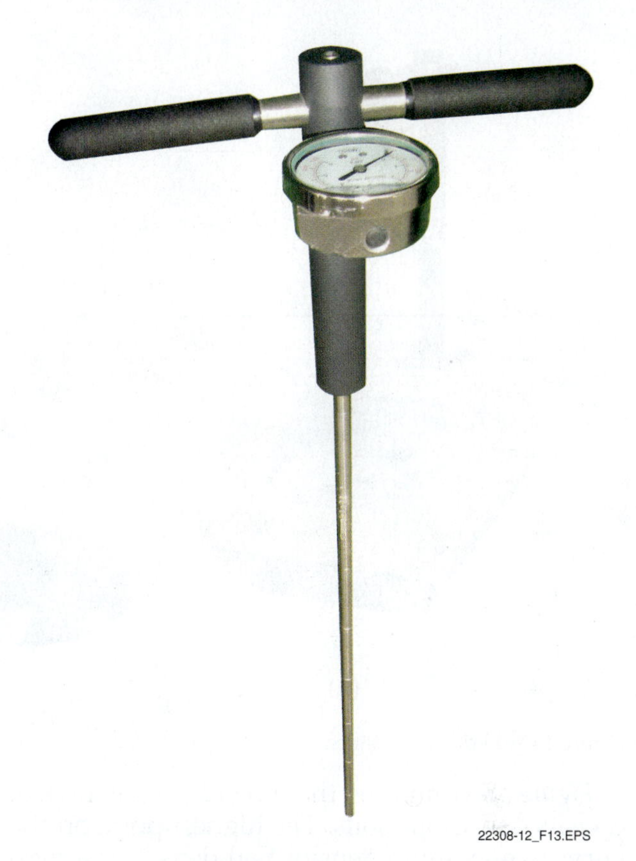

22308-12_F13.EPS

Figure 13 Portable penetrometer.

1.4.0 Effects of Moisture on Soils

The single greatest factor affecting a soil's behavior is its moisture content. Clay attracts and holds moisture, causing the clay to swell and increase in volume. Sand is very permeable to water, allowing quick drainage, so its volume changes very little with increased moisture.

Moisture also affects the compaction of soil. Moisture surrounds rough-surfaced particles of sand and allows them to slide across one another more easily than with dry particles, making compaction easier. Soil mixtures that are dry have a high degree of surface friction. This makes them difficult to compact, tending to leave air pockets and lower soil density. Soil mixtures that contain too much moisture can make the soil rubbery and resistant to compacting. They may even approach the point where they are fluid. This makes them impossible to compact until some of the moisture is removed. When a soil has just the right amount of moisture to achieve maximum density, it is said to have **optimum moisture** content.

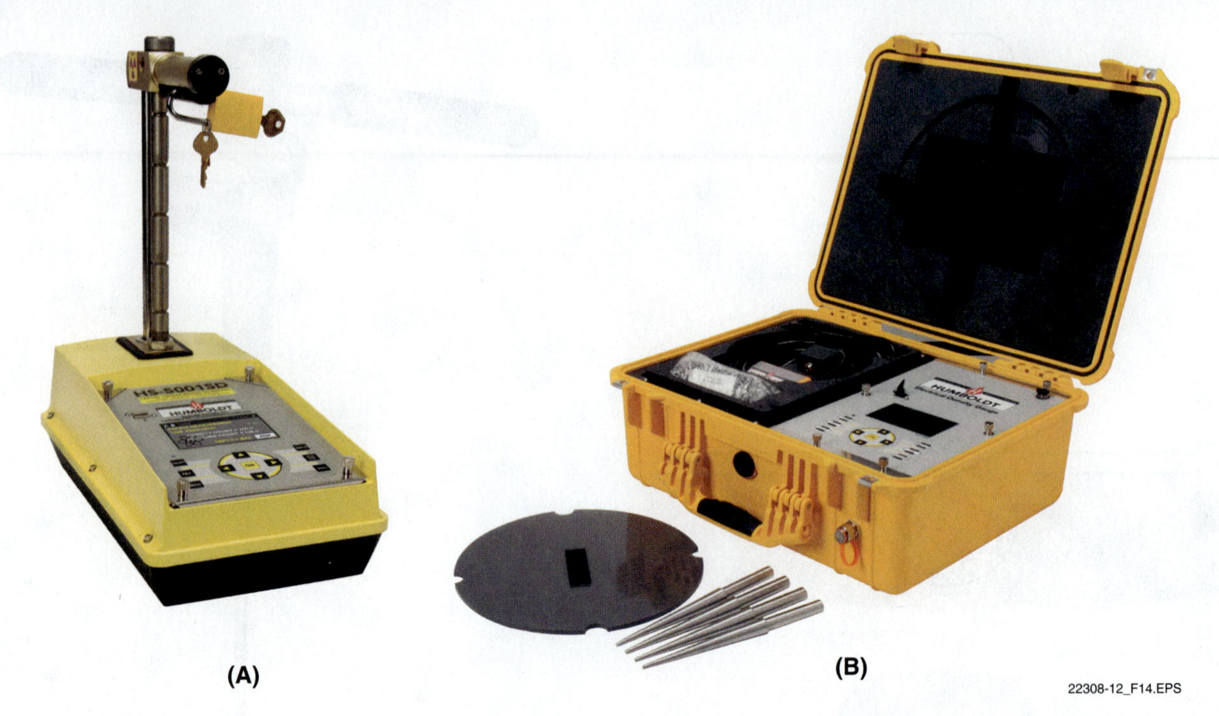

Figure 14 Soil density testers.

Figure 15 compares the moisture content and density of various soils. The highest point on the curve is maximum density. Soil density is a ratio of a soil's weight to its volume, so it is measured in pounds per cubic foot. When there are more soil particles in a cubic foot of space, the weight and density both increase. *Figure 15* shows that the soil with the highest density and lowest water content is well-graded sand, which is a mixture of predominantly sand and small amounts of clay and silt.

On the Beach

Think about a beach. Near the water's edge, the sand is tightly compacted. When you walk on it, the sand feels hard. As you move away from the water, the sand becomes softer, and your feet will sink into it. This condition demonstrates the relationship between moisture and compaction.

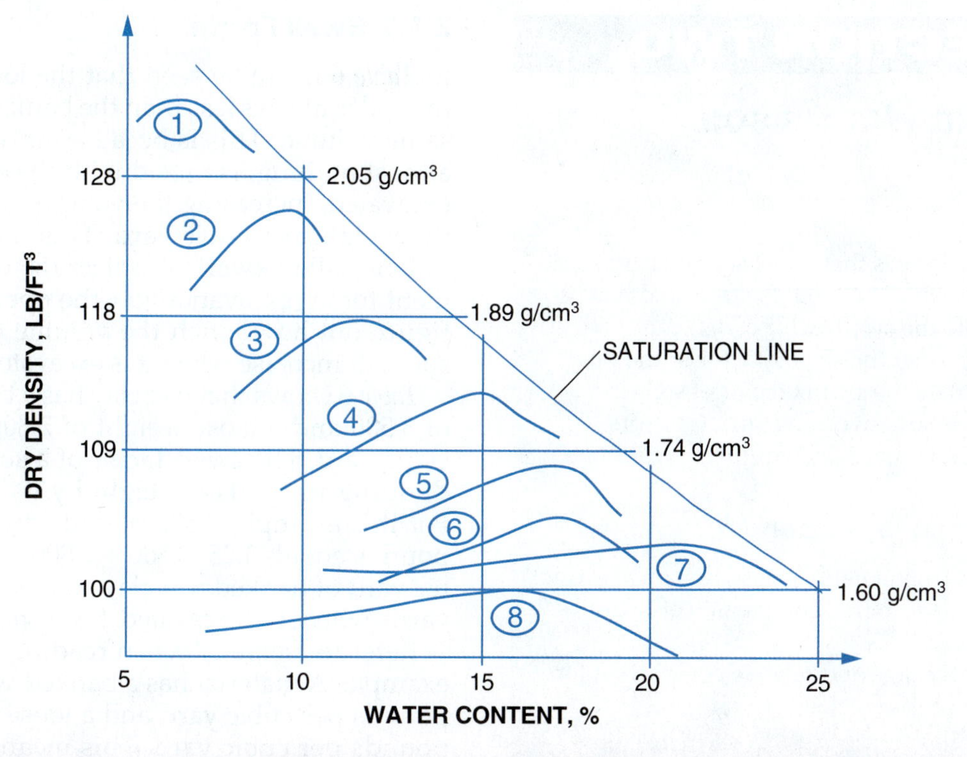

No.	Description	Sand %	Silt %	Clay %	w_L	I_P
1	Well-Graded Sand	88	10	2	16	–
2	Well-Graded Sandy Marl	72	15	13	16	–
3	Medium Sandy Marl	73	9	18	22	4
4	Sandy Clay	32	33	35	28	9
5	Silty Clay	5	64	31	36	15
6	Loess Silt	5	85	10	26	2
7	Clay	6	22	72	67	40
8	Poorly Graded Sand	94	6	–	–	–

22308-12_F15.EPS

Figure 15 How moisture affects soil density.

1.0.0 Section Review

1. Which of these soils has a tendency to ribbon?
   a. Sand
   b. Sandy loam
   c. Silt loam
   d. Clay

2. The grain size of soil is measured using a _____.
   a. sieve
   b. scale
   c. micrometer
   d. ruler

3. Soil density is measured as the _____.
   a. ratio of weight to volume
   b. ratio of moisture to volume
   c. pressure per cubic inch
   d. ratio of rock to sand

4. Which of these soils has the lowest percentage of clay content?
   a. Medium sandy marl
   b. Loess silt
   c. Sandy clay
   d. Silty clay

SECTION TWO

2.0.0 EXCAVATING SOIL

Objective 2

Describe the factors that affect soil excavation.

a. Explain what the swell factor is and how to calculate the swell factor of soils.
b. Explain what the shrink factor is and how to calculate the shrink factor of soils.
c. Describe how swell and shrink factors affect cycle times and equipment selection.

Performance Tasks 2 and 3

Read results from a field density test and explain what additional compaction effort is needed.
Compute shrinkage and relative compaction for two different types of soil.

Trade Terms

Banked: Any soil mass that is to be excavated from its natural position.

Consolidation: To become firm by compacting the particles so they will be closer together.

Settlement: To become firm by compacting the particles so they will be closer together.

Shrinkage: Decrease in volume when soil is compacted.

Swell: Increase in volume when soil is excavated.

Swell factor: The ratio of the banked weight of a soil to the loose weight of a soil.

As soil is excavated from its natural position, it is mixed with air and gives the appearance of swelling in volume. When the soil is used as fill, it is compacted, pushing air pockets out, and gives the appearance of shrinking in volume. This is important to know because heavy equipment is rated for maximum volumes and weights, so the operator needs to be able to select the proper equipment for the job.

Soil is measured in cubic units, with cubic yards being the most common measure. In the construction industry, a cubic yard is casually referred to as a yard. Undisturbed soil is called **banked**, while excavated soil is called loose. *Table 6* shows the typical loose and bank volumes and weights of many common soil mixtures.

2.1.0 Swell Factor

In *Table 6* it can be seen that the loose weight of the soil is always less than the bank weight for the same volume. This is because air, which weighs almost nothing, is mixed with the soil when it is excavated, increasing the volume and decreasing the weight per cubic yard. This increase in volume is called **swell**. To select the correct equipment for an excavation job, the operator needs to figure out how much the volume of compacted soil will increase when it is excavated.

Table 6 shows that wet clay has a banked weight of 3,500 and a loose weight of 2,800 pounds per cubic yard. The **swell factor** of a soil is found by dividing its banked weight by its loose weight. For this example, 3,500 pounds divided by 2,800 pounds equals 1.25 (3,500 ÷ 2,800 = 1.25). One cubic yard of banked wet clay equals 1.25 (1¼) cubic yards when it is excavated. See *Figure 16*.

Refer to *Figure 17* when reading the following example. A material has a banked weight of 1,000 pounds per cubic yard and a loose weight of 800 pounds per cubic yard. This means 800 pounds is used to make one loose cubic yard with 200 pounds of material left over. Calculate the swell factor by dividing the banked weight by the loose weight (1,000 ÷ 800 = 1¼ or 1.25). The calculation shows that one banked cubic yard equals 1¼ or 1.25 loose cubic yards.

Returning to the original problem (see *Figure 18*), wet clay weighs 3,500 pounds in a banked cubic yard and 2,800 pounds in a loose cubic yard, so after the first loose cubic yard is made, 700 pounds of material is left over. Those 700 pounds are enough

Effect of Soil Compaction

Water from such sources as rain and melting snow is absorbed by the soil. However, areas occupied by buildings, roads, and parking lots do not absorb water. Such areas are said to be impervious to water. The problem is that the water still needs to go somewhere. Populated areas need to have stormwater-runoff systems to compensate for the impervious area. For that reason, municipalities maintain stormwater-processing facilities to filter the water and return it to lakes, rivers, etc. As the amount of impervious area grows, the need to expand stormwater management grows with it. Many municipalities charge builders and developers an impact fee to cover the cost of stormwater management. Property owners may also be charged an annual fee for this purpose.

Table 6 Material Weights

Material	Loose Weight		Bank Weight	
	kg/m³	lb/yd³	kg/m³	lb/yd³
Clay – Natural Bed	1,660	2,800	2,020	3,400
Dry	1,480	2,500	1,840	3,100
Wet	1,660	2,800	2,080	3,500
Clay and Gravel – Dry	1,420	2,400	1,660	2,800
Wet	1,540	2,600	1,840	3,100
Decomposed Rock				
75% Rock, 25% Earth	1,960	3,300	2,790	4,700
50% Rock, 50% Earth	1,720	2,900	2,280	3,850
25% Rock, 75% Earth	1,570	2,650	1,960	3,300
Earth – Dry packed	1,510	2,550	1,900	3,200
Wet Excavated	1,600	2,700	2,020	3,400
Loam	1,250	2,100	1,540	2,600
Gravel – Dry (¼ to 2 inches)	1,690	2,850	1,900	3,200
Wet (¼ to 2 inches)	2,020	3,400	2,260	3,800
Sand – Dry, loose	1,420	2,400	1,600	2,700
Damp	1,690	2,850	1,900	3,200
Wet	1,840	3,100	2,080	3,500
Sand and Clay – Loose	1,600	2,700	2,020	3,400
Sand and Gravel – Dry	1,720	2,900	1,930	3,250
Wet	2,020	3,400	2,230	3,750

22308-12_T06.EPS

to make ¼ of a cubic yard more of loose material (700 ÷ 2,800). This material has a swell factor of 1.25 because 1 banked cubic yard makes 1.25 loose cubic yards. This is calculated by dividing the banked weight by the loose weight (3,500 ÷ 2,800).

2.2.0 Shrink Factor

Fill is brought onto the job site to bring the existing ground to the desired grade. When a fill site compacts naturally over time due to rain, freeze/thaw cycles, traffic, or the weight of a building, it is called **settlement** or **consolidation**; that is, the soil particles rearrange themselves so that they fit together more tightly. This is not desirable because the ground movement can damage foundations and road surfaces. To avoid settlement, the soil at construction sites is uniformly compacted before building begins.

When soil is compacted, its volume shrinks. The amount of **shrinkage** depends on the characteristics of the soil, the soil's moisture content, and the degree of compaction. The shrink factor can be calculated by dividing the compacted volume by the fill volume. The tricky part is determining the fill volume. Fill volume can be divided into the following three categories:

- *In situ* – There is no cut and no fill. The natural ground is cleared, grubbed, and then compacted.
- *Loose fill* – Fill is ordered and trucked in by the loose cubic yard, and then compacted.
- *Banked fill* – Fill is cut from another location, hauled to the building site, and compacted.

Fortunately, the method of calculating shrinkage is the same regardless of how the fill is obtained. The formula is: compacted volume divided by the loose volume (compacted volume/loose volume). For example, when 100 cubic yards of fill is compacted to 80 cubic yards, then the shrink factor is 0.8 (80 ÷ 100 = 0.8). See *Figure 19*.

This same formula works for weight, as well. If there are 2,800 pounds in a cubic yard of loose fill, and it takes 3,500 pounds of loose fill to make one cubic yard of compacted fill, the shrink factor is still 0.8, because 2,800 ÷ 3,500 equals 0.8. *Figure 20* shows that it takes 1¼ (1.25) cubic yards of loose fill to make a single cubic yard of compacted fill.

Now, how is the shrink factor used to determine how much fill is needed? When a material has a shrink factor of 0.8, then a compacted cubic yard equals 1¼ loose cubic yards. The amount of fill can be calculated by dividing the compacted

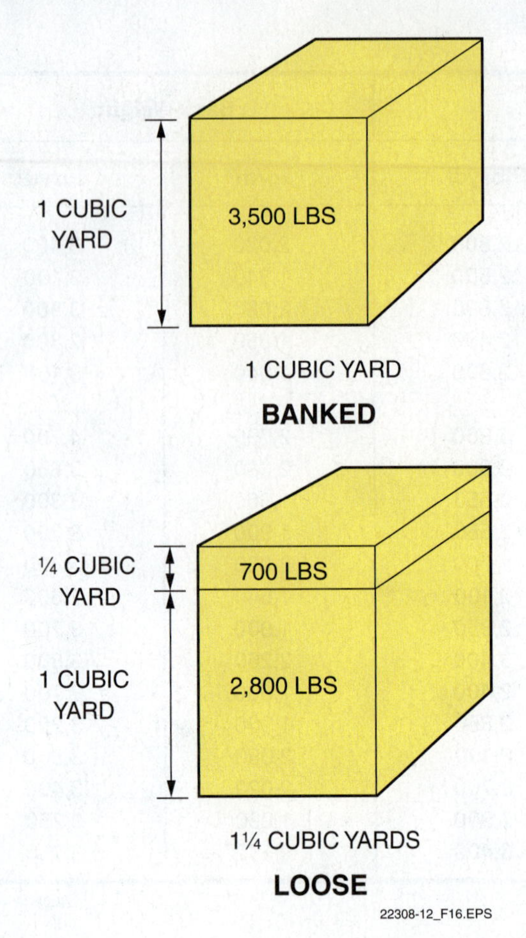

Figure 16 Swell factor.

volume by the shrink factor (1 ÷ 0.8 = 1¼ or 1.25). If an area needs to be filled with the volume of 10 cubic yards, using fill that has a shrink factor of 0.80, 12.5 cubic yards of fill (10 ÷ 0.8 = 12½ or 12.5) would need to be ordered.

These numbers should sound familiar, since they were used previously in the swell factor section. Shrinkage can be thought of as the opposite of swell. But keep in mind that when soil is being excavated, it may not be compacted to the density required at the building site. Therefore, assuming that a cubic yard of in situ banked soil will equal a cubic yard of compacted fill may get you in trouble.

2.3.0 Using Shrink and Swell Factors

In this section, you will learn to apply shrink and swell factors. You will also learn how these factors affect cycle times.

Overloading equipment can make it hard to control and places the operator at risk for an accident. Further, it can damage the equipment, resulting in costly down time and repairs. To ensure that equipment is not overloaded, the operator must apply the load and swell factors. Information about the equipment's weight and volume capacities can be found in the manufacturer's instructions.

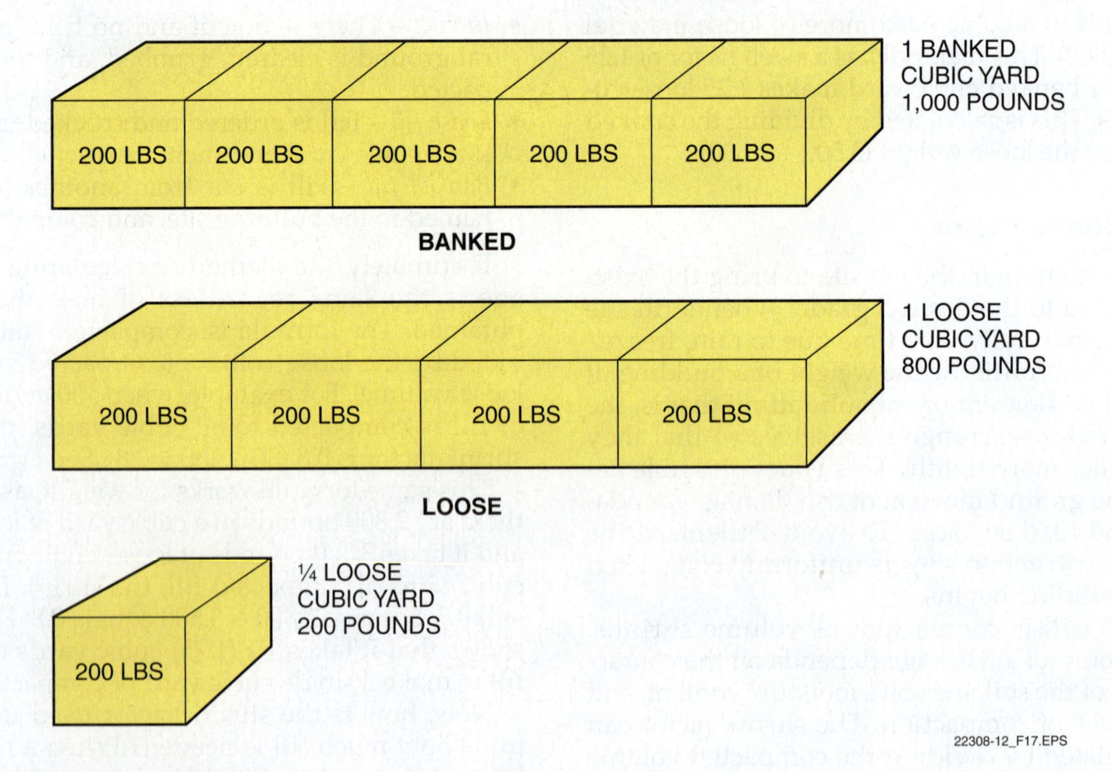

Figure 17 Swell factor example.

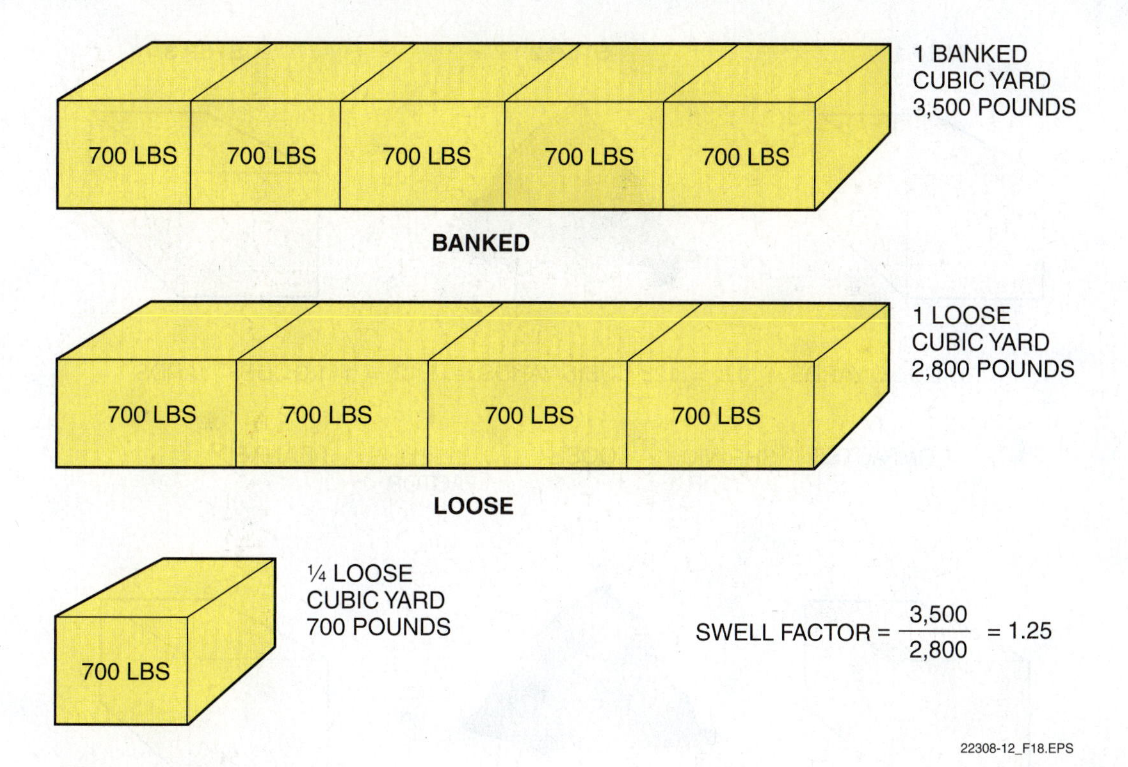

Figure 18 Wet clay swell factor.

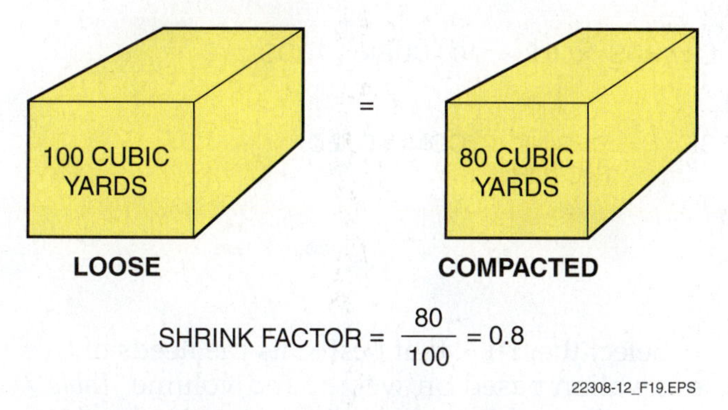

Figure 19 Shrink factor using volume.

2.3.1 Scenario

Let's assume that you need to excavate some damp sand from a borrow pit and take it to a building site. The sand has a banked weight of 3,200 pounds per cubic yard and a loose weight of 2,850 pounds per cubic yard (this information is from *Table 6*). You need to fill an area with a volume of 100 cubic yards. This material has a shrink factor of 0.8 when thoroughly compacted. This may be confusing, so refer to *Figure 21A* while carefully reading the following paragraph.

First, determine how much fill is needed. Since the compacted volume is 100 cubic yards and the shrink factor is 0.8, you need 125 cubic yards of loose fill (100 ÷ 0.8 = 125).

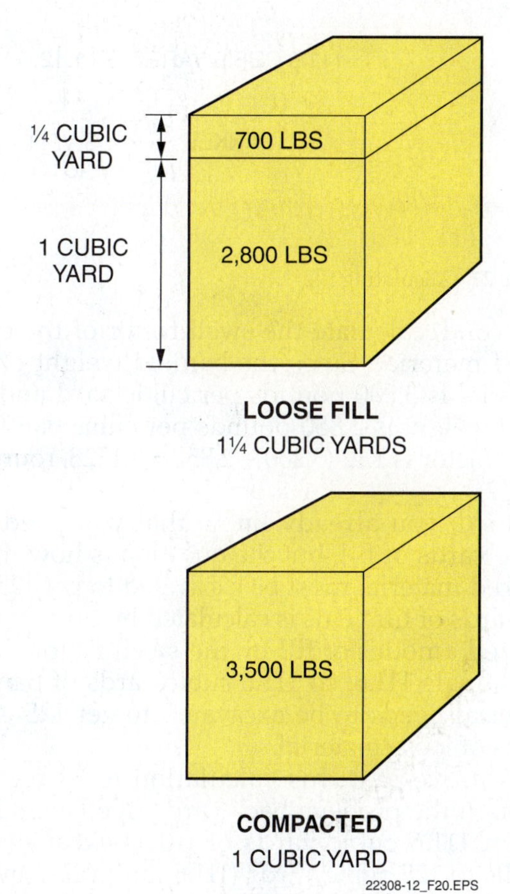

Figure 20 Shrink factor using weight.

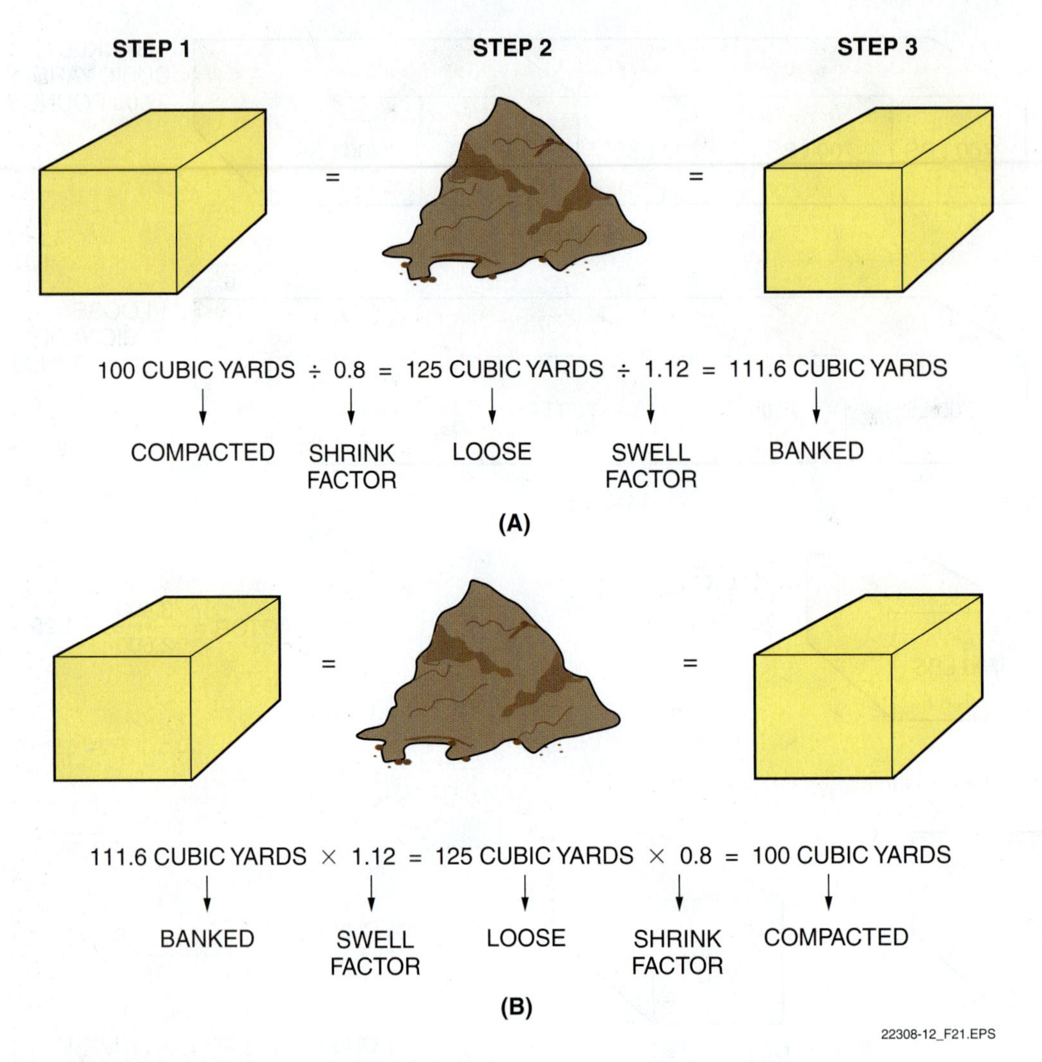

Figure 21 Calculating fill.

Second, calculate the swell factor of the excavated material. Since the banked weight of the material is 3,200 pounds per cubic yard and the loose weight is 2,850 pounds per cubic yard, the swell factor is 1.12 (3,200 ÷ 2,850 = 1.1228, rounded to 1.12).

Third, you already know that you need 125 cubic yards of fill, but the question is how much banked material must be excavated to get 125 cubic yards of fill? This is calculated by dividing the desired amount of fill by the swell factor. 125 ÷ 1.12 equals 111.6, so 111.6 cubic yards of banked material needs to be excavated to get 125 cubic yards of loose material.

To prove that this calculation is correct, go through the problem backwards (see *Figure 21B*). When 111.6 cubic yards of fill is excavated, it swells to 125 cubic yards (111.6 multiplied by the 1.12 swell factor equals 124.9, rounded to 125). When this material is compacted, it shrinks to 100 cubic yards because it has a shrink factor of 0.8 (125 multiplied by the shrink factor of 0.8, equals 100).

Select the truck that best suits the needs of this excavation based on weight and volume. *Table 7* shows the volume and weight capacity for three articulated trucks. Your instinct is probably telling you to select the truck with the largest payload, so you can get your work done quickly. This is a good instinct, but you need to check that the weight capacity is acceptable for the material you are going to haul, because the cost of operating a large truck is greater than that of using a smaller truck.

You already know that the weight of the loose material is 2,850 pounds per cubic yard, so you need to calculate the loaded weight of the material.

NOTE

The term *heaped* refers to material that has been heaped on the equipment. It is commonly assumed that material will heap in a two to one slope. The term *struck* refers to material level with the sides of the equipment.

Table 7 Articulated Trucks Payload Capacities

Truck Model	Volume Capacity (heaped SAE 2:1*, cubic yards)	Volume Capacity (struck, cubic yards)	Rated Payload (calculated, pounds)
TRUCK A	18.8	14.5	52,000
TRUCK B	22.1	17.1	62,000
TRUCK C	30	22.8	84,000

*SAE 2:1 refers to the angle of repose for the loaded material. For every 2 feet of horizontal run, there is 1 foot of rise. Source: Caterpillar® website.

22308-12_T07.EPS

Truck A:

- Heaped weight is 18.8 multiplied by 2,850 or 53,580 pounds. This exceeds the weight capacity of the vehicle.
- Struck weight is 14.5 multiplied by 2,850 or 41,325 pounds. This is within the weight capacity of the vehicle.

Truck B:

- Heaped weight is 22.1 multiplied by 2,850 or 62,985 pounds. This exceeds the weight capacity of the vehicle.
- Struck weight is 17.1 multiplied by 2,850 or 48,735 pounds. This is within the weight capacity of the vehicle.

Truck C:

- Heaped weight is 30 multiplied by 2,850 or 85,500 pounds. This exceeds the weight capacity of the vehicle.
- Struck weight is 22.8 multiplied by 2,850 or 64,980 pounds. This is within the weight capacity of the vehicle.

All of the available trucks can carry the load when it is struck. To get the job done as quickly as possible, you would select the largest truck because it can do the job in six trips, whereas Truck A would require nine trips and Truck B eight trips. (Divide the total quantity of 125 yards by the struck volume capacity.)

NOTE

This exercise demonstrated equipment selection based on speed. Large vehicles cost more to purchase, maintain, and operate. To decide which truck is the most cost efficient, you would need to compare operating costs.

2.0.0 Section Review

1. The banked weight of clay is less than its loose weight.
   a. True
   b. False

2. Given soil with a loose weight of 2,600 pounds per cubic yard and a bank weight of 3,100 pounds per cubic yard, the swell factor is _____.
   a. 1.19
   b. 1.25
   c. 1.30
   d. 1.33

3. The shrink factor of soil is found by _____.
   a. observing it for 1 hour after it is moistened
   b. dividing its compacted volume by its loose volume
   c. dividing its banked weight by its loose weight
   d. multiplying its weight by its compacted volume

SECTION THREE

3.0.0 WORKING IN VARIOUS SOIL CONDITIONS

Objective 3

Describe working in various soil conditions.

a. Describe the weight bearing and flotation properties of different soils.
b. Explain how soil characteristics affect machine performance.
c. Describe how soil conditions can affect trenching safety.

Different soils have different characteristics that affect the operation of heavy equipment. An operator must operate the equipment safely in different soil types and environmental conditions. Operators must know how to use the different types of heavy equipment to overcome problems caused by different soil characteristics. Three important characteristics that you need to keep in mind are the weight-bearing capability of a soil; the amount of traction a piece of equipment can get on a soil; and the resistance of the material to excavation.

> **NOTE**
> This section contains a number of tables that show technical information. Use the information in the tables to compare the characteristics of different soil types rather than trying to memorize the information.

3.1.0 Weight-Bearing and Flotation Properties

It is important to understand the relationship between the equipment and the weight bearing capacity of the soil on which the equipment is used. This section explains that relationship.

3.1.1 Weight Bearing

The weight-bearing capability of soil is how much weight it can safely support. It is related to soil density, so when a soil density is low, meaning it is not well compacted, its weight-bearing capacity is low. Compaction increases a soil's weight-bearing capability and makes it more stable. *Table 8* shows the approximate weight bearing capability of different types of soils, assuming a high degree of compaction. Weight bearing capability is usually measured in terms of pounds per square inch (psi), but it can be measured in any terms of weight per unit of measure, such as pounds per square foot. The abbreviation psi means the number of pounds that the soil can support on one square inch without sliding, excessively deforming, or collapsing.

Figure 22 illustrates the concept of pounds per square inch of pressure. In *Figure 22A*, a 200-pound person stands on one foot on a block of wood that is a 1-inch cube. The surface area in contact with the ground is 1 square inch, so the ground under the block is supporting 200 psi.

In *Figure 22B* the same person is standing with each foot on a 1-inch cube, so the ground under each cube is supporting 100 psi (200 pounds ÷ 2). This same concept may be applied to heavy equipment, because the total weight of the machine needs to be supported on the area of the tires or tracks in contact with the ground. When the weight on the tires exceeds the weight-bearing ability of the soil, the vehicle sinks into the ground until the pressure between the tires and the ground is equal.

3.1.2 Flotation

The ability of a tire or track to support the machine's weight on the ground is called flotation. To determine flotation capacity, consider the total weight of the machine, including its payload, the contact area (in square inches) of the tires or tracks on the ground, and the weight bearing ability of the ground. The total weight of the vehicle is divided by the total contact area of the tires or tracks with the ground to determine the pressure of the machine on the ground. If this pressure is greater than the load-carrying capability of the ground, the equipment will sink. Larger tires improve flotation because more tire surface is in contact with the ground. Reducing tire pressure can also increase flotation because more tire tread is in contact with the ground (*Figure 23*).

Table 8 Weight-Bearing Capability

Material	Weight-Bearing Capability in lb per in^2 (psi)
Rock (semi-shattered)	70
Rock (solid)	350
Clay (dry)	55
Clay (medium dry)	27
Clay (soft)	14
Gravel (cemented)	110
Sand (compact dry)	55
Sand (clean dry)	27
Quicksand	7

22308-12_T08.EPS

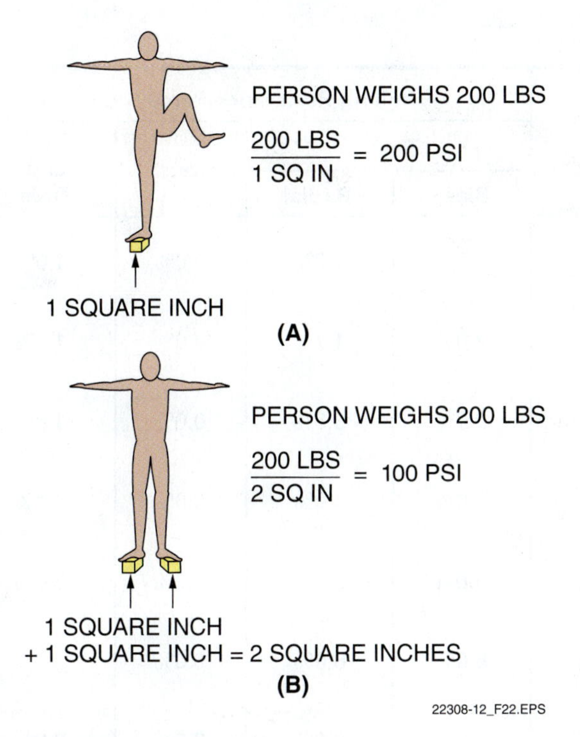

Figure 22 Determination of psi.

> **WARNING!** Always keep tire inflation pressure within the manufacturer's recommendation. Under-inflation may damage tires and cause a sudden tire failure, which may result in an accident.

3.2.0 Effects of Soil Conditions on Machine Performance

The condition of the soil on which a machine is working can have a significant effect on the performance of the machine. Soil conditions that affect machine performance include rolling resistance, digging resistance, and traction.

3.2.1 Rolling Resistance

Rolling resistance is the resistance of the tires to movement. You have probably experienced having your personal vehicle sink in mud or snow. In this situation, more power is needed to move the vehicle. The deeper the tires sink, the more power is needed to move the vehicle. This increases rolling resistance, and it seems as if the vehicle is always going uphill. Tires do not need to sink for the vehicle to experience rolling resistance. When the road surface flexes as the vehicle moves over it, as shown in *Figure 24*, the effect is the same as sinking. Rolling resistance is measured in terms of percentage of the machine's total weight. The effect of a rolling resistance of 20 percent is as if the vehicle is towing an additional 20 percent of its weight. *Table 9* shows some typical rolling resistances based on the type of tires or tracks.

3.2.2 Digging Resistance

All material has some degree of resistance to excavation. Some materials have more resistance than others. The type of material, along with its hardness, cohesiveness, weight, and degree of compaction, determine its resistance. Naturally, more resistant material will requires heavy-duty

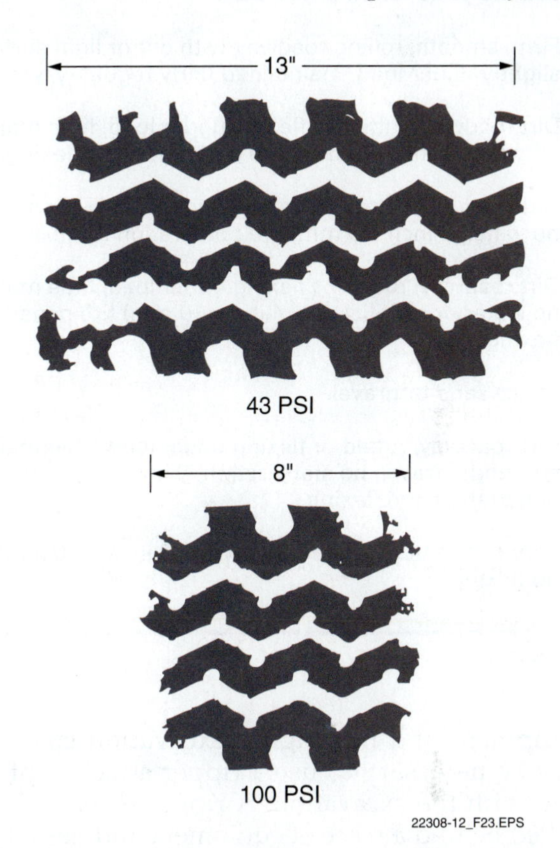

Figure 23 Reducing tire inflation improves flotation.

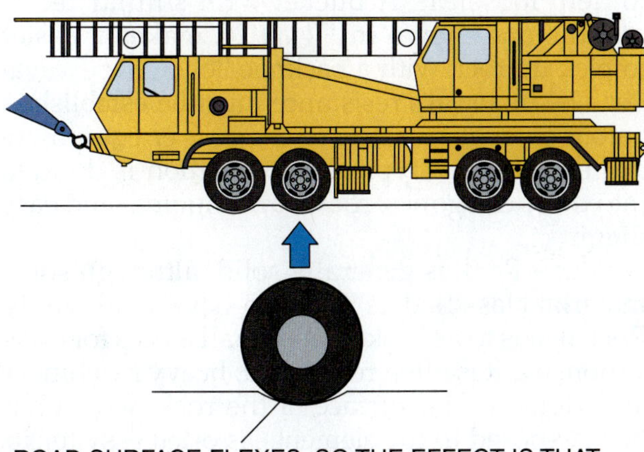

Figure 24 Flexing of road surface increases rolling resistance.

Table 9 Rolling Resistance

Surface Conditions	Rolling Resistance Percentage			
	Tires		Track	Track and Tires
	Bias	Radial		
Very hard, smooth concrete, cold asphalt, or dirt surface with no penetration or flexing.	1.5%	1.2%	0.0%	1.0%
Hard, smooth stabilized surface that is maintained and watered with no penetration under load.	2.0%	1.7%	0.0%	1.2%
Firm, smooth, rolling roadway with dirt or light surfacing, flexing slightly under load, maintained fairly regularly, watered.	3.0%	2.5%	0.0%	1.8%
Dirt roadway, rutted or flexing under load, little maintenance, no water, 1 inch (25 mm) tire penetration or flexing.	4.0%	4.0%	0.0%	2.4%
Dirt roadway, rutted or flexing under load, little maintenance, no water, 2 inch (50 mm) tire penetration or flexing.	5.0%	5.0%	0.0%	3.0%
Dirt roadway, rutted or flexing under load, little maintenance, no water or stabilization, 4 inch (10 mm) tire penetration or flexing.	8.0%	8.0%	0.0%	4.8%
Loose sand or gravel.	10.0%	10.0%	2.0%	7.0%
Dirt roadway, rutted or flexing under load, little maintenance, soft under travel, no stabilization, 8 inch (200 mm) tire penetration and flexing.	14.0%	14.0%	5.0%	10.0%
Very soft, muddy, rutted roadway, 12 inch (300 mm) penetration, no flexing.	20.0%	20.0%	8.0%	15.0%

SOURCE: CATERPILLAR

22308-12_T09.EPS

equipment and has higher excavation costs. It may be necessary to use a ripper attachment to help with the excavation. A ripper (*Figure 25*) is pulled behind a piece of equipment and has large teeth that comb through the excavation material to help loosen it. A bucket with similar teeth, like the one shown in *Figure 26*, makes it easier to pick up rock with a backhoe/loader or excavator. The material's resistance must be established before excavation starts so the proper equipment can be selected. Typically, excavation is divided into three categories: rock, hard digging, and easy digging.

Rock – Rock is generally solid, although some material classified as rock can have small voids. Rock needs to be broken into small pieces for excavation, but it is often resistant to heavy machinery and rippers. The surface of the rock, which has been exposed to the elements, is often easy to rip. But lower layers, which have been shielded from the weather, are more resistant. Sometimes rocks have been formed with fissures.. It can be broken along the fissures by approaching the rock mass from a different angle with the equipment. Often, rock can be broken by excavating beneath it and allowing the weight to shear off pieces of the rock, although this practice can be dangerous.

22308-12_F25.EPS

Figure 25 Ripper attachment.

22308-12_F26.EPS

Figure 26 Heavy-duty rock bucket.

Some material can be broken mechanically with a pneumatic jackhammer or a pneumatic drill (*Figure 27*), but sometimes it requires blasting. Blasting involves drilling a hole into the rock, packing it with explosives, and detonating them. Excavation of solid rock is the most expensive and dangerous type of earthwork.

WARNING!

Blasting requires advanced training and specialized knowledge of rock formations and explosives. A heavy equipment operator should never set explosives.

Hard digging – Hard digging doesn't require drilling or blasting, but the material is still very strong and resistant to digging. Materials that require hard digging include heavy material, such as crushed stone and gravel; or cohesive material, such as soft moist clay. These materials can be dug with heavy-duty machinery, such as front-end loaders, power buckets, bulldozers, and rippers.

Easy digging – Easy digging materials are usually light and dry, such as soft or fine loose earth or sand. Although these materials offer little resistance to digging, they may make it difficult for heavy equipment to move because they may lack traction.

3.2.3 Traction

Traction is the friction between the road surface and the tire or tracks of a vehicle. Low friction between the road surface and vehicle tires allows the tires to spin without moving the vehicle, forcing some of the engine power to be used to overcome the spinning of the tires rather than to move the vehicle. Some of the things that affect vehicle traction are the surface material coefficient of traction, the weight of the vehicle, and the type of tires or tracks.

Equipment weight distribution – Equipment weight distribution refers to the force of an engine's power on the drive wheels of a vehicle moves the vehicle across a surface. Ideally, all of the engine's power would be used to move the vehicle, but this is not possible because of traction and rolling resistance. The coefficient of traction and the amount of weight on the vehicle's drive wheels determine the amount of force a vehicle can deliver before the tires start to slip.

Another point to keep in mind is that when climbing a hill, gravity shifts some of the vehicle weight to the rear wheels, thus increasing traction for rear drive vehicles. Unfortunately, when driving down a steep grade, gravity shifts vehicle weight to the front wheels, thus decreasing traction of rear drive vehicles, which can make a vehicle difficult to handle.

Tire treads and traction – Tire treads are the working surface of a tire. The type of tread helps determine the flotation capacity of a vehicle. Many tires are designed to work effectively in many different environmental conditions. Generally, tires with a deep tread provide good traction under most conditions, but shallower treads are best on dry sand and ice, because they have better flotation. For tires to operate correctly, the operator needs to be sure that they are inflated according to manufacturer's specification and that the treads are free of foreign matter, such as mud.

22308-12_F27.EPS

Figure 27 Pneumatic rock drill.

Tracks – Tracked equipment is often used in muddy work because it has better flotation than wheeled equipment. Some tracks have especially wide track shoes to further increase flotation. All tracks are not equal, however; it is important to match the track shoe to the job that needs to be done.

Smooth or worn tracks are not suitable for mud work because they offer little traction on wet, slippery surfaces. It may be possible to improve the traction of worn tracks by replacing some of the shoes with new shoes, but this is not a good choice when the equipment must also be used on hard ground.

Grouser shoes (*Figure 28*) are often used to improve equipment footing on muddy ground, but they are not without problems. Grousers dig into loose soil very quickly and sometimes mud gets packed between them, giving the effect of a single, smooth surface with no traction.

3.2.4 Working in Mud

Mud is particularly troublesome for heavy equipment operators. Mud occurs when enough water has been mixed with a soil and to bring it to a liquid consistency. Mud has a very poor weight-bearing capability, so equipment sinks until it finds support. This increases rolling resistance on the tires and forces the engine to use extra power to overcome the resistance in order to move the vehicle. Some soils are more likely to turn muddy than others. Coarse-grained soils, such as sand, tend to drain water quickly and are unlikely to become muddy. Fine-grained soils, such as clay and silt, hold water for long periods and become muddy. Soil mixtures that contain a lot of fine-grained soil are also likely to become muddy.

Mud is often shallow. The vehicle tires can sink a small amount to reach solid ground, but when the mud is deeper, a vehicle with better floatation, such as a tracked vehicle or one with wide, soft tires, may need to be used. When the mud is too deep to support any vehicle, platforms, timbers, or mats will need to be laid on its surface in order to support the vehicle (*Figure 29*).

Even under the best of circumstances, driving on timbers or platforms is risky. Whatever is used must be wider than the vehicle tire or track and strong enough to support the weight of the vehicle. Both sides of the vehicle must be supported equally to avoid having one side sink into soft ground, causing the vehicle to tip over. Also, the timbers or platform must be moved as the vehicle is moved.

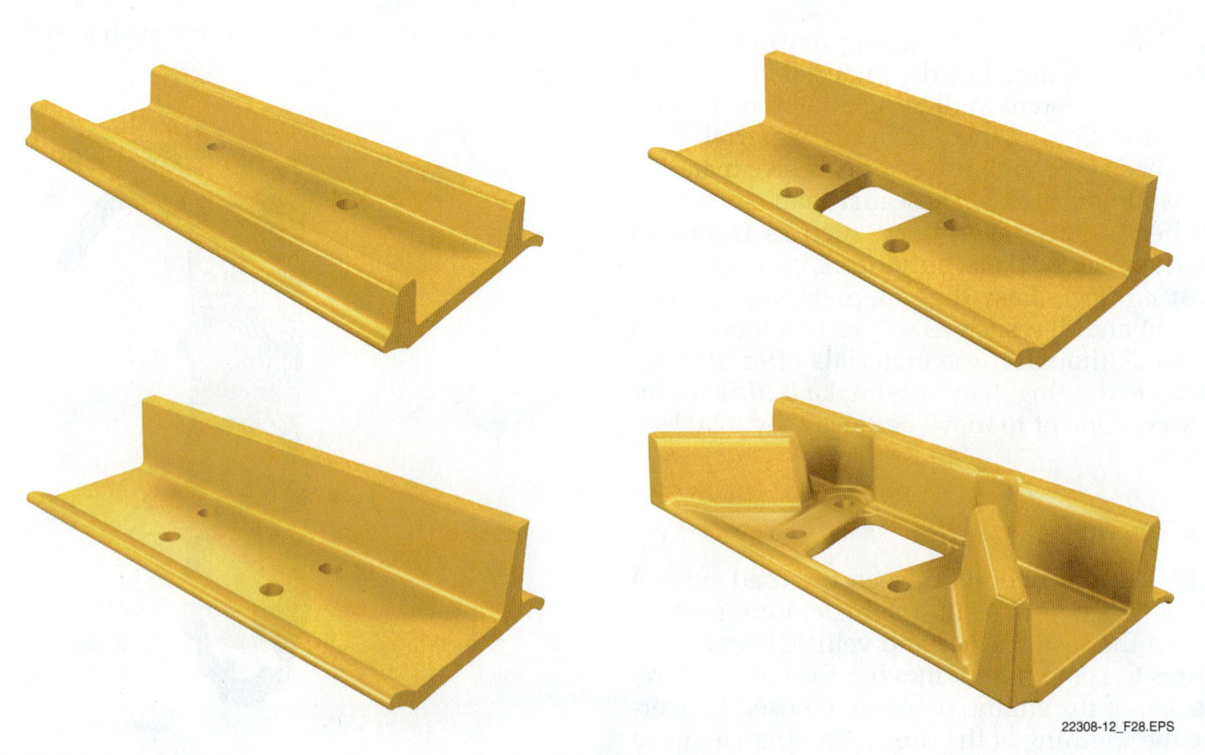

Figure 28 Grouser shoes.

22308-12_F29.EPS

Figure 29 Mud mat.

3.3.0 Trenching Hazards

The type of soil in and around a trench can contribute to the collapse of trench walls. Soil type is a major factor that must be considered in trenching operations. Only a competent person who has additional education, experience, and authority can decide if the soil in and around a trench is safe and stable. However, it is still the operator's responsibility to know some of the basics about soil and its hazards.

The soil found on most construction sites is a mixture of many types, including sand, loam, clay, and silt. It is the type of mixture that gives the soil its properties. For example, sand with a small amount of silt and clay may compact well and permit excavation of a stable trench.

Each of the various soil types, depending on their conditions at the time of excavation, behave differently. Sandy soils tend to collapse straight down, and wet clay and loams tend to slab off the sides of the trench (*Figure 30*). Firm, dry clays and loams tend to crack, and wet sand and gravel tend to slide.

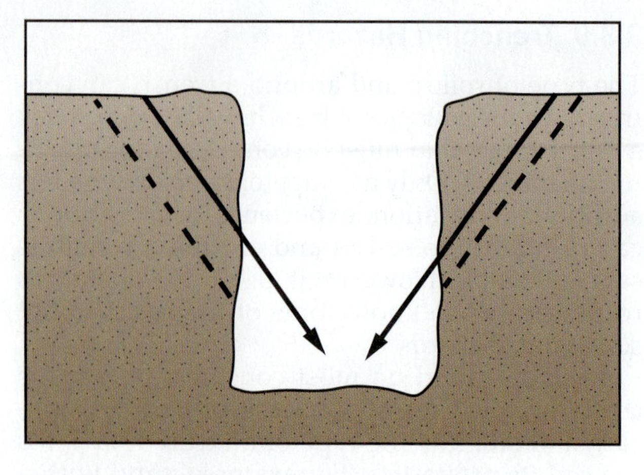
SANDY SOIL COLLAPSES

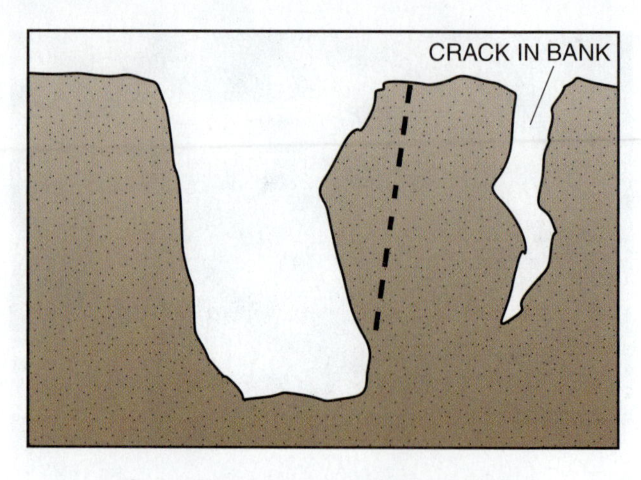

FIRM DRY CLAY AND LOAMS CRACK

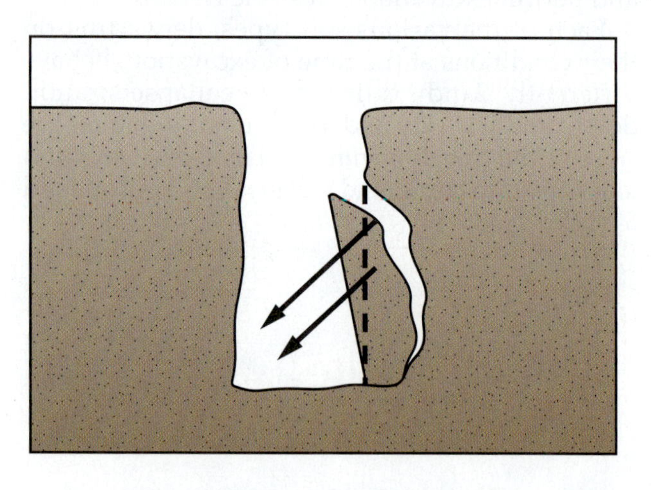
WET CLAY AND LOAMS SLAB OFF

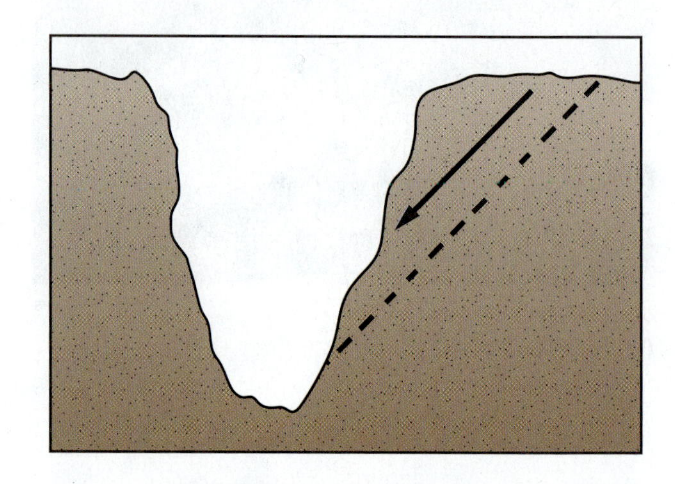
WET SANDS AND GRAVELS SLIDE

22308-12_F30.EPS

Figure 30 Behavior of different soils in trenches.

3.0.0 Section Review

1. Which of these soils has the greatest weight-bearing capacity?
   a. Dry clay
   b. Cemented gravel
   c. Compacted dry sand
   d. Medium dry clay

2. A ripper is used to _____.
   a. pick rocks out of soil
   b. loosen heavily compacted soil
   c. improve traction in damp soil
   d. break up rocks

3. A grouser shoe is used to _____.
   a. improve equipment footing on muddy ground
   b. compact soil that has been loosened by a ripper
   c. improve the traction of the tires on wheeled vehicles
   d. make it easier to walk on muddy ground

4. Which of these soil conditions is likely to cause a trench to fail by slabbing off?
   a. Sandy soil
   b. Firm, dry clay
   c. Wet sand
   d. Wet clay

SUMMARY

The term *soil* means any type of earthen material, including rock. It does not include bedrock, which is the base of Earth's surface. The soil forms on the bedrock in layers, which are called horizons. In general, there are three categories of soil: organic, inorganic, and rock. Organic material comes from living matter such as plants or animals. Inorganic material comes from rocks that have been broken down due to natural forces. Rock is any material that cannot be dug or loosened by available machinery.

The two soil classification systems that are used in construction are the ASTM Unified Soil Classification System and the AASHTO Soil Classification System. Soils are generally classified by particle size as granular and fine-grained. Sand, gravel, and rock are considered granular, while clay and silt are fine-grained. No single type of soil is ideal for construction; rather, it is a mixture of several types that will yield the ideal construction site.

Construction workers are mainly concerned with a soil's strength, which is related to its density. The density of soil is measured as a ratio of its weight to its volume, usually in pounds of wet soil or dry soil per cubic foot. Soil is compacted to improve its density. Compaction refers to the act of artificially increasing the density of the soil using vibration or weight. It is one of the most important activities in construction. To increase the density of soil, more soil particles must be moved into the same space by expelling air or water. A large amount of water makes it impossible to compact soil, but a small amount aids in compaction because it acts like a lubricant to rough-surfaced particles, allowing them to slip into voids more easily.

When working in various soil conditions, it is necessary to know how certain types of equipment will react to different soil conditions. The conditions that must be watched are the ability to rip rock, traction, and the flotation capacity of the equipment. Rocks have different levels of hardness and therefore cannot be broken up in the same way as soil.

It is important for heavy equipment operators to be able to identify different soil types and to understand how different environmental factors—especially rain—affect soil conditions. The equipment will behave differently in wet soil than it will in dry soil, so it is important to stay alert to soil changes to complete the job safely.

Review Questions

1. Air space or open pockets between soil particles are called ______.
   a. fines
   b. voids
   c. rakes
   d. spoils

2. Which of these factors has the most effect on the classification of soil?
   a. Size of particles
   b. Hardness of particles
   c. Behavior when moisture content varies
   d. Static weight of particles

3. Which of the following soil types has the finest grain size?
   a. Silt
   b. Gravel
   c. Clay
   d. Fine sand

4. The grain size of soil particles is measured by _____.
   a. compacting with a ram
   b. sieving
   c. measuring each particle and taking the average
   d. performing the sand cone test

5. When soil has enough small particles to fill the voids between large ones it is _____.
   a. well-graded
   b. sieved
   c. gradated
   d. poorly differentiated

6. Which of the following is *not* an engineering property of soil?
   a. Elasticity
   b. Conductivity
   c. Cohesion
   d. Plasticity

7. Soil density tests are performed to ______.
   a. see if a site needs additional compaction
   b. set a standard for measuring field tests
   c. test material for uniform gradation
   d. compare soils from different parts of the country

8. The ratio of a weight of a material to its volume is called ______.
   a. density
   b. lift
   c. mass
   d. specific gravity

9. The soil with the highest density and lowest water content is _____.
   a. poorly graded sand
   b. clay
   c. well-graded sand
   d. silty clay

10. The single most important factor in the compaction of soil is _____.
    a. surface texture
    b. chemical composition
    c. moisture content
    d. weight of the compaction equipment

11. When soil is removed from its natural (or banked) state, what typically happens to it?
    a. It shrinks.
    b. The chemical composition changes.
    c. It swells.
    d. Nothing happens.

12. The swell factor is the _____.
    a. number of loose yards in a banked yard
    b. amount of consolidation
    c. size of the soil particles
    d. small drainage slope between buildings

13. A material has a shrink factor of 0.8. In order to fill a 10 cubic yard trench, how many cubic yards of fill will be needed?
    a. 8
    b. 10
    c. 12.5
    d. 80

14. The purpose of compacting soil is to _____.
    a. produce a more stable surface
    b. make the surface smooth
    c. keep the excavation from failing
    d. squeeze out all of the moisture

15. Which of the following soils is difficult to dig because of strong, uniform cohesion?
    a. Soft clay
    b. Wet sand
    c. Gravel
    d. Boulders

Trade Terms Introduced in This Module

American Association of State Highway and Transportation Officials (AASHTO): An organization representing the interest of all state government highway and transportation agencies throughout the United States. This organization establishes design standards, materials-testing requirements, and other technical specifications concerning highway planning, design, construction, and maintenance.

American Society of Testing Materials (ASTM): A national organization that establishes standards for testing and evaluation of manufactured and raw materials.

Banked: Any soil mass that is to be excavated from its natural position.

Bedrock: The solid layer of rock under Earth's surface. Its solid-rock state distinguishes it from boulders.

Capillary action: The tendency of water to move into free space or between soil particles, regardless of gravity.

Cohesive: The ability to bond together in a permanent or semipermanent state. To stick together.

Consolidation: To become firm by compacting the particles so they will be closer together.

Density: Ratio of the weight of material to its volume.

Elasticity: The property of a soil that allows it to return to its original shape after a force is removed.

Expansive soil: A soil that expands and shrinks with moisture. Clay is an expansive soil.

Fines: Very small particles of soil. Usually particles that pass the No. 200 sieve.

Friable: Crumbles easily.

Horizon: Layers of soil that develop over time.

Humus: Dark swamp soil or decaying organic matter. Also called peat.

In situ: In the natural or original place on site.

Inorganic: Derived from other than living organisms, such as rock.

Liquid limit: The amount of moisture that causes a soil to become a fluid.

Loading: Applying a force to soil. A building can be a permanent load at a site, and a truck can be a passing load on a roadway.

Optimum moisture: The percent of moisture at which the greatest density of a particular soil can be obtained through compaction.

Organic: Derived from living organisms, such as plants and animals.

Peat: Dark swamp soil or decaying organic matter. Also called humus.

Plastic limit: The amount of water that causes a soil to become plastic (easily shaped without crumbling).

Plasticity: The range of water content in which a soil remains plastic or is easily shaped without crumbling.

Settlement: To become firm by compacting the particles so they will be closer together.

Shrinkage: Decrease in volume when soil is compacted.

Swell: Increase in volume when soil is excavated.

Swell factor: The ratio of the banked weight of a soil to the loose weight of a soil.

Voids: Open space between soil or aggregate particles. A reference to voids usually means that there are air pockets or open spaces between particles.

Water table: The depth below the ground's surface at which the soil is saturated with water.

Well-graded: Soil that contains enough small particles to fill the voids between larger ones.

Additional Resources

This module presents thorough resources for task training. The following resource material is suggested for further study.

"Arrest that Fugitive Dust," 2002. Roberta Baxter, *Erosion Control* magazine, Forester Communications, Inc.

Basic Equipment Operator, NAVEDTRA 14081, 1994 Edition. Morris, John T. (preparer), Naval Education and Training Professional Development and Technology Center.

Caterpillar Performance Handbook, Edition 33. A CAT® Publication. Peoria, IL: Caterpillar, Inc.

Moving the Earth, 3rd Edition, 1988. H.L. Nichols. Greenwich, CT: North Castle Books.

"Optimizing Soil Compaction and Other Strategies," 2004. Donald H. Gray, *Grading and Excavation Contractor* magazine. Forester Communications, Inc.

Soil Stabilization for Pavements Mobilization Construction Engineer Manual (EM 1110-3-137), 1984. Department of the Army, Corps of Engineers.

Temporary Stream and Wetland Crossing Options for Forest Management, 1998. Charles R. Blinn, Rick Dahlman, Lola Hislop, and Michael A. Thompson, USDA Forest Service, North Central Research Station, General Technical Report NC-202.

Using Lime for Soil Stabilization and Modification, 2001. National Lime Association website (**www.lime.org**).

Figure Credits

Reprinted courtesy of Caterpillar Inc., Module opener, Figures 25, 26, and 28

USDA Natural Resources Conservation Service, Figure 3

Topaz Publications, Inc., Figures 4, 5, 7, 11

Courtesy of American Educational Products, Figure 8

Courtesy of Durham Geo Slope Indicator, Figure 12B

Courtesy of Humboldt Mfg. Co., Figures 13 and 14

Courtesy of American Pneumatic Tools, Figure 27

Courtesy of SVE Portable Roadway Systems, Inc., Figure 29

Section Review Answers

Answer		Section Reference	Objective
Section One			
1	d	1.1.0, Table 3	1a
2	a	1.2.1	1b
3	a	1.3.0	1c
4	b	1.4.0, Figure 15	1d
Section Two			
1	b	2.1.0	2a
2	a*	2.1.0	2b
3	b	2.2.0	2b
Section Three			
1	b	3.1.0, Table 8	3a
2	b	3.2.2	3b
3	a	3.2.3	3b
4	d	3.3.0	3c

*3,100 ÷ 2,600 = 1.192

NCCER CURRICULA — USER UPDATE

NCCER makes every effort to keep its textbooks up-to-date and free of technical errors. We appreciate your help in this process. If you find an error, a typographical mistake, or an inaccuracy in NCCER's curricula, please fill out this form (or a photocopy), or complete the online form at **www.nccer.org/olf**. Be sure to include the exact module ID number, page number, a detailed description, and your recommended correction. Your input will be brought to the attention of the Authoring Team. Thank you for your assistance.

Instructors – If you have an idea for improving this textbook, or have found that additional materials were necessary to teach this module effectively, please let us know so that we may present your suggestions to the Authoring Team.

NCCER Product Development and Revision

13614 Progress Blvd., Alachua, FL 32615

Email: curriculum@nccer.org
Online: www.nccer.org/olf

❏ Trainee Guide ❏ AIG ❏ Exam ❏ PowerPoints Other ______________________________

Craft / Level: Copyright Date:

Module ID Number / Title:

Section Number(s):

Description:

Recommended Correction:

Your Name:

Address:

Email: Phone:

22212-13

Skid Steers

Overview

Skid steers are compact, highly maneuverable machines designed to accept a wide variety of attachments. Their standard configuration typically includes a bucket for moving soil, crushed stone, and related materials. With optional attachments, these machines provide a platform to accomplish an endless number of additional tasks. This module explores the basic care and proper operation of skid steer machines, as well as some of the many attachments that make it an indispensable tool in both construction and agriculture.

Module Seven

Trainees with successful module completions may be eligible for credentialing through NCCER's National Registry. To learn more, go to **www.nccer.org** or contact us at **1.888.622.3720**. Our website has information on the latest product releases and training, as well as online versions of our *Cornerstone* newsletter and Pearson's product catalog.

Your feedback is welcome. You may email your comments to **curriculum@nccer.org,** send general comments and inquiries to **info@nccer.org**, or fill in the User Update form at the back of this module.

This information is general in nature and intended for training purposes only. Actual performance of activities described in this manual requires compliance with all applicable operating, service, maintenance, and safety procedures under the direction of qualified personnel. References in this manual to patented or proprietary devices do not constitute a recommendation of their use.

22212-13
Skid Steers

Objectives

When you have completed this module, you will be able to do the following:

1. Identify and describe the components of a skid steer.
   a. Identify and describe chassis components.
   b. Identify and describe skid steer controls.
   c. Identify and describe skid steer instrumentation.
   d. Identify and describe skid steer attachments.
2. Describe the prestart inspection requirements for a skid steer.
   a. Describe prestart inspection procedures.
   b. Describe preventive maintenance requirements.
3. Describe the startup, shutdown, and operating procedures for a skid steer.
   a. State skid steer-related safety guidelines.
   b. Describe startup, warm-up, and shutdown procedures.
   c. Describe basic maneuvers and operations.
   d. Describe related work activities.

Performance Tasks

Under the supervision of the instructor, you should be able to do the following:

1. Complete a proper prestart inspection, maintenance, and housekeeping for a skid steer.
2. Demonstrate proper entrance and exiting of the operator cab on a skid steer.
3. Perform proper startup, warm-up, and shutdown procedures on a skid steer.
4. Execute basic maneuvers with a skid steer, including:
   - Grading
   - Removing stumps and boulders
   - Steering
   - Changing attachments
   - Loading a dump truck
   - Maneuvering on slopes
   - Utilizing a fork attachment for moving materials
5. Properly load and secure a skid steer for transport.

Trade Terms

Fixed displacement pump
Float mode
Hydrostatic drive
Rated operating capacity
Tipping load
Variable displacement pump

Industry Recognized Credentials

If you are training through an NCCER-accredited sponsor, you may be eligible for credentials from NCCER's Registry. The ID number for this module is 22212-13. Note that this module may have been used in other NCCER curricula and may apply to other level completions. Contact NCCER's Registry at 888.622.3720 or go to **www.nccer.org** for more information.

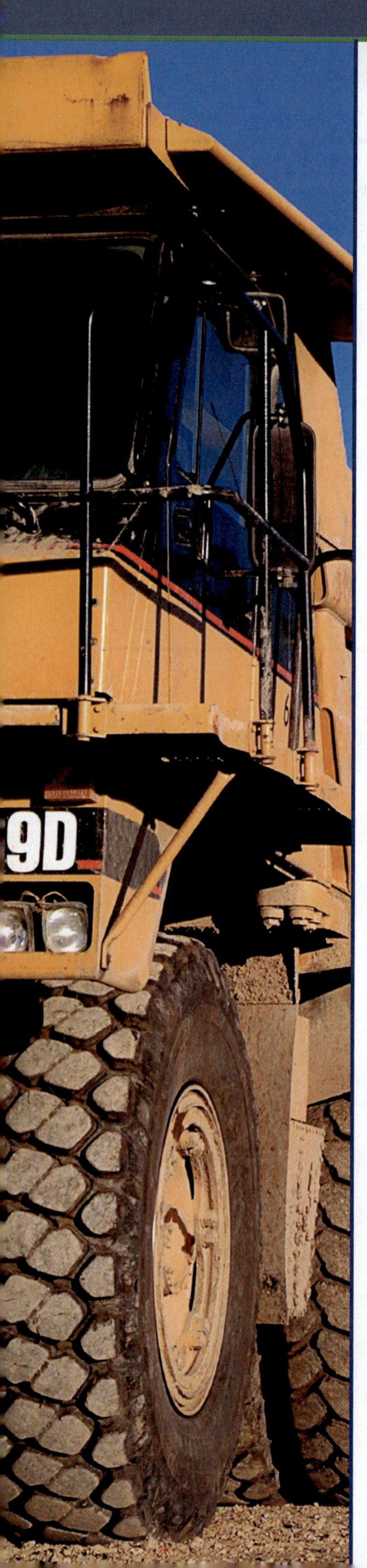

Contents

Topics to be presented in this module include:

Figures

Figures (continued)

Section One

1.0.0 Skid Steer Components

Objective 1

Identify and describe the components of a skid steer.

a. Identify and describe chassis components.
b. Identify and describe skid steer controls.
c. Identify and describe skid steer instrumentation.
d. Identify and describe skid steer attachments.

Trade Terms

Fixed displacement pump: A pump that cannot be adjusted to deliver more or less fluid volume in a pumping cycle.

Float mode: Placing the bucket or other attachment into the float mode allows the attachment to automatically follow the contour of the ground, remaining at the same elevation above it.

Hydrostatic drive: A system that relies on the flow and pressure of a fluid to transfer energy. In skid steers, the transfer of energy is from the engine, driven by diesel fuel, to the four drive wheels through hydraulic motors.

Variable displacement pump: A pump that can deliver a variable fluid volume in a pumping cycle to support the changing needs of a hydraulically operated device.

Skid steers are known by several names. They are commonly called skid-steer loaders or simply skid loaders. Although a number of companies manufacture and distribute them, the name Bobcat® has also become a common name used to describe any skid steer unit, much as Kleenex® has become synonymous with tissue.

Although skid steers may not be the first machine to come to mind when the term *heavy equipment* is heard, they are undoubtedly the most versatile piece of equipment found in many construction environments (*Figure 1*). Skid steers can be fitted with a vast array of attachments. They can break concrete, haul debris, and load it into a truck; saw and plane concrete or asphalt surfaces; mix fresh cement and pump it to an elevated form; auger holes in soil for fence posts and landscaping plants; and rake an area to remove debris and rocks before grass is planted. These are just a few of the tasks that can be done quickly and easily with a single piece of equipment.

The term *skid steer* is a very appropriate name, as it describes the way these units maneuver. Skid steers are built using tracks as well as wheels (*Figure 2*), but the wheeled versions are more common and the focus of this module. Although the earliest designs had three wheels, skid steers are typically four-wheel machines. However, unlike common loaders, skid steer units have no steerable wheels. Instead, the wheels on each side are driven together, always turning at the same speed and direction. By powering the pair of wheels on one side forward or backward while the opposite side wheels remain idle, the unit can be steered.

The Melroe Bobcat®

Founded in 1947, the Melroe Manufacturing Company was started by Edward G. Melroe, a Norwegian immigrant. Earlier compact loader models with three wheels were introduced, but the first true skid steer, with four powered wheels, was introduced in 1960. Here is an advertisement for one of the early Melroe Bobcats, bearing the name chosen for it in 1962 by a Melroe associate.

22212-12_SA01.EPS

22212-12_F01.EPS

Figure 1 A skid steer rakes a yard to prepare for turf grass planting.

22212-12_F02.EPS

Figure 2 A Bobcat® tracked loader uses a sod attachment to quickly install turf.

The idle wheels simply slide, or skid, in the direction they are forced by the driven wheels on the opposite side. There are compact loaders with four-wheel steering capability available, but they do not skid to steer.

One certain factor that makes skid steers popular and versatile is that they can be turned in extremely tight spaces. Powering the wheels on one side and allowing the wheels on the opposite side to slide does result in the ability to turn sharply, without creating a turning radius. However, these units can truly spin in a circle. By powering the wheels on the right backwards, while simultaneously powering the wheels on the left forward, the unit can literally spin as if it is on an axis. This special mobility is one of the primary features that make skid steers one of the most valuable tools on a job site.

Today's skid steers are far safer than their predecessors were. Earlier models offered little or no

protection for the operator in any direction. Required protection systems have reduced injuries remarkably. On the other hand, these units are so versatile, nimble, and responsive, they tend to make an operator feel somewhat invincible. This fact alone can lead to serious accidents.

This section introduces the basic components of a skid steer, along with the controls and instruments used in its operation. A number of the versatile attachments are also presented.

1.1.0 Main Chassis Components

A skid steer is built on an extremely rigid and strong frame known as the chassis. The extreme strength and rigid, compact construction of the chassis allows the machine to shrug off the stress imposed by skid steering with ease. All other major components associated with the operation of the machine are mounted to the chassis.

Figure 3 provides an overview of the primary chassis components and structure. Skid steers are very compact units by design. They have fewer and less complex parts than their larger relatives. Of course, each brand and model of skid steer is slightly different in its construction, capacity, and operation. It is essential that operators become familiar with the specific model of skid steer in use. When beginning to use a new or different machine, always review the operator manual.

One feature that allows their compact construction is the lack of a transmission. In fact, there is no direct relationship between the engine and the four wheels at all. The primary responsibility of the engine is to drive hydraulic pumps. All other functions, including movement of the machine, are then powered through hydraulic motors. Since the engine is not connected to the drive train, the differential is eliminated, along with axles that cross from one side of the frame to the other.

Figure 3 Skid steer chassis components.

1.1.1 Engine and Engine Area

Although a few skid steers use gasoline engines, the vast majority of engines are diesel-powered. The horsepower generally ranges from around 20 to 100 hp. As is clear in *Figure 4*, the engine and its accessories have been tightly packed into the compartment. However, serviceability is an important consideration. Although the compartment is quite crowded, most all items requiring regular attention are positioned to allow easy access. The engine coolant radiator as well as the fluid coolers, have been relocated in some models. The machine in *Figure 4* uses a tilt-up design, with the radiator above the engine. To service the engine, it is tilted up, as shown. This relocation is possible since the radiator is not cooled by an engine-driven fan blade. Instead, the fan blade is driven by a hydraulic motor. The fluid filters have been located near the access door to allow easy access to them as well.

The engine on all active models is located behind the operator. This keeps most of the weight at the rear of the machine to help balance the load.

1.1.2 Drive System

The drive system is one of the features that make skid steers unique. Instead of a traditional drivetrain with a gearbox and differential, the wheels are driven by powerful hydraulic motors. This type of drive system, one that relies on hydraulic fluid flow and pressure to transfer energy, is called a **hydrostatic drive**. Since there is no direct steering capability, there are no linkages or rods associated with turning the machine. In fact, there is no steering wheel for the operator. As a result, the drive system does not have to accommodate any change of direction in the wheels. They simply rotate in one direction or the other, and no more. The simplicity of the design adds to the reliability and reduces the maintenance requirements.

Figure 5 shows one example of the drive system common to skid steer machines. A single hydraulic motor drives both wheels on one side of the unit. Each of the two motors drives a dual sprocket, and the sprocket drives two chains. Each drive chain is then connected to a sprocket on a wheel. The sprockets on the wheels are

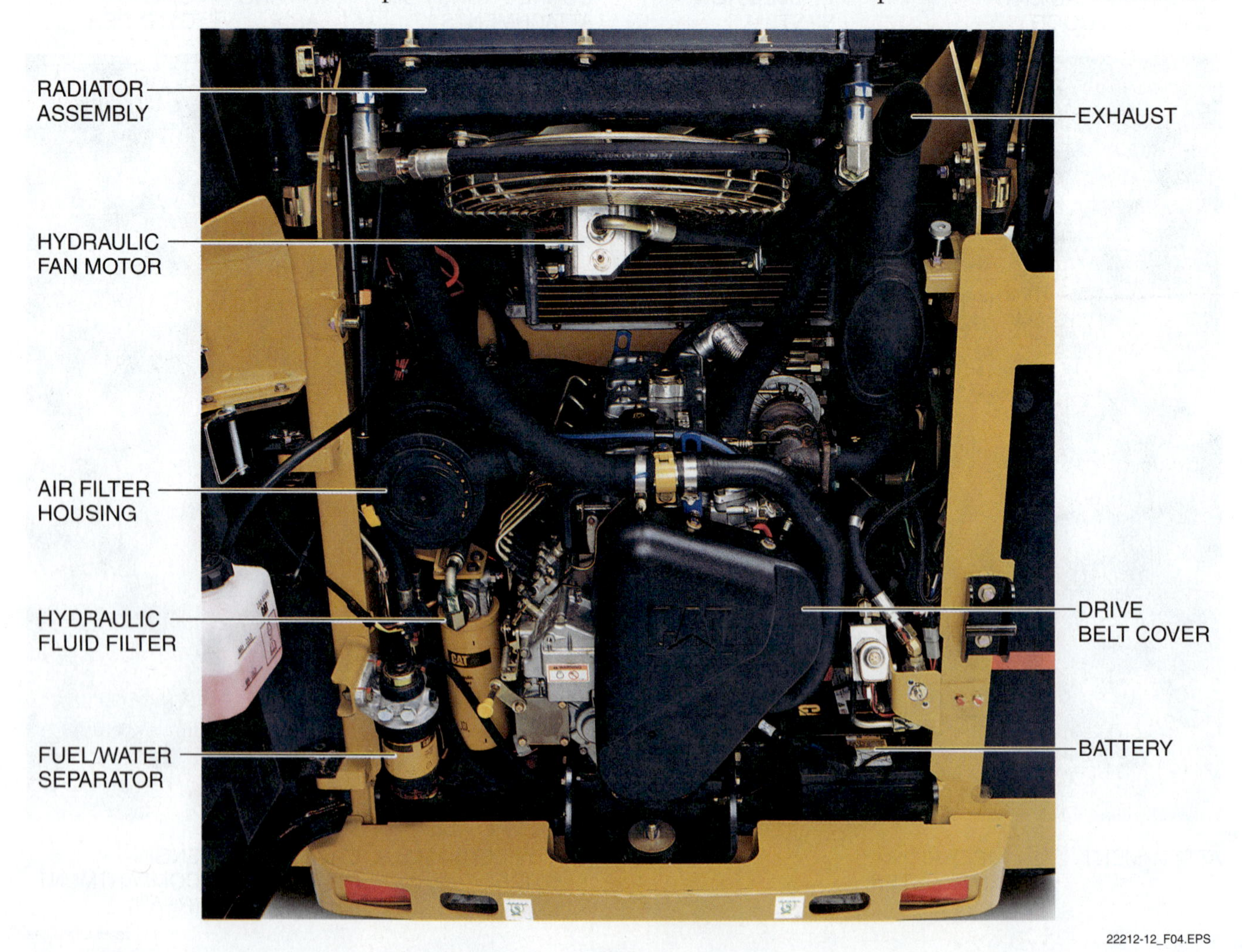

Figure 4 Engine and engine compartment.

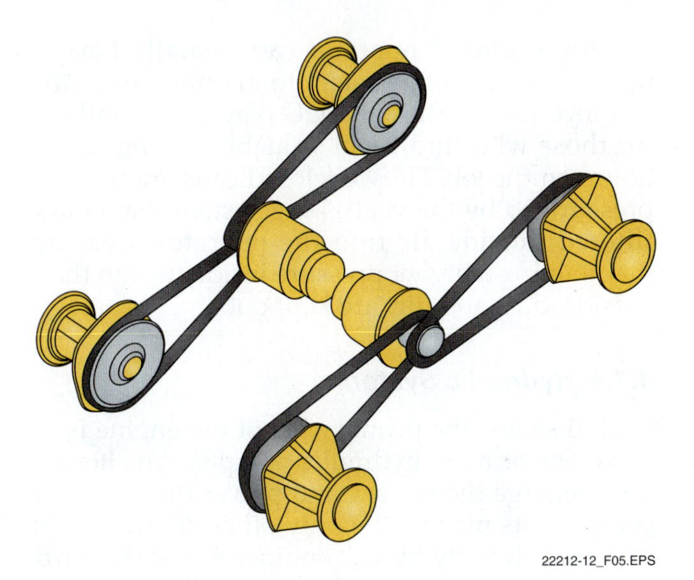

22212-12_F05.EPS

Figure 5 Skid steer drive system.

22212-12_F06.EPS

Figure 6 Bobcat® Model M600, introduced in 1966.

larger than the drive sprocket on the motor. This reduces the speed of the wheels compared to the motor, producing greater torque.

Since each motor is connected to the wheels by chains, the two wheels on each side remain synchronized at all times. If one of the wheels is turning, the second is turning as well. The two hydraulic motors are completely independent of each other. The drive sprockets and chains are sealed inside an enclosure and are bathed in oil to provide ideal lubrication and help transfer heat from the drive components.

1.1.3 Operator Cab

The design and functionality of the operator cab has changed a great deal over the years. Early models had no cab at all (*Figure 6*). This led to a number of injuries. The lack of any obstructions around the driver also encouraged unsafe behavior. Some injuries occurred as the result of falling loads that had been lifted overhead. Many others resulted from the up and down movement of the lifting arms, on both sides of the operator in very close proximity. One could argue that visibility was excellent, and that added to their popularity. Cabs and other safety features were continually added by manufacturers, some as the result of regulations. Today, the operator cab can be considered a self-contained capsule devoted to an operator's protection and comfort.

The operator cab of most models is constructed with great integrity. To maintain a compact size, every possible area is used to mount operating components. The area beneath the operator is no exception. The cab can be raised as a unit to gain access to the components below. *Figure 3* shows an example of a cab that tilts forward, while *Figure 7* provides an example of a cab that tilts toward the rear of the machine. This model also places the battery under the cab. The area under the operator is the typical location for the hydraulic pumps, driven directly from the engine drive shaft. The cab is hinged at one end, and is secured to the chassis for operation by pins or mechanical linkage. The only connections to the cab are electrical in nature, providing for the operation of cab electrical systems and control of the machine.

22212-12_F07.EPS

Figure 7 Operator cab tilted back for service access.

Note the location of the fuel tank. The area between the cab assembly and the engine compartment is a common location for the fuel tank.

Most of the structural features of the cab have resulted from requirements set forth by the US Occupational Safety and Health Administration (OSHA). The primary safety concerns for an occupied cab were related to objects falling from above, reaching or climbing in and out of the seat through the lifting arms, and rollovers. Standards for a Rollover Protective Structure (ROPS) and a Falling Object Protective Structure (FOPS) have answered these concerns effectively. The design of most ROPS and FOPS must be approved by a professional engineer, who often requires substantial testing for new designs or modifications to existing systems. ROPS and FOPS must be capable of withstanding a force equal to twice the weight of the machine at the point of impact. To increase safety further, a control interlock system is also built into many machines. In some cases, the machine controls are not enabled unless the hinged armrests are down in front of the driver. The machine in *Figure 8* is fitted with left and right armrests that fold down toward the center. Others have one-piece armrests that hinge straight up and over the operator's head for seat access. Some control systems may also be interlocked through the seat, where a certain amount of weight is required to enable the controls. Most people are familiar with this feature in passenger cars, where a seat switch disables the passenger air bag unless a significant amount of weight is present.

Figure 8 Skid steer with left and right armrests.

Fully enclosed operator cabs usually have a heating and cooling system to maintain operator comfort. Even radios and CD players are options for those who find them valuable during many hours on the job. Fully enclosed cabs are usually pressurized by the ventilating system that blows filtered outside air into the operator area. By maintaining a higher pressure inside the cab than outside, dust and dirt are kept out.

1.1.4 Hydraulic Systems

In skid steers, the primary job of the engine is to drive one or more hydraulic pump(s). An alternator to charge the battery and power the electrical components may be the only other device that is powered directly by the engine. All of the hard work is done by the engine indirectly, using the hydraulic system as a means of transferring energy.

The engine supplies power to the pumps that build pressure in the hydraulic system. The pumps are generally driven directly from the crankshaft of the engine (*Figure 9*). In most vehicles, the crankshaft is connected directly to a transmission or transaxle. Manufacturers save space by stacking the hydraulic pumps on top of each other. The crankshaft is connected to the first pump, and power is transmitted through its shaft on to the next pump, and so on.

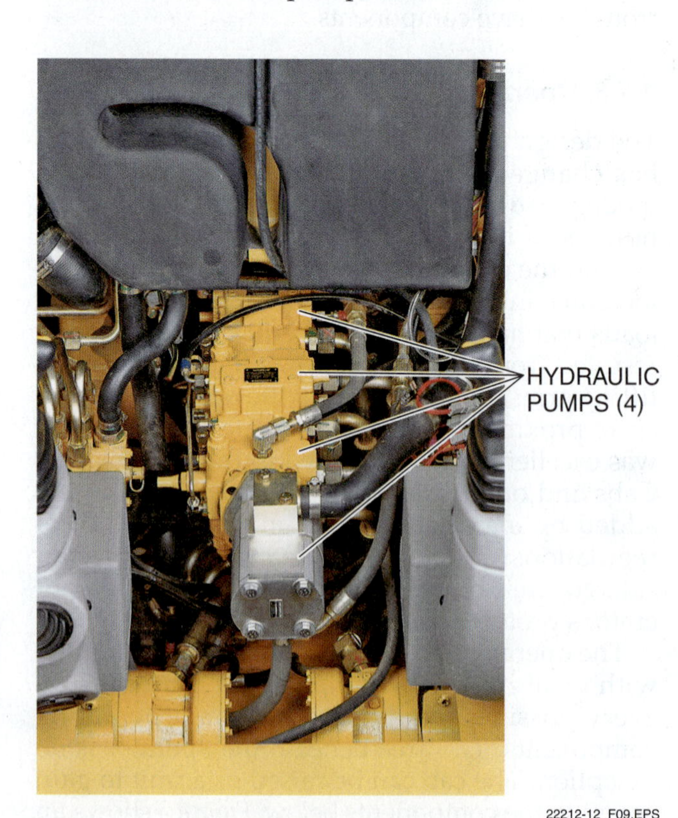

Figure 9 Hydraulic pumps.

Two hydraulic pumps are generally used to power the drive wheels. One pump provides power to the hydraulic motor on one side, and the other pump handles the opposite side. Another pump provides the operating pressure and flow for work tools and attachments. The smaller pump on the end of the assembly provides pressure to devices that require lower flow rates, and may also push hydraulic fluid for the entire system through one or more filters. There is only one reservoir for hydraulic fluid; the entire system shares the same fluid.

Some hydraulic pumps are designed for fixed displacement performance, while others provide variable displacement. In a **fixed displacement pump**, the flow rate of the fluid is based on the speed of the engine driving it. As the engine speed increases, the fluid flow rate increases. This type of pump is often used to power the lifting arms and attachments. Additionally, as long as the pump shaft is turning, flow is being produced whether it is needed or not. When a component is not being moved by the fluid, a relief valve must allow it to remain in motion and return to the pump inlet. Without this feature, the hydraulic fluid would soon overheat inside the pump.

The **variable displacement pumps** are typically used to power the drive wheels. These pumps can be controlled to output hydraulic fluid only on demand. If there is no demand, the pump just spins without creating any fluid flow. Since the engine provides all the energy to the pumps, too much of a hydraulic load can stall the engine. For example, when an attachment is in use and being operated, sending too much fluid to the drive wheels could cause this to happen. The control of the variable displacement pumps by the skid steer helps to prevent stalls from occurring. However, the operator must take an active role to avoid overloading and stalling the engine on many skid steers.

The operational controls open and close valves to the hydraulic lines connected to the loader's steering and lifting hydraulic cylinders and pistons. The lift arms mounted at the rear of the frame are connected to the attachment plate. When the hydraulic fluid is applied to the lifting cylinders, the pistons extend and raise the attachment. When the flow of hydraulic fluid is reversed, the cylinders retract their pistons and the attachment lowers. Another hydraulic cylinder connected to the attachment plate tilts the attachment forward and backward. Ensuring that the lift arms, and the hydraulic devices attached to them, are not damaged is critically important to the safe operation of the skid steer. The same goes for all the hydraulic lines connected to the hydraulic devices.

To power all of the possible attachments and their integral accessories, a number of connections to the hydraulic system are made available (*Figure 10*). Note that there is also an electrical connection, allowing the skid steer controls to operate features that are part of the attachment.

1.1.5 Tires

The tires of a skid steer must be able to withstand the punishment of constantly skidding to one side or the other, without rolling off of the rim. The stress on the tires to skid is even greater when the machine is carrying a load. Highly abrasive surfaces will cause a higher degree of wear and offer a higher resistance to skidding.

A variety of tires are available for skid steers. The least expensive type is the standard pneumatic (air-filled) tire (*Figure 11*). These tires need exceptionally heavy sidewalls. The tire tread is generally deep, allowing for a great deal of use before contact is made with the casing of the tire. Pneumatic tires offer a better ride and increased comfort for the operator in most cases. Several different categories are available, depending on the need and budget. The categories are regular duty, heavy duty, and severe duty. The duty level typically is related to the number of plies used in the construction of the tire.

22212-12_F10.EPS

Figure 10 Hydraulic system connection points for attachments.

Figure 11 Pneumatic skid steer tire.

Although they ride more comfortably and are less expensive, pneumatic tires are subject to punctures or other damage that may both ruin the tire and cause a work stoppage. The type of terrain and work dictates the tire choice. In areas where a lot of tire hazards exist, such as a building demolition site, a tire without air offers a more functional choice. These tires are sometimes called flat-proof tires (*Figure 12*). These tires are usually more expensive, but the savings over time from avoiding lost time on the job can be significant. The air cells in the tire help to cushion the ride somewhat. Note that this example has a completely smooth surface. This tire would be a good choice for hard surfaces such as concrete. However, with no tread, traction outdoors would likely be a serious problem. Flat-proof tires are available with tread as shown in *Figure 13*, and with more or less air cells for cushioning. With so many different tires to choose from, the best choice is made through consultation with tire specialists regarding the work environment. It is not unusual for an owner to have more than one type of tire on hand.

All skid steers have a tendency to damage the terrain. Two types of damage can occur. Damage occurs as the tire is skidded while the machine is turning, and damage occurs from the weight of the vehicle rolling across the ground. The wider the tire used, the more weight of the skid steer is displaced. Maximum displacement of the vehicle's weight is achieved by using tracks. Of course, skid steer machines are available as tracked models, without any wheels (*Figure 14*).

Figure 12 Flat-proof tires on a skid steer.

The drive system works much the same, with independent hydraulic pumps driving the tracks on each side forward or backward. The tracks help to more evenly distribute the weight and improve the stability of the machine. The tracks handle muddy conditions better than tires. For the best of both worlds, a skid steer with tires can be fitted with over-the-tire (OTT) tracks made from rubber or steel (*Figure 15*). Not all tracks are solid in construction. Some are made of steel bars several inches apart. They maximize traction but do not distribute the machine's weight as well. Depend-

Figure 13 Treaded flat-proof tire.

22212-12_F14.EPS

Figure 14 Tracked skid steer machine.

22212-12_F15.EPS

Figure 15 OTT tracks.

ing on the make and design of the machine, spacers may be needed between the wheels and their mounting hubs. This pushes the wheel farther away from the chassis, providing added clearance for the tracks.

1.2.0 Controls

All models of skid steer are either lever- or joystick-controlled; no steering wheel exists. Most current models are controlled by left and right joysticks. All of the most important and most common control actions are initiated by the joysticks. One distinct advantage of the joysticks is that they allow multiple control inputs to be made at the same time. A few other controls are spread around the cab in various locations. Review the operator manual to fully understand the controls available for the model in use.

1.2.1 Vehicle Movement Controls

The skid steer has no direct capability to steer. Turns occur when the drive wheels on one side are turning faster than the drive wheels on the opposite side. Since no input is required to perform steering, there is no need for a steering wheel.

Driving some units may require the use of two control levers. To engage and turn the drive wheels on one side, the lever on the corresponding side is pushed forward or backward only. The amount of travel in the lever controls the speed and force of the movement. If the right lever is very carefully and gently pulled toward the operator, the skid steer responds gently, and begins backing up to the left. This happens as the result of the right drive wheels turning in reverse, while the left wheels skid across the floor or ground. Pulling one lever back while pushing the other forward causes the machine to spin in a very tight circle. The drive wheels on one side are driving in the opposite direction from those on the other side.

Later models of skid steers are generally controlled with joysticks. For example, a Caterpillar machine is driven using only the left joystick (*Figure 16*). Pushing the joystick straight forward engages the drive wheels on both sides, and the machine moves forward. Pushing it forward and to the left executes a left turn. The farther left or right the joystick is pushed, the slower the drive wheels on that side turn in relation to the opposite side. If the tightest possible turn is desired, the left joystick is not pushed forward or backward at all. Pushing it directly to the right or left results in the machine spinning in place. If the normal engine speed is insufficient, the operator can use the foot throttle to increase the engine speed. The engine speed can also be controlled with a hand-operated throttle (*Figure 17*). It is not uncommon for both a foot pedal and a hand throttle to be on a single machine. However, remember that the ground speed of the machine is not dependent upon the accelerator pedal like most heavy equipment.

Some machines with advanced control schemes offer two or more control configurations. For example, Bobcat® units equipped with their Advanced Control System (ACS) can be switched from drive control exclusively through the left joystick (standard) to the older H-pattern of control. In the H-pattern, the left joystick controls the left wheels and the right joystick controls the right wheels.

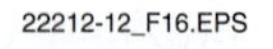

22212-12_F16.EPS

Figure 16 Left joystick for a Caterpillar skid steer.

1.2.2 Attachment Controls

Note the buttons on the joystick in *Figure 16*. The placard shown on the cab wall helps familiarize an operator with the function of the buttons and controls. One button on the left side controls the horn (top) and the other initiates continuous hydraulic fluid flow to an attachment, such as an auger. On the right side of the left joystick, the two buttons are used to control the position of three-way hydraulic valves. A trigger on the joystick is hidden from view. This skid steer has two speeds, allowing the operator to choose the top speed based on the work and terrain. The operator can enable high-speed operation for travel by pulling and holding the trigger.

Manufacturers try to place a number of control functions on the joystick. This allows the operator to control a variety of devices and systems without having to let go of the joystick. The operator should always try to maintain contact with both of the joysticks. When a joystick is released, it should always return to the neutral position.

The right joystick of the same machine provides control over the attachment as a whole. See *Figure 18*. Moving the joystick forward and backward lowers and raises the lifting arms, respectively. Moving it left and right tilts the bucket or other

22212-12_F17.EPS

Figure 17 Hand-operated throttle.

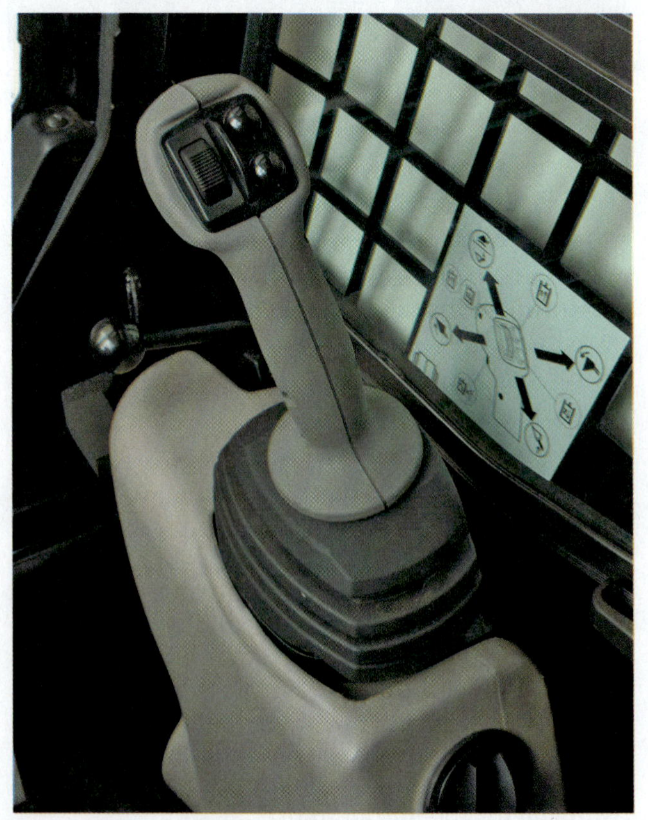

22212-12_F18.EPS

Figure 18 Right joystick.

attachment up and down. On some machines, the bucket or other work tool can be placed in a **float mode** by first pushing the joystick fully forward. It may stop in a slight detent and remain there to place the tool in the float mode. On other machines, a momentary trigger-pull while the joystick is fully forward is required.

On the top of the joystick, there are two buttons and a slide switch. The function of the buttons or switches depends upon the skid steer model and the attachment being used. If the attachment, such as an auger, also needs to be controlled, these switches and buttons may provide the input. The right joystick also has a trigger hidden from view. The black lever just forward of the joystick in the photo is a hand-operated throttle control.

Older and more basic skid steers may not provide many control options for the attachments. For these situations, attachments may have their own controls that are simply routed into the operator cab and mounted in a convenient location (*Figure 19*).

Note that this is only one example of how a skid steer may be operated. There are many different control schemes and layouts, sometimes within the same model family. Operators must take the time to read and understand the operator manual for the skid steer in use and be familiar with all the features and controls available. If a simple attachment, such as a bucket, is in use, then many of the other attachment controls may not be needed.

1.2.3 Other Controls

Since skid steers are entered from the front, there is no room for a traditional dashboard. Some models may provide a control or instrument on the armrest for convenient access. Most controls, other than those on the joystick, must be located to the left and right of the operator's head (*Figure 20*), or forward of the operator to the left and right (*Figure 21*).

Figure 20 shows the controls for the heating and cooling system on an overhead console. Many of today's skid steers have fully enclosed cabs that can be heated and cooled. Models that are not fully enclosed typically have no heating or cooling system. *Figure 20* also shows two other control switches. The work speed control allows the operator to choose how quickly the work tool moves. The operator seat ride control allows air-cushioned seats to provide shock-absorbing motion. Compressed air is used to adjust the seat height, but the ride remains firm unless this switch is turned On.

Figure 22 provides a clear view of the ignition switch on the left console. The switch directly above the ignition switch operates the parking

22212-12_F19.EPS

Figure 19 Snow blower control accessory, mounted in the skid steer cab.

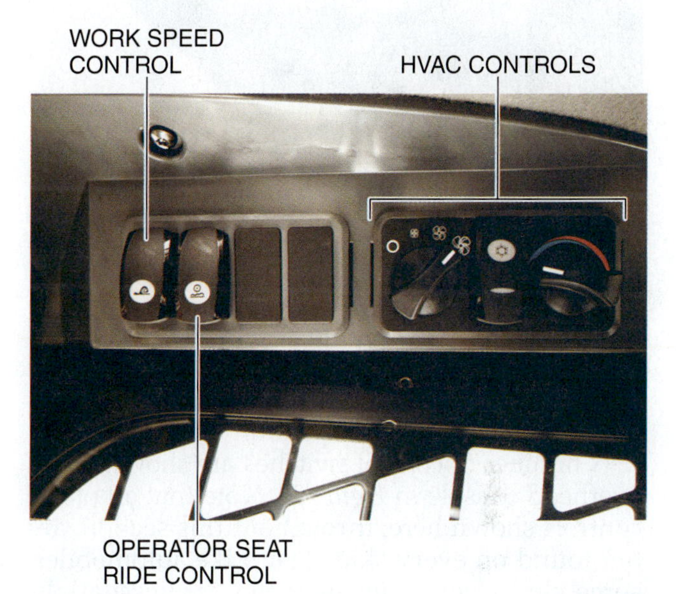

22212-12_F20.EPS

Figure 20 Left overhead control console.

22212-12_F21.EPS

Figure 21 Left and right forward consoles.

22212-12_F22.EPS

Figure 22 Left forward console.

brake. Note that the parking brake may automatically engage on some models when the armrest is raised. This ensures that it is not forgotten, and that it is engaged as the operator leaves the cab.

A number of control switches are shown on an overhead console in *Figure 23*. Note that all of the controls shown here, throughout this section, are not found on every skid steer. Like automobiles, some skid steers offer only the bare essentials, while others are far more feature-rich. The difference in cost between a simple unit and a feature-rich model, assuming they both have equal capacity, can easily be $20,000. Moving from left to right, the controls shown in *Figure 23* are as follows:

- *Auxiliary hydraulic control for hydraulic valves* – This control changes the position of a three-way hydraulic valve associated with an attachment.
- *Hydraulic lockout and interlock override* – The hydraulic lockout control at the top of the switch enables or disables the hydraulic system. This switch would be used when driving the skid steer on a roadway for a distance, disabling the hydraulics for the trip. The interlock override bypasses the safety devices that normally prevent the machine hydraulic systems from operating unless the armrests are down, or some similar interlock is satisfied. There are some attachments, such as some backhoe attachments, that require the operator to leave the normal seat and operate from an alternate position.

22212-12_F23.EPS

Figure 23 Overhead console switches.

WARNING!

The operator Override should never be used for any other purpose beyond allowing skid steer operation from an alternate seating position. Using the switch to intentionally bypass safety interlocks can result in serious injury or death.

- *Rear work light controls*
- *Front work light controls*
- *High-volume hydraulic device enable switch* – Some attachments require a higher volume of hydraulic fluid flow to operate properly. To support these attachments, some skid steer models provide an auxiliary hydraulic pressure source to support the higher flow. This switch allows this auxiliary source to provide the higher volume of flow or a normal flow rate. When it is enabled, the light remains illuminated. This switch may also change the behavior and use of the left trigger switch.
- *Auxiliary electric control* – This switch enables electrical controls on attachments that require an electrical interlock as well as a hydraulic pressure source.
- *Level bucket control* – This switch instructs the skid steer to automatically maintain the bucket opening level, to keep the contents from spilling out as the lifting arms are raised. Note that this switch only affects the level of the bucket as it is raised; it has no effect while the bucket is being lowered.
- *Attachment locking feature* – This actuates an automatic locking mechanism for the attachment, locking it to the coupler.

One additional operator-accessible control is used in case of a shutdown or malfunction in the

machine while the lifting arms are in the raised position. A manual control that opens a hydraulic circuit valve and allows the lifting arms to drop in a controlled manner is available to the operator. It is typically to the left or right of the seat and has a red handle (*Figure 24*).

1.3.0 Instruments

Most skid steers offer very few instruments related to the machine's operation. This is true for two reasons. First, skid steers are relatively simple and are built with much smaller engines than their heavier cousins. Since the engines may be as small as 15 to 20 hp, there is a limited need for instrumentation to monitor their operation. The other reason is limited space. Skid steer cabs are quite compact, and visibility outside of the machine is very important. On very basic machines, only a few common lights to alert the operator to a problem are present.

The service hour counter is one instrument common to all models. Although some models use a type with a liquid crystal diode (LCD) display (*Figure 25*), older models use a simple mechanical version. The counter is a valuable tool used to ensure that maintenance is done on a consistent schedule. The value on the counter should never be changed.

1.3.1 Gauges

The fuel gauge is another common instrument (*Figure 26*). This particular gauge captures an operator's attention since the white face contrasts sharply with the dark surroundings. It is extremely important that diesel engines not be allowed to run out of fuel. Running the engine completely out of fuel can cause engine damage or significantly slow down the job while the matter is resolved. Once the fuel is depleted, fuel must be added and the fuel lines must be bled of air before the engine will restart. Other possible gauges on a skid steer are an engine oil temperature gauge and a hydraulic fluid temperature gauge. Beyond these three gauges, skid steers typically use lights to alert the operator to a problem or warn of a developing fail-

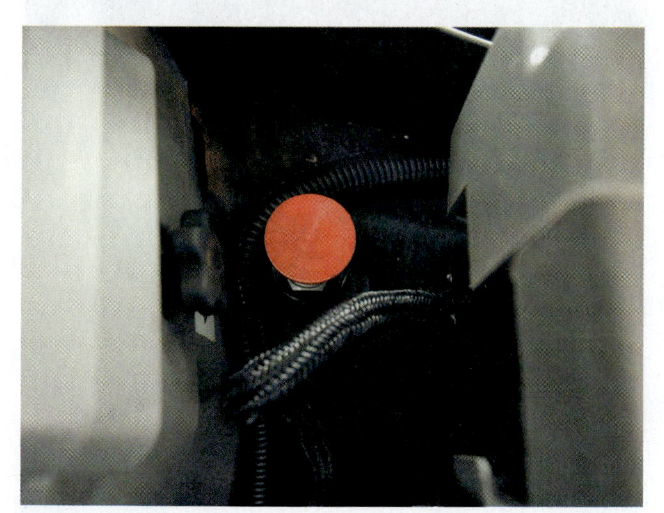

22212-12_F24.EPS

Figure 24 Dead engine valve control to lower the lifting arms.

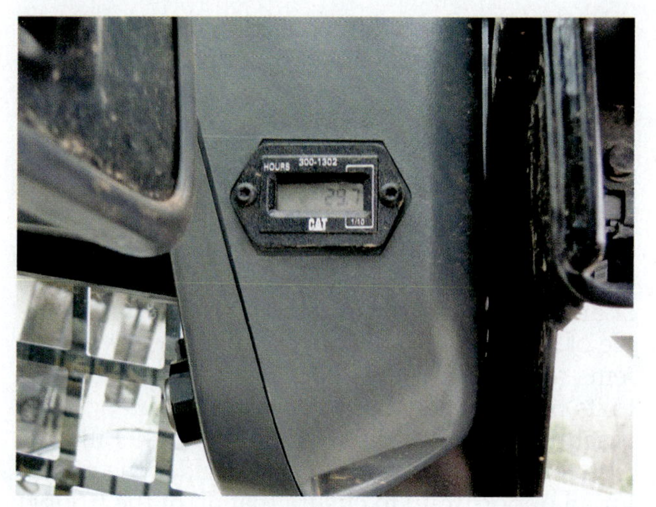

22212-12_F25.EPS

Figure 25 Service hours counter.

22212-12_F26.EPS

Figure 26 Fuel gauge.

ure. Some feature-rich models do provide an LCD display that shows a variety of settings (*Figure 27*).

1.3.2 Information and Warning Lights

Some of the lights in the operator cab simply provide information to the operator that a particular switch is enabled or a selection has been made. Other lights warn of problems that are developing or an unsafe condition exists. *Figure 28* is an example of a warning light panel with a variety of systems represented. Once the ignition key is turned to its first position, battery power is routed to the machine systems and all the panel lights illuminate for about 3 seconds. This provides the operator an opportunity to ensure that all of the trouble lights are working. Failed lamps or circuits should be repaired immediately. Without these lights operating, the operator may not be aware of a serious problem developing. Note that there are three different colors of lights. Green lights simply provide information; yellow lights indicate a situation that requires caution; and red lights indicate a serious problem that should be addressed immediately.

Beginning with the top row and moving from left to right, the lights shown in *Figure 28* provide the following indications:

Top Row:

- *Yellow square with exclamation point* – Indicates a driver alert and lights up when any other warning light is on; also warns of low fuel.
- *Battery* – Warns that the alternator output to the electrical system is too low or too high.
- *Circle with a P* – Warns that the parking brake is engaged.
- *Person seated with seat belt* – Warns that the seat belt has not been fastened and/or the armrest has not been lowered.

Middle Row:

- *Hydraulic vessel with a swipe* – Indicates that the hydraulic high-volume flow feature is enabled.
- *Hydraulic vessel with arrows around it* – Informs the operator that the continuous flow feature is in use. This feature may be used to operate a constantly rotating device like a broom.
- *Hydraulic vessel with dashed lines* – Warns of a problem with the work tool hydraulic system.
- *Hydraulic vessel with exclamation point* – Warns that the hydraulic fluid temperature is too high.
- *Engine crankcase with a coil* – Indicates that the engine is cold and the glow plugs are warming up the cylinders for cold-start situations. Other machines may require that the glow plugs be operated manually.

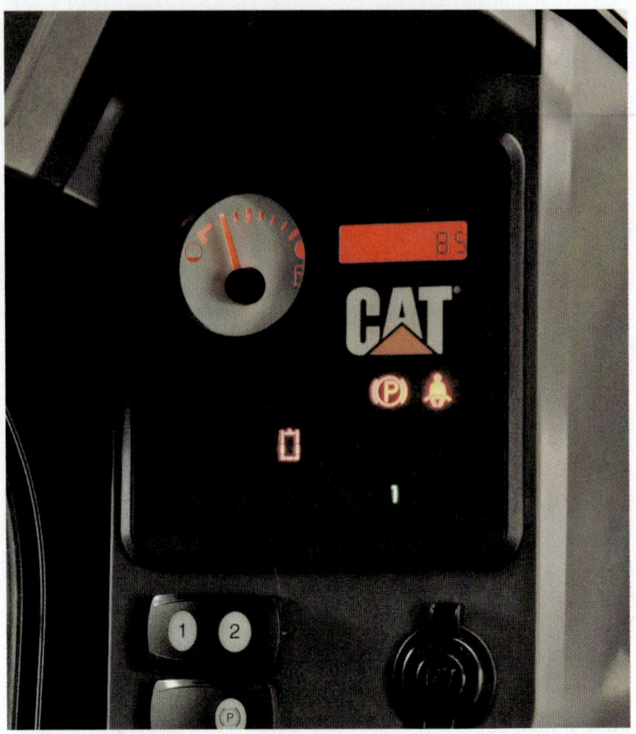

22212-12_F27.EPS

Figure 27 Back-lit LCD display provides information on various settings.

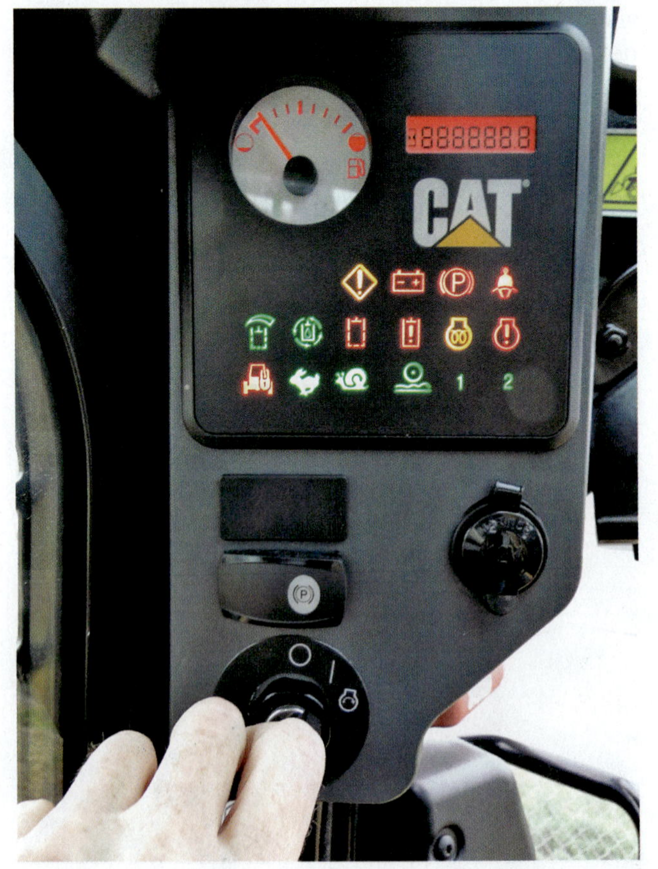

22212-12_F28.EPS

Figure 28 Information and warning light panel during a prestart test.

- *Crankcase with an exclamation point* – Warns of an engine problem.

Bottom Row:

- *Skid steer with a padlock* – Warns that the machine security system is active.
- *Rabbit* – Informs the operator that high speed has been selected.
- *Snail* – Informs the operator that a low work tool speed is in use.
- *Wheel rolling over rough terrain* – Informs the operator that the air-ride shock-absorbing feature of the seat has been enabled.
- *Numbers 1 and 2* – Informs the operator which of two joystick control patterns is in use.

At the top of the panel, the digital service hour counter is also testing its individual digit display and backlight.

1.4.0 Attachments

If there is one single feature that defines the skid steer among its users, it is the vast array of available attachments. There are many different types, and many different versions within each type. New or improved attachments are consistently being developed.

Skid steers have been developed around a simple yet strong and versatile system to connect attachments quickly and easily. Not only are attachments available through the skid steer machine manufacturer, but a number of other companies specialize in making attachments to fit any make of skid steer. Skid steer owners and users have also developed their own attachments to resolve specific situations. Note, however, that using these attachments and the attachments of another manufacturer may affect an existing warranty on the machine. It is also important to note that most attachments go through a rigorous series of tests, and a great deal of data regarding their weight, capacity, and other specifications must be made available. None of this testing or data is available for home-built or site-built attachments. As a result, their safe use may be questionable.

A typical Cat skid steer attachment coupler is shown in *Figure 29*. Once the attachment is properly mounted on the coupler, the two locking levers are pushed toward the center, locking the

The Ultimate Attachment?

Tired of sitting in the cab? Are dusty work conditions causing too much discomfort? Are there hazardous materials possibly in the area? All of these may be reasons to try your hand at operating a skid steer without actually getting in.

The remote control system receiver developed by Bobcat® can be installed in minutes using magnets. It is then connected to the machine control system through a cable and plug. The transmitter mimics the commands of the machine's joysticks, allowing the operator to drive and operate the attachments. If the operator falls or becomes incapacitated and drops the control transmitter, the machine automatically shuts down. The transmitter also has an Emergency Stop button, and a second button is provided in the kit for the rear of the machine. Not only are there military applications, but the system also allows a single operator to do jobs that normally would require both a spotter and an operator.

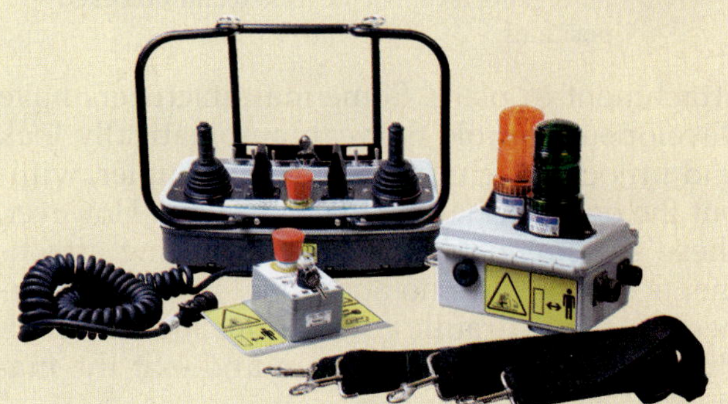

22212-12_SA02.EPS

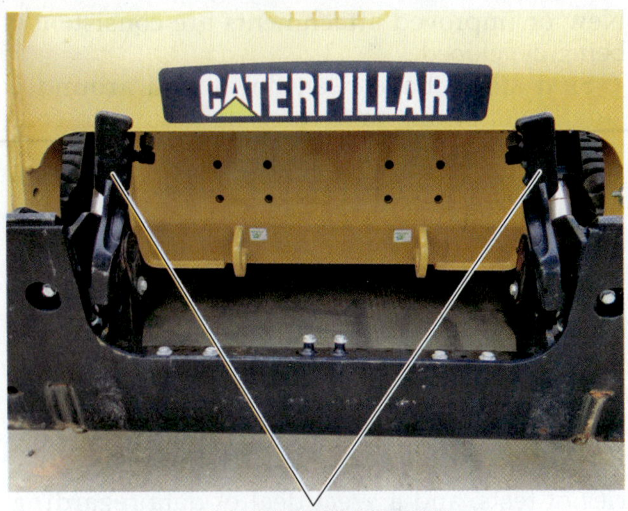

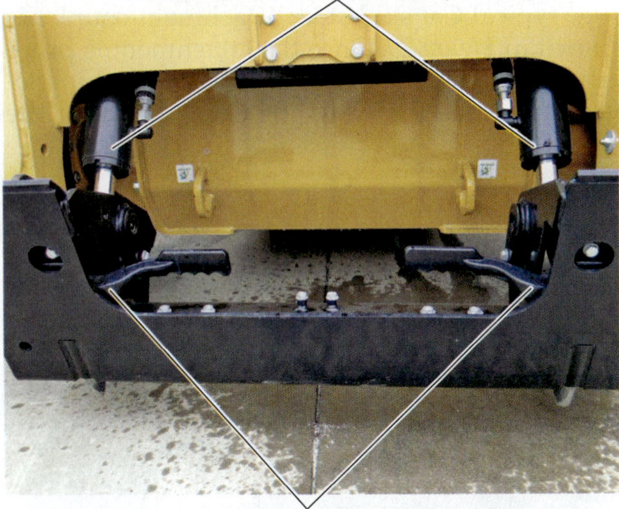

Figure 29 Cat skid steer attachment coupler, showing the locking levers in the locked and unlocked positions.

attachment in place. Some manufacturers have developed controls that can automatically lock and unlock attachments from the coupler, without the operator having to exit the cab. However, these systems only allow for changing attachments that require no additional hydraulic connections. If hydraulic connections are required, the operator must shut down and exit the machine to complete the task.

The most common attachment is the standard bucket (*Figure 30*). The bucket can be used to move soil and rock, doze, level soil or gravel, dig, or load/unload materials. There are a number of different bucket configurations and sizes available. Teeth can be added to the leading edge of most buckets. It is important that the bucket (or any other attachment) be properly matched to the capacity of the skid steer. Since the coupler specifications remain the same from the smallest to the largest skid steer models, an attachment that is beyond the capacity of the machine can be installed. It is the operator's responsibility to ensure that the attachments are appropriate for the machine in use.

In *Figure 30*, note the two pieces of steel welded to the upper edge. These are step points provided to assist the operator in accessing the cab. Without any attachment on a machine, the first step up to the cab area is long and clumsy. It is quite common for attachments to provide this point for the operator's first step toward the cab.

One modification to the standard bucket is the grappler (*Figure 31*). This style of grappler provides a relatively flat surface on the bottom, and a clamshell-like moving jaw above it. The lifting arms and tilt cylinders move the complete attachment up or down, or tilt the forward edge up or down. The two cylinders on the attachment operate the jaw, closing it around material to be taken away. Since there is an additional hydraulic feature, this attachment requires a connection to the hydraulic system (*Figure 32*). The grappler jaw is then controlled by buttons on one of the joysticks that is in control over the chosen hydraulic circuit. This particular machine was prepared for duty in a recycling facility. Note the smooth tires, which indicate the unit will be driven almost exclusively on concrete or a similar surface. Grapplers are available in a number of styles.

Another very popular attachment is a set of forks (*Figure 33*) allowing the skid steer to function as a forklift. This set of forks, and the related bale forks (*Figure 34*), do not require additional

Figure 30 Standard bucket attachment.

22212-12_F31.EPS

Figure 31 Grappler attachment.

22212-12_F32.EPS

Figure 32 Hydraulic connections to support grappler jaw operation.

hydraulic connections. Bale forks are more slender than pallet forks, and the forks are well above the ground. Bales are then picked up by simply driving into them with the forks, since they are positioned near the center of the bale.

Figure 35 provides some examples of the many other attachments available for a skid steer. It is beyond the scope of this module to cover all of the possibilities. However, a variety of them are presented here, with a brief description of them as needed.

- *Laser-guided grading box (A)* – This box levels loose soil, collecting it from high spots and spreading it in low spots, much like a grader. The model depicted interfaces with a laser system to provide elevation information to the operator. Laser guidance systems are available for both single-slopes and dual-slopes (slopes in two directions). When fully automated, the task requires less operator involvement, improved accuracy, and faster results. The Bobcat® operator can also see the exact elevation of the box blade at any time.

22212-12_F33.EPS

Figure 33 Common fork attachment (pallet forks).

22212-12_F34.EPS

Figure 34 Bale forks.

- *Cement mixer (B)* – Simply add water and the desired mix. Shown mounted on a Bobcat® skid steer, the mixer becomes very mobile and can be moved around at will.
- *Drop hammer (C)* – This attachment is primarily designed to break up large expanses of concrete or asphalt for demolition. The Bobcat® model shown here delivers 3,600 foot-pounds of energy per blow, with a three-second delay between blows. This delay provides the operator enough time to slightly reposition the machine for the next blow. The drop hammer is significantly quieter than using a concrete breaker or chipping hammer.

(A) LASER-GUIDED GRADING BOX

(B) CEMENT MIXER

(C) DROP HAMMER

(D) CEMENT PUMP

22212-12_F35A.EPS

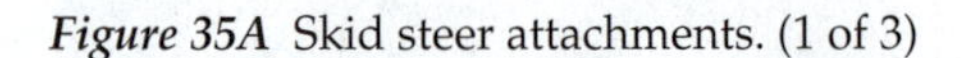

Figure 35A Skid steer attachments. (1 of 3)

- *Cement pump (D)* – Cement is accepted into the hopper from a truck, and is then pumped to a distant or elevated location.
- *Power rake (E)* – The power rake smoothes soil surfaces and collects rocks and sticks, just as a common rake would. They are mostly used on loose soil where new turf grass seed or sod is to be installed.
- *Enclosed sweeper (F)* – Sweepers are popular attachments. They do an excellent job of cleaning pavement after other work has been completed. Sweepers, which are enclosed, create less airborne dust and dirt. Sweepers like the one shown here also collect the debris onboard, so it can be disposed of elsewhere. Power broom attachments do not collect the debris—they are usually angled and simply sweep the debris off to one side of the surface.
- *Snow blower (G)* – The snow blower attachment provides a very effective means of deep snow removal, carried by a machine that performs well in adverse weather. The discharge chute rotates and adjusts so the snow can be placed as desired.
- *Concrete breaker/chipping hammer (H)* – These attachments are also known as hydraulic breakers. A variety of bit shapes and sizes are available for different surfaces and tasks. The model shown can be mounted just as easily on an excavator.
- *Planer attachment (I)* – The planer is capable of cutting or milling concrete and asphalt surfaces. The depth, width, and slope of the cutting device can be controlled. Milled material can be collected and recycled. They are excellent for cleaning up areas where large, dedicated planing and milling equipment cannot reach, establishing drainage in parking areas, and leveling uneven pavement.
- *Excavator (J)* – The excavator attachment turns the skid steer into a backhoe. Depending on the make and model, the attachment may require

(E) POWER RAKE

(F) ENCLOSED SWEEPER

(G) SNOW BLOWER

(H) CONCRETE BREAKER/CHIPPING HAMMER

22212-12_F35B.EPS

Figure 35B Skid steer attachments. (2 of 3)

the operator to change seating positions and operate the backhoe from outside the normal cab, seated on the attachment itself. This type of excavator attachment requires the operator to use the Operator Override switch to allow the skid steer to function without an operator in the cab seat. The model shown, however, extends the controls right into the cab, requiring the front door to be removed.

- *Earth auger (K)* – Augers powered by a skid steer are excellent for boring post holes accurately and dependably. Since they are power by hydraulic motors that are relatively small for their output, the head of the auger can follow the bit down into the hole for greater depth. This is another attachment that can also be attached to an excavator.
- *Brush cutter (L)* – The brush cutter, shown here mounted to a Bobcat® unit, is as simple as it sounds. It creates a powerful hydraulically driven mower, driven by a very powerful and maneuverable skid steer.

There are still many other possible attachments, including saws, trenchers and trench compactors, underground bores, tree spades, and tillers. If you can conceive it, it can probably be made to operate with a skid steer acting as a carrier. It is important to repeat that, even though an attachment may mount to a machine with no problem, it is not an indication that it is an appropriate match for that machine. Before connecting and using any attachment, be sure that it is approved for use on the machine and does not exceed its capacity in any way. Attachments that perform the same function may have different operating methods and precautions associated with them. Always read the documentation for the attachment as well as that of the skid steer to protect the equipment investment and ensure the safety of nearby personnel.

(I) PLANER ATTACHMENT

(J) EXCAVATOR

(K) EARTH AUGER

(L) BRUSH CUTTER

22212-12_F35C.EPS

Figure 35C Skid steer attachments. (3 of 3)

1.0.0 Section Review

1. Skid steer engines are mounted behind the operator _____.
   a. for easier service access
   b. to help balance the load
   c. for improved engine cooling
   d. to keep the operator cooler

2. Most, if not all, skid steers manufactured today are controlled by _____.
   a. levers
   b. a yoke
   c. joysticks
   d. a steering wheel

3. The service hour meter is reset after each significant maintenance inspection action is performed.
   a. True
   b. False

4. The drop hammer is primarily used to _____.
   a. drive pilings
   b. install fence posts
   c. pack soil in a trench
   d. break up concrete slabs

SECTION TWO

2.0.0 PRESTART INSPECTIONS AND MAINTENANCE

Objective 2

Describe the prestart inspection requirements for a skid steer.

a. Describe prestart inspection procedures.
b. Describe preventive maintenance requirements.

Performance Task 1

Complete a proper prestart inspection, maintenance, and housekeeping for a skid steer.

Skid steers require very limited care, since they are small and use smaller diesel engines than most heavy equipment. However, they still require a proper prestart inspection at the beginning of the workday. Consistently putting off common maintenance activities can lead to significant and expensive damage. Any time the machine is not operating smoothly on a job site, revenue is being lost. Operators have a responsibility to operate their machine as efficiently as possible, and that efficiency begins with timely inspections and maintenance.

2.1.0 Prestart Inspections

A thorough walk-around inspection should be conducted at the beginning of each shift or when a different operator is taking over. The skid steer should always be equipped with the appropriate operator manual. Whenever there is any doubt about the tasks to be done, refer to the manual. Many machines are also equipped with a logbook so that inspections and other actions can be documented.

The prestart walk-around, also called the daily inspection, may occasionally require some additional actions. For example, greasing specific components is done on a schedule based on the operating hours. When greasing is due, it should be accomplished as part of the walk-around. If an area is to be greased every 50 hours, then the operator will need to do it once each week as part of the walk-around, assuming a normal 40- to 50-hour workweek.

Maintenance time intervals for most machines are established by the Society of Automotive Engineers (SAE) and adopted by most equipment manufacturers. Instructions for preventive maintenance are usually in the operator manual for each piece of equipment. Typical time intervals are: 10 hours (daily); 50 hours (weekly); 100 hours, 250 hours, 500 hours, and 1,000 hours. The operator manual also includes lists of inspections and servicing activities required for each time interval. For specific preventive maintenance actions, refer to the operator manual associated with the machine being used.

2.1.1 Prestart Inspection Process

The daily inspection is also called a prestart inspection for a very good reason—it should be accomplished before the machine is started at the beginning of the workday. This helps to identify any potential problems that could cause a breakdown and indicate whether the machine can be operated. The items to be attended to on the daily inspection are often listed as 10-hour inspection items in the operator manual. If the skid steer is in operation around the clock, with three different operators working normal shifts, then the daily inspection would be conducted three times each day.

The operator should walk completely around the machine checking various items. The machine should be parked on a stable, flat surface with the parking brake engaged and the lifting arms down. The key should be removed from the ignition.

Items to be checked and serviced on a daily inspection are identified below. The operator can begin wherever it is convenient. In this review, the inspection begins at the right front wheel. It is always a good idea to carry a rag along to clean accumulated dust and dirt from grease points, hydraulic connections, and similar locations. Keep in mind that this walk-around example cannot possibly cover every possible machine and configuration. Always follow the procedures for the specific machine in use.

WARNING!

Do not check for hydraulic leaks with your bare hands. Use cardboard or another device. Pressurized fluids can cause severe injuries to unprotected skin.

- Check the right front wheel and tire assembly. Look for missing or loose lug nuts, damaged air valve stems, and cut or damaged tires. Check the tire pressure to ensure it conforms to the operator's manual.
- As the inspection continues down the right side, check the hydraulic hoses routed along the lifting arms for damage or leakage. Look for spots that may be chafing due to vibration or movement.
- Check the hydraulic fluid reservoir level (*Figure 36*). Some may be a little difficult to see. The reservoir level is typically checked by observing a vertical or round sight glass. The machine should already be parked with the lifting arms down. If the lifting arms are raised, a great deal of the hydraulic fluid will be in the operating cylinders, and the reservoir may appear to be low. Always check hydraulic fluid levels with the lifting arms down.
- Check the fuel tank cap for security (*Figure 37*). Note that the cap can usually accept a padlock for security if needed.

22212-12_F36.EPS

Figure 36 Hydraulic fluid sight glass.

22212-12_F37.EPS

Figure 37 Fuel tank cap.

NOTE

It is best to refuel the skid steer at the end of the work day rather than waiting until the next day. As temperatures change overnight, water may condense inside the fuel tank. This problem is significantly reduced if the tank is left full of fuel.

- Check to be sure that lifting arm locks, used to safely lock the lifting arms in the upright position, are properly secured to the side of the machine (*Figure 38*).
- Check the right rear wheel and tire condition and pressure in the same manner as the right front.
- Open the engine compartment, and lock the door in the open position. Most machines have a retaining pin or other device used for this purpose.
- Make sure that all required guards are in place, such as the one covering the drive belt(s). Remember that the engine in a skid steer is typically backwards to its position in a rear-wheel drive auto.
- Inspect the cooling system for leaks and damaged hoses. The machine in *Figure 39* has the radiator positioned above the engine. The top guard is shown in its normal position, which keeps debris such as gravel from striking the radiator face. Remove any debris, then remove the retaining pin holding the guard down and lift it up (*Figure 40*).
- Check the face of the radiator for debris. Note that this radiator is split into two separate sections. The upper section is filled with engine coolant. The lower section contains the hydraulic fluid coolant. Some models also have a fill port on the hydraulic fluid cooler coil so that small amounts of fluid can be added easily (*Figure 41*). However, the hydraulic reservoir itself also has a fill port.

22212-12_F38.EPS

Figure 38 Lifting arm locks.

22212-12_F39.EPS

Figure 39 Radiator assembly with the guard in place.

- This model also has a sight glass to make checking the engine coolant level easy (*Figure 42*). The coolant should be visible in the sight glass. Since this radiator hinges up for engine service (*Figure 43*), be sure to check this glass with the radiator in the operating position.

WARNING!

The radiator may be under pressure, and the coolant may be at or near the boiling point. When the cap is removed and the radiator returns to atmospheric pressure, steam and extremely hot water may be expelled. Use extreme caution and the proper PPE (safety glasses, heavy rubber gloves) when removing the cap from a radiator.

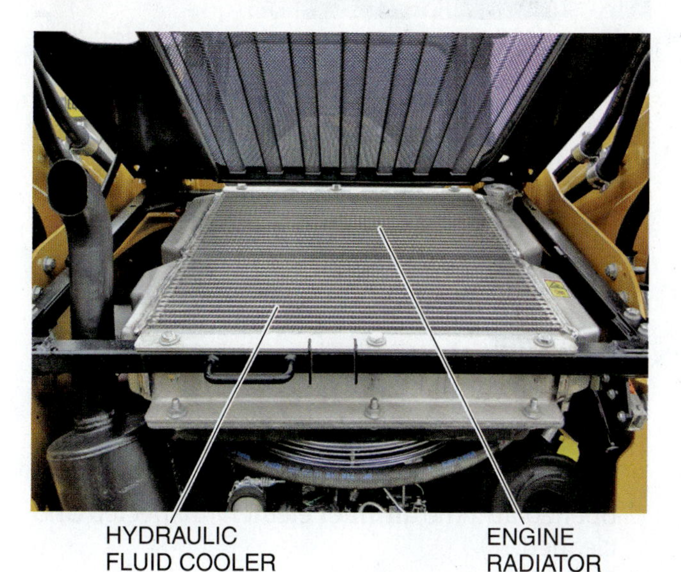

22212-12_F40.EPS

Figure 40 Radiator assembly exposed (guard up).

Figure 41 Engine coolant and hydraulic fluid fill caps.

Figure 42 Engine coolant sight glass.

- Check the oil using the provided dipstick (*Figure 44*). Add oil if necessary. Operating the machine with a low oil level threatens both lubrication and cooling, since the oil also serves to carry heat away from moving parts.
- Check the air filter. Most machines have a way to make a quick check without disassembling the air filter housing. This machine, as seen in *Figure 44*, has a trap. This soft rubber piece has a slot at the bottom that is opened by simply twisting it slightly. If a great deal of debris falls out when it is disturbed, the housing should be opened and the air filter element inspected and cleaned.

Figure 43 Radiator assembly raised for engine service.

Figure 44 Oil dipstick and air filter trap.

- Drain any accumulated water from the fuel/water separator (*Figure 45*). Note the instructions shown here to remove the entire bowl for cleaning if required. Drain fuel until all the water (which is clear in color, compared to the fuel) is gone. Dispose of the drained fluid in an approved container and according to policy. It should not be drained on the ground.
- Make sure the radiator and guard are returned to their normal position and secured. Make a final visual inspection of the engine compartment to look for obvious defects, leaks, or objects that do not belong there. Wipe down any surface that has grease or oil, to reduce the buildup of dust and dirt.
- Before closing and securing the door, check the lubrication schedule on the placard (*Figure 46*), and compare the hours to the operator's log. Note on this placard that items to be serviced at equal intervals are connected by a line around the machine. Further, all grease points are connected by the 10-hour interval line. Therefore, greasing at these points is required every 10 hours of operation. Remember that this means they are daily inspection items. Many owners leave a grease gun with the machine at all times.

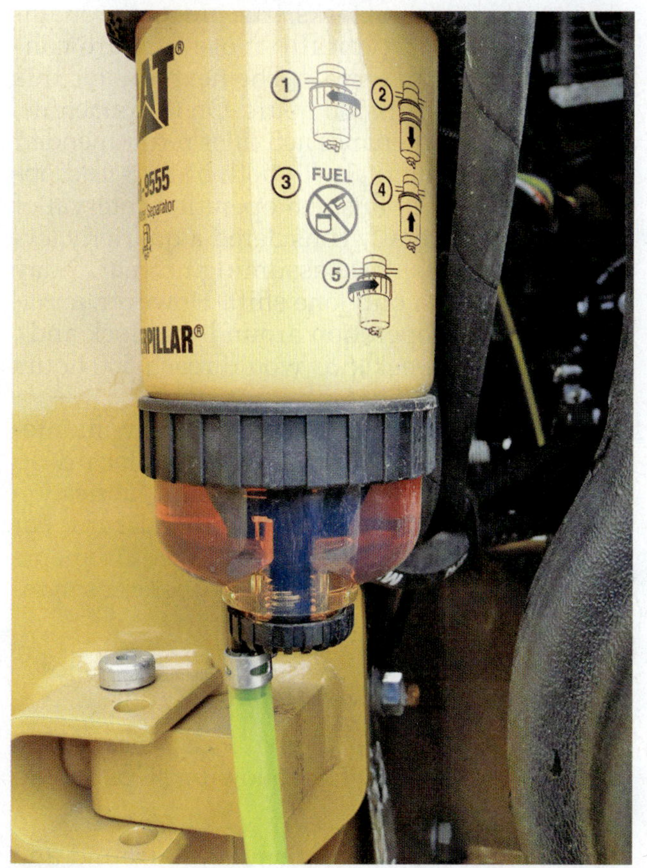

22212-12_F45.EPS

Figure 45 Fuel/water separator.

NOTE

In cold weather, it may be preferable to grease pivot points at the end of a work shift when the machine is warm. Warm the grease gun before using it for better grease penetration.

- Once the engine door is secure, continue the walk-around to the left side of the machine. Inspect the hydraulic lines along the lifting arms and other accessible or visible locations.
- Check both left side wheels and tires for damage and proper pressure, as was done for the right side.
- Examine and clean any visible fluid from the attachment hydraulic line connections (*Figure 47*). Make sure any protective caps provided are installed on unused lines.
- At the front of the machine, look for and remove any debris trapped between the coupler and the attachment, or the coupler and skid steer chassis. Examine the entire attachment or work tool for signs of damage or fluid leakage.
- Make a final visual check of light assemblies and windshield wipers to ensure they are serviceable. On this machine, the windshield washer reservoir is checked as the cab is entered, just inside the door (*Figure 48*).

Before entering the cab to start the machine, redirect attention from the machine to the surrounding environment. Look for unusual obstructions or changes in the work site that may interest a heavy equipment operator. Once inside the cab, the operator does not have the same visibility. Look up as well as around the area.

2.2.0 Maintenance

Preventive maintenance is an organized effort to regularly perform periodic lubrication and other service work in order to avoid performance problems and breakdowns. By performing preventive maintenance on the skid steer, the operator keeps the equipment in a safe and reliable operating condition, and reduces the possibility of breakdowns or expensive repairs.

Accurate, up-to-date maintenance records are essential for documenting the history of the equipment. Each machine should have a record that describes any inspection or service that was performed and the corresponding time intervals. Typically, an operator's manual and some sort of inspection log are kept with the equipment at all times.

Preventive maintenance of equipment is essential and is not that difficult if the right tools and equipment are used. The leading cause of

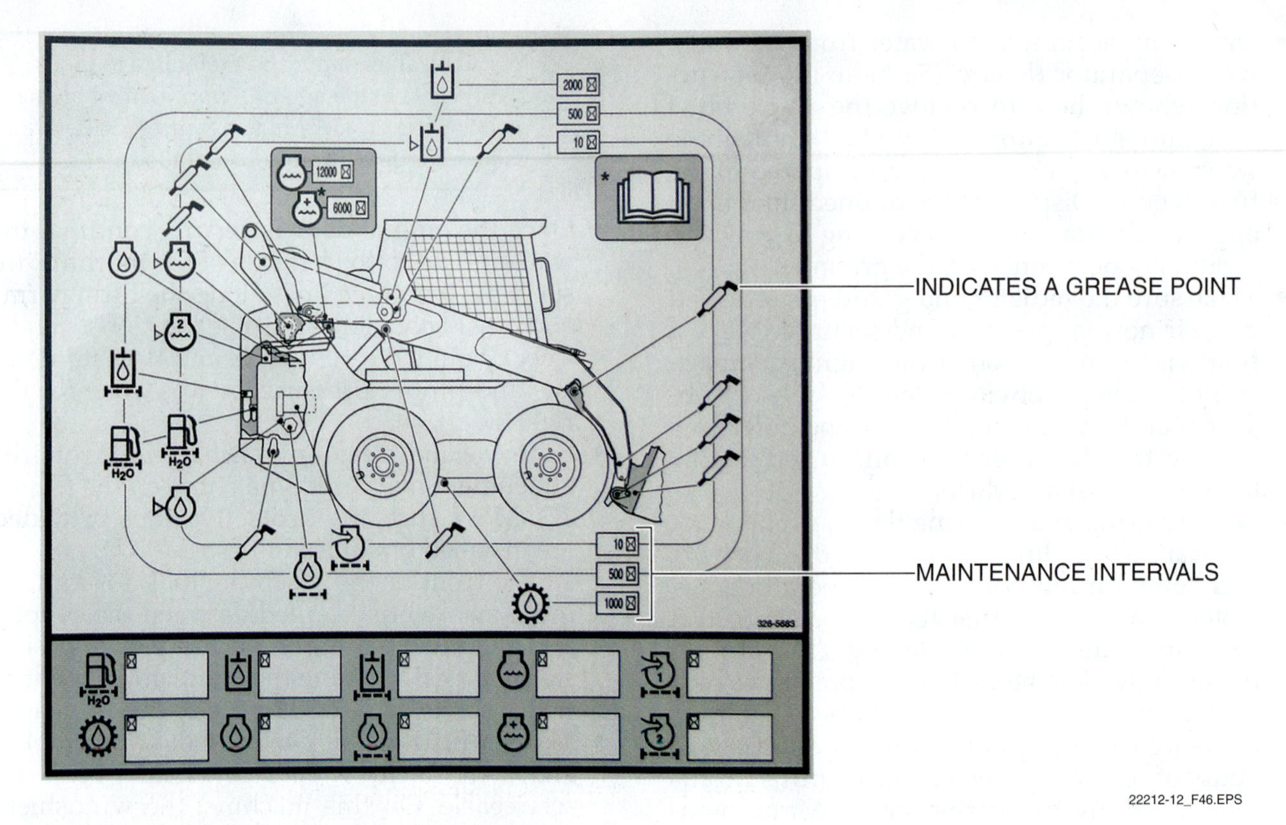

Figure 46 Lubrication and service schedule placard.

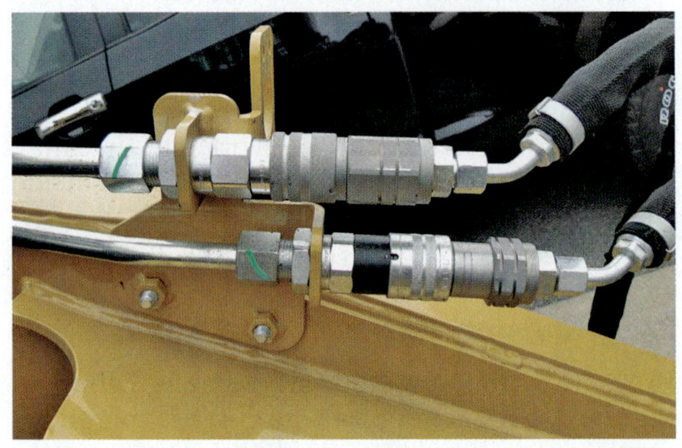

Figure 47 Hydraulic line connections.

premature equipment failure is putting things off. Preventive maintenance should become a habit, performed on a regular basis. Refer to the operator manual for the location of any specific component or lubrication point, and perform the prescribed maintenance activities.

Figure 49 shows a maintenance schedule as it appears on an engine compartment placard. Many skid steers require very limited maintenance. Note that this machine uses lines around the outline of the machine to show items that are to be accomplished on the same schedule. All the grease points shown are connected by the 10-hour interval line. Since inspection items identified as 10-hour requirements are considered daily or prestart inspection tasks, true preventive maintenance tasks are typically those that are conducted at longer intervals. The next line for this skid steer represents 500 hours. On this machine, there are no maintenance activities recommended between those intervals. To both owners and operators, that is a bonus. An operating interval of 500 hours is generally considered a quarterly (every three months) task, assuming normal, 5-day work weeks with only one shift. However, a machine that is in operation around the clock and/or through the weekends would reach 500 hours much more quickly.

It is important to note that not every maintenance requirement is shown on the placard. In this example, only common fluid and lubrication maintenance items are noted on the placard. For complete preventive maintenance requirements and guidance, always refer to the operator's manual.

2.2.1 Preparing for Maintenance

The first step in the maintenance process is to prepare the machine. The following steps should be taken:

- Park the skid steer on firm, level ground and apply the parking brake.

22212-12_F48.EPS

Figure 48 Windshield wiper and washer reservoir.

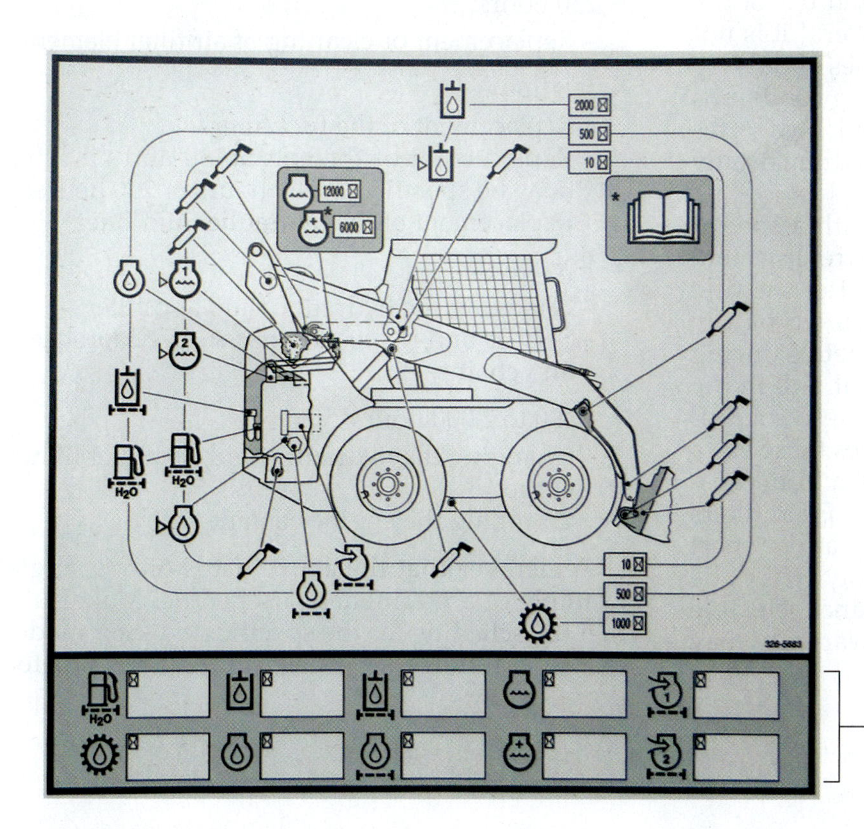

22212-12_F49.EPS

Figure 49 Maintenance schedule placard.

- Lower the attachment or coupler to the ground; the ground should support any attachment, rather than relying on the hydraulic system to hold it up. If the maintenance activity requires the lifting arms to be up, then the safety locks must be installed.

WARNING!

The operator should never exit the machine while the lifting arms are raised unless the safety locks are installed. A second person should install the guards while the operator raises the arms and waits inside the cab. Never raise the lifting arms and then exit the machine to install the locks.

- Operate the engine at idle speed for three to five minutes, if it has been shut down long enough to cool off.
- Shut the engine down and take the key out of the ignition.
- Firmly chock the wheels to prevent any movement.

2.2.2 Long Interval Maintenance

Although some machines do not require any specific maintenance beyond daily tasks for many operating hours, the environment and use of the machine may dictate otherwise. Even if it is not specifically called for, some activities should be carried out weekly, or as the operating conditions may indicate. The operator manual likely states clearly that operating conditions can dramatically alter the maintenance schedule.

New skid steers may start out with a special maintenance schedule. For example, replacement of the engine oil is often required after the first 50 hours of operation. The new engine typically produces significantly more wear metals during the first 50 hours of operation than it will in the future. Replacing the lubricant in these components after this short interval removes these extra metal bits and flakes, preventing them from causing damage or plating other surfaces. Fluid filters may also be changed for the first time after a short interval. If a new skid steer is being driven, be sure to follow the required maintenance schedule for new machines. Otherwise, the warranty may not remain valid.

The battery is a common example of a component that may require inspection weekly or monthly. The case should be inspected for any signs of damage from an external source, casing cracks, or buckling. Check the electrolyte solution level if the battery is not sealed. There is typically a ring or other visual indicator around the bottom of the fill opening on each cell to indicate the proper level. Sealed batteries often have an eye showing the charge condition. Remember that ventilated batteries should be filled with distilled water only. Common tap water contains too many minerals that eventually form deposits on the battery plates, causing the battery to fail. Sealed batteries require only a visual inspection.

WARNING!

Proper PPE should be worn at all times when working near batteries. The electrolyte solution is a strong acid that can cause severe burns or blindness.

A number of tasks are required at longer intervals. Generally, the tasks become more complex and are done in the maintenance shop environment. Tasks that may be required at the intervals shown include the following:

- 250 hours:
  - Replacement or cleaning of air filter elements
- 500 hours:
  - Replacement of the fuel filter
  - Replacement of the engine oil and oil filter (may be specified at 500 hours or 250 hours)
  - Replacement of the hydraulic fluid filter
- 1,000 hours:
  - Changing the hydraulic fluid and filter
  - Service and inspection of the drive sprockets and chains
- 1,500 to 2,000 hours:
  - Changing the hydraulic fluid and filter (if not earlier)
  - Changing the engine coolant

Remember that the above list is only a single example of a maintenance schedule. Make sure that the schedule for the specific skid steer model in use is followed. Service intervals can differ dramatically.

Hydraulic fluid maintenance is addressed in all maintenance schedules. However, the hydraulic fluid should be changed whenever it becomes dirty or breaks down due to overheating. Continuous operation of the hydraulic system in hot environments can heat the hydraulic fluid to the boiling point and cause it to break down. In dusty areas, the fluid cooler may become clogged or blocked, causing the fluid to remain hot. As a result, the hydraulic fluid condition should be observed consistently, and replaced if it shows signs of overheating and breakdown. Filters should also be replaced whenever the fluid is changed.

2.2.3 Preventive Maintenance Records

Accurate, up-to-date maintenance records are essential for knowing the history of your equipment. Each machine should have a record that describes any inspection or service that is to be performed and the corresponding time intervals. Typically, an operator manual and some sort of inspection sheet are kept with the equipment at all times. Actions taken, along with the date, are recorded on the log. Various operators can then share information about fluids that have been added, for example, so that patterns in fluid loss are noted.

The operator manual usually has detailed instructions for performing periodic maintenance. If you find any problems with your machine that you are not authorized to fix, inform the foreman or field mechanic before operating the machine.

2.0.0 Section Review

1. The hydraulic fluid level is typically checked by _____.
   a. removing a dipstick
   b. observing a sight glass
   c. observing a floating indicator arm
   d. removing the fill cap and observing the level

2. Where should an operator look for specific information regarding preventive maintenance on the skid steer?
   a. On the company's maintenance bulletin board
   b. On the service decals inside the operator cab
   c. In the operator manual for the machine
   d. On the machine's maintenance record

SECTION THREE

3.0.0 STARTUP AND OPERATING PROCEDURES

Objective 3

Describe the startup, shutdown, and operating procedures for a skid steer.

a. State skid steer-related safety guidelines.
b. Describe startup, warm-up, and shutdown procedures.
c. Describe basic maneuvers and operations.
d. Describe related work activities.

Performance Tasks 2, 3, 4, and 5

Demonstrate proper entrance and exiting on a skid steer.

Perform proper startup, warm-up, and shutdown procedures on a skid steer.

Execute basic maneuvers with a skid steer, including:

- Grading
- Removing stumps or boulders
- Changing attachments
- Loading a dump truck
- Maneuvering on slopes
- Utilizing a fork attachment for moving materials

Properly load and secure a skid steer for transport.

Trade Terms

Rated operating capacity: The amount of weight that a skid steer is projected to handle safely through common maneuvers. The rated operating capacity is typically derived from the tipping load test, and is equal to 50 percent of the tipping load value.

Tipping load: The maximum load that a skid steer can hold in the air, with the coupler outstretched, before the rear wheels begin to leave the ground.

Now that all the basic skid steer components have been covered, along with the operator-performed inspections and preventive maintenance activities, the next step is to actually start and operate the machine. Before that can happen, operators must fully understand the safety issues associated with this versatile piece of equipment. The operator manual for a skid steer contains safety information about that specific machine. All skid steers do have similarities in their operation. However, they also all have slight differences. Some may be a bit clumsier to operate than others are. The type of steering modes available and the job-site terrain also change how a machine handles.

3.1.0 Safety Guidelines

Safe skid steer operation is the responsibility of the operator. Operators must develop safe working habits and recognize hazardous conditions to protect themselves and others from injury or death. Always be aware of unsafe conditions to protect the load and the skid steer from damage. Become familiar with the operation and function of all controls and instruments before operating the equipment. Read and fully understand the operator manual. Operators must be properly trained and certified. Take note of and read the many safety hazard placards on the machine.

3.1.1 Operator Training and Certification

Not everyone is allowed to operate a skid steer on a job site. However, skid steers are often rented by homeowners or other non-professionals that have no training in the machine's operation. This often leads to property damage or personal injury. In spite of their size, skid steers are powerful and dangerous units in unskilled hands. Due to liability and insurance guidelines, only trained and qualified personnel are allowed to operate skid steers on a job site. Operators may receive a license certifying them to operate specific pieces of equipment. They also have to be re-evaluated at different times over the years. Since operators often move from one job site to another or from one company to another, they may be required to requalify at each new location.

3.1.2 Operator Safety

Nobody wants to have an accident or be hurt. There are a number of things workers can do to protect themselves and those around them from getting hurt on the job. Be alert and avoid accidents.

The most frequent type of fatality with skid steers involves workers in the immediate area, not the operator. Operators are not killed nearly as often as are workers in the surrounding area, especially with ROPS and other safeguards now in place for operators. When operators are seriously injured or killed, the seat belt is usually not being used.

Know and follow your employer's safety rules. Your employer or supervisor will provide you with the requirements for proper dress and safety equipment. The following are recommended safety procedures for all occasions:

- Only operate the machine from the operator's cab.
- Wear the seat belt and tighten it firmly. The ROPS depends heavily on the operator remaining in the seat in the event of an accident. The ROPS should never be removed or modified in any way—no drilling, cutting, or welding should ever be done to a ROPS.
- Mount and dismount the equipment carefully using three points of contact. Never dismount a running machine, and never dismount when the lifting arms are up, unless the proper locks are in place.
- Wear a hard hat, safety glasses, and safety shoes when operating the equipment. Hearing protection should also be on hand and used with machines that are not enclosed.
- Do not wear loose clothing or jewelry that could catch on controls or moving parts.
- Keep the windshield, windows, and mirrors clean at all times.
- Never operate equipment under the influence of alcohol or drugs.
- Never smoke while refueling. Avoid any potential sources of static electricity when refueling.
- Do not use a cell phone or text while the machine is in operation.
- Never remove protective guards or panels.
- Never attempt to search for leaks with your bare hands. Hydraulic and cooling systems operate at high pressure. Fluids under high pressure can cause serious injury.
- Always lower the lifting arms and attachments to the ground before performing any service or when leaving the skid steer unattended.

3.1.3 Safety of Co-Workers and the Public

The operator is not only responsible for his/her own personal safety, but also for the safety of others who are working around in the area. Sometimes, an operator may be working in areas that are very close to pedestrians or motor vehicles. In

these areas, take time to be aware of what is going on around the site. Create a safe work zone using cones, tape, or other barriers. Remember, it is often difficult to hear when operating a machine. Use a spotter and a radio in crowded conditions.

The main safety points when working around other people include the following:

- Walk around the equipment to make sure that everyone is clear of the equipment before starting and moving it.
- Always pay attention to the direction of travel.
- Do not drive the skid steer up to anyone standing in front of an object or load.
- Make sure that personnel are clear of the rear area before turning.
- Know and understand the traffic rules for the area in which you are operating.
- Use a spotter when landing an elevated load or when you do not have a clear view of the landing area.
- Exercise particular care at blind spots, crossings, and other locations where there is other traffic or where pedestrians may step into the travel path.
- Do not swing loads over the heads of workers. Make sure you have a clear area to maneuver.
- Travel with the attachment or coupler no higher than 12 to 18 inches above the ground; always drop the coupler or attachment to the ground when parking the machine.
- Do not allow other workers to ride anywhere in or on the machine.

3.1.4 Equipment Safety

The skid steer has been designed with certain safety features to protect you as well as the equipment. For example, it has guards, canopies, shields, rollover protection, falling object protection, and seat belts. Know your equipment's safety devices and be sure they are in working order.

A Tragic Month

In the period between mid-February and mid-March 2012, four skid steer accident fatalities occurred in the agricultural environment alone. One operator died after being crushed between the bucket and the machine. The other three fatalities resulted from rollovers. One was crossing uneven ground, while the other two resulted from loads being carried too high during travel. The seat belts were not in use.

All of these losses could have been prevented easily. Don't be the next statistic.

Skid steer rollovers are a significant cause of deadly accidents. In too many cases, the seat belt was not in use. However, the lifting arms tend to be involved in the majority of fatalities. Operators have tried to take too many shortcuts by leaving the machine running and exiting the cab. The lifting arms, due to their close proximity to the machine and tight clearances, are the most dangerous part of the machine. Use the following guidelines to maintain a safe environment in and around the skid steer:

- Perform prestart inspection and lubrication daily.
- Look and listen to make sure the equipment is functioning normally. Stop if it is malfunctioning. Correct or report trouble immediately.
- Always travel with the coupler or load low to the ground.
- Never exceed the manufacturer's limits for speed, lifting, or operating on inclines.
- Know the weight of all significant loads before attempting to lift them. Review the appropriate load chart. Do not exceed the rated capacity of the skid steer.
- Always lower the coupler or work tool, engage the parking brake, turn off the engine, and secure the controls before leaving the equipment.
- Never park on an incline.
- Maintain a safe distance between your skid steer and other equipment that may be on the job site.
- Balance has everything to do with the operation of a skid steer. With no load, roughly two-thirds of the machine's weight is over the rear axle. The balancing point makes a dramatic shift forward with a significant load, changing the behavior of the machine.
- Skid steer stability decreases sharply during travel if the lifting arms are raised.
- Always move up and down inclines with the heavy end of the machine at the high end. If the machine is unloaded, the rear is the heaviest end. Therefore, the skid steer should be backed up a hill when it is unloaded. This may still hold true for lightweight attachments and work tools, such as an auger.
- Under no circumstances should an operator attempt to reach into the cab and operate the machine from the outside. Many serious injuries and fatalities on older, less-safe skid steers happened this way.

Know your equipment. Learn the purpose and use of all gauges and controls as well as your equipment's limitations. Never operate your machine if it is not in good working order.

3.1.5 Spill Containment and Cleanup

Skid steers are powered by combustible fuels. Fuels must always be contained during refueling activities. Most companies have designated refueling areas specifically to contain any fuel spills that may occur.

Skid steers also use hydraulic systems to operate all aspects of the machine. With so many of the operational controls relying on the hydraulic system, it is critical that hydraulic fluid be contained without spills or leaks. Anyone who has ever worked around any hydraulic system knows that it is almost impossible to stop all leaks. Leaks need to be cleaned up and repaired whenever reasonably possible. Hydraulic fluid spills, on the other hand, are much more of a problem. If a hydraulic hose ruptures under pressure, there may be gallons of hydraulic fluid spilled. To protect the environment, companies maintain an oil spill kit easily accessible to the operator. A typical oil spill kit may include the following:

- A large (20-gallon or more) salvage drum/container used to store spill kit materials until needed. It is used for disposal after materials are used for cleanup.
- An emergency response guide for spills of petroleum-based products.
- Personal protective equipment (goggles and unlined rubber gloves).
- One or more 3" × 10' absorbent socks for containment.
- Half dozen 3" × 4' absorbent socks for containment.
- 20 or so absorbent pads for cleanup.
- 50 or so wiper towels or rags for cleanup.
- 3 to 5 pounds of powdery absorbent material.
- Several disposable bags with ties.

Although a flat-tipped shovel may not fit into a spill kit, having one handy is a good idea for spill cleanups.

3.2.0 Startup, Warm-Up, and Shutdown Procedures

Before starting skid steer operations, make sure that you are familiar with the task and the area of operations. Check the area for both vertical and horizontal clearances, and for obstructions that interfere with normal turns or pivots. Make sure that the path is clear of electrical power lines and other hazardous obstacles.

The following suggestions can help improve operating efficiency:

- Observe all safety rules and regulations.
- Analyze the tasks to be done and ensure the skid steer is capable of accomplishing the tasks safely. Visualize the work being done.
- Use a spotter if you cannot see the area where a load will be placed. A spotter may also be needed for work such as milling or cutting concrete.

3.2.1 Preparing to Work

Preparing to work involves getting organized in the cab, fastening the seat belt, and starting the machine. Mount your equipment using the grab rails and foot rests. Getting in and out of equipment can be dangerous. Always face the machine and maintain three points of contact when mounting and dismounting a machine. That means you should have three out of four of your hands and feet on the equipment. That can be two hands and one foot or one hand and two feet. Skid steers have grab bars and secure, slip-resistant foot holds to help.

WARNING!

OSHA requires that approved seat belts and a ROPS be installed on virtually all heavy equipment. Old equipment must be retrofitted and the seat belt must be used at all times. Do not use heavy equipment that is not equipped with these safety devices.

Adjust the seat to a comfortable operating position. Ensure that the throttle pedal on the floor is within comfortable reach. In a skid steer, the seat position is not quite as critical to operation, since there is no steering wheel. Skid steers do not have a service brake. Without these two concerns, positioning the seat is primarily related to maintaining a comfortable grip on the joysticks or levers, with sufficient legroom for comfort. The seat height is adjustable on most machines to provide comfort. The seat height will affect the shoulders and how the arms lay on the arm rests. Some models also offer the ability to adjust the position of the joystick controls forward or backward (*Figure 50*). Make sure you can see clearly and reach all the controls. Also adjust the mirrors before starting the engine. The operator in *Figure 51* is properly prepared for startup, with the seat belt fastened and armrests down. Remember that late-model skid steers have an interlock in the armrests that will not allow operation unless they are in position.

The startup and shutdown of an engine is very important. Proper startup lengthens the life of the engine and other components. A slow warm up is essential for proper operation of the machine

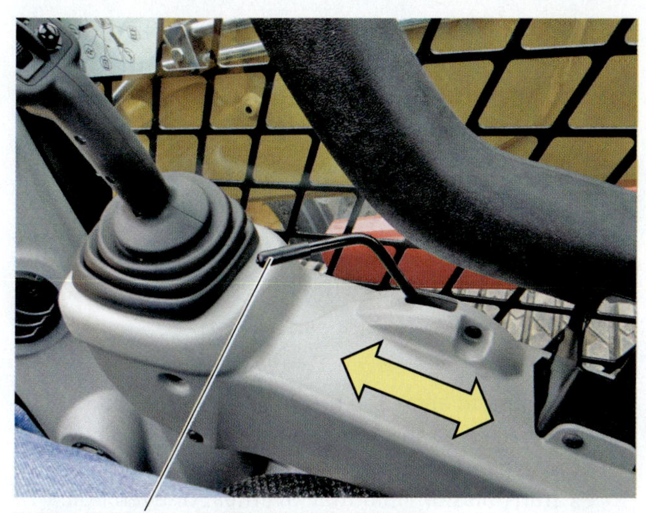

THIS LEVER ALLOWS THE JOYSTICK MODULE TO MOVE FORWARD OR BACKWARD TO ACCOMMODATE THE OPERATOR.

22212-12_F50.EPS

Figure 50 Joystick control module adjustment.

under load. Similarly, the machine must be shut down properly to cool the hot fluids circulating through the system. These fluids must cool so that they can help cool the metal parts of the engine before it is switched off.

3.2.2 Startup

There may be specific startup procedures for the piece of equipment you are operating, but in general, the startup procedure should follow this sequence. However, it is important to always follow the sequence offered in the operator manual. Before beginning the startup sequence, the operator should be seated in the cab, with the seat belt fastened and the armrest(s) pulled down into position.

Step 1 Be sure the parking brake is engaged.

Step 2 Turn the ignition key to the first detent, or the On position. This should enable the electrical system, and the instrument lights should illuminate for several seconds. Watch the lights to ensure that they are all functional. If lights that alert the operator to serious problems (typically red lights) are not working, do not operate the machine until the problem is addressed. The parking brake lamp should remain On as the others turn Off.

Step 3 Set the throttle control to the middle of its travel, about half-throttle.

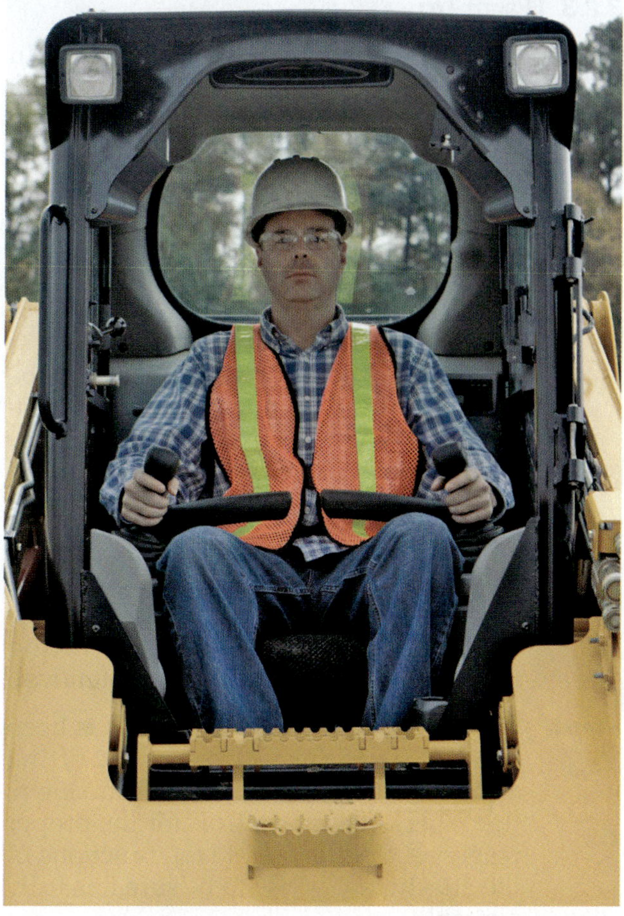

22212-12_F51.EPS

Figure 51 Operator properly prepared for startup.

Step 4 Turn the ignition switch to the Start position. The engine should turn over. Never operate the starter for more than 30 seconds at a time. If the engine fails to start, wait two to five minutes before cranking again. The lesser amount of time typically applies in cold weather, which allows the starter to cool off more quickly.

Step 5 As soon as the engine starts, release the key; it should return to the On position on its own.

Maintain the engine speed at one-third to one-half throttle for the warm-up period. Always keep the engine speed low until the oil pressure light is extinguished. The oil pressure light should initially light and then go out. If the oil pressure light does not turn off within 10 seconds, stop the engine, investigate, and correct the problem.

There are special procedures for starting a diesel engine in cold temperatures. Many diesel engines have glow plugs that preheat the engine before starting. Follow the manufacturer's instructions for starting the machine in cold weather.

3.2.3 Warm Up

Warm up a cold engine for at least 5 minutes. In colder temperatures, warm up the machine for a longer period of time. While allowing time for the engine to warm up, complete the following tasks.

Step 1 Check all the gauges and instruments to make sure they are working properly. Watch the temperature gauge as the coolant temperature climbs if the machine has one.

Step 2 Determine what operating modes will be needed for the work ahead and set the controls accordingly. For example, the operator may or may not need to provide a high-flow hydraulic source for an attachment.

Step 3 Release the parking brake. This is often necessary to enable use of the controls. Ensure that, with the joysticks in their neutral position, the skid steer, lifting arms, and any attachments do not move.

Step 4 Once a sufficient warm-up time has been allowed, cycle the controls for the lifting arms. Lift them up and bring them down, then check the tilt controls for the proper response. Briefly operate the functions of the attachment in use to be sure it is fully operable.

Step 5 Position the lifting arms and attachment in the travel position, and use the controls to move the skid steer slightly forward and backward. Then execute a turn in each direction.

> **CAUTION**
>
> Do not pick up a load if the hydraulics are sluggish. Allow the machine to warm up until the hydraulic components function normally. The hydraulics can fail if not warmed up completely.

Step 6 Reset the parking brake while the machine continues to warm up.

If there are any problems that have no obvious cause, shut down the machine and investigate or get a mechanic to look at the problem.

3.2.4 Shutdown

Shutdown should also follow a specific procedure. Proper shutdown reduces engine wear and possible damage to the machine.

Step 1 Find a dry, level spot to park the skid steer. Stop the machine by simply allowing the driving controls to return to a neutral position.

Step 2 Lower the coupler or attachment to the ground and engage the parking brakes.

Step 3 Place the speed control in low idle and let the engine run for approximately five minutes.

> **CAUTION**
>
> Failure to allow the machine to cool down can cause excessive temperatures in the engine or systems. The engine oil, hydraulic fluid, or transmission fluid may overheat or even boil.

Step 4 Turn the engine start switch to the Off position.

Step 5 Release hydraulic pressures by moving the control levers until all movement stops.

Step 6 Remove the ignition key. If you must park on an incline, chock the wheels.

Lock the cab door, and secure and lock the engine enclosure. Always engage any available security systems and devices when leaving the skid steer unattended.

3.3.0 Basic Maneuvers

To maneuver a skid steer, you must be able to move forward, backward, and turn. The following guidelines assume that the left joystick has full control over vehicle movement.

Remember that allowing the joystick to return to its neutral position stops all vehicle movement. There is no foot brake on today's models. With the skid steer's hydrostatic drive system, once the control returns to neutral, all hydraulic flow in and out of the hydraulic motors is stopped. The machine will not typically lock up the wheels as the joystick is released and it comes to a stop. However, it should stop rather quickly and surely. The machine will also come to a stop on an incline that is within its specifications, with only slightly more travel. They do not coast when the joystick is moved to the neutral position. The lack of a positive, dedicated brake in the skid steer is probably the most confusing feature to a new operator. However, the inherent stopping ability of the drive system has proven to be an effective and safe design.

3.3.1 Moving Forward

The first basic maneuver is learning to drive forward. To move forward, follow the steps below.

CAUTION

Always travel with the coupler or attachment low to the ground (12 to 18 inches). Be aware of the forward edge of the attachment and avoid hitting things with it. On very rough terrain, a slightly higher travel position may be required.

Step 1 Disengage the parking brake.

Step 2 Before starting to move, use the joystick to raise the coupler or attachment roughly 15 inches above the ground. This is the travel position.

Step 3 Push the left joystick straight forward, slightly and easily. All four drive-wheels should engage and move the machine forward. To reach full speed at the current control settings, slowly push the joystick fully forward. The farther forward the joystick is pushed, the faster the machine will go, within its current or preset limits.

Step 4 To stop, allow the joystick to return to neutral.

3.3.2 Moving Backward

To back up or reverse direction, the only action required is to move the left joystick backwards. As the joystick is moved closer to the backward limit of its travel, the skid steer moves faster, up to its top speed for the current settings.

3.3.3 Turning

The concept of skid steering has been presented but not yet practiced. Remember that skid steer wheels are fixed on their axle and do not turn left or right. Turns happen as a result of the drive wheels on one side turning faster than the wheels on the opposite side. The skid steer can be made to pivot in place by driving the wheels on each side in opposite directions.

The sequence, using a single joystick to control machine movement, is as follows:

- With the joystick pushed forward and to the left, the machine executes a relatively gradual turn to the left. The drive wheels on the right rotate at a higher speed than those on the left.
- With the joystick pushed forward and to the right, the machine executes a relatively gradual turn to the right. The drive wheels on the left rotate at a higher speed than those on the right.
- Pushing the joystick from its neutral position directly to the left causes the skid steer to pivot in place to the left. It continues to spin in a tight circle in this position. The drive wheels on the left will be pushing forward, while the drive wheels on the right will be pushing backwards, thus causing the unit to pivot in place.
- Pushing the joystick from its neutral position directly to the right causes the skid steer to pivot in place to the right. It continues to spin in a tight circle in this position. The drive wheels on the right will be pushing forward, while the drive wheels on the left will be pushing backwards, thus causing the unit to pivot in place.
- With the joystick pushed backward and to the left, the machine gradually turns the front of the unit toward the left while backing up. The drive wheels on the left rotate at a higher speed than those on the right.
- With the joystick pushed backward and to the right, the machine gradually turns the front of the machine toward the right while backing up. The drive wheels on the right rotate at a higher speed than those on the left.

3.4.0 Work Processes

The operation of skid steers is fairly straightforward, but it requires some practice. They can start and stop quickly, and new operators may feel a bit clumsy at first. The lack of a steering wheel and brake makes skid steers a very different vehicle than those that people are used to driving. Although many pieces of heavy equipment are quite different to drive and manipulate than a car, the vast majority retain a steering wheel and a service brake in common with standard motor vehicles.

With practice, good skills in operating the skid steer will be developed. Some basic work activities performed with a skid steer are described in this section.

NOTE

The controls on specific skid steer machines may be different from those described in the procedures. Check your operator manual for information about the controls and limitations of your equipment.

3.4.1 Skid Steer Capacities and Limitations

Skid steer machines do not typically have load charts as extensive as a forklift may have. There are a few important specifications to consider about the machine itself before putting it to work.

A lot of the information an operator may need to know comes from the attachment or work tool documentation.

Skid steers are rated first by their **rated operating capacity**. This is generally the first specification noted. The rated operating capacity is considered the amount of weight the machine can safely lift and maneuver. However, the rated operating capacity is actually derived from another value that is based on testing.

The **tipping load** of a skid steer machine is a specification that is based on testing. Full-size loaders may have a variety of tipping load values in their specifications. Straight static tipping load and full turn tipping load are two examples. Skid steers usually have only one tipping load value in their specifications. The tipping load is simply defined as the amount of weight that can be placed on the coupler with the lifting arms at their maximum forward reach, without causing the skid steer to lift the rear tires off of the ground.

The rated operating capacity of a skid steer is based on the tipping load. It represents 50 percent of the tipping load. Therefore, a skid steer with a tipping load of 2,800 pounds (1,270kg) would have a rated operating capacity of 1,400 pounds (635kg). The rated operating capacity is considered a safe limit in comparison to the tipping load, indicating the machine can both lift and maneuver with this much weight. The tipping load test is done with the skid steer stationary. Note that skid steers can be fitted with counterweights (*Figure 52*) that increase their rated operating capacity somewhat. However, the increase in capacity is usually only 100 to 300 pounds (45 to 136kg).

22212-12_F52.EPS

Figure 52 Counterweights at the rear of the skid steer, to add mass and increase its rated operating capacity.

It is important to be fully aware of the machine's operating limitations. It is equally important that the weight of a load also be known. Knowing the machine's limitations alone is simply not enough. Knowledge of the machine has little value to the operator if he or she does not pay attention to the weight of the load. Remember that a load that can be lifted off of the ground without tipping the skid steer is not necessarily a load that can be safely maneuvered or carried to another location.

Lift and take control of all loads carefully. If the rear wheels of the skid steer begin to rise off of the ground when trying to lift a load, stop and slowly lower the load back to its starting point.

3.4.2 Using a Bucket

The most common attachment for a skid steer is the bucket. It may be one of several different designs, but the bucket is the most popular overall. Unlike full-sized loaders though, skid steers cannot typically reach over the sides of a large dump truck for loading. However, they are often used to load material, such as mulch or topsoil, into a smaller vehicle.

The procedure for carrying out a loading operation from a stockpile of material is as follows:

Step 1 Travel to the work area with the bucket in the travel position.

Step 2 Position the bottom of the bucket parallel to and just skimming the ground.

Step 3 Drive the bucket straight into the stockpile. Once the bucket stops penetrating, ease off of the joystick to avoid spinning the wheels.

Step 4 Adjust the controls and raise the bucket to fill it, as shown in *Figure 53*.

22212-12_F53.EPS

Figure 53 Filling the bucket.

Step 5 Work the tilt control lever back and forth if necessary to move material to the back of the bucket. Keeping the contents closer to the skid steer helps keep the machine stable for maneuvering.

Step 6 Use the joystick to back the skid steer away from the stockpile.

Step 7 Place the bucket in the travel position and move toward the destination.

Step 8 Raise the bucket high enough to clear the side of the truck or other obstruction.

Step 9 Move the bucket over the truck bed and tilt the bucket forward to dump the bucket. Moving the tilt control back and forth a few times may help move the contents out. *Figure 54* shows the proper position for dumping from a standard bucket.

Step 10 Tilt the bucket back to level the bottom again and back the skid steer away from the truck.

Step 11 Return the bucket to the travel position and return to the stockpile.

Step 12 Repeat the cycle until the truck is loaded.

Loading material into a truck with a skid steer requires that the operator have good reflexes and distance judgment. The loader must be placed close to the side of the truck in order to get the bucket positioned to dump properly.

There are two main points where accidental contact is the most common. The first area of contact is between the bucket and the side of the truck. Either the operator has misjudged the height of the truck bed or approached the truck too quickly, not allowing sufficient time to raise the bucket before getting to the truck. The second contact point is between the front wheels or tracks of the skid steer and the side of the truck body. Again, this is due to the operator misjudging the distance between the front of the loader and the side of the truck. Contact from these situations can cause severe damage to both pieces of equipment.

22212-12_F54.EPS

Figure 54 Dumping the bucket.

When loading from a stockpile or bank, the placement of both the truck and the skid steer are variable and must be adjusted to local operating conditions. Conditions to be considered are the weight of material, the gradient of the loading area, and available traction. If the skid steer works too close to the truck, it is necessary to pause during each cycle for the bucket to clear the side of the truck. If the machine works too far from the truck, the cycle is excessively long, with a resulting waste of time. The operator, by experience, must determine the most efficient arrangement for the particular operation and direct the trucks accordingly.

While there are many ways to maneuver a skid steer, the two most common patterns for a truck loading operation are the I-pattern and the Y-pattern.

For the I-pattern, both the skid steer and the dump truck move in only a straight line, backward and forward (*Figure 55*). This is a good method for small, cramped areas. The operator fills the bucket and backs approximately 20 feet away from the pile. The dump truck backs up between the skid steer and the pile. The operator dumps the bucket into the truck. The truck moves out of the way and the cycle repeats. This pattern requires constant truck movement however, and is not generally needed with the quick and compact movements of the skid steer.

To perform this I-pattern loading maneuver, position the skid steer so that it is on the truck driver's side of the truck. That way, eye contact can be made with the driver. Fill the bucket and back far enough away from the pile to allow the truck to back into position. Signal the truck driver with the horn. The truck backs up to a predetermined position, as shown in *Figure 55B*. Move the machine forward and center it on the truck bed. Raise the bucket to clear the side of the truck and place it over the truck bed. Dump the bucket. When the bucket is empty, move the tilt control to shake out the last of the material. Back the skid steer away from the truck and signal the truck driver to move. When the truck is out of the way, lower the bucket and position the machine to return for another bucket of material.

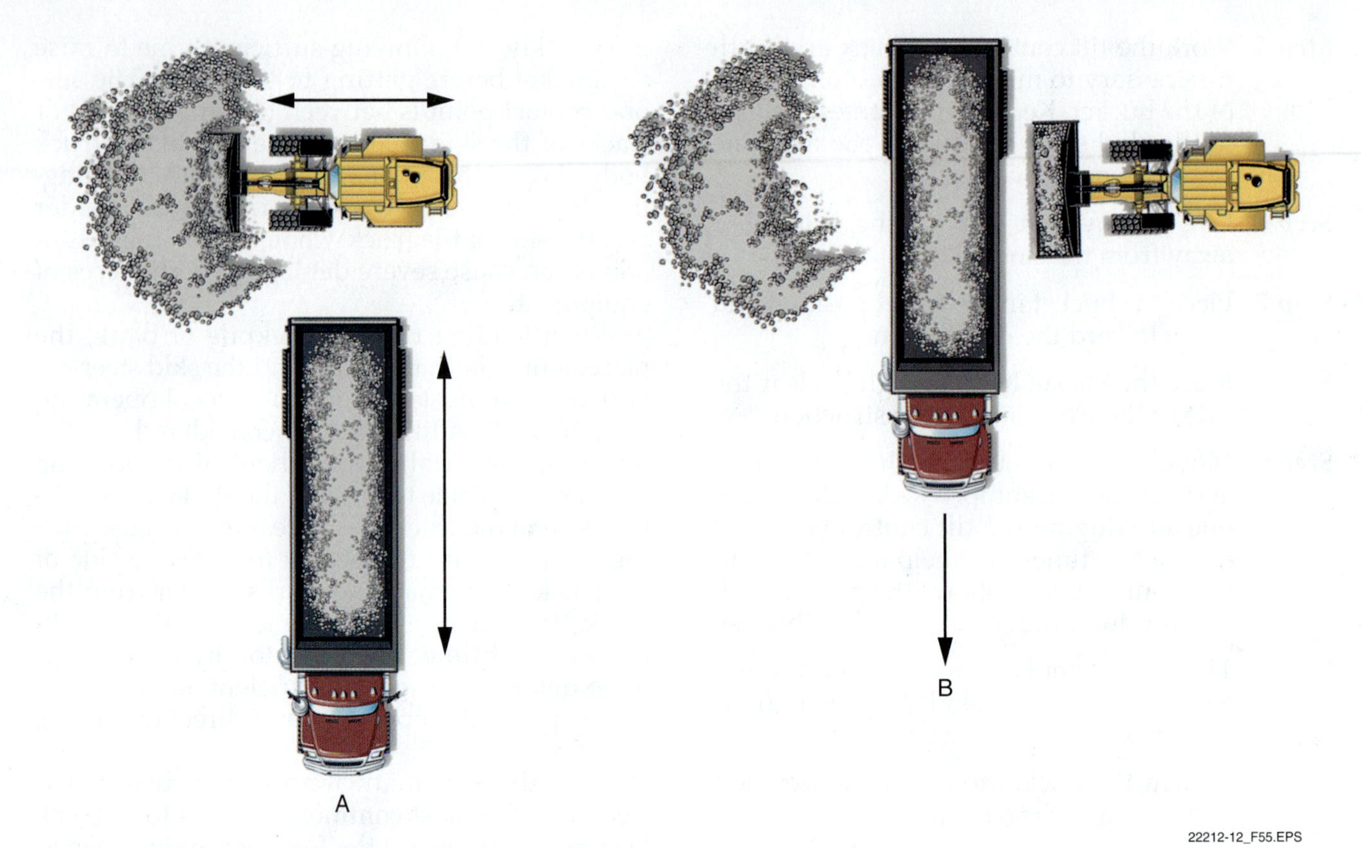

Figure 55 The I-pattern for loading.

The other loading pattern is the Y-pattern (*Figure 56*). This method is used when larger open areas are available. The dump truck remains stationary and as close as possible to the pile. The skid steer does all the moving in a Y-shaped pattern.

Position the truck so that eye contact can be made with the driver. Fill the bucket with material. While backing up, turn the skid steer to the right or left, depending on the position of the truck. Shift forward and turn the machine while approaching the truck slowly. Stop when the skid steer is lined up with the truck bed. Dump the bucket in the same way done for the I-pattern. When the bucket is empty, back away from the truck while turning toward the pile. Drive forward into the pile to repeat the pattern. Repeat the cycle until the truck is full.

3.4.3 Leveling and Grading

Minor leveling and grading operations can be performed with the skid steer under most conditions. A multipurpose bucket is well suited for grading. Leveling can be done by tilting the bucket down and placing the cutting edge on the ground surface. Backing up with the bucket in this position will smooth out loose material.

Figure 56 The Y-pattern for loading.

For grading operations, the bottom of the bucket should be parallel to the ground surface (*Figure 57*). While maintaining this position, material from high spots will be scraped into the bucket.

To perform a grading operation with the skid steer, follow these steps:

Step 1 Line up the bucket and skid steer on the area to be graded.

Step 2 Move the joystick to position the bucket at the desired height, with the bottom level.

Step 3 Begin moving forward through the area.

Step 4 Maintain a low steady speed and keep the bucket at a constant height. As high spots are encountered, they will be trimmed by the blade and loaded into the bucket.

Step 5 When low spots are encountered, tilt the bucket forward and dump some of the material. Back up over the dumped material and smooth it out with the back of the blade, then return to grading while moving forward.

Step 6 Reposition the machine and make another pass to the left or right of the area just graded.

3.4.4 Fork Attachments

Use a fork attachment with pallet-style forks (*Figure 58*) to move palletized materials. They come in several different sizes depending on the lifting weight and reach required. Operating the skid steer as a forklift is basically the same procedure for all models.

22212-12_F57.EPS

Figure 57 Grading with a skid steer.

22212-12_F58.EPS

Figure 58 Fork attachment.

The following steps are used to unload material from a flatbed:

Step 1 Position the skid steer at either side of the truck bed.

Step 2 Manipulate the control levers in order to obtain the appropriate fork height and angle.

Step 3 Drive the forks into the opening of the pallet or under the loose material. Use care not to damage any material.

Step 4 Manipulate the controls as required to lift the material slightly off of the bed.

Step 5 Tilt the forks back slightly to keep the pallet or other material from sliding off the front of the forks (*Figure 59*).

Step 6 Use the joystick to back the machine away from the truck.

Step 7 Lower the forks to the travel position and travel to the stockpile area.

Step 8 Position the skid steer so that the material can be placed in the desired area.

Step 9 Lower the lifting arm slowly until the material is set on the required surface.

Step 10 Adjust the forks to relieve the pressure under the pallet.

Step 11 Back the machine away from the pallet.

Step 12 Repeat the cycle until the truck is unloaded or loaded.

Safe operation of the skid steer as a forklift requires extreme caution. Make sure to follow these safety requirements:

22212-12_F59.EPS

Figure 59 Lifting the load.

- Do not swing loads over the heads of workers. Make sure that there is enough clear area to maneuver.
- Do not allow workers to ride on the forks.
- The forks should always be lowered to the ground when the skid steer is parked.
- Using the fork attachment usually requires operation in a confined area. Always know what clearance is available for maneuvering.

3.4.5 Other Attachments

The skid steer can accommodate many different attachments. Each one, even those that do the same job but are different in brand or model, has different operating characteristics. For many attachments, the skid steer is merely a carrier. As far as the skid steer is concerned, the operator need only worry about the capacity of the skid steer to carry and maneuver the attachment into position. Operating the attachment, though, may be more complex. It is essential that the operator reads and understands the operating requirements and safety guidelines of the attachment before beginning work with one. Test the controls in a safe setting before beginning, to ensure that it can be operated safely and efficiently.

3.4.6 Transporting a Skid Steer

Transporting a skid steer is a very common task. It is not unusual for a skid steer to be on a different job site each day.

The skid steer should be transported on a properly equipped trailer or other transport vehicle. Before beginning to load the equipment for transport, make sure the following tasks have been completed:

- Check the operator manual of the transportation vehicle or trailer to determine if the loaded equipment complies with height, width, and weight limitations for over-the-road hauling.
- Check the operator manual to identify the correct tie-down points on the equipment.
- Plan the loading operation so that the loading angle is at a minimum.

Once these tasks are complete and the loading plan has been determined, carry out the following procedures:

Step 1 Position the trailer or transporting vehicle. Always block the wheels of the transporter after it is in position, before loading is started.

Step 2 Place the bucket in the travel position and carefully back the skid steer onto the transporter. Remember that the heavy end of the skid steer should always be uphill. With the skid steer unloaded, the heavy end is always at the rear.

Step 3 Lower the bucket to the floor of the transporter.

Step 4 Engage the parking brake, shut down the engine, and remove the ignition key.

Step 5 Lock the door to the cab as well as any access covers. Attach any vandalism protection.

Step 6 Secure the machine with the proper tie-down equipment as specified by the manufacturer (*Figure 60*). Chocks are recommended at the front and back of all four tires, and may be required. If so, make sure they are secure and cannot leave the trailer while the trip is underway. Use all four tie-down points for the machine as documented by the manufacturer.

Step 7 Cover the exhaust and air intake openings with tape or a plastic cover.

Step 8 Place appropriate flags or markers on the equipment if needed for height and width restrictions.

22212-12_F60.EPS

Figure 60 Front tie-down points for transporting the skid steer.

3.0.0 Section Review

1. Fatalities related to skid steers most often _____.
   a. occur during a rollover
   b. involve the operators themselves
   c. involve workers in the immediate area
   d. occur when loading or unloading the machine

2. When do the instrument lights typically light up for a brief test?
   a. When they key is first turned to the On position.
   b. When the key is first turned to the Start position.
   c. Within the first 30 seconds of starting the engine.
   d. When the key is turned to the Off position for shutdown.

3. Skid steers can be stopped by _____.
   a. using the service brake pedal
   b. using the service brake lever
   c. using the trigger switch on a joystick
   d. letting go of the joystick that controls machine movement

4. What would the rated operating capacity of a skid steer be if it has a tipping load of 2,480 pounds (1,125kg)?
   a. 1,240 pounds (562.5 kg)
   b. 1,637 pounds (742.5 kg)
   c. 2,180 pounds (988.8 kg)
   d. 3,720 pounds (1,687.4 kg)

SUMMARY

The introduction of the skid steer years ago has changed the speed and manner in which work is accomplished. No machine has more versatility, thanks to the wide variety of attachments available. Although skid steers may be limited in capacity, they are nearly unlimited in the tasks that can be accomplished quickly and effectively. As the name implies, skid steers turn by the drive wheels on one side turning more or less than those on the opposite side.

The vast majority of skid steers operate on diesel fuel, and have engines rated at less than 100hp. Instead of a traditional drivetrain with a transmission, skid steers are moved by hydraulic pumps and motors. The drive wheels on each side are powered by separate hydraulic motors. Additional hydraulic pumps serve the lifting arms and various attachments. Skid steers are not as complex and large as their heavier relatives are, and are easier to maintain and inspect.

Preparing a skid steer for operation is relatively simple. Thanks to ROPS and FOPS, injuries and fatalities resulting from skid steer operations have dropped significantly. Although earlier machines were driven by a set of levers, today's skid steers are operated by joysticks. The joysticks are equipped with a variety of buttons and switches that take on different roles, based on the type of attachment or work tool in use. Some work tools require additional hydraulic pressure sources and control interlocks. Others, such as the bucket, require no additional support for operation other than a sound connection to the coupler.

Operating a skid steer is very different from operating a standard vehicle. However, with practice, an operator can soon become highly skilled at maneuvering and completing tasks in one of the most versatile pieces of heavy equipment ever developed.

Review Questions

1. One feature that makes the skid steer more compact is the absence of _____.
   a. a radiator
   b. a transmission
   c. hydraulic pumps
   d. a hydraulic fluid reservoir

2. On a skid steer, there is *not* a direct relationship between the engine and the drive wheels.
   a. True
   b. False

3. The hydraulic pumps of a skid steer are typically located _____.
   a. at each drive wheel
   b. under the operator's seat
   c. in the engine compartment
   d. under the chassis, with one on each side

4. ROPS systems must be able to withstand an impact of _____.
   a. the weight of the machine
   b. twice the weight of the machine
   c. three times the weight of the machine
   d. four times the weight of the machine

5. With a fixed displacement hydraulic pump, _____.
   a. the flow rate of fluid is dependent on the engine speed
   b. a relief valve to allow continuous fluid flow is not required
   c. the flow rate of the fluid is controlled by valves inside the pump
   d. the amount of energy transferred through the fluid is greatly reduced

6. When the H-pattern of control is selected in a skid steer, the _____.
   a. left joystick controls the machine movement
   b. right joystick controls the machine movement
   c. joysticks only control the coupler and attachments while the skid steer sits still
   d. left joystick controls the left wheels while the right joystick controls the right wheels

7. It is common for skid steers to have both a foot- and hand-operated throttle.
   a. True
   b. False

8. Compared to pallet forks, bale forks are _____.
   a. more slender
   b. much heavier
   c. wider and longer
   d. vertical instead of horizontal

9. How might cold weather affect the skid steer lubrication schedule?
   a. Points requiring grease must be serviced half as often.
   b. Points requiring grease must be serviced twice as often.
   c. It is best done in the morning when the grease is coldest and thickest.
   d. It is best done at the end of the day when the grease and grease points are warmer.

10. Assuming standard five-day workweeks with normal working hours, what is the interval for a 500-hour inspection?
    a. Monthly
    b. Bi-monthly
    c. Quarterly
    d. Semi-annually

11. Which of the following items can be done to the ROPS by an operator?
    a. Drill a small hole and mount a cup holder.
    b. Weld a small bracket on it to mount something.
    c. Cut a small opening in one spot to improve visibility
    d. Attach a copy of the inspection checklist using a spring clamp.

12. At shutdown, the skid steer should be allowed to continue operating to cool the fluids for approximately _____.
    a. 1 minute
    b. 2 minutes
    c. 5 minutes
    d. 15 minutes

13. When the skid steer is running, it will coast down most any incline unless the parking brake is used.
    a. True
    b. False

14. When a single joystick is used to control skid steer travel, pushing the joystick forward and to the right will result in _____.
    a. the machine pivoting in place to the left
    b. the machine pivoting in place to the right
    c. a gradual turn to the left while moving forward
    d. a gradual turn to the right while moving forward

15. The operator can help the contents of the bucket move out when dumping by _____.
    a. bumping the tilt control back and forth
    b. bumping the machine forward and backward
    c. gently nudging the side of the truck tires a few times
    d. running the engine at idle speed when the bucket is tilted forward

Trade Terms Introduced in This Module

Fixed displacement pump: A pump that cannot be adjusted to deliver more or less fluid volume in a pumping cycle.

Float mode: Placing the bucket or other attachment into the float mode allows the attachment to automatically follow the contour of the ground, remaining at the same elevation above it.

Hydrostatic drive: A system that relies on the flow and pressure of a fluid to transfer energy. In skid steers, the transfer of energy is from the engine, driven by diesel fuel, to the four drive wheels through hydraulic motors.

Variable displacement pump: A pump that can deliver a variable fluid volume in a pumping cycle to support the changing needs of a hydraulically operated device.

Figure Credits

Reprinted courtesy of Caterpillar Inc., Module opener, Figures 1, 3, 4, 7, 9, 10, 12–14, 16, 18–21, 23, 24, 26, 27, 35B (e, f, g, h), 35C (i, j, k), 43, 51–54, 57–59

Courtesy of Bobcat®. Bobcat®, the Bobcat logo and the colors applied to the Bobcat vehicle are registered trademarks of Bobcat Company in the United States and various other countries., Figures 2, 6, 30, 33, 34, 35A (a, b, c, d), 35C (l), SA01, SA02

Topaz Publications, Inc., Figures 8, 17, 22, 25, 28, 29, 31, 32, 36, 37–42, 44–50, 60

Courtesy of Titan Tire Corporation, Figure 11

Courtesy of Skid Steer Solutions, Figure 15

Section Review Answers

Answer	Section Reference	Objective
Section One		
1 b	1.1.1	1a
2 c	1.2.0	1b
3 b	1.3.0	1c
4 d	1.4.0	1d
Section Two		
1 b	2.1.1	2a
2 c	2.2.0	2b
Section Three		
1 c	3.1.2	3a
2 a	3.2.2	3b
3 d	3.3.0	3c
4 a	3.4.1	3d

NCCER CURRICULA — USER UPDATE

NCCER makes every effort to keep its textbooks up-to-date and free of technical errors. We appreciate your help in this process. If you find an error, a typographical mistake, or an inaccuracy in NCCER's curricula, please fill out this form (or a photocopy), or complete the online form at **www.nccer.org/olf**. Be sure to include the exact module ID number, page number, a detailed description, and your recommended correction. Your input will be brought to the attention of the Authoring Team. Thank you for your assistance.

Instructors – If you have an idea for improving this textbook, or have found that additional materials were necessary to teach this module effectively, please let us know so that we may present your suggestions to the Authoring Team.

NCCER Product Development and Revision

13614 Progress Blvd., Alachua, FL 32615

Email: curriculum@nccer.org

Online: www.nccer.org/olf

❑ Trainee Guide ❑ AIG ❑ Exam ❑ PowerPoints Other ______________________________

Craft / Level: Copyright Date:

Module ID Number / Title:

Section Number(s):

Description:

Recommended Correction:

Your Name:

Address:

Email: Phone:

22205-13

Loaders

Overview

Loaders are one of the more popular pieces of equipment used in construction. They are characterized by a large bucket mounted on two lift arms on the front of the machine. For this reason, they are often called front-end loaders. Loaders are mounted either on rubber tires or on metal tracks. Loaders mounted on tracks are often called track loaders or crawler loaders.

Module Eight

Trainees with successful module completions may be eligible for credentialing through NCCER's National Registry. To learn more, go to **www.nccer.org** or contact us at **1.888.622.3720**. Our website has information on the latest product releases and training, as well as online versions of our *Cornerstone* newsletter and Pearson's product catalog.

Your feedback is welcome. You may email your comments to **curriculum@nccer.org**, send general comments and inquiries to **info@nccer.org**, or fill in the User Update form at the back of this module.

This information is general in nature and intended for training purposes only. Actual performance of activities described in this manual requires compliance with all applicable operating, service, maintenance, and safety procedures under the direction of qualified personnel. References in this manual to patented or proprietary devices do not constitute a recommendation of their use.

22205-13 LOADERS

Objectives

When you have completed this module, you will be able to do the following:

1. Identify and describe the components of a loader.
   a. Identify and describe chassis components.
   b. Identify and describe loader controls.
   c. Identify and describe loader instrumentation.
   d. Identify and describe loader attachments.
2. Describe the prestart inspection and preventive maintenance requirements for a loader.
   a. Describe prestart inspection procedures.
   b. Describe preventive maintenance requirements.
3. Describe the startup and operating procedures for a loader.
   a. State loader-related safety guidelines.
   b. Describe startup, warm-up, and shutdown procedures.
   c. Describe basic maneuvers and operations.
   d. Describe related work activities.

Performance Tasks

Under the supervision of your instructor, you should be able to do the following:

1. Complete a proper prestart inspection and maintenance on a loader.
2. Perform proper startup, warm-up, and shutdown procedures on a loader.
3. Execute basic maneuvers with a loader, including proper movements and curling the bucket.
4. Carry out basic earthmoving operations with a loader, load a truck (to capacity, if possible), and build a storage pile.

Trade Terms

Accessories
Blade
Breakout force
Bucket
Debris
Dozing
Grouser
Grubbing
Roading
Rubble
Spot
Stockpile
Stripping
Tipping load

Industry Recognized Credentials

If you are training through an NCCER-accredited sponsor, you may be eligible for credentials from NCCER's Registry. The ID number for this module is 22205-13. Note that this module may have been used in other NCCER curricula and may apply to other level completions. Contact NCCER's Registry at 888.622.3720 or go to **www.nccer.org** for more information.

Contents

Topics to be presented in this module include:

Figures and Tables

SECTION ONE

1.0.0 LOADER COMPONENTS

Objective 1

Identify and describe the components of a loader.

a. Identify and describe chassis components.
b. Identify and describe loader controls.
c. Identify and describe loader instrumentation.
d. Identify and describe loader attachments.

Trade Terms

Blade: An attachment on the front end of a loader for scraping and pushing material.

Breakout force: The maximum vertical upward force created by the curling action of a bucket attached to an excavator, backhoe, or loader, measured 4 inches behind the tip of the bucket's cutting edge.

Bucket: A U-shaped closed-end scoop that is attached to the front of the loader.

Debris: Rough broken bits of material such as stone, wood, glass, rubbish, and litter after demolition.

Tipping load: The loaded weight that will lift the rear wheels off the ground with the machine in a static (not moving) condition.

Loaders are manufactured in various sizes and configurations by several different companies. There are two main categories of loaders, the wheel loader (*Figure 1*) and the track or crawler loader (*Figure 2*). The wheel loaders have rubber tires, while the crawler loaders have tracks instead of wheels. Additionally, some manufacturers make a small tracked loader similar to a skid steer, which can rotate in place.

The primary operating components of the loader are the **bucket**, lift arms, and dump rams. Operators use a steering wheel on older machines to turn wheel loaders, but newer loaders often use joysticks. Levers, pedals, or joysticks are used to turn the track loaders. The controls for the lift arms and bucket are usually the same on either style of loader.

Loaders are versatile and are easy to handle and maneuver. They are primarily used for lifting and moving materials, but can be used for other tasks such as loading, excavating, demolition, grading, leveling, hauling, and stockpiling. When loaders are equipped with attachments such as brooms, rakes, and forks (tool carriers), they can be used for a variety of other functions. *Figure 3* shows a loader equipped with a set of forks (tool carrier) instead of its normal bucket.

Operators must understand the general properties of the materials being moved by a loader. This is especially true when using a loader for earthmoving or other construction-site jobs. Here are a few examples of materials that may be lifted and moved by a loader:

- One cubic yard of excavated dry clay weighs roughly 1,836 pounds.
- One cubic yard of wood bark (refuse) weighs just over 400 pounds.
- One cubic yard of common red brick weighs 3,240 pounds.

WARNING!

Overloading the lifting devices on a loader can cause damage to the lifting devices. Overloading may also cause the loader to tip over.

1.1.0 Chassis Components

The frame used in any vehicle is often called its chassis. The frame or chassis is the backbone of the vehicle. The engine, transmission, axles, operator cab, and the main body parts are attached directly to the frame. Loaders with solid frames are referred to as rigid-frame loaders. Track loaders have rigid frames. Most wheel loaders are articulated, which means that there is a joint in the frame that allows the front and back to move independently. Articulated loaders are easier to reposition because they have a tighter turning radius. *Figure 4* shows a typical articulated loader.

Loaders are powered either by gasoline or diesel engines. On most loaders, the engine is behind the operator's cab. The rear-mounted engine provides added weight over the rear wheels, which provides better traction and helps balance heavy loads in the bucket.

As noted earlier, loaders exist in various sizes and configurations. Some loaders have specialized designs to be used in a specific industry. For example, loaders are frequently used at waste transfer stations to move small amounts of waste. Several manufacturers make specialized loaders in various sizes for waste handling. They are designed to keep **debris** out of the machine. Large service doors allow for easy cleaning. Additional guards keep debris out of the engine, hydraulics, and other compartments. The fan reverses to blow debris out of the radiator cores and screens. These

Figure 1 A smaller wheel loader.

Figure 2 A track loader.

features help minimize the amount of waste that gets caught in the cooling system which helps keep the machine from overheating.

Operating a loader is not limited to moving the machine and materials. An operator must also perform daily checks and maintenance to keep the machine in good working order. Such activities are also an important part of safe operations. If operators are unsure of where the components of a machine are located, and what they are supposed to be doing, it is hard to ensure that the daily checks are being properly completed. The machine's operator manual covers such information. *Figure 5* shows how the service doors on a typical loader can be opened to give access to the engine and radiator areas.

Figure 3 Loader with forks attachment.

1.1.1 Engine Area

Even though loaders are physically different sizes, many use the same type and size of engine. Most manufacturers try to standardize the location of the major engine components. Operators must use the assigned operator manual to locate any specific components on a loader engine. *Figure 6* shows a left-side view of a loader engine with a few components identified.

1.1.2 Loader Transmissions and Axles

A loader's engine provides power for loader operations. The power from the engine is transmitted through some form of transmission to the drive axle(s) of the loader. The engine and transmission systems on loaders are similar to those found in most tractors. Most modern loaders use an electronically controlled hydrostatic transmission. It provides infinitely variable speed within the speed range of the machine. Many have a load-sensing feature that automatically adjusts the speed and power to changing load conditions.

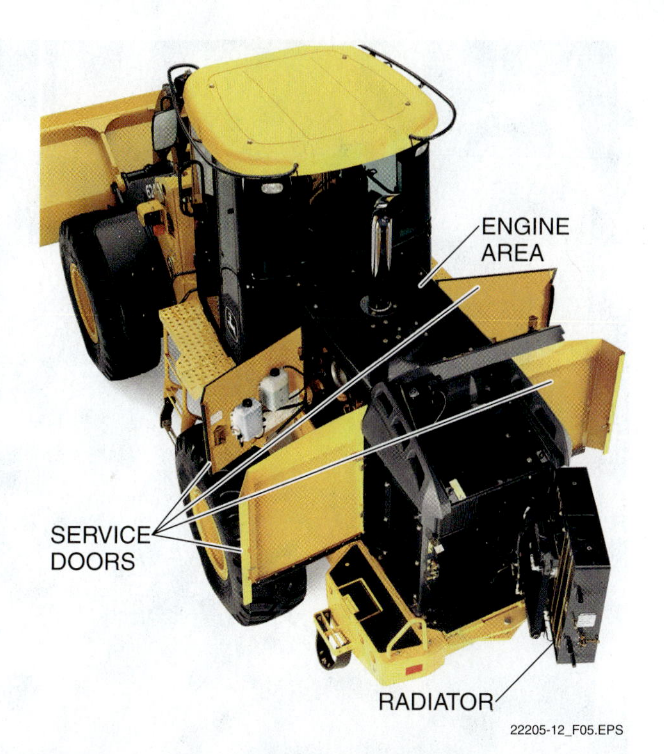

Figure 5 Service doors to engine and radiator areas.

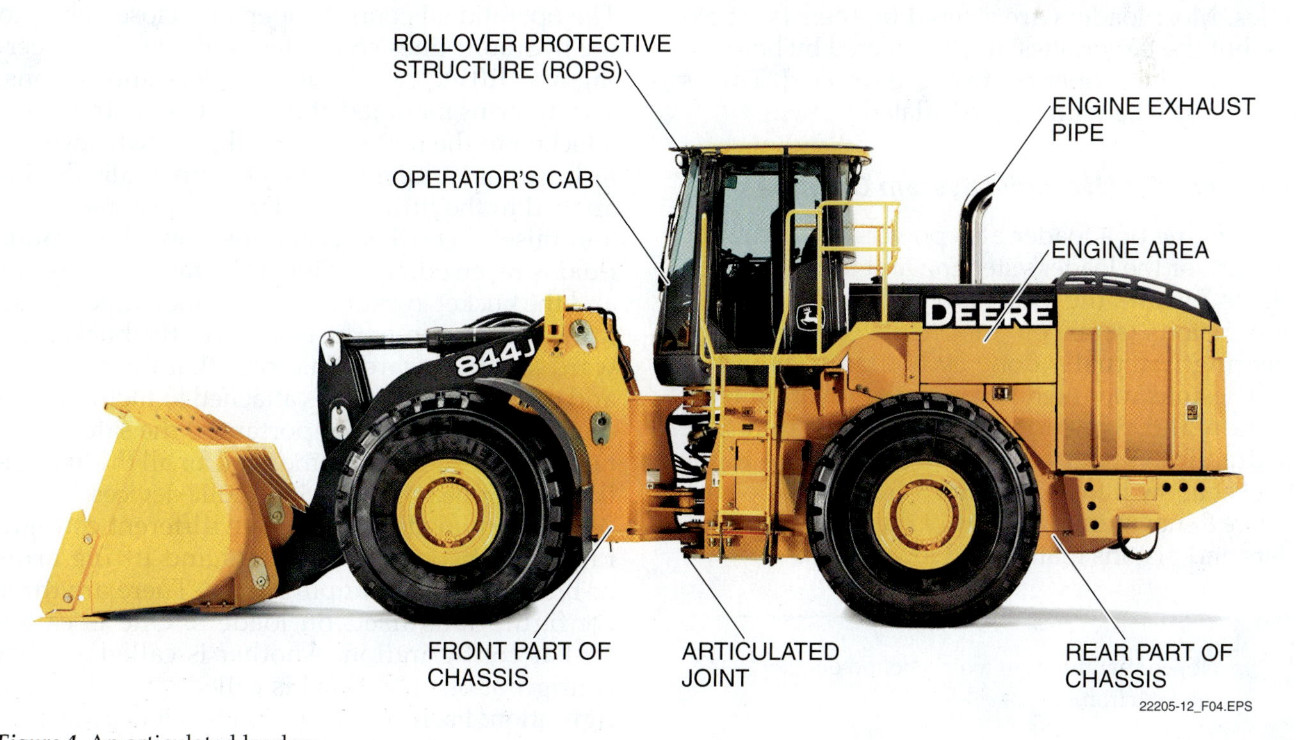

Figure 4 An articulated loader.

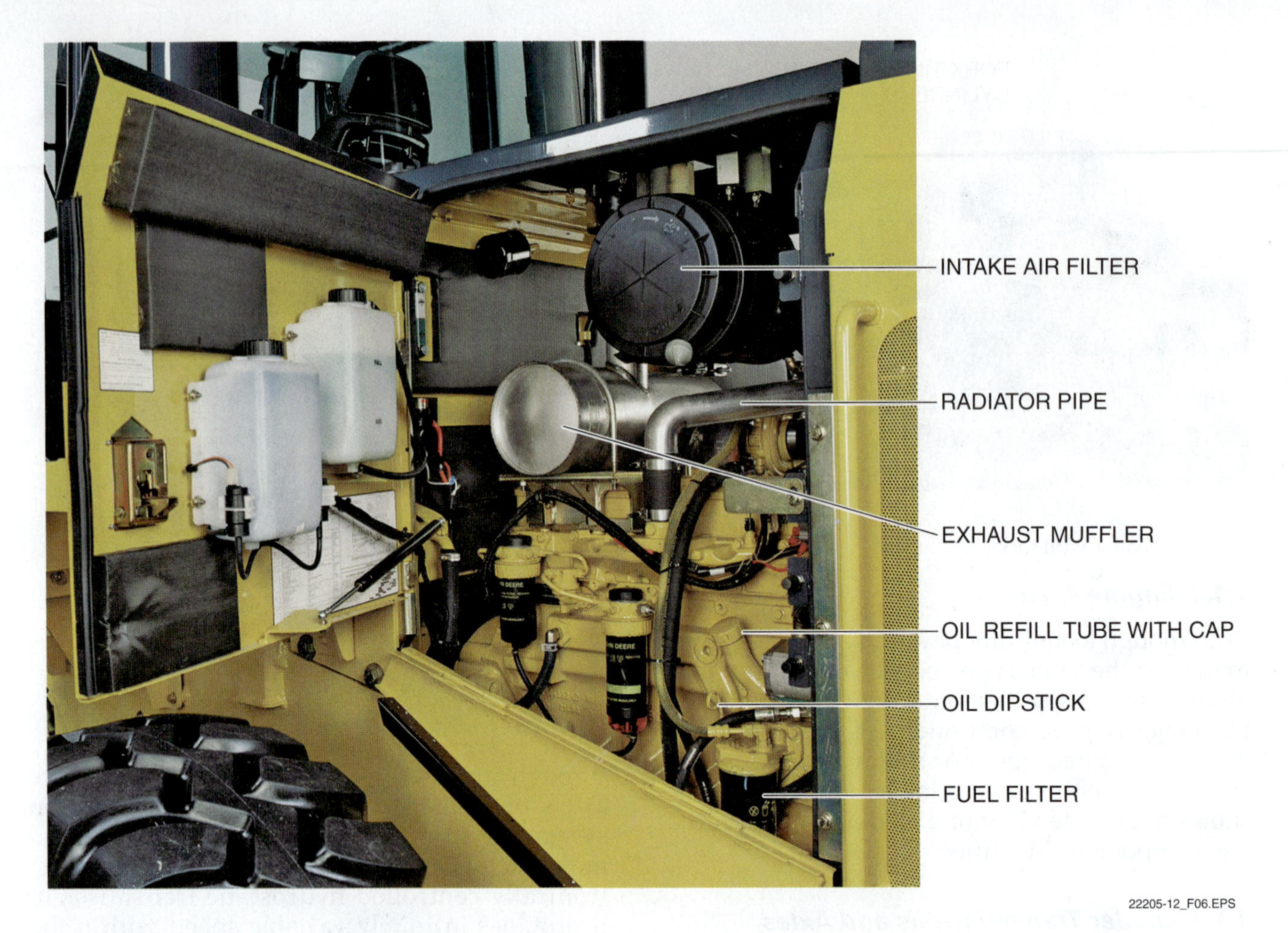

Figure 6 Engine compartment components.

Some loaders are driven only by a single axle while others are driven by both the front and rear axles. Most loaders are steered by their front axles, but the larger ones may be steered by both axles when the proper controls are selected. This is especially true with the articulated loaders.

1.1.3 Loader Hydraulic System Components

The engine of a loader also powers the hydraulic system for the loader's steering and lifting devices. *Figure 7* shows the location of hydraulic cylinders and pistons (rams) on an articulated loader. The steering is actually done by hydraulic cylinders and pistons mounted between the pivoting front section and the loader's main rear section. The hydraulic cylinders and pistons connected to the bucket allow it to be tilted forward and backward. *Figure 8* shows another view of the hydraulic cylinders and pistons that lift and lower a load.

> **NOTE**
> Notice that the front axle is bolted directly to the loader's frame.

The engine supplies power to one or more pumps that build pressure in the hydraulic system. The operational controls open and close valves to the hydraulic lines connected to the loader's steering and lifting hydraulic cylinders and pistons. The lift arms mounted at the front of the frame are attached to the bottom the bucket (or whatever attachment is installed). When the hydraulic fluid is applied to the lifting cylinders, the pistons extend and raise the bucket. When the flow of hydraulic fluid is reversed, the cylinders retract their pistons and the bucket lowers. The hydraulic cylinder(s) attached to the top of the bucket tilt the bucket forward and backward. Ensuring that the lift arms, and the hydraulic devices attached to them, are not damaged is critically important to the safe operation of the loader. The same goes for all the hydraulic lines connected to the hydraulic devices.

Different loaders often have different configurations of hydraulic cylinders and lifting arms or bucket tilting/dumping arms. There are three configurations used on loaders. One is called a J-bar configuration. Another is called a Z-bar configuration. The third is called a parallel configuration. Each configuration has its advantages

Figure 7 Location of some hydraulic cylinders and pistons.

Figure 8 Location of lifting and lowering hydraulic cylinders and pistons.

when it comes to lifting and controlling a loaded bucket or attachment. *Figure 9* shows examples of these different configurations.

Loaders are rated by **breakout force**, which is the amount of force their arms and hydraulic cylinders can apply to a load as the bucket is being pushed into a load such as dirt or gravel. Loaders are also rated on the amount of load they can lift before the rear wheels begin to lift off the ground. The term used here is **tipping load**. After a load is lifted, the term *Straight Static Tipping Load* is used for a machine's ability to move a load in a straight line without tipping over forward. The amount of load carried in a straight line is more than the machine can carry when making turns. The term used for machine's load-carrying ability while making turns is *Full Turn Tipping Load*. The operator manual for each machine should list the machine's Straight Static Tipping Load and Full Turn Tipping Load capabilities or limitations. Both of these amounts are critically important to the safe operation of the machine. The type of arms and hydraulic components used on the machine play major roles in determining the breakout force and tipping load abilities of the machine.

1.1.4 Operator Cab

The operator cab of a loader serves several purposes. The first is that it provides a controlled environment in which the operator can comfortably work. The cab also provides some operator

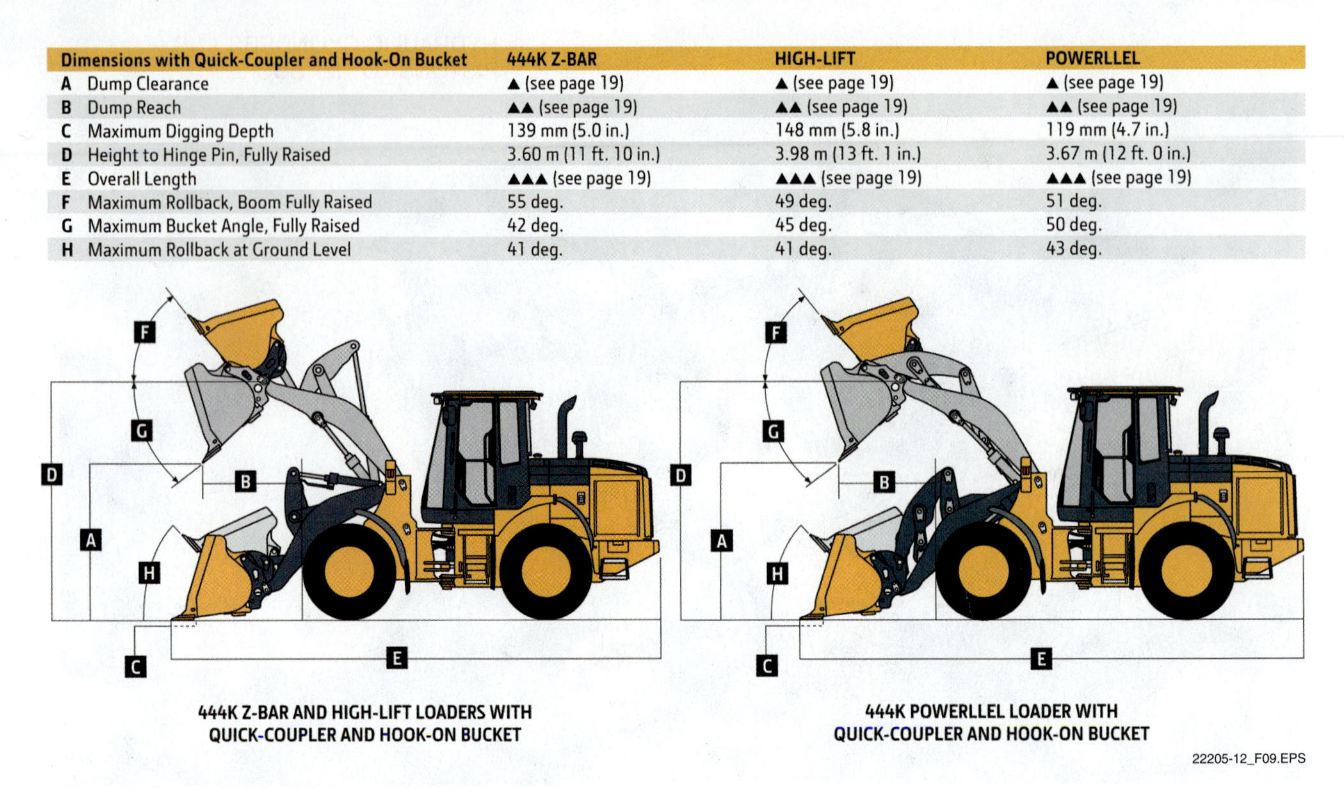

Dimensions with Quick-Coupler and Hook-On Bucket		444K Z-BAR	HIGH-LIFT	POWERLLEL
A	Dump Clearance	▲ (see page 19)	▲ (see page 19)	▲ (see page 19)
B	Dump Reach	▲▲ (see page 19)	▲▲ (see page 19)	▲▲ (see page 19)
C	Maximum Digging Depth	139 mm (5.0 in.)	148 mm (5.8 in.)	119 mm (4.7 in.)
D	Height to Hinge Pin, Fully Raised	3.60 m (11 ft. 10 in.)	3.98 m (13 ft. 1 in.)	3.67 m (12 ft. 0 in.)
E	Overall Length	▲▲▲ (see page 19)	▲▲▲ (see page 19)	▲▲▲ (see page 19)
F	Maximum Rollback, Boom Fully Raised	55 deg.	49 deg.	51 deg.
G	Maximum Bucket Angle, Fully Raised	42 deg.	45 deg.	50 deg.
H	Maximum Rollback at Ground Level	41 deg.	41 deg.	43 deg.

Figure 9 Different loader lift bar configurations.

protection in case the loader was to overturn. And finally, the cab serves as the central hub for loader operations. An operator must understand the controls and instruments before operating a loader. Big or small, the operator cabs on most wheel loaders are laid out in a similar manner. The steering wheel is in the center of the cab. The operator's seat is at the back of the cab and centered on the steering wheel. A dash panel with indicators and some controls is below the steering wheel. The brake and accelerator pedals are near the floor and under the steering wheel. A controls console, usually on the right of the seat, houses most of the controls used on the loader. Most modern loader cabs also have good heating and cooling systems, and adjustable and comfortable seating for the operator. *Figure 10* shows an operator cab of a John Deere 544J loader, as seen from the outside.

1.2.0 Controls

Depending on the manufacturer and model, a loader may be controlled with a joystick or steering wheel. Foot pedals are often used to steer track loaders. If a joystick is used to control the loader's steering, the joystick is typically located in front of the left armrest. The bucket can be con-

Buckle Up!

Two incidents in recent years—just the tip of the iceberg—demonstrate the importance of using the required seat belt.

In a Michigan incident, an operator was clearing trees from an area adjacent to a steep hill. The loader came too close, and rolled over the hill. The operator was not wearing the seat belt, and was thrown from the cab. The same roll-over protection cage that could have protected him rolled over the operator as the loader tumbled.

In a second incident, an operator was driving a loader up a dirt ramp onto a lowboy trailer. The loader was not well-positioned, and one tread began to slip off the side of the trailer. Lacking a seat belt, the operator was thrown off, and the loader rolled across him. In this case, a seat belt was not even installed.

These are just two of many unnecessary incidents that resulted in fatalities. There is no reason to believe that these operators would not have weathered the accidents without serious injury if a seat belt was used. Do not operate a loader or similar equipment that lacks the proper safety equipment. Buckle up. Do not be the next example.

Figure 10 Operator cab of a loader.

trolled with levers or a joystick. Bucket controls (levers or joystick) are typically located in front of or slightly to the right of the right armrest. Review the operator manual to fully understand other types of vehicle movement and bucket controls.

1.2.1 Vehicle Movement Controls

If a steering wheel is used, steering the loader is pretty much like steering a truck or car. Rotating the steering wheel counterclockwise guides the machine to the left. Rotating the steering wheel clockwise guides the machine to the right. If the loader's steering is controlled with a joystick, the stick is moved to the right to make the loader move to the right. When the joystick is moved to the left, the loader goes to the left. When the joystick is released, it will return to the central position. The machine will maintain the direction it was moving in before the joystick was released. Some loaders may be controlled by both a steering wheel and a joystick. *Figure 11* shows a typical steering joystick.

The steering joystick on loaders is aided by computer systems built into the machine. The computer system(s) detects the ground speed of the machine and then adjusts the reaction speed of the joystick to ensure that the steering is smooth and controlled.

As noted earlier, the power of a loader's engine is sent through the loader's transmission to the drive axle(s). The loader's transmission is

Figure 11 Example of a steering joystick.

controlled by a forward/neutral/reverse (F-N-R) control (*Figure 12*) that is normally mounted either on the steering wheel shaft or on the dash panel under the steering wheel (for loaders with steering wheels). If a joystick is used for steering, the F-N-R control is usually built into the joystick. The selected transmission direction is indicated on one of the instrument panels or digital readouts in the cab.

F-N-R
CONTROL

22205-12_F12.EPS

Figure 12 F-N-R transmission control on steering column.

> **CAUTION**
>
> Always come to a complete stop before changing travel direction. Changing travel direction while the machine is moving can damage some machines.

Most loaders have multiple speed options. There are usually two gear selection switches. One is an upshift switch that shifts into higher gears. The other switch downshifts. Typically, loaders have three speed ranges: first, second, and third. Use the upshift switch to move the machine into the next higher speed. Use the downshift to move the machine into the next lower speed.

An automatic shift function is available on some machines. When this function is activated, the control selects the proper gear according to the speed of the machine. An indicator light on the dash panel shows that this feature is activated.

1.2.2 Lift and Tilt Movement Controls

Two lift arms attach the bucket to the front frame of a loader. This adds stability to the bucket and improves material handling. The arms are raised and lowered by one or two hydraulic cylinders. The bucket can be tilted forward and backward. One or two more hydraulic cylinders are used to move the bucket. Inside the operator cab, the controls for these lift arm and bucket movements are usually mounted to the right side of the operator's seat. The controls may be levers or joysticks. *Figure 13* shows a set of lift and tilt controls used on a newer model John Deere K-series loader.

Different manufacturers may have slightly different terms that apply to the control settings for the lift arms and the bucket of a loader. The terms may also vary depending on whether the loader is equipped with a single joystick that controls both the lift arms and the bucket, or if the loader is equipped with individual levers for the lift arms and the bucket. When the control lever for the arms is pulled back, the arms lift. When the lever is pressed forward, the arms lower. Joysticks work the same way. When the control lever for the bucket is pulled back, the bucket tilts up. When the lever is pressed forward, the bucket tilts down. When a joystick controls both the arms and the bucket, moving it to the left (on most joysticks) tilts the bucket up, and moving it to the right tilts the bucket into a dump position. Moving the joystick between a backward or forward direction and a left or right

On-Board Weighing Systems

Highly advanced weighing systems are available on new loaders or as an aftermarket addition. Such systems can weigh the load while the bucket and vehicle remain in motion. They eliminate the need for constant stops at weigh stations. Load after load of similar or different materials can be internally logged. At the end of the work day, a precise report can be printed or downloaded via a portable drive or memory card, providing a comprehensive summary of an operator's performance, sales figures for various materials, and other important data. Loading times and times in transit are also logged. In addition, the system provides protection against overloads.

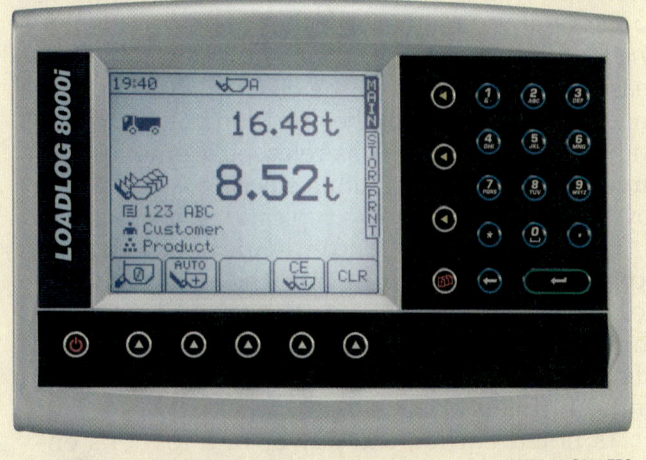

22205-12_SA01.EPS

Figure 13 Joystick and pushbutton controls for bucket lift and tilt.

direction allows the joystick to control the movements of both the lift arms and the bucket at the same time.

The following are control terms associated with the movement of loader lift arms:

- *Float* – A control position that allows the hydraulic fluid to the lift arm hydraulic cylinders to flow in and out both ends of the cylinders so that the bucket can follow the contour of the ground as the loader moves forward or backward.
- *Hold* – This is the neutral position. The controls return to this position unless locked into the float position. The arms hold the bucket in a selected position.
- *Lower* – A control position that causes the arms to lower the bucket. When released, the arms return to the hold position and the bucket stops.
- *Raise* – A control position that causes the arms to lift or raise the bucket. When released, the arms return to the hold position and the bucket stops.

CAUTION

Do not use the float position to lower a loaded bucket. Such action can cause seriously damage to the machine if the load is lowered too quickly.

The following are terms associated with tilting the bucket:

- *Dump* – A control position that causes the bucket to rotate forward to dump its load.
- *Hold* – A control position that causes the bucket to remain in a selected position.
- *Tilt back* – A control position that causes the bucket to rotate backward to a position with its cutting edge pointing up.

1.3.0 Instruments

Modern loaders are heavily loaded with electronic systems that include a digital display or monitor (*Figure 14*) that is mounted in front of the operator or slightly off to the operator's right side when the operator is looking out the front windshield. Software on these newer electronic systems provides all sorts of machine-related information. Some newer loaders have keyless security systems that only allow the machine to be started if the proper start code is entered through the computer controls on the digital monitor.

NOTE

Since heavy equipment is being sold worldwide, some computer systems allow the operator to select different languages.

Computer controls, selected by pushbutton controls, can provide an operator with a wealth of machine information such as:

- Vital and general operating information about the loader (transmission mode, gear, engine speed, and ground speed)
- Troubleshooting information for speed, pressure, temperature, and switch settings
- Payload scale information (useful in determining how much load has been placed into a truck)

Figure 15 shows some of the options possibly available on the diagnostic screen.

To improve safety, some loaders are equipped with a rear-view camera that allows the operator to see what is directly behind the machine without twisting around in the seat. A simple radar system may also be installed to alert the operator when anything comes too close to the rear of the loader.

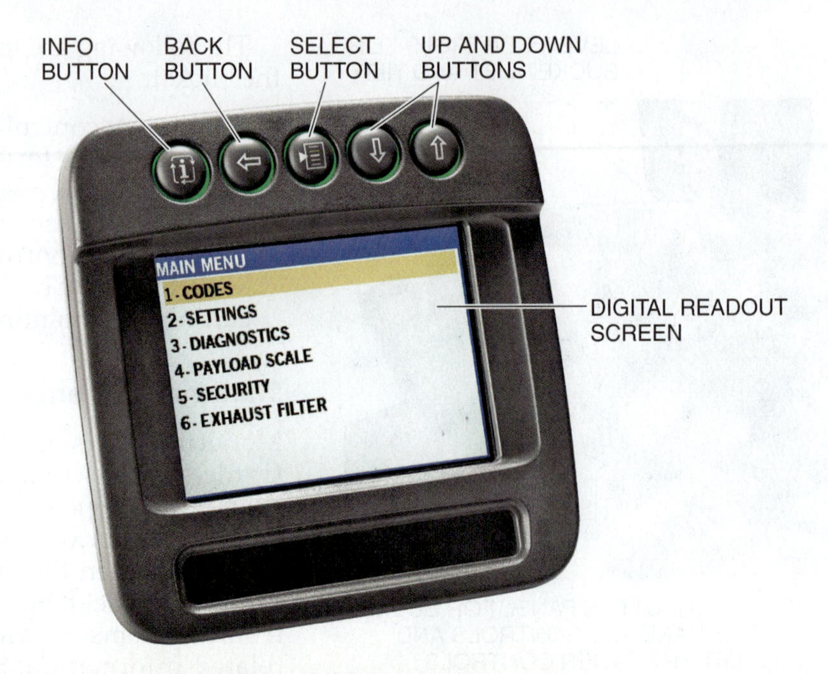

MAIN MENU DEFINITIONS	
1. CODES	Allows service or operator personnel to view active or stored diagnostic trouble codes.
2. SETTINGS	Allows the operator to change various operating characteristics of the machine.
3. DIAGNOSTICS	Provides a limited set of troubleshooting tools for operators and service personnel.
4. PAYLOAD SCALE	If the loader is so equipped, this option provides the operator with weight information about the material being loaded. Also provides a means of changing certain operating parameters of the payload weighing system.
5. SECURITY	Provides a way for the machine owner to assign personal identification numbers (PINs) to only authorized operators. Helps prevent theft or unauthorized use of the machine.
	Pressing the UP or Down buttons allows the selection of different items on the MENU. Pressing the SELECT button activates the highlighted item on the display.

22205-12_F14.EPS

Figure 14 Digital readout screen with Main Menu options.

Regardless of the equipment being used, operators must pay attention to the machine's instrument readouts (gauges or digital). The following are some of the key readouts to monitor:

- Engine temperature
- Transmission temperature
- Hydraulic oil temperature
- Fuel level
- Speed
- Alerts or warnings

Figure 16 shows a Caterpillar indicator panel containing both analog gauges and digital readouts.

1.3.1 Engine Temperature

The coolant temperature gauge or digital readout indicates the temperature of the coolant flowing through the engine's cooling system. Refer to the operator manual to determine the correct operating range for normal loader operations. Temperature gauges normally read left to right with cold on the left and hot on the right. Digital readouts simply show numbers that increase or decrease. Most gauges have a section that is red. If the needle is in the red zone, the coolant temperature is excessive. Some machines may also activate warning lights if the engine overheats.

CAUTION

Operating equipment when temperature gauges are in the red zone may severely damage it. Stop operations, determine the cause of problem, and resolve it before continuing operations.

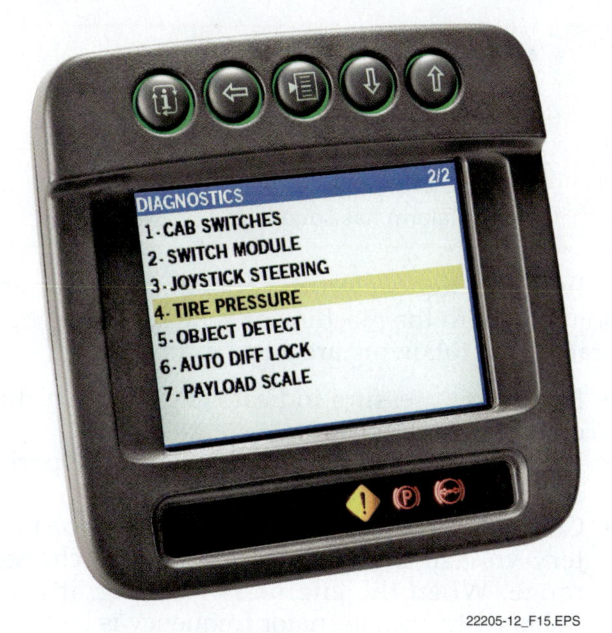

22205-12_F15.EPS

Figure 15 Diagnostic screen with list of things that may be checked on digital display.

WARNING!

To prevent personal injury, never open the cap on a hot radiator. Always allow the engine to cool before opening the radiator cap to check the coolant level in the radiator.

1.3.2 Transmission Temperature

The transmission oil temperature gauge or digital readout indicates the temperature of the oil flowing through the transmission. This gauge also reads left to right in increasing temperature, and digital readouts increase or decrease. The gauge has a red zone that indicates excessive temperatures. Refer to the operator manual to determine the correct operating range for normal loader operations.

1.3.3 Hydraulic Oil Temperature

The hydraulic oil temperature gauge or digital readout indicates the temperature of the hydraulic fluid used in the loader's hydraulic system. The gauge has a red zone that indicates excessive temperatures. Refer to the operator manual to determine the correct operating range for normal loader operations. When hydraulic equipment is being forced to work harder than normal, the temperature of the hydraulic fluid increases. The hydraulic fluid is supposed to be cooler after it passes through the hydraulic system reservoir. If the hydraulic fluid temperature increases, ensure that the reservoir has enough fluid in it, and that the fluid is passing easily through the reservoir.

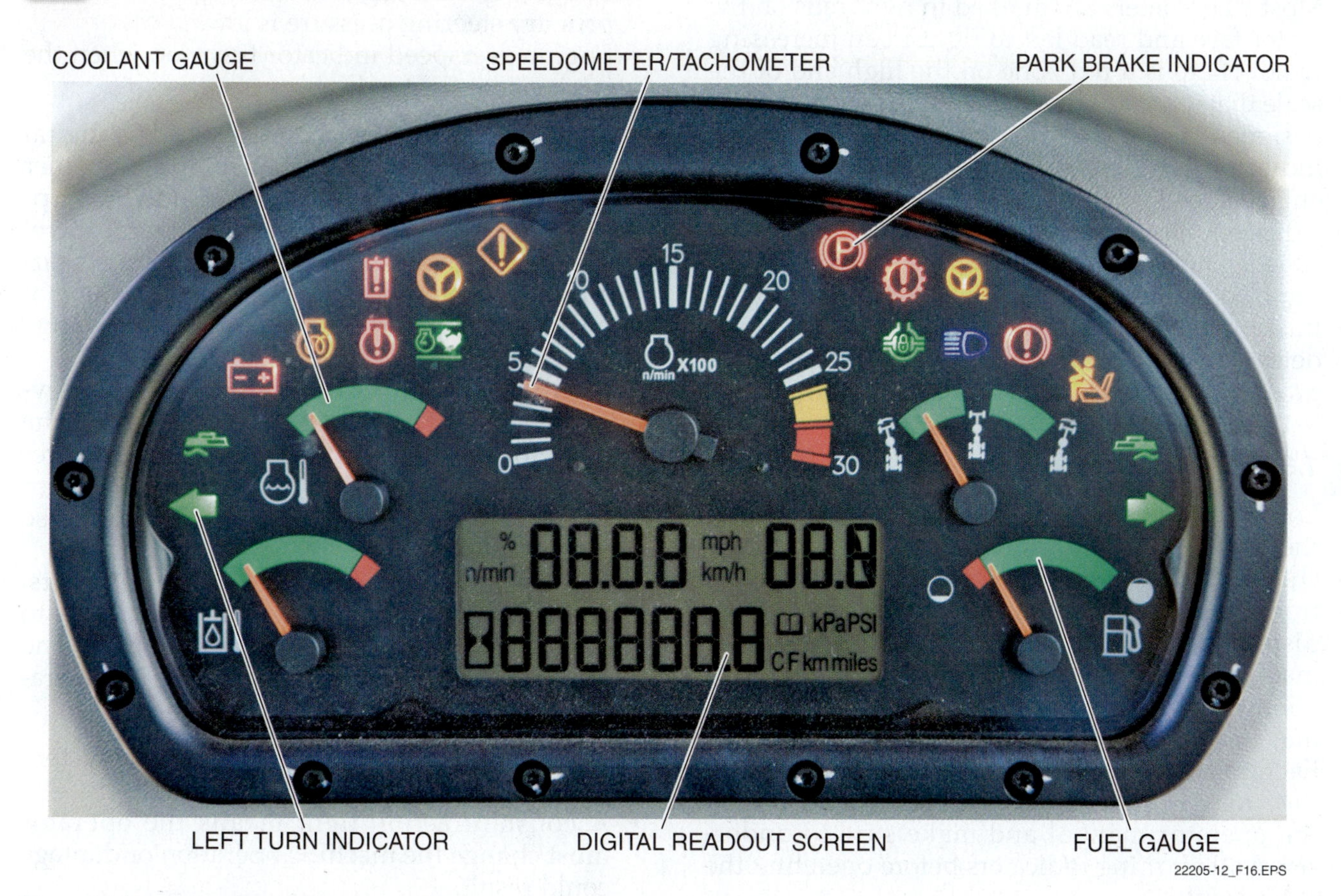

22205-12_F16.EPS

Figure 16 Example of an indicator panel.

> **CAUTION**
>
> Do not operate the loader under a load until the hydraulic oil has reached the correct operating temperature. Operating with a heavy load before the hydraulics have adequately warmed can damage the equipment and may cause failure.

1.3.4 Fuel Level

Fuel gauges or digital readouts indicate the amount of fuel in the loader's fuel tank. On diesel engine loaders, the gauge may contain a low-fuel warning zone. Some models have a low-fuel warning light.

> **CAUTION**
>
> Avoid running out of fuel on diesel engine loaders because the fuel lines and injectors must be bled of air before the engine can be restarted.

1.3.5 Speed

Most loaders show their speed on both a speedometer and a tachometer. Tachometers indicate the engine speed in revolutions per minute (rpm). Most tachometers are marked in hundreds on the meter face and read left to right in an increasing scale. There is a red zone on the high end of the scale that indicates that the engine is exceeding its designed speed. On some loaders there is also an indicator light that will warn the operator if the engine speed is too high.

Speedometers show the machine's ground speed. They can be set for either miles per hour (mph) or kilometers per hour (kph). The transmission gear selected and the rpm of the engine determine the machine's overall speed across the ground.

1.3.6 Alerts or Warnings

Loader systems have sensors in various parts of the engine and at various locations on the machine. These sensors are connected to an electronic monitoring system. The monitoring system display has alert indicators that light when the machine is not functioning properly. If the situation needs immediate attention, the indicators may flash, remain lit, or an audible alarm sounds. Each machine has different indicator lights, or they may be arranged differently. Always read the operator manual and make sure to understand all warning indicators before operating the equipment.

> **CAUTION**
>
> A flashing indicator light may require immediate action by the operator. To prevent possibly serious damage to the machine, or possible personal injury, stop and look up the meaning of an indicator after it is flashing, or after an alarm has sounded.

Indicator lights alert the operator to make an adjustment to the machine or how it is being operated. The following are a few examples:

- Engine oil pressure indicator lights when the oil pressure is too low.
- Parking brake indicator lights when the parking brake is engaged.
- Charging system indicator lights when the battery voltage is outside of the battery charge range. When the engine is running, it also lights when the alternator frequency is low.
- Engine coolant flow indicator comes on when the engine coolant is too low, or when there is no coolant flow to the engine.
- Hydraulic oil filter indicator turns on when the hydraulic oil filter is plugged and the bypass valve is open.
- Transmission oil filter indicator lights when the transmission oil filter is restricted.
- Secondary steering indicator lights when the primary steering pressure is low.
- Engine overspeed indicator turns on when the engine speed is too high.

A digital readout can be set to show a digital reading of any of the gauges or set to show other readings. These other readings include total operating hours, current engine speed, total travel distance, or active diagnostic codes. As noted earlier, newer machines have computer screens that allow the operator to bring up all sorts of information about the machine.

Many monitoring systems have several levels of urgency. The categories are based on the severity of the problem. Different warnings require different actions by the operator. Higher warning levels demand an immediate response by the operator.

Some control panels have indicator lights, warning action lights, and an action alarm to show increasing levels of potential danger. The latter two require immediate action by the operator. Typical warning levels are as follows:

- A flashing indicator light means the system needs attention soon.
- A constant action light means the operator must change the machine operation or damage could result.

- When an action alarm sounds, the operator must change machine operations immediately or the machine will be damaged.
- When an action alarm becomes intermittent the operator must immediately perform a safe shutdown. Otherwise, the machine will be damaged or the operator will be injured.

1.3.7 Other Indicators

All loaders have other indicators that may show when various features are activated under normal operating conditions. Generally, they do not indicate that there is something wrong with the machine, merely that a function is active. The exception is the action light, which indicates that there is a fault in the monitoring system.

Typical indicator lights show that the following features are activated:

- Turn signals
- Reduced rimpull
- Autoshift
- Loose material mode
- Automatic ride control
- Torque converter lockout clutch
- Throttle lock

Reduced rimpull allows the operator to control driveline torque while the engine is operating at a high idle. The operator can reduce driveline torque while giving full power to the bucket hydraulics. Activating the reduced rimpull function changes the pressure that must be applied to the foot pedal to activate the brakes. This is useful when inching into a pile to scoop a load. Reduced rimpull reduces wheel slipping.

The reduced rimpull function provides an impeller clutch pressure that limits rimpull to the desired level when the brakes are fully released. The reduced rimpull function is engaged with a knob that can be set in five positions. The positions represent a maximum allowable percentage of total rimpull. Typically, the intervals are off or 100 percent, 90 percent, 80 percent, 70 percent and 60 percent. The reduced rimpull function can only be activated when the machine is operating in first gear.

The loose material mode is activated by a switch. This provides maximum hydraulic pump flow and provides for faster loading of loose materials. Do not activate this switch if digging conditions are rough. The normal mode provides a reduced hydraulic flow for better digging.

Automatic ride control is used when traveling at higher speeds over rough terrain. The system acts as a shock absorber, dampening the forces from the bucket. This minimizes bucket movement and swinging motion and helps stabilize the machine.

The torque converter lockout is also activated with a switch on the side of the operator's cab. This feature enables the lockup clutch. When the engine is in a certain speed range, the torque converter locks up. This provides more efficient operations for higher speeds and is used when carrying loads.

The throttle lock is engaged with a switch on the left side of the instrument panel. This switch keeps the engine at the current rpm. Some machines have an indicator light that shows when the throttle lock is active. To disengage the throttle lock, either turn off the switch or depress the brake pedal slightly.

1.4.0 Attachments

The bucket is the main attachment for the loader, although there are several other useful attachments. Different buckets are available to maximize productivity under different circumstances. In a quarry setting, the operator may need a stronger bucket with teeth to scrape a wall face. A larger bucket with a higher back plate aids in handling loose material like dry soils.

In addition to lifting and loading, loaders are sometimes used for other operations such as sweeping and scraping. Brooms, forks, rakes, and snow blades can be attached to a loader. Many loaders are equipped with a hydraulic quick-connect that allows the operator to change attachments from the cab. In other models, the attachments are fixed to the lift arms with pins (*Figure 17*). The operator must dismount and manually secure the attachment.

WARNING!

Do not leave the operator's cab while the machine is running. Place the controls in park and shut down the engine before dismounting to adjust attachments.

1.4.1 Buckets

Several types of buckets are made to accommodate different work situations. Working with different materials may require different bucket designs. Some buckets have teeth for scraping or digging, while others have a flat **blade** for scraping on flat surfaces. Regardless of the design, the bucket width is usually the same as that of the back wheels or tracks of the loader. *Figure 18* shows different types of loader buckets.

Figure 17 Attachment quick-disconnect for loaders.

General-purpose buckets are used for a variety of activities from loading to clearing. These buckets are generally larger and sturdier than standard buckets. They have a straight edge, but allow for a toothed edge to be bolted on.

Loose material buckets have a straight edge in front. The back plate is elevated to minimize how much material spills out of the back when the bucket is filled and lifted. These buckets are designed for loose soils, wood chips, light gravel, and similar material.

Heavy-duty, high-abrasion rock buckets are designed for use in quarries. They are used in bank or face loading or in abrasive low-to-moderate impact conditions. An upper rock guard aids in load retention. The bucket is available with either a straight or toothed spade edge. The straight edge offers higher breakout force and increased dump clearance. The toothed spade edge is designed to dig efficiently.

A multipurpose bucket is designed to allow the bucket to function in four different ways. It is also known as a four-in-one bucket. It can be used as a standard bucket, dozer blade, clamp, or controlled-discharge bucket. The operator can load material with the bucket, doze with the straight blade, grab items with the hydraulically operated front clamp, or meter out the load with the clamshell. Teeth can also be attached to the flat front blade for digging. *Figure 19* shows one form of multipurpose bucket that is often called a grapple bucket because of its ability to clamp whatever has been loaded into the bucket.

The auxiliary hydraulic lines are used to open and close the multipurpose bucket. When the bucket is fully opened, it can function as a dozer blade. The auxiliary control lever also opens and closes the bucket for clamshell and clamp features. When the auxiliary controls are completely closed, the multipurpose bucket functions like a standard bucket.

There are many types of specialty buckets. Several specialty buckets are designed to maximize load capacity for common materials such as coal or wood chips. These buckets have the maximum size given the average density of the material and the capacity of the loader. This ensures maximum operator and machine efficiency.

Some buckets are designed to achieve maximum dump height. They are also called rollout buckets. Finally, there are buckets with a top clamp that can grab and hold loose material. These are often used to move refuse, trash, or brush.

1.4.2 Other Attachments

Other types of attachments for a loader include dozer blades, snowplows, brooms, rakes, rippers, augers, lift booms, and forks. Some of these attachments are designed to meet the needs of specific industries, such as forestry or waste handling. Other attachments, such as snow blowers or snowplows, enable the loader to be used in the winter when it would otherwise be idle. *Figure 20* shows some other attachments that can be used with a loader.

GENERAL PURPOSE BUCKET

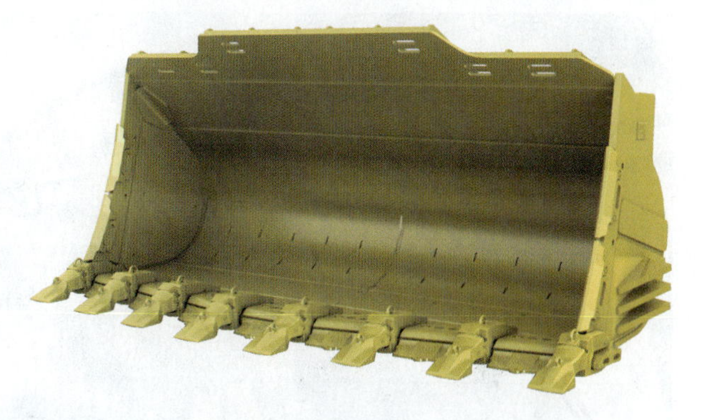

HIGH-ABRASION ROCK BUCKET

LOOSE MATERIAL BUCKET

22205-12_F18.EPS

Figure 18 Different loader buckets.

There are several loader attachments used in the forestry industry, including log forks (*Figure 21*) and mill yard forks. Log forks are used to load, deck, and sort lumber, logs, or palletized material. The clamp can be raised to almost 90 degrees to allow the operator to scoop and load the forks. The clamp is then closed to hold the logs in place for transport. Mill yard forks a have a similar design, but the upper tine closes between the lower two tines. This allows the operator to hold a single log or several logs.

Various kinds of dozer-type blades can be attached to the loader. Snowplow blades can be either box shaped, V-shaped, or have the traditional concave blade. Angle blades can be hydraulically adjusted to move the material straight ahead or to either side. This allows the operator to place material without performing multiple passes. The angle blade is used for side casting, backfilling, cutting ditches, snow removal, pioneering, and maintaining rural or logging roads.

The broom is another popular attachment. The broom can be angled horizontally up to 30 degrees to either side. It is adjusted hydraulically or manually. They are used for clearing parking lots, industrial plants, mill yards, airport runways, streets, driveways, and lanes. The brooms are powered with one or two hydraulic motors. An optional water spray can be added for dust control.

Forks are a common attachment. They can be either attached to the lift arms in place of the bucket or attached to the bucket. With forks, the loader can be used to pick up and move large pieces of rigid material such as lumber, pallets, pipe, or concrete block.

The rake attachment is used for clearing brush. It rakes up brush and other vegetation. It can also lift up and dump it into a haul unit or burn pile.

22205-12_F19.EPS

Figure 19 Grapple bucket.

Some attachments are made for special applications by various manufacturers. The function and operation of an attachment is described in the manufacturer's literature and the operator manual. Always review the instructions before using any equipment or attachments.

CAUTION

Only use attachments that are compatible with the machine. Using nonstandard attachments could damage the equipment or cause injury.

LIFT BOOM ATTACHMENT

AUGER ATTACHMENT

22205-12_F20.EPS

Figure 20 Other attachments.

22205-12_F21.EPS

Figure 21 Loader with log fork attachment.

1.0.0 Section Review

1. Why are loader engines mounted behind the operator?
   a. For cooling
   b. For more power
   c. To improve operator visibility
   d. For additional weight over the rear wheels

2. What is the term used for a control position that causes the bucket to remain in a selected position?
   a. Dump
   b. Swing
   c. Hold
   d. Tilt

3. What instrument shows the engine speed in revolutions per minute?
   a. The tachometer
   b. The speedometer
   c. The ground speed indicator
   d. The operational hours indicator

4. Which loader bucket is designed to be used with wood chips?
   a. A multipurpose bucket
   b. A loose material bucket
   c. A general purpose bucket
   d. A high-abrasion rock bucket

Section Two

2.0.0 Prestart Inspections

Objective 2

Describe the prestart inspection requirements for a loader.

a. Describe prestart inspection procedures.
b. Describe preventive maintenance requirements.

Performance Task 1

Complete proper prestart inspection and maintenance for a loader.

Trade Term

Grouser: A ridge or cleat across a track that improves the track's grip on the ground.

Part of any machine operator's job, regardless of machine type, is to do a thorough walk-around inspection of the machine before starting and after using it. Most companies have inspection checklists that must be followed, and in some cases signed off on, prior to starting the machine. Machine manufacturers such as Caterpillar, John Deere, Komatsu, as well as others, have such inspection lists posted in the operator manual assigned to a given machine. These inspections are often called daily checks, but when a machine is being used during multiple shifts, such checks may be shift checks.

In addition to the daily or shift checks, operators may also be required to do some preventive maintenance work, such as cleaning operator cab windows, greasing moving parts, checking and changing air filters, or checking and adding fuel, coolant, or hydraulic fluid.

Maintenance time intervals for most machines are established by the Society of Automotive Engineers (SAE) and adopted by most equipment manufacturers. Instructions for preventive maintenance are usually in the operator manual for each piece of equipment. Typical time intervals are: 10 hours (daily); 50 hours (weekly); 100 hours, 250 hours, 500 hours, and 1,000 hours. The operator manual also includes lists of inspections and servicing activities required for each time interval. For specific preventive maintenance actions, refer to the operator manual associated with the machine being used.

2.1.0 Inspection Procedures

The first thing a loader operator must do each day is to conduct the required daily inspections. This should be done before starting the engine. This identifies any potential problems that could cause a breakdown and indicate whether the machine can be operated. The equipment should be inspected before, during, and after operation.

The daily inspection is often called a walk-around. The operator should walk completely around the machine checking various items. Items to be checked and serviced on a daily inspection are as follows:

NOTE

If a leak is observed, located the source of the leak and fix it. If leaks are suspected, check fluid levels more frequently.

WARNING!

Do not check for leaks with your bare hands. Use cardboard or a similar material. Pressurized fluids can cause severe injuries to unprotected skin. Long-term exposure can cause cancer or other chronic diseases.

- Inspect the cooling system for leaks and faulty hoses.
- Inspect the engine compartment and remove any debris. Clean and secure access doors.
- Inspect the engine for obvious damage. Look for fuel or oil leaks.
- Check the condition and adjustment of drive belts on the engine.
- Inspect tires for damage and replace any missing valve caps.
- Inspect the axles, differentials, wheel brakes, and transmission for leaks.
- Inspect the hydraulic system for leaks, faulty hoses, or loose clamps.
- Inspect all attachments and the linkage for wear and damage. Check this hardware to make sure there is no damage that would create unsafe operating conditions or cause an equipment breakdown. Make sure the bucket is not cracked or broken. If an implement such as a roller is attached, make sure the hitch is properly set and that the safety pin is in place.
- Inspect and clean steps, walkways, and handholds.
- Inspect the ROPS for obvious damage.
- Inspect the lights and replace any broken bulbs or lenses.
- Inspect the operator's compartment and remove any trash.

- Inspect the windows for visibility and clean them if they are fouled.
- Adjust the mirrors.
- Test the backup alarm and horn. Put equipment in reverse gear and listen for backup alarm.

Some manufacturers require that daily maintenance be performed on specific parts. These parts are usually those that are the most exposed to dirt or dust and may malfunction if not cleaned or serviced. For example, the service manual may recommend lubricating specific bearings every 10 hours of operation, or always cleaning the air filter before starting the engine. *Figure 22* shows the location of points to be inspected on a wheel-type loader.

Most loaders are designed so that the operator enters and exits the cab on the left side of the machine. Because of that, most of the areas an operator must check on the engine are accessible from the left side of the machine. Some loader manufacturers install a checklist inside an access door to the motor area. Such a checklist shows recommended items to check; recommended lubricants or replacement parts; and capacity limits for coolant, hydraulic fluid, and fuel.

For track-type loaders, there are several other items to check in the same manner as required for bulldozers. These include the following:

- *Idlers* – These components keep tension on the tracks to prevent them from coming off the sprockets. Check for broken or cracked rollers, tension springs, or other damage.
- *Sprockets* – The teeth on a worn sprocket are not well-defined and may even be broken. This may cause the loader to throw a track and badly damage the track, drive shaft, and bearings.

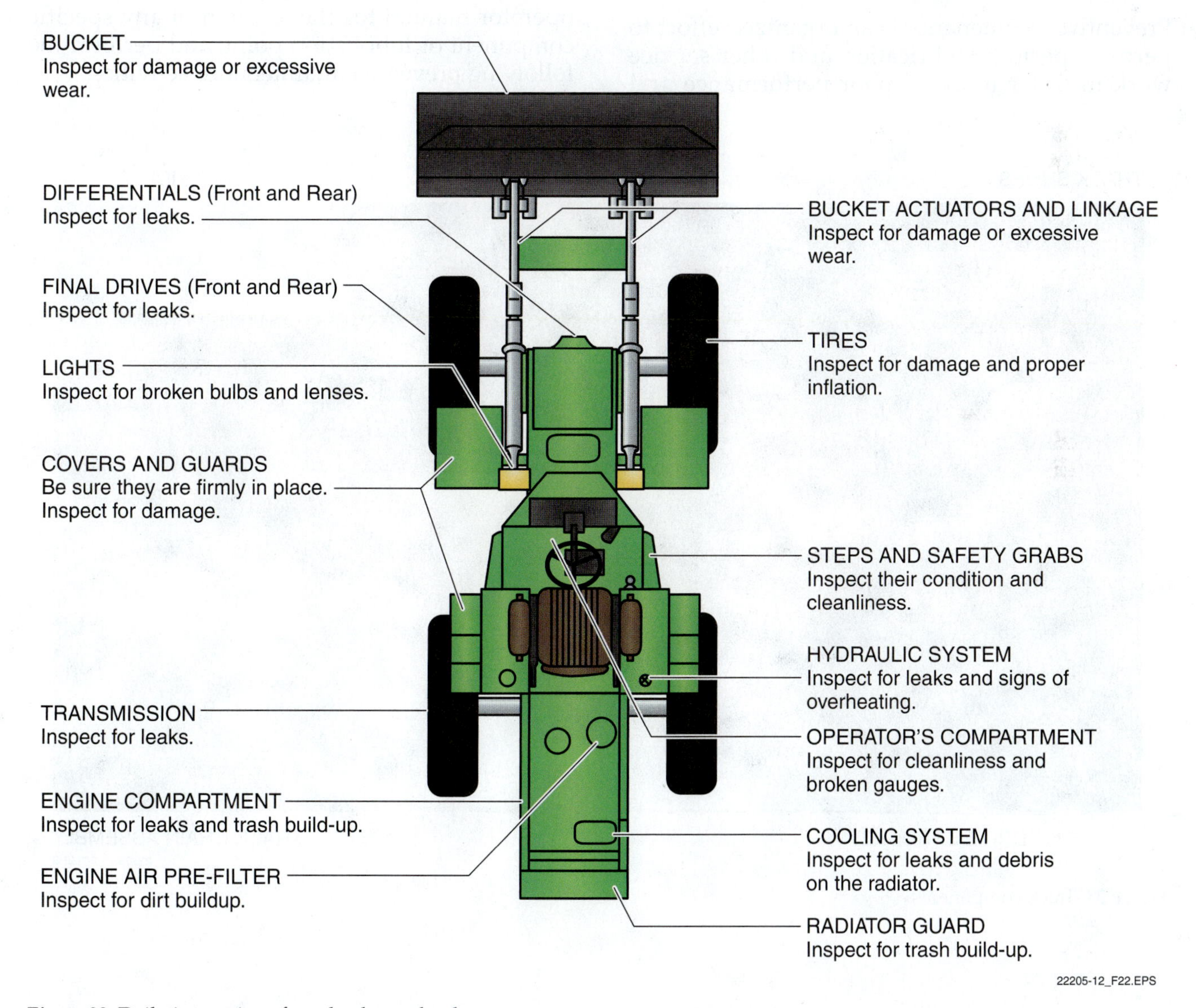

Figure 22 Daily inspections for wheel-type loaders.

- *Tracks* – Tracks wear out very fast because they are the one part of the equipment that is constantly in use. Check for broken or missing shoes and bolts. Also, inspect the **grousers** to see if they will need replacing soon. If the grousers are worn down too far, the loader will not get good traction. Clean accumulated mud and debris from the tracks.

Figure 23 shows the key components of a track loader or bulldozer.

The operator manual usually has detailed instructions for performing periodic maintenance. If an operator finds any problems with the machine that he or she is not authorized to fix, inform the foreman or field mechanic for correction before operation.

2.2.0 Preventive Maintenance

Preventive maintenance is an organized effort to perform periodic lubrication and other service work in order to avoid poor performance and breakdowns at critical times. By performing preventive maintenance on the loader, the operator keeps the equipment operating efficiently and safely and avoids the possibility of costly failures in the future.

Accurate, up-to-date maintenance records are essential for knowing the history of the equipment. Each machine should have a record that describes any inspection or service that is to be performed and the corresponding time intervals. Typically, an operator manual and some sort of inspection sheet are kept with the equipment at all times.

Preventive maintenance of equipment is essential and is not that difficult if the right tools and equipment are used. The leading cause of premature equipment failure is putting things off. Preventive maintenance should become a habit, performed on a regular basis. Refer to the operator manual for the location of any specific component or lubrication point, and perform the following preventive maintenance activities:

Figure 23 Track components.

- *Battery* – Check the battery cable connections. If loose, tighten them. If corroded, remove and clean them before re-installing and tightening them.
- *Crankcase oil* – Check the crankcase oil level and make sure it is in the safe operating range.
- *Cooling system* – Check the coolant level and make sure it is at the level specified in the operating manual.
- *Fuel level* – Check the fuel level in the fuel tank(s). Do this manually with the aid of the fuel dipstick or marking vial. Do not rely on the fuel gauge at this point. Check and clean the fuel pump sediment bowl if one is fitted on the machine.
- *Hydraulic fluid* – Check the hydraulic fluid level in the reservoir.
- *Transmission fluid* – Measure the level of the transmission fluid to make sure it is in the normal operating range.
- *Pivot points* – Clean and lubricate all pivot points (*Figure 24*).

> **NOTE**
> In cold weather it is sometimes preferable to lubricate pivot points at the end of a work shift when the grease is warm. Warm the grease gun before using it for better grease penetration.

LUBRICATION POINTS IDENTIFIED BY ARROWS

WARNING: NEVER LUBRICATE ARMS WHEN THEY ARE RAISED. LUBRICATE ONLY WHEN ARMS ARE LOWERED WITH BUCKET ON THE GROUND.

22205-12_F24.EPS

Figure 24 Pivot points to lubricate on a loader arms.

2.0.0 Section Review

1. The typical time interval for weekly inspections on a loader is at least once every _____.
   a. 24 hours
   b. 36 hours
   c. 40 hours
   d. 50 hours

2. Where should a loader operator look for specific times to do preventive maintenance on the loader?
   a. On the company's maintenance bulletin board
   b. On the service decals inside the operator cab
   c. In the operator manual for the machine
   d. On the machine's maintenance record

SECTION THREE

3.0.0 LOADER STARTUP AND OPERATING PROCEDURES

Objective 3

Describe the startup and operating procedures for a loader.

a. State loader-related safety guidelines.
b. Describe startup, warm-up, and shutdown procedures.
c. Describe basic maneuvers and operations.
d. Describe related work activities.

Performance Tasks 2, 3, and 4

Perform proper startup, warm-up, and shutdown procedures.

Execute basic maneuvers with a loader, including proper movements and curling the bucket.

Carry out basic earthmoving operations with a loader; load a truck (to capacity, if possible), and build a storage pile.

Trade Terms

Accessories: Attachments used to expand the use of a loader.

Dozing: Using a blade to scrape or excavate material and move it to another place.

Grubbing: Digging out roots and other buried material.

Roading: Driving a piece of construction equipment, on a public road, from one job site to another.

Rubble: Fragments of stone, brick, or rock that have broken apart from larger pieces.

Spot: To line up the haul unit so that it is in the proper position.

Stockpile: Material put into a pile and saved for future use.

Stripping: Removal of overburden or thin layers of pay material.

Now that all the basic loader components have been covered along with the operator-performed inspections and preventive maintenance activities, the next step is to actually start and operate the machine. Before that can happen, operators must fully understand the safety issues associated with a loader. The operator manual for a given loader contains safety information about that specific machine. Keep in mind that while loaders in general are similar, operating a smaller loader is different than operating a very large loader. The same goes for a loader mounted on a solid frame verses one mounted on an articulated frame. Wheel loaders are also different than track loaders in some respects. This module focuses on the safe operation of a wheel loader.

3.1.0 Safety Guidelines

Safety can be divided into three areas: safety of the operator, safety of others, and safety of the equipment. Loader operators are responsible for performing their work safely, protecting the public and co-workers from harm, and protecting the equipment from damage. Knowing the equipment and the work area, and operating with safety in mind are the easiest ways to keep a job site safe.

3.1.1 Operator Safety

Nobody wants to have an accident or be hurt. There are a number of things an operator can do to protect one's self and those around the loader from getting hurt on the job. Be alert and avoid accidents.

The first rule is to know and follow the employer's job-site safety rules. The employer or the immediate supervisor will provide the requirements for proper dress and safety equipment. The following are recommended safety procedures for all loader operators:

- Clean steps, grab irons, and the operator's cab.
- Mount and dismount the equipment carefully using three-point contact and facing the machine.

- Wear the required PPE when operating or working around the equipment.
- Do not wear loose clothing or jewelry that could catch on controls or moving parts.
- Keep the windshield, windows, lights, mirrors, and air-intake openings clean at all times.
- Never operate equipment under the influence of alcohol or drugs.
- Never smoke while refueling or checking batteries or fluid.
- Never remove protective guards or panels.
- Never attempt to search for leaks with bare hands. Hydraulic and cooling systems operate at high pressure. Fluids under high pressure can cause serious injury.
- Always lower the bucket or other attachment to the ground before performing any service or when leaving the loader unattended, and turn off the machine.

CAUTION

Getting in and out of equipment can be dangerous. Always face the machine and maintain three points of contact when mounting and dismounting a machine. That means you should have three out of four of your hands and feet on the equipment. That can be two hands and one foot or one hand and two feet.

3.1.2 Safety of Co-Workers and the Public

Operators are not only responsible for their personal safety, but also for the safety of other people who may be working around them. Sometimes, an operator may be working in areas that are very close to pedestrians or motor vehicles. In these areas, take time to be aware of everything going on around the loader. Remember, it is often difficult to hear when operating a loader. Use a spotter and a radio in crowded conditions.

The main safety points when working around other people include the following:

- Walk around the equipment to make sure that everyone is clear of the equipment before starting and moving it.
- Before beginning work in a new area, take a walk to locate any cliffs, steep banks, holes, power or gas lines, or other obstacles that could cause a hazard to safe operation.
- When working in traffic, find out what warning devices are required. Make sure to know the rules and the meaning of all flags, hand signals, signs, and markers.
- Maintain a clear view in all directions. Do not carry any equipment or materials that obstructs any view.
- Always look before changing directions.
- Never allow riders on the loader.
- Maintain eye contact with other workers.
- Maintain a safe speed.

3.1.3 Equipment Safety

The loader being used has been designed with certain safety features to protect the operator as well as the equipment. For example, it has guards, canopies, shields, roll-over protection, and seat belts. Know the equipment's safety devices and be sure they are in working order.

Use to the following guidelines to keep the equipment in good working order:

- Perform prestart inspection and lubrication daily.
- Use your natural senses to make sure the equipment is functioning normally. Pay attention to the normal smells of the machine. Unusual smells like burned oil or antifreeze indicate something is wrong. Feel the machine's vibrations as it normally runs, and listen or sense any knocking noise or vibration. Stop if it is malfunctioning. Correct or report trouble immediately.
- Always travel with the bucket low to the ground.
- Never exceed the manufacturer's limits for speed, lifting, or operating on inclines.
- Always lower the bucket, engage the parking brake, turn off the engine, and secure the controls before leaving the equipment.
- Never park on an incline.
- Use a rigid-type coupler when towing the loader.
- Make sure clearance flags, lights, and other required warnings are on the equipment when **roading** or moving.
- When loading, transporting, and unloading equipment, know and follow the manufacturer's recommendations for loading, unloading, and tie-down.

The basic rule is to know the equipment. Learn the purpose and use of all gauges, indicators, and controls as well as all of the equipment's limitations. Never operate the machine if it is not in good working order. Some basic safety rules of operation include the following:

- Do not operate the loader from any position other than the operator's seat.

- Do not coast or make sudden steering changes. Neutral is for standing still only.
- Maintain control when going downhill. Do not shift gears and keep the load low.
- Whenever possible, avoid obstacles such as rocks, fallen trees, curbs, or ditches.
- If an obstacle must be crossed, reduce the loader's speed and approach at an angle to reduce the impact on the equipment and yourself.
- Use caution when undercutting high banks, backfilling new walls, and removing trees.

3.2.0 Startup, Warm-Up, and Shutdown Procedures

The operation of a loader requires constant attention to the controls and the surrounding environment. Operators must plan their work and movements in advance and be alert to the other operations going on around the equipment. Do not take risks.

If there is doubt concerning the capability of the machine to perform some work, stop the equipment and investigate the situation. Discuss it with the foreman or engineer in charge. Whether it is a slope that may be too steep, or an area that looks too unstable to work, know the limitations of the equipment. Decide how to do the job before starting operations. Once a job is started, it may be too late to do something about an unusual situation that may cause extra work for others or possibly an unsafe condition.

The following suggestions can help improve operating efficiency:

- Keep equipment clean. Make sure the cab is clean so nothing affects the operation of the controls.
- Calculate and plan operations before starting.
- Do not move the machine until the brake air system reaches 100 pounds per square inch (psi).
- Set up the work cycle so that it will be as short as possible.
- Observe all safety rules and regulations.
- **Spot** trucks properly.
- Level off the work area if necessary.
- Keep transport distances as short as possible.
- Keep the transmission range lever in low range during loading operations and when transporting any load.
- Keep the wind to the back of the loader when dumping into a truck.

While traveling, maintain good visibility and loader stability by carrying the bucket low, approximately 15 inches above the ground. When loading trucks from a **stockpile**, as shown in *Figure 25,* use the wait time to clean and level the work area. Cleanup of spillage around the stockpile will smooth loader cycles and lessen operator fatigue. Maintain traction when loading the bucket by not putting excessive down pressure on the bucket. Excessive down pressure forces the front wheels or the front portion of the tracks to raise up off the ground.

To control dumping, move the bucket tilt control lever to the dump position. Repeat this operation until the bucket is empty. When handling dusty material, try to dump the material with the wind to your back. This will keep dust from entering the engine compartment and the operator's cab.

Use the appropriate size bucket. Check the work to be done and choose the right bucket for the job. Using the wrong bucket increases the wear on the bucket. Using the wrong bucket also increases the potential to exceed the machine's operating limits, which reduces its service life.

3.2.1 Preparing to Work

Preparing to work involves getting organized in the cab and starting the machine. Mount the equipment using the grab rails and foot rests. Adjust the seat to a comfortable operating position. The seat should be adjusted to allow full brake pedal travel with your back against the seat back. This will permit the application of maximum force on the brake pedals. Make sure that all the controls can be seen clearly and are within easy reach.

NOTE

Always maintain three points of contact when mounting equipment. Keep grab rails and foot rests clear of dirt, mud, grease, ice, and snow.

WARNING!

OSHA requires that approved seat belts and a roll-over protective structure (ROPS) be installed on all heavy equipment. Old equipment must be retrofitted. Do not use heavy equipment that is not equipped with these safety devices.

Operator areas vary depending on the manufacturer, size, and age of the equipment. However, all operator areas have gauges, indicators and switches, levers, and pedals. Gauges and indicators show the status of critical items such as water temperature, oil pressure, battery voltage, and fuel level. Indicators and alarms alert the operator

22205-12_F25.EPS

Figure 25 Loading a truck from a stockpile.

to low oil pressure, engine overheating, clogged air and oil filters, and electrical system malfunctions. Switches are for activating the glow plugs, starting the engine, and turning **accessories**, such as lights, on and off. Typical instruments and controls were described previously. Review the operator manual so that the specifics of the machine being operated are fresh in your mind.

The startup and shutdown of an engine is very important. Proper startup lengthens the life of the engine and other components. A slow warm up is essential for proper operation of the machine under load. Similarly, the shutdown of the machine is critical because of all the hot fluids circulating through the system. These fluids must cool so that they can cool the metal parts before the engine is switched off.

3.2.2 Startup

There may be specific startup procedures for the piece of equipment being operated, but in general, the startup procedure should follow this sequence:

Step 1 Sit in operator seat and fasten seat belt.

Step 2 Turn battery disconnect switch to On.

Step 3 Ensure that the machine's parking brake is engaged. Use a lever or knob to engage the parking brake depending on the loader make and model.

NOTE

When the parking brake is engaged an indicator light on the dash will light up or flash. If it does not, stop and correct the problem before operating the equipment.

Step 4 Be sure all controls are in neutral (F-N-R control to N) and the bucket is on the ground.

NOTE

The next step is to start the engine, which is a two-step process on modern loaders. The first step is to apply power to the control and display units (the computer system and its displays). The second is to actually apply power to the starting motor.

Step 5 Enter the security code (if applicable).

Step 6 Press and release the engine start switch, and then wait for the computer system to activate.

Step 7 Sound the machine's horn to alert bystanders that the machine is being started.

Step 8 Set the throttle control as indicated in the operator manual.

Step 9 Press and hold the starter button until the engine starts. Release the starter button when the engine starts.

> **NOTE**
> Never operate the starter for more than 30 seconds at a time. If the engine fails to start, wait two to five minutes before cranking again. If the engine fails to start after two or three tries, press the engine stop switch and wait two minutes before trying again. If engine still does not start, call for a tow to maintenance.

Step 10 Warm up the engine for at least five minutes.

> **NOTE**
> Warm up the machine for a longer period in colder temperatures.

Step 11 Check all the gauges and instruments to make sure they are working properly.

Step 12 Shift the gears to low range.

Step 13 Release the parking brake and depress the service brakes.

Step 14 Check all the controls for proper operation.

Step 15 Check service brakes for proper operation.

Step 16 Check the steering for proper operation.

Step 17 Manipulate the controls to be sure all components are operating properly.

> **CAUTION**
> Do not pick up a load if the hydraulics are sluggish. Allow the machine to warm up until the hydraulics function normally. The hydraulics can fail if not warmed up completely.

Step 18 Shift the gears to neutral and lock.

Step 19 Reset the brake.

Step 20 Make a final visual check for leaks, unusual noises, or vibrations.

If the machine being used has a diesel engine, there are special procedures for starting the engine in cold temperatures. Many diesel engines have glow plugs that heat up the engine for ignition. See *Table 1* for recommended glow plug heating times.

To use the glow plug heater, push in and turn the heat start switch or depress the glow plug button for the indicated time. After holding that position for the indicated time, push in and turn the start switch to start. Some units have a small sight glass to observe the glow plugs. When they stop glowing, the engine is ready to start. Some units are also equipped with ether starting aids. Review the operator manual to ensure that the procedures for using these aids are fully understood.

Table 1 Recommended Glow Plug Heating Times

Starting Aid Chart	
Starting Temperature	**Glow Plug Heat Time**
Above 60°F (16°C)	None
60°F (16°C) to 32°F (0°C)	1 Minute
32°F (0°C) to 0°F (–18°C)	2 Minutes
Below 0°F (–18°C)	3 Minutes

22205-12_T01.EPS

As soon as the engine starts, release the starter switch and adjust the engine speed to approximately half throttle. Let the engine warm up to operating temperature before moving the loader.

Let the machine warm up for a longer period of time when it is cold. If the temperature is at or above freezing, 32°F (0°C), let the engine warm up for 15 minutes. If the temperature is between 32°F (0°C) and 0°F (–18°C), warm the engine for 30 minutes. If the temperature is less than 0°F (–18°C) or hydraulic operations are sluggish, additional time is needed. Follow the manufacturer's procedure for cold starting.

3.2.3 Checking Gauges and Indicators

Keep the engine speed low until the oil pressure registers. The oil pressure light should come on briefly and then go out. If the oil pressure light does not turn off within 10 seconds, stop the engine, investigate, and correct the problem.

Check the other gauges and indicators to see that the engine is operating normally. Check that the water temperature, ammeter, and oil pressure indicator are in the normal range. If there are any problems, shut down the machine and investigate or get a mechanic to look at the problem.

3.2.4 Shutdown

Shutdown should also follow a specific procedure. Proper shutdown reduces engine wear and possible damage to the machine. Perform the following steps to shut down a loader.

Step 1 Find a dry, level spot to park the loader. Stop the loader by decreasing the engine speed, depressing the clutch, and placing the direction lever in neutral. Depress the service brakes and bring the machine to a full stop.

NOTE

If the loader must be parked on an incline, block the tires.

Step 2 Place the transmission in neutral and engage the brake lock. Lock out the controls if the machine has a control lock feature.

Step 3 Lower the bucket so that it rests on the ground. If any other attachment is being used, be sure it is also lowered.

Step 4 Place the speed control in low idle and let the engine run for approximately five minutes.

CAUTION

Failure to allow the machine to cool down can cause excessive temperatures in the turbocharger. This can cause the oil to overheat. Oil that has been overheated must be replaced.

Step 5 Turn the start switch off.

Step 6 Release the hydraulic pressure by moving the control levers until all movement stops.

Step 7 Turn the disconnect switch to off and remove the key.

Some machines have additional disconnect switches for added security. If the loader being used has a battery disconnect, disconnect it. Other machines have a fuel shutoff switch. These controls provide an additional safety feature and deter unauthorized users. Always engage any additional security systems when leaving the loader unattended.

3.3.0 Basic Maneuvers/Operations

To maneuver the loader, the operator must be able to move forward, backward, and turn. Although basic maneuvering was covered in detail in *Heavy Equipment Operations Level One*, this section serves as a review.

On wheel-type loaders, direction is controlled using a steering wheel or a joystick. Crawler-type loaders run on tracks that are controlled by foot pedals and levers or joysticks. For loaders with joystick steering controls, it takes some practice to coordinate the control of the hand levers and foot pedals to steer the machine, while at the same time operating the bucket or other attachment. This section highlights joystick steering as it may be unfamiliar.

3.3.1 Moving Forward

The first basic maneuver is learning to drive forward. To move forward, follow these steps:

Step 1 Before starting to move, raise the bucket assembly by pulling the boom and bucket control lever. Raise the bucket to about 15 inches above the ground. This is the travel position.

Step 2 Put the shift lever in low forward. Release the parking brake, and press the accelerator pedal to start moving the loader.

Step 3 Steer the machine using a steering wheel or joystick for wheel-type loaders or levers and pedals or joystick for crawler-type loaders.

Step 4 Once underway, shift to a higher gear to drive on the road. To shift from a lower to a higher gear, move the shift lever forward. Remember, high gear is used only for traveling on the road.

NOTE

Always travel with the bucket tilted back and low to the ground (12 to 18 inches). This will provide better visibility.

3.3.2 Moving Backward

To back up or reverse direction, always come to a complete stop before placing the machine into reverse gear. Move the shift lever to reverse and apply some acceleration to begin moving backwards.

NOTE

Although it is possible to change directions while moving if the machine has a hydrostatic transmission, it is not recommended. Changing direction while in motion can cause a sudden jolt to the operator and the load.

For articulated loaders, it may take additional practice to be able to back in a straight line. Steering one of these machines is about the same as trying to back up a trailer attached to truck. If a turn is made too sharply, the machine may jack-knife. If that happens, stop and pull forward to straighten out before continuing to move backward.

3.3.3 Steering and Turning

How a loader is steered depends on whether it has a steering wheel or a joystick. Some wheel-type loaders will have a joystick for one-hand use. This allows the operator to keep the other hand on the bucket controls for simultaneous operations.

Steering wheels on wheel-type loaders operate in the same manner as steering wheels on cars and trucks. Moving the wheel to the right turns the loader to the right. Turning the wheel to the left moves the wheels to the left.

Some loaders use a joystick for steering instead of a wheel. The joystick may also incorporate a gear-change lever. Moving the steering lever to the left steers the machine to the left. The further the steering lever is moved, the faster the machine steers to the left. Moving the lever to the right accomplishes the same action to the other side.

The turning radius of a rigid-frame wheel-type loader is greater than that of an articulated loader. Therefore, articulated equipment can be used in tighter work areas without having to pull forward and backward to be repositioned. *Figure 26* shows an example of an articulated loader with a turning capability of 40 degrees left and right.

Crawler-type loaders require the use of foot pedals and levers or a joystick to steer the machine. This allows the operator to maintain at least one hand on the bucket controls at all times. Crawler-type loaders are steered in the same manner as bulldozers.

3.3.4 Operating the Bucket

Generally, the bucket on a loader is attached to the front frame by two lift arms. This adds stability to the bucket and improves material handling. Loaders may be equipped with one or two control levers. These controls operate the bucket lift and bucket tilt. Joysticks have replaced the lever operation on most new loaders. All movements for both bucket tilt and lift are incorporated into one joystick.

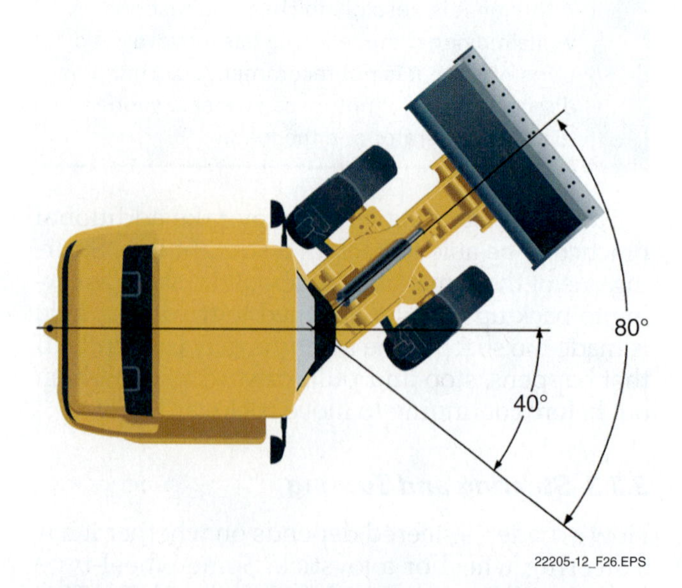

Figure 26 Articulated loader turn radius.

A loader with one control lever can accomplish all the functions of a two-lever unit. This operation would be similar to the use of a joystick, except that the joystick may have additional controls in the form of buttons or switches. Additional controls perform functions such as load metering or operation of special attachments.

All loaders have some type of bucket level indicator. The purpose of the indicator is to show the operator the position of the bucket as it is being raised or lowered. This helps keep the bucket from being rolled back during a high lift. A rollback can cause the contents of the bucket to spill out, possibly causing injury to the operator and damage to the machine.

Mechanical bucket level indicators have several different designs. One type uses two pointers on the bucket links. The bucket is level to the ground when the two pointers are opposite each other. Another type has an indicator rod that travels back and forth inside a tube attached to the dump ram. When the end of the indicator rod is flush with the end of the tube, the bucket is level in any boom position. Newer models of loaders may have an automatic self-leveling feature. The bucket is leveled throughout the hoist cycle. This reduces spillback and maintains better load control.

3.4.0 Work Activities

Operation of the loader is not as complex as some machines, but it does require constant attention and planning. By thinking ahead, an operator should have no trouble operating the loader. The basic work activities performed with either the wheel-type or crawler-type loader are described in this section.

NOTE

The controls on specific loaders may be different than those described in the procedures. Check your operator manual for information about the controls and limitations of your equipment.

The basic activities performed by a loader are all accomplished using one of the bucket attachments. Most loaders can perform several different activities with the same bucket. However, the bucket may not be the most effective or efficient for the job. Always check your operator manual to make sure the bucket is designed for the intended purpose. Exceeding the model's design limits reduces the service life of the equipment.

3.4.1 Loading Operations

Loaders are frequently used to load trucks, bins, and other containers. Usually, this loading is done by taking material from a stockpile. The procedure for carrying out a loading operation from a stockpile is as follows:

Step 1 Travel to the work area with the bucket in the travel position.

Step 2 Position the bucket parallel to and just skimming the ground.

Step 3 Drive the bucket straight into the stockpile.

Step 4 Adjust the controls and raise the bucket to fill it, as shown in *Figure 27*.

Step 5 Place the bucket and boom control levers on hold when the bucket is filled.

Step 6 Work the tilt control lever back and forth to move material to the back of the bucket. This is referred to as bumping. When the bucket is full, move the tilt control lever to the tilt back position.

Step 7 Shift the gears to reverse and back the loader away from the stockpile.

Step 8 Place the bucket in the travel position and move the loader to the truck.

Step 9 Center the loader with the truck bed and raise the bucket high enough to clear the side of the truck.

Step 10 Move the bucket over the truck bed and shift the bucket control lever forward to dump the bucket. At the same time, pull the boom control lever to the rear in order to retract the bucket. *Figure 28* shows the proper position for dumping from a standard bucket.

22205-12_F27.EPS

Figure 27 Filling loader bucket.

22205-12_F28.EPS

Figure 28 Proper dumping position.

Step 11 Pull the bucket control lever to retract the empty bucket and back the loader away from the truck as soon as the bucket is empty.

Step 12 Lower the bucket to the travel position and return to the stockpile.

Step 13 Repeat the cycle until the truck is loaded.

As the truck fills, the material needs to be pushed across the truck bed to even the load. As the leading edge of the bucket passes the sideboard of the truck, roll the bucket down quickly. Dump the material in the middle of the bed. Back up and push the load across the truck as the bucket is raised. By raising the bucket and backing up slowly, the material is distributed evenly across the bed.

Loading material into a truck with a loader requires that the operator have good reflexes and distance judgment. The loader must be placed close to the side of the truck in order to get the bucket positioned to dump properly.

There are two main points where accidental contact is the most common. The first area of contact is between the bucket and the side of the truck. Either the operator has misjudged the height of the truck bed or approached the truck too quickly, not allowing sufficient time to raise the bucket before getting to the truck. The second contact point is between the front wheels or tracks of the loader and the side of the truck body. Again, this is due to the operator misjudging the distance between the front of the loader and the side of the truck. Contact from these situations can cause severe damage to both pieces of equipment.

When loading from a stockpile or bank, the placement of both the truck and the loader are variable and must be adjusted to local operating conditions. Conditions to be considered are weight of material, gradient of the loading area, traction, and turning capability of the loader. If the loader works too close to the truck, it is necessary to pause during each cycle for the bucket to clear the side of the truck. If the loader works too far from the truck, the cycle is excessively long, with a resulting waste of time. The operator, by experience, must determine the most efficient arrangement for the particular operation and direct the trucks accordingly.

While there are many ways to maneuver a loader, the two most common patterns for a truck-loading operation are the I-pattern and the Y-pattern.

For the I-pattern, both the loader and the dump truck move in only a straight line, backward and forward (*Figure 29*). This is a good method for small, cramped areas. The loader fills the bucket and backs approximately 20 feet away from the pile. The dump truck backs up between the loader and the pile. The loader dumps the bucket into the truck. The truck moves out of the way and the cycle repeats.

To perform this I-pattern loading maneuver, position the loader so that it is on the truck driver's side of the truck. That way, eye contact can be made with the driver. Fill the bucket, as shown in *Figure 29A*. Back far enough away from the pile. Signal the truck driver with the horn. The truck will back to a predetermined position, as shown in *Figure 29B*. Move the loader forward and center it on the truck bed. Raise the bucket to clear the side of the truck and place it over the truck bed. Move the boom and bucket control lever to the left to dump the bucket. At the same time, raise the boom to make sure the bucket clears the truck bed. When the bucket is empty, move the boom and bucket control from side to side to shake out the last of the material. Back the loader away from the truck and signal the truck driver to move. When the truck is out of the way, lower the bucket and position the loader to return for another bucket of material.

The other loading pattern is the Y-pattern (*Figure 30*). This method is used when larger open areas are available. The dump truck remains stationary and as close as possible to the pile. The loader does all the moving in a Y-shaped pattern.

Position the dump truck or trucks so that eye contact can be made with the driver(s). Fill the bucket with material. While backing up, turn the loader to the right or left, depending on the position of the truck. Shift forward and turn the loader while approaching the truck slowly. Stop when the loader is lined up with the truck bed. Dump the bucket in the same way done for the

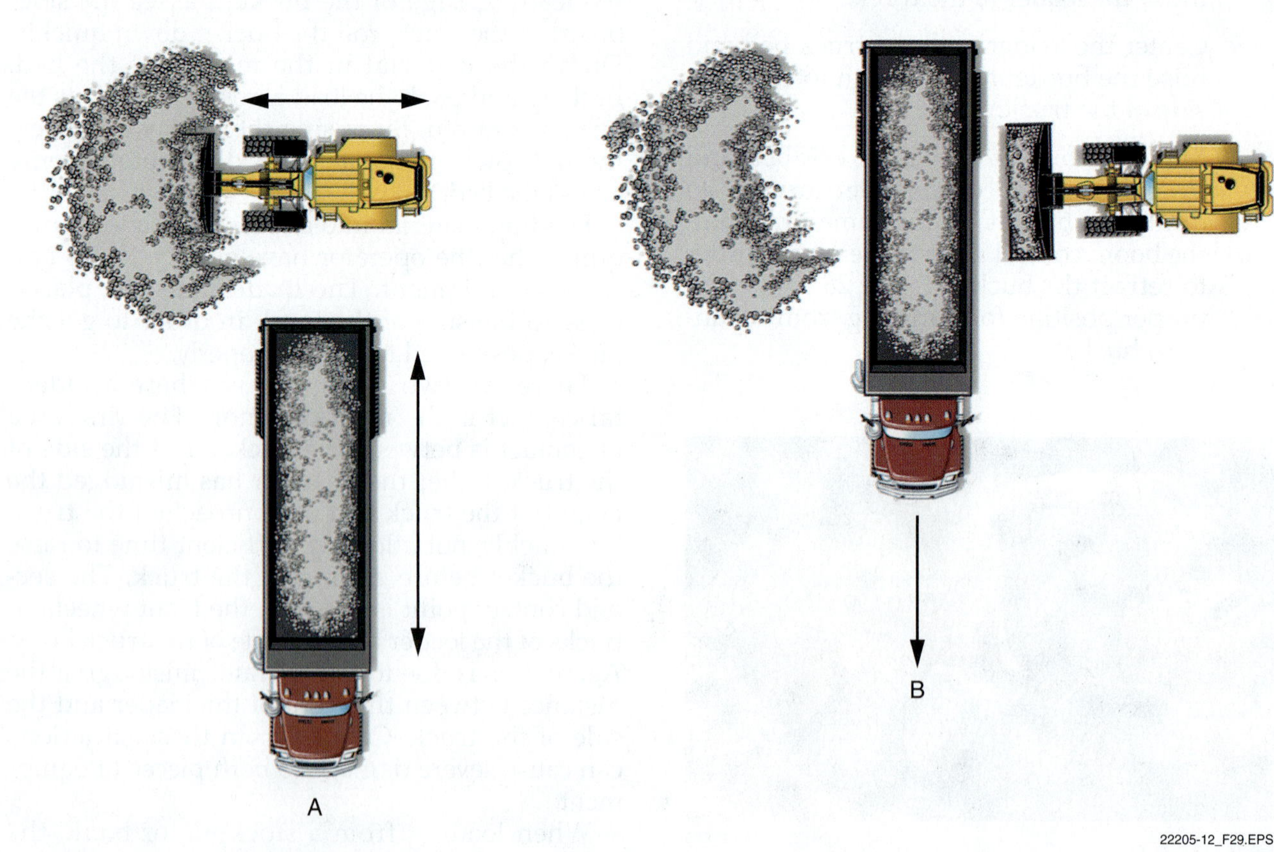

Figure 29 I-pattern for loading.

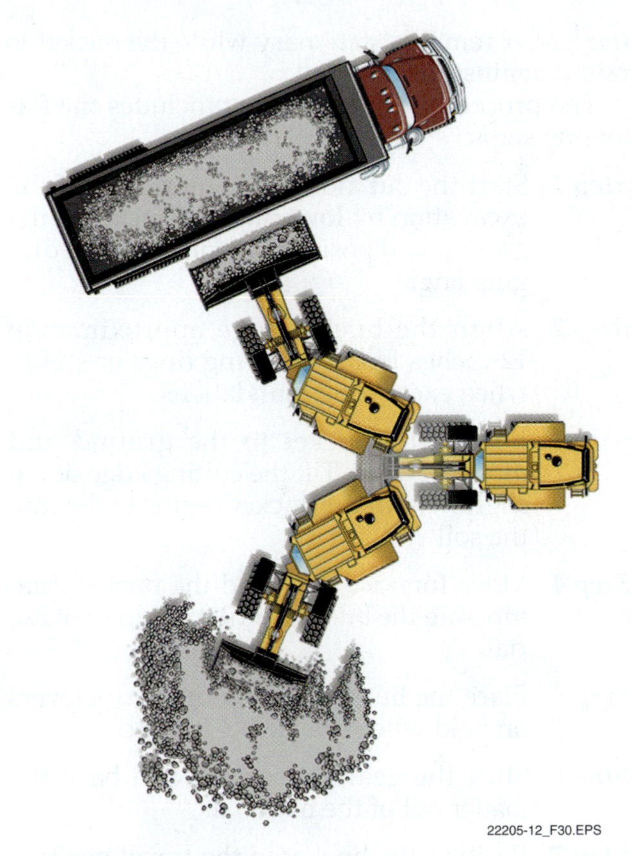

Figure 30 The Y-pattern for loading.

I-pattern. When the bucket is empty, back away from the truck while turning toward the pile. Drive forward into the pile to repeat the pattern. Repeat the cycle until the truck is full.

3.4.2 Leveling and Grading

Leveling and grading operations can be performed with the loader under most conditions. The multipurpose bucket is well suited for grading. The general purpose bucket is also used for this operation. Leveling can be done by tilting the bucket down and placing the cutting edge on the ground surface. Backing up with the bucket in this position will smooth out loose material.

For grading operations, the bottom of the bucket should be parallel to the ground surface. While maintaining this position, load material into the bucket and use the loaded bucket as the main **dozing** blade.

> **NOTE**
> Do not grade material with the bucket in the dump position. This puts stress on the structure and the hydraulics.

To perform a grading operation with the loader, follow these steps:

Step 1 Line up the bucket and loader on the area to be graded.

Step 2 Move the boom control lever forward and position the bucket on the ground or at the desired height.

Step 3 Shift to low gear and press the accelerator to begin moving forward.

Step 4 Maintain a low steady speed and keep the bucket at a constant height. As high spots are encountered, they can be trimmed by the blade and loaded into the bucket (*Figure 31*).

Step 5 When low spots are encountered, tilt the bucket forward and dump the material. Back up over the dumped material and smooth it out with the back of the blade.

Step 6 Reposition the loader and make another pass to the left or right of the area just graded.

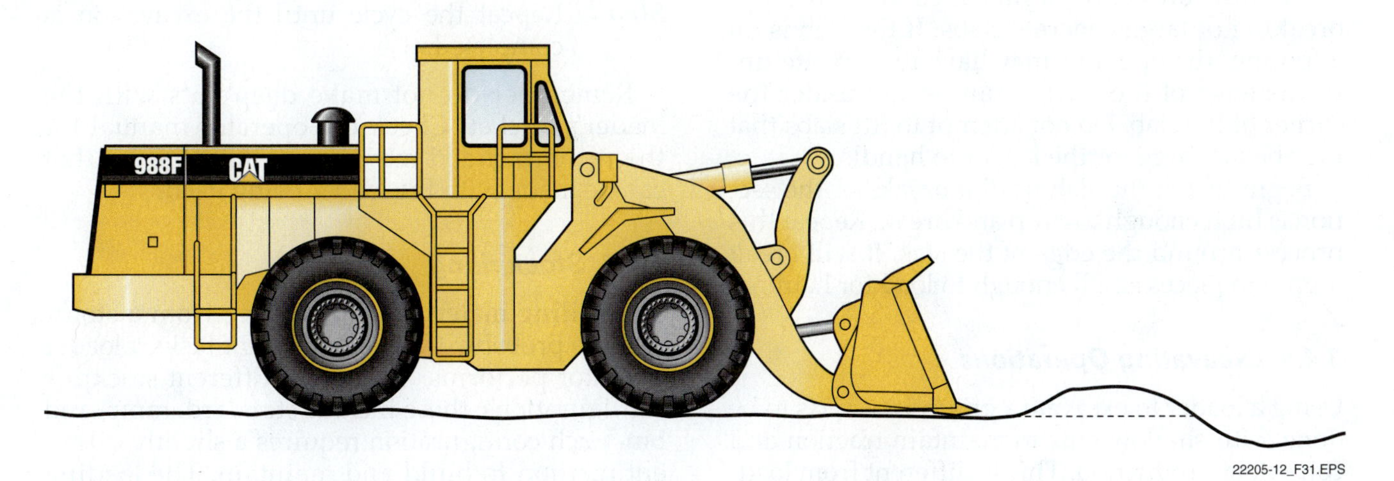

Figure 31 Trimming a high spot with a loader.

3.4.3 Demolition

Demolition work is a good activity for a track-type loader. Track loaders have the power and size to demolish structures such as walls and small buildings while at the same time being able to load and haul the rubble for removal from the area. Since track loaders operate on metal tracks instead of inflated rubber tires, they are less likely to be damaged by the debris of the demolition process.

Use the following procedure to perform a demolition activity with a track- or crawler-type loader:

Step 1 Travel to the building site with the bucket raised in the travel mode.

Step 2 Downshift into first gear and raise the boom to an overhead position.

Step 3 Approach the building cautiously and place the bucket on the top of the building wall.

Step 4 Very slowly, move forward to push the top of the wall in.

Step 5 As the building begins to fall, select the reverse gear and back away.

Step 6 Continue placing the bucket on the wall until all of the building is down.

Step 7 Load the bucket and approach the dump area or haul unit.

Step 8 Raise the bucket to clear the side of the haul unit or truck.

Step 9 Dump the material slowly in order to avoid jarring or damaging the truck.

Step 10 Continue loading until all the debris is removed from the site.

Demolition work sometimes requires the breaking of large concrete slabs. If the slab is on a footing, the operator may have to excavate until the edge of the bucket can be slid under the corner of the slab. Do not attempt to lift slabs that may be too large for the loader to handle.

Begin lifting the slab until it breaks or the section is high enough to drop and break. Repeat this process around the edge of the slab. It will break it up into pieces small enough to load for hauling.

3.4.4 Excavating Operations

Using a loader to excavate a ground area has to be done with shallow cuts to maintain traction and consistent production. This is different from loading the bucket from a bank or stockpile where the loader remains stationary while the bucket is raised against the material.

The procedure for excavating includes the following steps:

Step 1 Start the cut along the outer edge of the excavation by lowering the bucket to the ground and positioning for a straight digging angle.

Step 2 Align the bucket edge approximately 12 inches from the string lines or stakes when excavating foundations.

Step 3 Lower the bucket to the ground and move forward. Tilt the cutting edge down slightly until the bucket begins to dig into the soil.

Step 4 Move forward and load the bucket. Manipulate the bucket until it is full of material.

Step 5 Place the bucket and boom control levers on hold when the bucket is filled.

Step 6 Shift the gears to reverse and back the loader out of the material.

Step 7 Position the bucket in the travel position and proceed to the dump site. The travel position is about 12 to 18 inches above ground with the bucket tilted back.

Step 8 Move the bucket control lever forward to dump the bucket. At the same time, pull the boom control lever to the rear to raise the bucket.

Step 9 Pull the bucket control lever to retract the empty bucket. At same time, back the loader away from the pile.

Step 10 Position the bucket in the travel position and return to the excavation.

Step 11 Repeat the cycle until the excavation is completed.

Remember do not make deep cuts with the loader's bucket. Check the operator manual for the recommended maximum allowable cut that can be made with the loader being used.

3.4.5 Stockpiling

Stockpiling materials and loading from a stockpile are probably the most frequent tasks a loader operator performs. There are different stockpile configurations; these include standard, ramp, and bin. Each configuration requires a slightly different method to build and maintain. The loading operation was described earlier. The procedures

for maintaining the stockpiles and working the material are described in this section.

To make a standard stockpile as shown in *Figure 32*, use the basic bucket-loading technique. Pick up the material from the bottom of the pile while moving forward and raising the bucket to the top of the pile. Raise the bucket high enough to clear the pile of material. Move forward so the raised bucket is over the top of the pile. Dump the bucket and allow the material to spread from the top of the pile downward.

Put the loader in reverse, back away from the pile, and start over. Work this way in a pattern all around the stockpile, moving the material continually toward the center. Always start at the point furthest from the center and work the area smooth.

A ramp stockpile is used to store large quantities of material in a small area with a larger loader. The shape of the stockpile is a high, long, and narrow ramp.

To make a ramp stockpile, start the ramp close to the work area. Dump the material, then lower the bucket to approximately 15 inches above the ground. Use it to push the highest area of the pile forward. Dump the next load of material beside the first dump area and spread this material out the same way. Repeat the steps of dumping onto the pile and dragging the bucket back to build the ramp with a gradual slope. The base of the stockpile should be twice as wide as the loader so that when the ramp is complete, it is still wide enough to support the loader.

Follow these tips to be more efficient in maintaining a ramp stockpile:

- After each dump on the ramp, push only the top half of the material off the end. The other half will add to the ramp.
- Do not run the loader in the same track all the time. Move the loader from one side of the ramp to the other to help compact the material.
- As the ramp grows, level it off and continue dumping until the limit of the stockpile area is reached.
- To remove material from a ramp stockpile, reverse the process used to make the ramp. As more material is taken from the ramp, the ramp is reduced layer by layer.
- Begin loading the bucket at the point where the last material was placed

A bin stockpile is contained in some sort of three-sided enclosure. The bin's walls are usually made of wood, concrete, or metal. The bin is usually rectangular in shape, with a floor of concrete or asphalt. Bin stockpiles provide good storage and keep different types of materials separated. They are usually found at materials manufacturing plants and maintenance facilities. Use extreme care when working with material in a bin. Bumping the side of the bin or pushing material into the wall can cause damage to the structure as well as the loader.

To fill a bin, place the material close to the wall at the farthest point from the entrance. Dump the material along this wall first. Next, dump material on top of this first row, allowing it to fall forward.

22205-12_F32.EPS

Figure 32 Standard stockpile.

As the back of the bin begins to fill, place the additional material in front of the last pile of material. Do not let any material run over the sides of the bin. Always enter the bin from the center of the opening and work out from the back of the bin.

To consolidate material or clean up the floor of the bin, pick up the material and place it on the top of the pile. Do not push material forward. This will put pressure on the walls and damage the bin.

3.4.6 Clearing Land

Clearing land involves the removal of vegetation, trees, and other obstructions above the surface. It may also include **stripping** away the top soil and doing some rough grading. Wheel-type loaders normally do not do this type of work because the rubber tires do not perform well under the conditions surrounding clearing and **grubbing**.

Track-type loaders are better suited for this activity because of their tracks. The tracks give them the stability and firmness to clear brush and small trees and move small boulders. Track or crawler loaders are able to maneuver the rough terrain and are free of problems associated with rubber-tired equipment.

3.4.7 Backfilling

Backfilling can be accomplished with either the wheel-type loader or crawler-type loader. The first method requires that materials be loaded from a stockpile, carried to the site, and dumped. A second method involves using spoils material located close to the area to be backfilled (*Figure 33*). When backfilling a trench, have someone observe the operation from a safe vantage point and direct the operation so that the loader does not come too close to the edge of the trench or collapse the side of the trench from the weight of the loader.

When backfilling from the spoils pile, use the bucket to push the material toward the area. Do not overfill the bucket or try to push too much material up against a structure. This may put too much pressure on the structure, causing it to fall over.

3.4.8 Excavating Work in Confined Areas

Sometimes, loader operators are required to work in small spaces or confined areas. This requires careful planning and execution.

When working in a confined area, the type of loader used has a great impact on both the time it takes to do the job and the quality of the work. Different types of confined areas require different approaches. The first type of confinement is below grade, where the space is limited and enclosed by walls or other vertical restrictions. The second type is at grade, where space is limited by some obstruction such as a trench, wall, building, or other equipment. *Figure 34* shows a loader working through the door of a storage building.

When working below grade, exercise extreme caution and do not run into or damage any wall supports or shoring. This may cause a collapse of the embankment, burying people and equipment. Usually, loading out of a deep excavation such as a foundation or pit requires workers on the ground to direct truck traffic and spot equipment for loading. Always watch the spotter for signals when loading trucks in a confined area.

Loading trucks from within an excavation is basically the same as the standard loading operation. Loading is normally from a stockpile of spoils placed in an open area by other equipment,

22205-12_F33.EPS

Figure 33 Backfilling.

22205-12_F34.EPS

Figure 34 Working in confined spaces.

such as hydraulic excavators or bulldozers that are digging out the excavation. The loader is used for loading from the stockpile into the truck. When the excavation is complete, the loader can drive out on the remaining ramp. After all equipment has been removed, the ramp can be removed by equipment from outside the excavation.

Working in limited space between vertical structures may restrict the bucket height and turning ability of the loader. Articulated loaders are a better choice under these conditions because they have sharper turning and better maneuverability. Make sure the view in all directions is unobstructed. If raising the bucket limits the view, have someone stand off to the side and assist by spotting your loading operation.

3.4.9 Working in Unstable Soils

Working in mud or unstable soils that do not support the loader can be aggravating and dangerous. This is a problem even for experienced operators.

When entering a soft or wet area, go very slowly. If the front of the loader feels like it is starting to settle, stop and back out immediately. That settling is the first indication that the ground is too soft to support the equipment. The engine will lug slightly and the front end of the loader will start to settle.

After backing out, examine how deep the wheels or tracks sank into the ground. If they sink deep enough that the material hits the bottom of the loader, the ground is too soft to work in a normal way.

To work in soft or unstable material, follow this procedure:

Step 1 Start from the edge and work forward slowly.

Step 2 Push the mud ahead of the bucket and be sure the ground below is firm.

Step 3 Don't try to move too much material on any one pass.

Step 4 Keep the wheels or tracks from slipping and digging in.

Partially stable material can also be a hazard because an operator may drive in and out over relatively firm ground many times, while it slowly gets softer because the weight of the wheels or tracks pumps more water to the surface. If this happens, the wheels or tracks will sink a little more each time until the loader finally gets high-centered. Then, the machine must be pulled out with a winch.

To keep this from happening, do not run in the same track each time entering or leaving an area. Move over slightly in one direction or the other so the same tracks are not pushed deeper into the unstable material each time.

3.4.10 Using Special Attachments

In addition to the various types of buckets used with a loader, there are several special attachments that expand the loader's operational capability. The three main attachments are the multipurpose bucket, ripper, and forklift. The buckets and forks were covered earlier in this module. Rippers are usually used only on graders and dozers. But if a ripper is to be used on a loader, it is more likely to be used on a track loader because of the track loader's good traction and heavier weight. *Figure 35* shows a ripper mounted behind a tracked loader.

Ripping aids in breaking up hard material for ease of loading. After the material is broken loose, it can be removed to the haul unit or stockpiled. Good traction is required to pull the ripper teeth through the material without spinning the tracks.

When ripping, the loader should be operated at a low travel speed and only one or two ripper teeth should be used. If the material breaks up easily, teeth can be added. As the ripping operation continues, keep some of the loose material on the ground to cushion the loader and improve traction. The basic procedure for using a ripper with a track loader is the same as with a bulldozer. Refer to the operator manual for specific instructions on location and use of controls.

3.4.11 Forklift Attachments

Use a forklift attachment on the loader (*Figure 36*) to move palletized materials. Some models have forks that attach to a general-purpose bucket, while other models attach directly to the lift arms. Forks come in several different sizes depending on their lifting weight and reach. Operating the loader as a forklift is basically the same procedure for all models.

The following steps are used to unload material from a flatbed truck:

Step 1 Position the loader at either side of the truck bed.

Step 2 Manipulate the control levers in order to obtain the appropriate fork height and angle.

22205-12_F35.EPS

Figure 35 Ripper in use.

Step 3 Drive the forks into the opening of the pallet or under the loose material. Use care to not damage any material.

Step 4 Manipulate the controls and brake pedal as required for lifting the material slightly off the bed.

Step 5 Tilt the forks back to keep the pallet or other material from sliding off the front of the forks.

22205-12_F36.EPS

Figure 36 Forklift attachment for a loader.

Step 6 Shift the gear to reverse and back the loader away from the truck.

Step 7 Lower the forks to the travel position and travel to the stockpile area.

CAUTION

Do not lower the forks with the boom and bucket control lever in the float position. This could cause equipment damage.

Step 8 Position the loader so that the material can be placed in the desired area.

Step 9 Lower the boom until the material is set on the required surface.

Step 10 Adjust the forks with the boom lever in order to relieve the pressure under the pallet.

Step 11 Back the loader away from the pallet.

Step 12 Repeat the cycle until the truck is unloaded.

Safe operation of the loader as a forklift requires extreme caution. Make sure to follow these safety requirements:

- Do not swing loads over the heads of workers. Make sure that there is enough clear area to maneuver.
- Do not allow workers to ride on the forks or the bucket.
- Using the forklift attachment usually requires operation in a confined area. Always know what clearance is available for maneuvering.

3.4.12 Transporting a Loader

If the loader needs to be transported from one job site to another, it may either be driven if it is a short distance, or loaded and hauled on a transporter.

When roading a loader from one site to another, make sure the necessary permits for traveling on a public road have been obtained. Flags must be mounted on the left and right corners of the machine. Lights and flashers should be switched on. Depending on the location, a scout vehicle may also be required. Because the wheel-type loader is top heavy and prone to bouncing, drive at a slow rate of speed, especially around corners and over rough terrain. Keep the bucket in the travel position of 12 to 18 inches above the ground.

If the equipment must be moved a long distance, it should be transported on a properly equipped trailer or other transport vehicle. Before beginning to load the equipment for transport, make sure the following tasks have been completed:

- Check the operator manual to determine if the loaded equipment complies with height, width, and weight limitations for over-the-road hauling.
- Check the operator manual to identify the correct tie-down points on the equipment.
- Be sure to get the proper permits, if required.
- Plan the loading operation so that the loading angle is at a minimum.

Once these tasks are complete and the loading plan has been determined, carry out the following procedures:

Step 1 Position the trailer or transporting vehicle. Always block the wheels of the transporter after it is in position but before loading is started (*Figure 37*).

Step 2 Place the loader bucket in the travel position and drive the loader onto the transporter. Whether the loader is facing forward or backward will depend on the recommendation of the manufacturer. Most manufacturers recommend backing the loader onto the transporter.

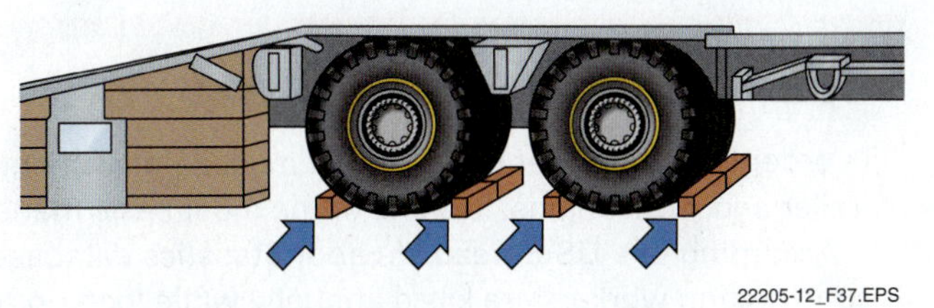

Figure 37 Block trailer wheels.

Step 3 If the loader is articulated, connect the steering frame lock link to hold the front and rear frames together (refer to the operator manual for lock link location). The steering frame lock is secured by a pin when not in use. Remove the cotter pin and holding pin. Swing the lock to the transport position and secure it with a pin and cotter pin.

Step 4 Lower the bucket to the floor of the transporter.

Step 5 Move the transmission lever to neutral, engage the parking brake, and turn off the engine.

Step 6 Manipulate the bucket raise and tilt controls to remove any remaining hydraulic pressure.

Step 7 Remove the engine start key and place the fuel valve in the Off position.

Step 8 Lock the door to the cab as well as any access covers. Attach any vandalism protection.

Step 9 Secure the machine with the proper tie-down equipment as specified by the manufacturer. Place chocks at the front and back of all four tires.

Step 10 Cover the exhaust and air-intake openings with tape or a plastic cover.

Step 11 Place appropriate flags or markers on the equipment if needed for height and width restrictions.

NOTE

Unloading the equipment from the transporter would be the reverse of the loading operation.

Transporting Loaders

In order to get a loader on the job, it must be driven onto a trailer. Upon arrival, the loader is driven off of the trailer and work begins. It is one of the most basic maneuvers related to heavy equipment operation.

According to a US Bureau of Labor Statistics database covering an 11-year period, an average of nine construction workers are killed annually while loading and unloading heavy equipment. Of those, 70 percent occurred during the loading process. About 75 percent of the incidents occurred due to the equipment turning over. However, only 35 percent of those killed were operators.

There is no question that the very narrow amount of room for the equipment on a trailer is a key danger. To minimize the chances for an incident, the following is recommended:

- The trailer should be on flat, level, firm ground.
- Operators should personally ensure that the trailer is wide enough.
- Nearby workers should remain clear and out of the range of a possible rollover incident.
- Position and use a spotter during loading and unloading.

22205-12_SA02.EPS

Additional Resources

The Occupational Safety and Health Administration (OSHA) publishes safety requirements for loaders and related material-handling equipment in *Standard 1926.602*, found at *www.osha.gov*.

3.0.0 Section Review

1. When getting into or out of a loader, the operator must do what at all times?
   a. Keep both hands on the machine
   b. Maintain visual contact with the ladder
   c. Maintain three-point contact with the machine
   d. Keep both hands and both feet on the machine

2. When traveling a loader, maintain good visibility and loader stability by carrying the bucket _____.
   a. approximately 15 inches above the ground
   b. approximately 30 inches above the ground
   c. approximately 36 inches above the ground
   d. approximately 60 inches above the ground

3. To change movement direction from forward to reverse, the operator should stop the loader and then change the gears from forward to reverse.
   a. True
   b. False

4. When doing demolition work with a loader, approach a wall slowly and place the bucket _____.
   a. against the base of the wall
   b. approximately half way up the wall
   c. against the top of the wall
   d. approximately 15 inches above the base of the wall

SUMMARY

Loaders are used primarily for loading material from the ground or from a stockpile. They can also be used for digging, grading, hauling, and light clearing work. There are two basic types of loaders: the wheel-type loader and the crawler-type loader. There are two basic configurations for the wheel-type loader: a rigid-frame machine and an articulated machine.

All loaders have a large steel bucket mounted on two rigid arms extending from the frame. The bucket can dig, scoop, and curl upward in order to pick up large quantities of material. A loader can be equipped with a standard bucket, a multipurpose bucket, or other specialty buckets. Other attachments include forks, plows, rakes, and brooms.

Safety considerations when operating a loader include keeping the loader in good working condition, obeying all safety rules, not taking chances, and being aware of other people and equipment in the same area where the loader is being operated. If unsure of the terrain or operation, stop the machine, get down, look around, and discuss the work with your supervisor or the resident engineer. When starting the dig, be sure to start in the right place and that the area is clear of any underground utilities or other structures.

Use caution when loading into dump trucks. Do not use excessive speed when making turns with a loaded bucket. Always carry the bucket low to the ground.

The two common loading patterns are the I-pattern and Y-pattern. They are named after the configuration the loader makes when performing the loading operation. With the I-pattern, the loader moves up to fill the bucket and back to allow the dump truck to move in between the loader and the stockpile. The Y-pattern is used when there is more room for the loader to maneuver. In this operation, the loader moves up to load the bucket, then moves back and turns right or left approximately 45 to 90 degrees to approach the truck waiting to the side of the loader. The loader creates a Y-pattern when making one complete loading cycle.

Use caution when working in unstable soils. If the front of the loader feels like it is starting to settle, stop and back out immediately. This is an indication that the ground is too soft. When working in soft material, work from the edge and move forward slowly. Push any soft material ahead of the bucket and be sure the ground below is firm. Keep the wheels or tracks from slipping and digging in.

Review Questions

1. Loaders are grouped into how many main categories?
   a. One
   b. Two
   c. Three
   d. Four

2. What plays a major role in the breakout force and the tipping load abilities of a loader?
   a. The type and size of the engine
   b. The type and size of the loader's frame
   c. The type of lift arms and hydraulic components
   d. The size of the bucket and the arms that lift and tilt it

3. Bucket controls (levers or joystick) are typically located _____.
   a. on the left armrest
   b. on the right armrest
   c. in front of or slightly to the left of the left armrest
   d. in front of or slightly to the right of the right armrest

4. What is the term used for a control position that allows the hydraulic fluid to the lift arm hydraulic cylinders to flow in and out both ends of the cylinders so that the bucket can follow the contour of the ground as the loader moves forward or backward?
   a. Tilt
   b. Hold
   c. Float
   d. Skim

5. What control term used with a loader's arm controls is considered to be a neutral position?
   a. Hold
   b. Float
   c. Stay
   d. Neutral

6. If a loader's diesel engine runs out of fuel, refuel the loader and _____.
   a. restart the engine
   b. clean the injectors
   c. allow the engine to completely cool before restarting the engine
   d. bleed the air out of the fuel lines and injectors before restarting the engine

7. A flashing indicator light means _____.
   a. that the system associated with the indicator needs attention
   b. to recheck the machine system associated with the indicator
   c. to check all machine temperatures and levels
   d. to stop all operations immediately

8. The width of a loader's bucket is normally the same as the _____.
   a. loader's back wheels
   b. loader's front wheels
   c. width of the loader's chassis or frame
   d. length of the loader's chassis or frame

9. A broom attachment can be angled horizontally up to how many degrees on either side?
   a. 10 degrees
   b. 15 degrees
   c. 20 degrees
   d. 30 degrees

10. A grouser is part of the _____.
    a. track
    b. idler
    c. sprocket
    d. ROPS

11. On a track loader, what component(s) keep tension on the tracks to keep them from jumping off?
    a. The idlers
    b. The sprockets
    c. The grousers
    d. The tensioning springs on the axles

12. The fuel level in a loader should be checked _____.
    a. manually by opening the fuel fill cap and looking inside the tank
    b. manually using a fuel dipstick or marking vial
    c. visually using the fuel gauge on the dash
    d. visually using the digital readout

13. When traveling a loader, always travel with the bucket _____.
    a. in the float mode
    b. in the hold mode
    c. low to the ground
    d. raised above the front wheels

14. Before shutting down a loader's engine, place the transmission into _____.
    a. neutral and place chocks under the wheels
    b. low gear and disengage the brake lock
    c. high gear and engage the brake lock
    d. neutral and engage the brake lock

15. When using a loader to excavate foundations, align the bucket edge approximately how far from the string lines or stakes?
    a. 6 inches
    b. 12 inches
    c. 18 inches
    d. 24 inches

Trade Terms Introduced in This Module

Accessories: Attachments used to expand the use of a loader.

Blade: An attachment on the front end of a loader for scraping and pushing material.

Breakout force: The maximum vertical upward force created by the curling action of a bucket attached to an excavator, backhoe, or loader, measured 4 inches behind the tip of the bucket's cutting edge.

Bucket: A U-shaped closed-end scoop that is attached to the front of the loader.

Debris: Rough broken bits of material such as stone, wood, glass, rubbish, and litter after demolition.

Dozing: Using a blade to scrape or excavate material and move it to another place.

Grouser: A ridge or cleat across a track that improves the track's grip on the ground.

Grubbing: Digging out roots and other buried material.

Rubble: Fragments of stone, brick, or rock that have broken apart from larger pieces.

Roading: Driving a piece of construction equipment, on a public road, from one job site to another.

Spot: To line up the haul unit so that it is in the proper position.

Stockpile: Material put into a pile and saved for future use.

Stripping: Removal of overburden or thin layers of pay material.

Tipping load: The loaded weight that will lift the rear wheels off the ground with the machine in a static (not moving) condition.

Additional Resources

This module presents thorough resources for task training. The following resource material is suggested for further study.

The Occupational Safety and Health Administration (OSHA) publishes safety requirements for loaders and related material-handling equipment in *Standard 1926.602*, found at **www.osha.gov**.

Figure Credits

Courtesy of Deere & Company, Module opener, Figures 1–15, 17, 18B, 19–21, 24, 25, 27, 28, 32–36

Courtesy of Loup Electronics, SA01

Reprinted courtesy of Caterpillar Inc., Figures 16, 18A, 18C, 23

Courtesy of Michael Cereghino, SA02

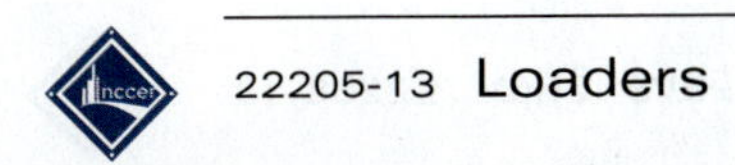

Section Review Answers

Answer	Section Reference	Objective
Section One		
1 d	1.1.0	1a
2 c	1.2.2	1b
3 a	1.3.5	1c
4 b	1.4.1	1d
Section Two		
1 d	2.0.0	2a
2 c	2.2.0	2b
Section Three		
1 c	3.1.1; Caution	3a
2 a	3.2.0	3c
3 a	3.3.2	3c
4 c	3.4.3	3d

NCCER CURRICULA — USER UPDATE

NCCER makes every effort to keep its textbooks up-to-date and free of technical errors. We appreciate your help in this process. If you find an error, a typographical mistake, or an inaccuracy in NCCER's curricula, please fill out this form (or a photocopy), or complete the online form at **www.nccer.org/olf**. Be sure to include the exact module ID number, page number, a detailed description, and your recommended correction. Your input will be brought to the attention of the Authoring Team. Thank you for your assistance.

Instructors – If you have an idea for improving this textbook, or have found that additional materials were necessary to teach this module effectively, please let us know so that we may present your suggestions to the Authoring Team.

NCCER Product Development and Revision

13614 Progress Blvd., Alachua, FL 32615

Email: curriculum@nccer.org
Online: www.nccer.org/olf

❑ Trainee Guide ❑ AIG ❑ Exam ❑ PowerPoints Other ______________________________

Craft / Level: Copyright Date:

Module ID Number / Title:

Section Number(s):

Description:

Recommended Correction:

Your Name:

Address:

Email: Phone:

22204-13

Scrapers

Overview

Bulk earthmoving is the primary purpose of scrapers. Layers of earth are scraped and collected, and then transported for use in another location. A scraper represents a compromise in vehicle design between a machine that loads well and one that can haul a load a significant distance. Scrapers can accomplish a lot of work alone, but they often work with bulldozers, excavators, and other scrapers as a team. An understanding of their components, controls, and maintenance requirements will provide a basis for safe and efficient operation.

Module Nine

Trainees with successful module completions may be eligible for credentialing through NCCER's National Registry. To learn more, go to **www.nccer.org** or contact us at **1.888.622.3720**. Our website has information on the latest product releases and training, as well as online versions of our *Cornerstone* newsletter and Pearson's product catalog.

Your feedback is welcome. You may email your comments to **curriculum@nccer.org**, send general comments and inquiries to **info@nccer.org**, or fill in the User Update form at the back of this module.

This information is general in nature and intended for training purposes only. Actual performance of activities described in this manual requires compliance with all applicable operating, service, maintenance, and safety procedures under the direction of qualified personnel. References in this manual to patented or proprietary devices do not constitute a recommendation of their use.

22204-13
Scrapers

Objectives

When you have completed this module, you will be able to do the following:

1. Identify and describe the components of a scraper.
   a. Identify and describe chassis components.
   b. Identify and describe scraper controls.
   c. Identify and describe scraper instrumentation.
2. Describe the prestart inspection and preventive maintenance requirements for a scraper.
   a. Describe prestart inspection procedures.
   b. Describe preventive maintenance requirements.
3. Describe the startup, shutdown, and operating procedures for a scraper.
   a. State scraper-related safety guidelines.
   b. Describe startup, warm-up, and shutdown procedures.
   c. Describe basic maneuvers and operations.
   d. Describe related work activities.

Performance Tasks

Under the supervision of your instructor, you should be able to do the following:

1. Complete a proper prestart inspection and preventive maintenance on a scraper.
2. Perform proper startup, warm-up, and shutdown procedures on a scraper.
3. Execute basic maneuvers with a scraper, including forward/backward movement, turning, loading, and unloading.

Trade Terms

Apron
Bowl
Cycle
Ejector
End shoes
Governor
Grade checker
Lug down
Pay material
Rimpull
Ripping
Stockpile
Stripping

Industry Recognized Credentials

If you are training through an NCCER-accredited sponsor, you may be eligible for credentials from NCCER's Registry. The ID number for this module is 22204-13. Note that this module may have been used in other NCCER curricula and may apply to other level completions. Contact NCCER's Registry at 888.622.3720 or go to **www.nccer.org** for more information.

Contents

Topics to be presented in this module include:

Figures

SECTION ONE

1.0.0 SCRAPER COMPONENTS

Objective 1

Identify and describe the components of a scraper.

a. Identify and describe chassis components.
b. Identify and describe scraper controls.
c. Identify and describe scraper instrumentation.

Trade Terms

Apron: A movable metal plate in front of the scraper bowl that can be raised or lowered to control the flow of material into or out of the bowl.

Bowl: The main component at the center or back of a scraper where the material is collected and hauled.

Cycle: One complete trip made by a scraper starting from the load point and returning to the same or adjacent point. A cycle typically includes scraping, hauling, spreading the contents, and returning to the origin to begin the next load.

Ejector: A large metal plate inside the bowl of a scraper that can be activated to push the material forward, causing it to fall from the front of the bowl.

Governor: A control that limits the maximum speed of an engine or vehicle.

Rimpull: The torque applied to the axle from the engine that is then converted into pulling force at the face of the tire; more commonly known as gross traction in mathematical calculations.

Stockpile: Material put into a pile and saved for future use.

The scraper is undoubtedly the most efficient piece of equipment that has been developed for bulk, wide-area excavation. Scrapers are highly mobile excavators that can dig, carry, and spread loads. They may also be called carryalls or pans. Scrapers are used on many different types of excavation and construction projects, such as road building, mining, dam construction, and industrial plant development. Scrapers are also used to build canals, dikes, and levees. They can be used efficiently where any large amount of material must be moved. They are used to remove dirt and other materials from a site by scraping it into a **bowl**. They are also used to spread dirt and other material that has been loaded into the bowl with a separate piece of equipment, such as a loader.

Scrapers are used in various construction projects for the following tasks:

- Cut, load, and spread granular material
- Strip and **stockpile** topsoil
- Perform finish grading

There are many variations of the scraper. Self-propelled scrapers were invented in the late 1930s. Some of the earlier scrapers were mounted on four wheels and drawn by a tractor or bulldozer. A mechanical winch on the dozer powered the cables operating the scraper. Today, many scrapers are self-propelled, although towed units remain popular. Four-wheeled, self-propelled units are preferred if any length of haul is required. When scraping in dense soils such as clay, the additional resistance can be overcome by using bulldozers as pushers during the loading operation.

Scrapers are manufactured in different sizes for performing different kinds of work. There are small scrapers that are used for small jobs, such as industrial land development or for work in tight places. Smaller scrapers can be towed by a dozer or tractor to economically complete a small project. Larger scrapers are typically used for rough grading on large excavation or construction projects for roads, airports, and dams.

It is nearly impossible to keep track of the size and capacity range of scrapers over the years. The smallest self-propelled scrapers handle around 11 cubic yards of material, driven by 150 to 200hp engines. The large end of the scale is a moving target. For single bowl, self-propelled scrapers, the Caterpillar 657 Series is certainly in the running with twin engines developing roughly 950hp.

With the larger scrapers, tremendous amounts of material can be cut and moved in one **cycle**. Because of their size, scraper operation requires good coordination and quick reflexes.

A number of manufacturers make several different types of scrapers. Generally they fall into the following four categories based on the arrangement of the bowl and the power unit:

- Standard self-propelled
- Tandem
- Elevating
- Towed

A common self-propelled (*Figure 1*) scraper consists of a bowl unit and a tractor. A cutting edge is mounted at the bottom of the bowl to scrape the earth into the bowl. On the largest models, the bowl can hold more than 40 cubic yards of material, and the cutting edge width spans nearly 13 feet.

Figure 1 Standard self-propelled scraper.

Figure 2 Tandem scraper.

Figure 3 Tandem scrapers working together.

The tractor is connected to the bowl unit by an arched gooseneck. A standard single-engine self-propelled scraper has two axles. The front axle, under the tractor, is the drive axle. The engine projects forward of the drive axle to balance the weight of the tractor over the drive wheels for better traction. This is known as an overhung configuration.

In some conditions, a standard single-engine scraper does not have sufficient power or traction to fully load the bowl. There are several methods used to supply more power to the scraper to accomplish this. Tandem scrapers have two engines: one in the tractor and a second at the rear of the scraper unit. These machines can work on steeper grades and rougher terrain.

Even with tandem engines, additional power may still be needed. In certain conditions, another machine is used to push the scraper to load it completely. A tractor or bulldozer can be used, or two push-pull scrapers can work together. The two scrapers are connected by a bail and hook (*Figure 2*). One machine pushes the other and the power of the two machines is used to load one bowl at a time. Once both bowls have been filled, the scrapers unhook and travel separately to unload. *Figure 3* shows two tandem scrapers working together.

An elevating scraper has an independently powered auger or elevator mounted in the bowl. The elevator lifts the material to the top of the pile and evenly distributes it into the bowl. As the material is lifted, it is conditioned, or broken up. This blends the material, reduces air voids, and creates a more consistent payload. These scrapers do not need additional power to obtain a full load.

There are some drawbacks to this type, however. In most models, the elevator replaces the **apron** in the front of the bowl. Large rocks or tree stumps can damage the elevator or get stuck in the bowl. The weight of the elevator or auger can slow the scraper during the hauling phase. The elevator must also be maintained, which increases the operating cost. However, these models are economical to use in many situations.

Figure 4 shows an elevating scraper loading soil. The elevator on this machine is a paddle wheel, much like the paddle wheel of river steamboat. Other elevators have a single spiral blade that rotates like an auger (*Figure 5*). Typically the elevator is operated with hydraulic pressure or an electric motor.

The pull-type or towed scraper does not have its own engine. It must be pulled by a tractor, bulldozer, or other heavy equipment. Towed scrapers predate self-propelled models, but were replaced by them for a time on most construction projects. Recently, however, towed units have regained popularity and are again common. Self-propelled

Figure 4 Paddle wheel elevator scraper.

Figure 6 Towed scraper.

models remain popular for use in rocky soils. Towed models offer the convenience of having a separate tractor that can be used for other tasks as well. Some contractors use a large crawler tractor to pull two towed scraper units to move dirt economically (*Figure 6*). Several companies produce construction-grade towed scrapers in sizes from 7 to 24 cubic yards.

Towed scrapers must depend on the hydraulic system of the tow vehicle for the operation of the apron and other moveable components. Many are also equipped with brakes to reduce the stopping distance, instead of relying solely on the brakes of the tow vehicle. To allow two units to be towed by a single vehicle, towed scrapers are equipped with hitches and hydraulic connections on both the front and rear.

Towed scraper models have chassis components that are very similar to self-propelled models. However, the controls and instrumentation vary widely based on the tow vehicle being used. It is unlikely that the instrument icons or control labels in the tow vehicle are related specifically to scrapers. This is because the hydraulic and electric connections on the tow vehicle can be used for many different purposes and accessories. As a result, the tow vehicle operator must be inti-

Towed Scrapers

Towed scrapers were on the job many years before self-propelled models were conceived. The earliest models were pulled by horses, mules, and oxen. Movable components, like the apron, were added later and originally controlled by cables or ropes.

Robert LeTourneau is generally considered the father of modern earthmoving. Among other accomplishments, he developed the first towed scraper to be pulled by a tracked vehicle in 1923. Roughly 20 years later, there were nearly 30 manufacturers of scrapers in the United States alone.

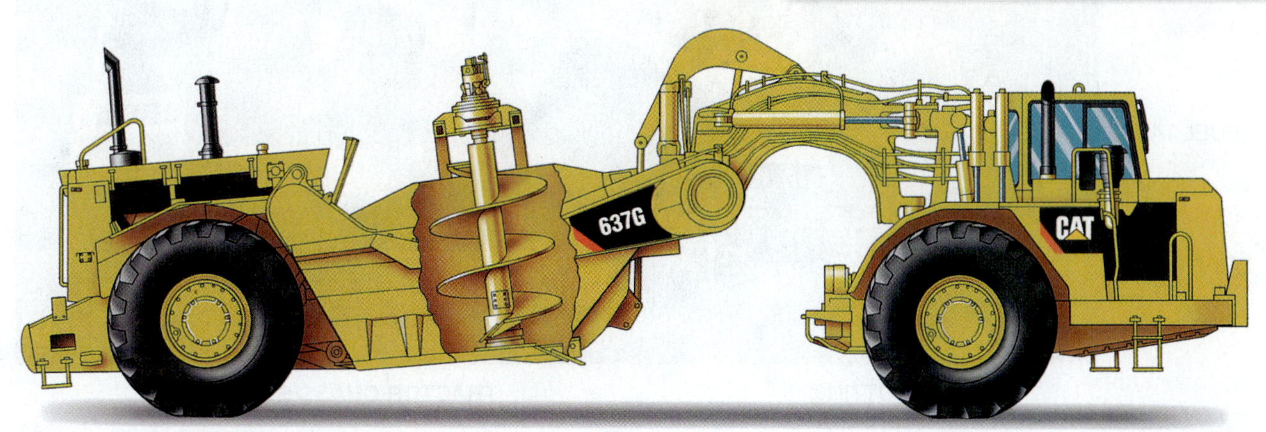

Figure 5 Auger-type elevating scraper.

mately familiar with the operation of the accessory hydraulic and electric controls in the cab to successfully control the towed scraper.

1.1.0 Scraper Chassis Components

This section reviews and describes the components of a single-engine scraper. Although there are differences among all scrapers, there are many similarities as well.

The basic chassis components of a single-engine scraper are shown in *Figure 7*. The chassis is the basic frame of the unit. All other components, either directly or indirectly, mount to the chassis. Since scrapers are articulated, there are two distinct chassis sections. The front section is the tractor chassis, and the rear section is the bowl chassis.

The tractor chassis carries the engine and the operator's compartment. The engine is mounted over, and slightly forward of, the two drive wheels. This helps to balance the tractor. The rear wheels of a single-engine scraper are dolly wheels only; they do not have any drive or steering capability. The hydraulic fluid reservoir is located on the tractor chassis, as are the hydraulic pumps. The pumps are driven by the engine. Note that this model does not have a pull bar on the tractor section. There is a small hitch, but this model is designed to be pushed by a dozer. It does not have the capability of pushing another scraper, or of being pulled by another unit.

The rear wheels of tandem scrapers are driven by the rear engine, and they sometimes do have steering capability as well. Another operator may be positioned at the rear of the scraper for this purpose. The two engines are coordinated via an electronic control module. Both engines have numerous sensors which monitor engine conditions and alert the operator to potential problems. The transmission is electronically controlled. Gears one and two are used for loading operations, and the higher gears are used for hauling.

The bowl chassis is pulled by the tractor. The draft arms, or draft frame, connect the bowl to the hitch of the tractor. A towed scraper is little more than a bowl chassis section with a hitch. The front of the bowl has a wide opening at ground level to gather the soil. It hangs from a frame supported on rubber-tired wheels. The forward edge of the bowl can be tilted vertically about 20 degrees. As the scraper moves forward, a thin layer of earth is cut and forced back into the bowl.

The cutting edge of the bowl is made of steel. It is usually assembled in sections, and is replaceable. It must endure a great deal of punishment, especially in rocky soils. The bowl is lowered by hydraulic cylinders until the cutting edge penetrates the soil to the desired level. For harder surfaces, teeth are fitted onto the edge to improve

Figure 7 Scraper chassis components.

digging. They are replaced when they become dull or worn. The cutting edge shown in *Figure 8* is known as a stinger. It is identified by the center piece that protrudes forward. When using a cutting edge that is even across the front, the sections can be moved around to accommodate greater wear at the outside edges. This increases the life of the edge assembly.

Elevator scrapers have paddles (*Figure 9*) or augers at the bowl opening. The cutting edge still separates a layer of soil and delivers it to the elevator system. The elevator then carries the soil to the upper back of the bowl. The elevator helps to evenly and fully load the bowl, and helps break up the soil.

When the bowl is full, the operator tilts it upward. This brings the cutting edge above the ground surface so that cutting stops. The apron is dropped down over the open end of the bowl and rests on the cutting edge. The apron closes the bowl and prevents spillage during hauling. The apron can also be lowered and used like a clamp for bulky objects like tree stumps or boulders.

The **ejector** is used to unload material from the bowl. When the scraper reaches the unloading area, the bowl is lowered until the cutting edge is one to six inches above the ground surface. The tilt alone is not sufficient to force the material to spill out of the bowl. The ejector is used to push the material out from the back as the apron is raised. The ejector plate, which is the width of the bowl, is forced forward slowly against the material in the bowl. The apron is lifted gradually as the ejector moves forward to provide a uniform discharge of the material. The design of the scraper does not allow it to simply dump the contents in a pile. The scraper must remain in motion while the bowl unloads, spreading the contents.

The front and back chassis sections are connected by the hitch. The hitch of many self-propelled units is designed with shock-absorbing capability. These are referred to as cushioned hitches. Since the tractor chassis has only two wheels, the two chassis sections rely on each other for support. The hitch is a relatively complex assembly. It must be able to move and turn in different directions, while still being strong enough to handle the immense loads and stress of scraping.

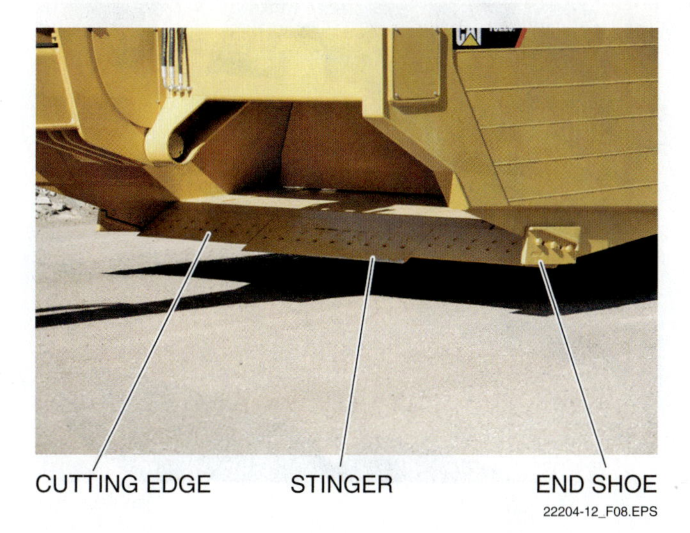

Figure 8 Cutting edge.

Figure 9 Elevator paddles.

In addition to these physical components, warning labels are found all around the scraper. Warning labels should always be given the proper attention when working on or around the scraper, and they should be properly maintained like all other functioning parts.

1.1.1 Engine and Drivetrain

Scrapers require a lot of horsepower. They are propelled by diesel engines. New diesel engines designed for use in heavy equipment must conform to US Environmental Protection Agency (EPA) guidelines for emissions. Scrapers have become cleaner and more efficient as a result.

The engine shown in *Figure 10* is used in single-engine scrapers, and is rated at nearly 500hp. The rear engine of a tandem scraper has less horsepower than the tractor engine. The most notable advancements in these engines in recent years are related to the electronic ignition and fuel controls.

Scraper transmissions and differentials must be rugged as well. Electronic controls have become a major part of transmissions and other driveline components. The synchronization of engine and

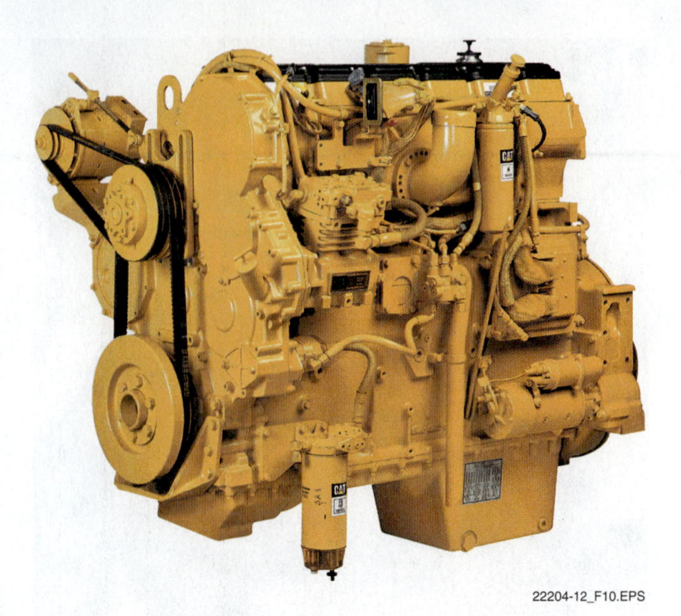

Figure 10 Scraper engine.

driveline controls has eliminated many of the possible operator errors that can be made. The controls are constantly evaluating performance and operating characteristics, then applying logic to determine the proper gear and drive mode. The first two gears are typically used for cutting and filling the scraper. This is when maximum torque is required. The remaining gears, usually four to eight more, are used for hauling the load to the dumpsite.

One variation in scrapers is the difference in the chosen tire type. The four tire patterns typically used are based on tread pattern and tire profile. Considerable thought must be given to the selection of tires for a particular job, as they are expensive and wear differently depending on the particular operation.

The four primary types of tires are as follows:

- *Traction tire* – A self-cleaning, directional-bar-type tire that provides maximum traction.
- *Hard-lug or rock-rib tire* – Used if the tires are subject to rough operation over rocky terrain.
- *Button-type tire* – Generally used on the non-driving or trailing wheels and is particularly desirable in sand.
- *Balloon tire* – Used where work is done in soft areas and where maximum flotation is more important than traction.

1.1.2 Hydraulic System

The hydraulic system of the scraper must provide the power to operate the bowl, tilting it up or down. It also powers the ejector and the apron. It may power the bowl's elevator system if it is equipped with one. The power steering system is hydraulic, and even cushioned hitch systems often rely on power from the hydraulic system.

Obviously, the hydraulic system is a crucial part of the scraper. They do not always rely on a single hydraulic pump. A number of scrapers use several separate hydraulic pumps. Each pump is driven from the main engine. Over 100 gallons of hydraulic fluid may be moving through the hydraulic lines each minute on a large scraper.

The hydraulic reservoir is conveniently located on the tractor chassis. A typical reservoir holds roughly 50 gallons. The larger units have reservoirs that hold 80 gallons or more.

The operator's controls position hydraulic valves that allow fluid to flow in one direction or another. Although there are a number of systems on the scraper that rely on hydraulic pressure, the operation of the individual components is relatively simple. Actuators (*Figure 11*) that move in or out are used to move the apron and the ejector. Actuators are also part of the steering system. For elevator systems, a continuous flow of fluid is used to turn a hydraulic motor. The hydraulic motor, through a gearbox, is used to turn the auger or paddle wheel. The gearbox is needed to reduce the hydraulic motor shaft speed to the proper speed for the driven component.

1.2.0 Controls

The scraper should be operated only from the operator's cab. Indeed, there is little the scraper can do if it is not moving. Although the location of the individual controls varies with manufacturer and model, all of the primary controls are located within easy reach of the operator's seat.

Some typical controls found in the cab of a scraper are shown in *Figure 12*. Always check the

Figure 11 Hydraulic actuator.

Figure 12 Scraper cab.

operator manual for the model in use to become familiar with the controls. There are many controls relevant to the operation of a scraper. It is critical that operators know their location when a quick response or action is required. Remember that tandem scrapers will have additional controls related to the rear engine as well.

Simply driving a scraper is similar to that of other large trucks. The scraper, apron, and ejector are controlled with either a joystick or levers (*Figure 13*). This module describes joystick controls. Review the operator manual before operating a machine to understand how each of the controls functions. Joystick buttons have different functions in different machines.

1.2.1 Disconnect Switches

Some loader models have shutoff switches that must be moved to the On position before the ma-

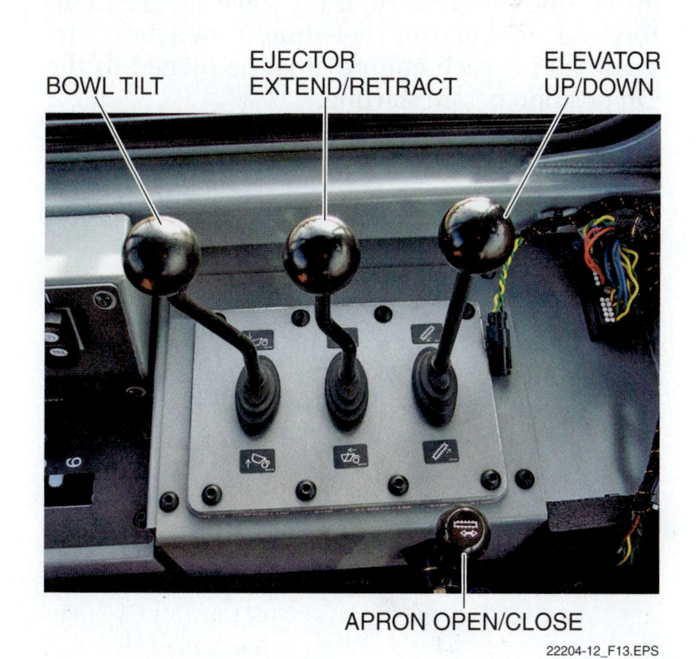

Figure 13 Bowl control levers.

chine can be operated. These switches may be outside of the cab near the engine compartment. Before entering the cab, the switches should be located and switched On.

Many machines have a battery or electrical system disconnect switch (*Figure 14*). This switch is also known as a master switch, since nothing works without it being on. When the master switch is turned off, the entire electrical system is disabled. This switch should be turned Off when the machine is left overnight or longer to prevent a short circuit or battery drain. Before entering the cab, check that the switch is in the On position. Tandem scrapers with two engines may have two switches, one on the scraper engine and one on the tractor engine.

> **CAUTION**
>
> Never switch the battery disconnect switch to the Off position while the engine is running. This could seriously damage the electrical system.

Some older machines have a fuel-shutoff switch that physically prevents fuel from flowing from the fuel tank. This prevents unwanted fuel flow during idle periods or when transporting the machine, significantly reducing the potential for fuel leaks. The switch controls an electric fuel valve. If equipped, this switch must also be moved to the On position. Note that the valve does not operate unless the electrical system disconnect switch is on.

Remember that tandem scrapers have two engines. Both engines typically have separate battery disconnect and fuel-shutoff switches. The switches for each engine must be turned to the On position before starting.

22204-12_F14.EPS

Figure 14 Electrical system disconnect switch.

1.2.2 Seat and Steering Wheel Adjustment

Upon entering the cab, the operator should first adjust the seat and steering wheel. While seat and steering wheel adjustments are not directly involved in scraper operations, correct positioning of these items can affect safe operation as well as comfort. The operator should adjust the seat and steering wheel position and then fasten the seat belt before operating the scraper.

Most seats can be moved up or down and forward or backward. Some provide an adjustable shock absorber function. They may even have their own air system for complete control over shock absorption and seat height. The seat should be adjusted so that the operator's legs are almost straight when the clutch or brake pedals are fully depressed and the operator's back is flat against the back of the seat. The knees should remain slightly bent.

After the seat is correctly positioned, adjust the steering wheel. Most scrapers have a lever on the steering column to adjust the wheel. On most models the lever returns to the locked position when released. Move the lever up and tilt the steering column to position it correctly. The steering wheel can also be moved higher or lower through a telescoping function. Push the lever down to move the wheel up or down. Because seat and steering wheel adjustment devices vary widely for various makes and models, refer to the operator manual for specific instructions.

1.2.3 Engine Start Switch

The ignition switch functions can vary widely between makes and models of scrapers. Some only activate the starter and ignition system. Others may activate fuel pumps, fuel valves, and starting aids. In some cases, the starter and starting aids are engaged by other manual controls. The engine start switch for the tractor engine is located on the console instrument panel.

Tandem scraper engines are started separately. Typically, the tractor engine is started first and then the scraper engine is started.

Common engine start switches have the following three positions:

- *Off* – Turning the key to the Off position stops the engine. It also disconnects power to electrical circuits in the cab itself. However, several lights often remain active when the key is at the Off position, including the hazard warning light, the interior light, and the parking lights.
- *On* – Turning the key to the On position activates all of the electrical circuits except the starter motor circuit. When the key is first

turned to the On position, it may initiate an instrument panel and indicator bulb check. If so, all instrument and indicator lamps should illuminate momentarily. Note any that are not working and have the lights replaced.

- *Start* – The key is turned to the Start position to activate the starter, which starts the engine. This position is spring-loaded to return to the On position when the key is released. If the engine fails to start, the key must be returned to the Off position before the starter can be activated again. To reduce battery load during starting, the ignition switch of some scrapers may be configured to shut off power to accessories and lights when the key is in the start position.

CAUTION

Activate the starter for a maximum of 30 seconds. If the machine does not start, turn the key to the Off position and wait two minutes before activating the starter again.

Start the scraper engine after starting the tractor engine. There should be a separate switch on the instrument panel to start the scraper engine. The key switch for the tractor engine must be in the On position before the scraper engine can be started.

1.2.4 Vehicle Movement Controls

A steering wheel is used in combination with foot pedals to control vehicle movement. The throttle foot pedal is used to control the engine speed. In tandem scrapers, there are two throttle pedals. One controls the tractor engine speed, while the other controls the scraper engine speed. Depress the tractor throttle to increase the engine rotations per minute (rpm) and travel speed, release it to decrease the rpm and speed. Both throttles can be operated with one foot simultaneously. Many scrapers can automatically synchronize the two engines, so the operator does not need to work both pedals. The steering wheel is used to turn the vehicle. Due to the immense size and weight of the vehicle, hydraulically assisted steering is a standard feature.

Another pedal is for the service brakes. The brakes are typically powered by air from an onboard air-compressor system. As an added safety feature, the parking brake may be applied by spring pressure, and released by air pressure. The primary brake system works in the opposite way—air-applied and spring-released. With this arrangement, the sudden loss of air pressure will not result in a complete loss of braking ability. In fact, on many units, if the air pressure drops below a preset value, the parking brakes automatically engage and stop the scraper.

The transmission is controlled by a gear-selection lever on the right side of the operator's seat. The gear selector should be set to N or neutral when parking or starting the engine. The transmission is not designed to hold the scraper in a parked position. The parking brake is used to hold the scraper's position when the scraper is parked. To further prevent movement, the pan is lowered to the ground when the vehicle is parked. In addition, the engine will not start unless the gear-selection lever is set to neutral.

There are a number of forward speeds. On some models, the number of available forward gears is set by the service personnel. To change gears, squeeze the trigger to unlock the shifter and move the lever to the desired gear. Release the trigger to mechanically lock the control into the current gear. Move the lever to R to move the machine in reverse.

Advanced scrapers have the capability of using a torque converter for increased torque during scraping and loading. The transmission then shifts to a direct-drive mode for improved efficiency for hauling in the higher gears. This automatic feature can generally be overridden when increased torque is required in the higher gears and using the torque converter is preferred.

On some machines, the transmission automatically shifts between second gear and any higher gear that is selected. Manual shifting may only be required between first and second gear. First and second are the gears used for scraping and loading. Once the engine is running and the transmission is in neutral, the operator may select the highest gear appropriate for the site and project. This top gear will be the highest gear chosen by the system logic. All the lower gears will remain available, and may be automatically chosen. Be sure to review the transmission controls before operating the machine. In many of today's scrapers, a great deal of automation and logic has been incorporated into engine speed control and gear selection.

1.2.5 Scraper Controls

Scraper functions are controlled with a joystick (*Figure 15*) or a series of levers. This section describes joystick controls. The operation and control scheme of the joystick varies from one unit to another. The control scheme is often different within the same model family. Therefore, it is critical that operators familiarize themselves with the specific scraper being driven. The layout shown in *Figure 15* is only a single example.

Unlike some other heavy equipment, the joystick of the scraper does not control any vehicle movement. As a general rule, it controls only the moving components of the bowl. It may also control the function of a cushioned hitch or the tow bail. The joystick combines the functions of several levers and switches into one control unit. This makes it much easier to operate several functions at the same time with one hand.

The joystick is located on the right side of the operator's seat. The function of each control has been identified in *Figure 15*. However, it cannot be stressed enough that this represents only one of many possible configurations.

Moving the joystick forward extends the ejector blade in the bowl, pushing out the load. Pulling it back retracts the ejector to the back of the bowl. Moving the joystick to the left tilts the nose of the bowl down, positioning it for scraping and loading. Moving the joystick to the right tilts the nose of the bowl up for travel. The speed of movement is increased as the joystick is moved farther away from the central position. Some joysticks also have some automatic features. For example, when the joystick is pulled all the way back to the stop, it will remain there until the ejector has fully retracted, then spring back to the neutral position.

Most joysticks have additional buttons or slide switches on the top. Most are designed to be operated with the thumb. The tow bail control raises or lowers the tow bail onto the scraper ahead. This allows scrapers to connect and disconnect easily and quickly, without risk to a worker being between

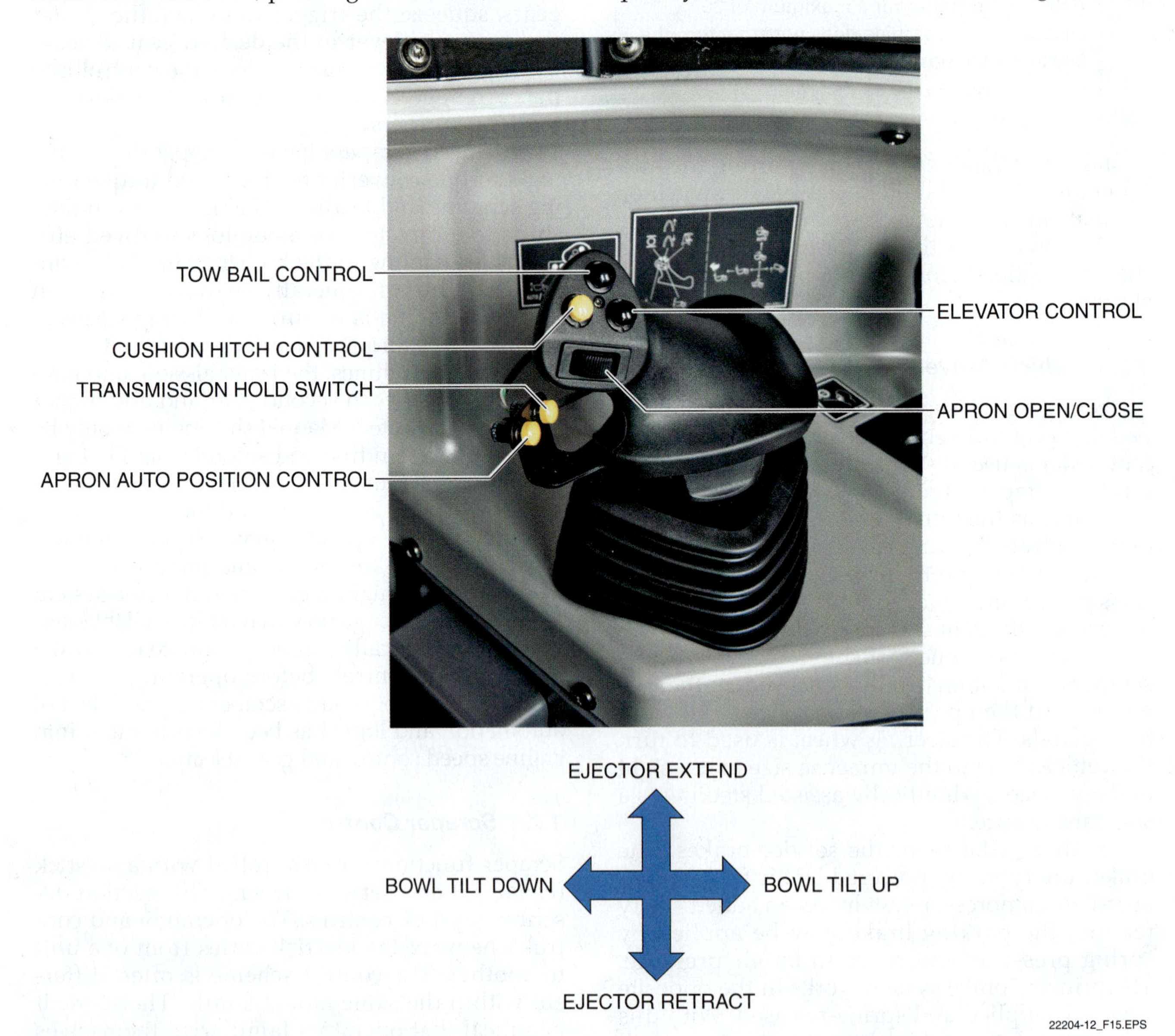

Figure 15 Joystick control.

the vehicles. An elevator control button starts and stops the elevator, if the scraper has one.

In the example in *Figure 15*, the slide switch raises and lowers the apron. The apron auto control button allows it to float as necessary, maintaining its position relative to the bowl opening. This function is valuable when hauling a load. It allows the bowl position to be changed without concern for the position of the apron.

The transmission hold switch prevents the transmission from shifting. Push the button in to engage this feature. Push the button again to disengage this feature. The transmission hold allows the operator to maintain the torque converter drive for increased **rimpull** during scraping. If the scraper is already in third gear or higher, and is in its direct-drive mode, the present gear is maintained as long as the hold is engaged. Note that, on tandem scrapers, the tractor and scraper transmissions both respond at the same time.

The cushion hitch control locks the cushion hitch for improved control of the cutting edge during both scraping and spreading. Remember that the cushion hitch functions much like a shock absorber. Although it does improve comfort, it also allows the bowl to move up and down over bumpy terrain. To ensure that the cutting edge stays at a consistent distance from the ground, the action of the cushion hitch can be disabled. Except when scraping and spreading, the cushion system should generally be enabled.

1.2.6 Differential Lock

The differential lock is engaged with a foot-operated button on the floor of the cab next to the brake and throttles (*Figure 16*). To engage the differential lock, depress and hold the button. Release the button to disengage it. On some models there are two buttons, one on either side of the pedals. Both buttons have the same function, but are used from different seat positions.

Figure 16 Differential lock switch.

The differential is the gearbox that accepts power from the engine driveline and transfers it to the two drive axles. It is not unusual to have one drive wheel in slippery material while the other has reasonable traction. Normally, the wheel without traction tends to rob power from the drive wheel that still has traction. This causes the traction-free wheel to spin rapidly, while the other wheel is not receiving enough power to pull the vehicle. The differential lock overrides the normal tendency of the front axle differential. An equal amount of torque is transmitted to both wheels, even though one wheel may not have traction. This helps maintain improved traction when ground conditions are soft or slippery, preventing wheel spin.

The differential lock should be engaged while the wheels are not spinning. If the wheels start to spin, release pressure on the throttle until the wheels stop spinning. Engage the differential lock and then increase pressure on the accelerator. Once clear of the problem area, release pressure on the accelerator, and release the switch to disengage the differential lock.

CAUTION

Do not engage the differential lock at high speeds or while the wheels are spinning. This could damage the differential. The scraper should not be turned while the differential lock is engaged – release the lock before making a turn.

1.2.7 Additional Controls

There are several other important controls in addition to those described above. Some of these features offer added control of the machine in specialized situations. These controls are not available on all models. The controls and their functions are as follows:

- *Fuel ratio override* – Changes the mixture of fuel to the injectors.
- *Governor override button* – Allows the operator to exceed the **governor** setting of the engine speed.
- *Heated-start switch* – Control panel switch used to activate glow plug and start engine. Newer machines have an automatic feature that operates glow plugs or admits an ether starting fluid to the injectors.
- *Individual wheel brake lever* – Controls braking action to individual wheels.

- *Parking brake control* – Secures the scraper when stopped or parked (*Figure 17*).
- *Retarder lever* – The hydraulic retarder acts like an internal brake on the driveline. It reduces the need to apply the service brakes. The retarder takes a few seconds to engage and must be activated three to four seconds before it is needed. Using the retarder decreases wear on the service brakes and enhances machine control.
- *Stop button* – A kill switch for the engine.
- *Engine speed lock* – Operates somewhat like a cruise control, but sets and holds an engine speed rather than a ground speed.

1.3.0 Instrumentation

An operator must pay attention to the instrument panel, as well as to guiding the scraper. The instrument panel includes the gauges that indicate engine and transmission temperature. There are several warning lights and indicators that must also be monitored. An operator can seriously damage the equipment if the instrument panel is not closely monitored. Like good drivers, scraper operators do well to develop the habit of quickly scanning the control panel regularly.

The instrument panel varies on different makes and models of scrapers. Generally the panels include the instruments and indicators covered in the following sections. Advanced scraper instrument panels may incorporate all the gauges into a single color display (*Figure 18*). Newer scrapers may also have built-in cameras that allow the operator to clearly see behind the scraper or into the mouth of the bowl (*Figure 19*). A more traditional instrument panel is shown in *Figure 20*. Most of these instruments and indicators are similar to those in other machines.

22204-12_F17.EPS

Figure 17 Parking brake control.

22204-12_F18.EPS

Figure 18 Digital instrument panel.

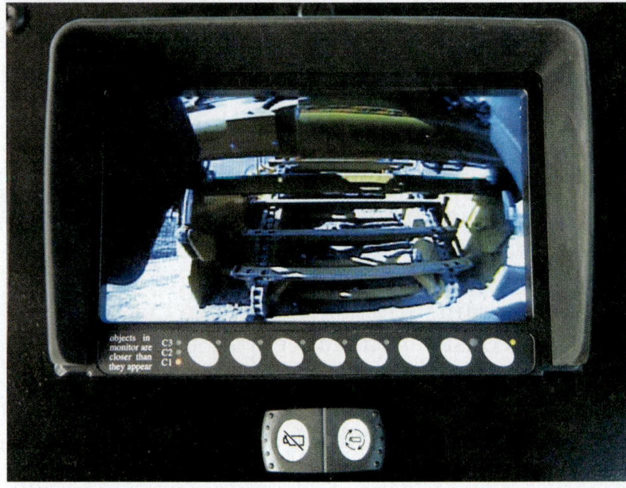

22204-12_F19.EPS

Figure 19 Camera view of elevator operation.

The quad gauges typically include the engine coolant temperature, transmission/torque converter oil temperature, fuel level, and air supply pressure. On dual-engine scrapers, the display can be switched between the tractor and scraper engine monitoring systems. The function of these gauges is described in the sections that follow. Other types of scrapers may have different gauges. Refer to the operator manual for the specific gauges on the machine in use.

The instrument panel includes indicator and warning lights, which are also described in the sections that follow. Indicator lights show that various machine functions are enabled or disabled. Warning lights indicate that the machine's systems are not functioning properly. Frequently used switches and indicator lights are located on

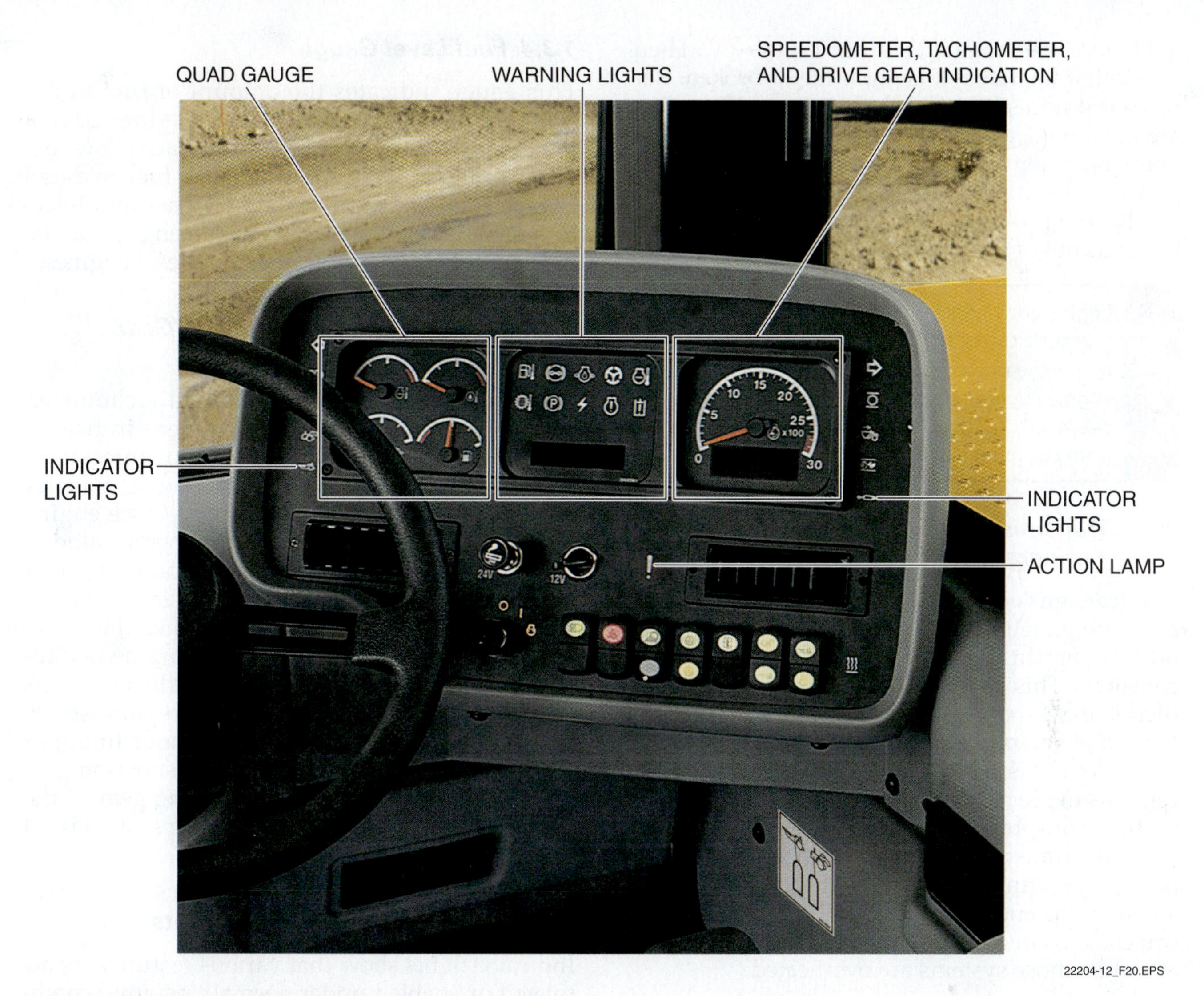

Figure 20 Traditional instrument panel.

the instrument panel. An operating hour meter for the engine may also be on the main panel. Other lights and switches may be located on an overhead console.

On tandem scrapers with advanced displays, if one engine is experiencing trouble, the display automatically shows the data for that engine. When the engine coolant or torque converter oil temperature exceeds the maximum operating temperature, warning lights come on and the action lamp flashes. The action lamp, if equipped, is a larger, brighter warning light that lights up in conjunction with any other warning light. This helps to ensure that the operator takes notice of the problem.

1.3.1 Engine Coolant Temperature Gauge

The engine coolant temperature gauge indicates the temperature of the coolant flowing through the cooling system. Refer to the operator manual to determine the correct operating range for normal scraper operations. The acceptable range is usually shown on the gauge as a green or white band.

Most gauges also have a section that is red. If the needle is in the red zone, the coolant temperature is excessive. Stop the machine immediately and investigate the problem. Most scrapers also activate warning lights if the engine overheats. The lights are more likely to gain the attention of a busy operator than a gauge alone.

CAUTION

Operating a scraper when the coolant temperature gauge(s) is in the red zone may severely damage the engine. Stop operation, determine the cause of problem, and resolve it before continuing.

Get out of the machine and follow the proper guidelines for investigating and solving the problem. There are several checks that the operator can perform. First, check the engine coolant level.

Add more water or antifreeze if it is too low. Then check that the fan belt is not loose or broken. Replace it if necessary. Check that the radiator fins are not fouled, and clean them if necessary. These items represent the three primary causes of engine overheating. If initial troubleshooting fails to resolve the problem, stop operation and take the machine out of service.

WARNING!

Engine coolant is extremely hot and is under pressure. Check the operator manual and follow the procedure to safely check and fill engine coolant. Removing the radiator cap of an overheated engine may result in severe burns from escaping steam.

1.3.2 Transmission/Torque Converter Oil Temperature Gauge

The transmission and torque converter oil temperature gauge indicates the temperature of the oil flowing through the transmission or torque converter. This gauge also has a red zone that indicates excessive temperatures. If the gauge is in the red zone, immediately reduce the load on the machine. This should reduce the temperature. If it remains in the red zone, stop the machine and investigate the problem. When the weather is colder, allow the transmission oil to warm up sufficiently before operating the machine. Placing too much stress on the engine and drivetrain when it is cold can cause as much damage as continuing to operate when those systems are overheated.

1.3.3 System Air Pressure Gauge

The air supply pressure gauge indicates the air pressure in the air tanks. On tandem scrapers, there is likely to be an air compressor located on each engine. When the monitoring system is in tractor mode, the gauge shows the air pressure in the tractor's air tanks. When the monitoring system is in scraper mode, it shows the air pressure in the scraper's air tanks.

The brakes are actuated with air pressure. Thus, it is critical that the air pressure remain within the normal operating range. A pushbutton releases a spring-applied mechanism that engages the parking brake. On many models, the brakes lock on if the air pressure decreases below a certain level. This forces the scraper to stop before the air supply is insufficient to do so. However, the operator must monitor the air pressure in the system and take action if it drops below safe operating levels.

1.3.4 Fuel Level Gauge

This gauge indicates the amount of fuel in the scraper's fuel tank. Most diesel engines have a low-fuel warning zone. Some also have a low-fuel warning light. Avoid running out of fuel on diesel engine scrapers because the fuel lines and injectors must be bled of air before the engine can be restarted. This is typical of many diesel engines.

1.3.5 Speedometer/Tachometer/Gear Indicator

Most scrapers provide a combined tachometer, speedometer, and transmission gear indicator. They may also be separated, but all three indicators are commonly provided. The tachometer indicates the engine speed in rpm. Since engine speed is generally the most important value to monitor, the tachometer usually dominates this part of the display. The tachometer can be set to display the engine speed of either the front-tractor engine or the rear-scraper engine on tandem units. The speedometer shows the machine's ground speed. Digital speedometers can usually be set to display either miles per hour (mph) or kilometers per hour (kph). The transmission gear indicator shows the actual operating gear of the tractor engine unless the monitoring system is set in scraper mode.

1.3.6 Indicator and Warning Lights

Indicator lights show that various features are activated or enabled under normal operating conditions. They do not indicate that there is something wrong with the machine. This is what separates indicator lights from warning lights. Indicator lights are often a part of the control button or switch.

Typical indicator lights include the following:

- Turn signals
- Headlights/high beams
- Hazard lights
- Windshield wipers
- Apron float
- Tractor or scraper engine display mode
- Transmission hold
- Ejector return
- Throttle lock

Warning lights are usually grouped together. When one is lit, the related system is not functioning properly. For example, if the oil pressure light is lit, the oil pressure on the selected engine is too low. The operator can set the digital display to show data on several different machine systems as needed.

Typical alert indicators or warning lights include the following:

- Low fuel level, clogged fuel filter, or water in fuel
- Air pressure
- Engine oil pressure
- Secondary steering
- Engine coolant temperature
- Hydraulic system failure
- Power train system
- Charging system
- Parking brake On
- Torque converter oil temperature

CAUTION

The parking brake warning lights up when the parking brake is engaged. It should go out when the parking brake is disengaged. If the light comes on when the parking brake is not engaged, shut the machine down immediately and investigate the problem. Any issues related to braking must be given immediate attention.

There are often increasing levels of warning alarms on a machine. At the first level, the alert indicators light up. The operator must take action in the near future to correct the problem. The problem or the related system may not represent any imminent danger to the operator or the equipment. At the second level, the alert indicator and the action light may both come on. The operator must take immediate action and change how the machine is being operated to correct the problem and avoid machine damage. At the third level, the alert indicator and the action light come on, and an audible alarm sounds. The operator must immediately shut down the machine to avoid machine damage or operator injury.

1.0.0 Section Review

1. The component that is located at the front of the bowl and can be raised and lowered is called the _____.
   a. lug
   b. apron
   c. ejector
   d. end shoe

2. The differential lock is used _____.
   a. to help prevent wheel spin
   b. lock the scraper into one gear
   c. to help the scraper turn more sharply
   d. set the height of the bowl above the ground

3. New scrapers may be equipped with cameras that monitor the back of the scraper and the _____.
   a. mouth of the bowl
   b. engine area
   c. top of the bowl
   d. hydraulic pumps

SECTION TWO

2.0.0 SCRAPER INSPECTION AND MAINTENANCE

Objective 2

Describe the prestart inspection requirements for a scraper.

a. Describe prestart inspection procedures.
b. Describe preventive maintenance requirements.

Performance Task 1

Complete a proper prestart inspection and preventive maintenance on a scraper.

Preventive maintenance is an organized effort to regularly perform periodic lubrication and other service work. The goal is to reduce instances of poor performance and breakdowns at critical times, and to extend the life of the equipment. Performing preventive maintenance on the scraper allows it to operate efficiently and safely. Maintenance helps avoid the possibility of costly failures in the midst of a project. Falling behind on the project schedule can even be more costly than the equipment repair.

Preventive maintenance of equipment is essential and relatively easy with the right tools and equipment in hand. The leading cause of premature equipment failure is putting things off. Preventive maintenance should become a habit, and be performed on a regular basis.

CAUTION

Scraper service is normally based on the number of operating hours. A service schedule is contained in the operator manual. Failure to perform scheduled maintenance could result in damage to the machine.

2.1.0 Prestart Inspection Checks

The first thing to be done before work is to conduct a prestart inspection. This should be done before starting the engine. The inspection is intended to identify any potential problems that could cause a breakdown and to determine whether or not the machine should be operated. It is wise to inspect the equipment both before and after operation. Follow the manufacturer's instructions for a prestart inspection on the specific scraper model in use.

Before beginning an inspection, the scraper must be left in a safe condition. Prepare it for an inspection with these basic steps:

- Park the scraper on level ground.
- Set the bowl, the apron, and the ejector to an appropriate position for inspection per the manufacturer's guidance. For example, the manufacturer may instruct the operator to lower the bowl to the ground, lower the apron, and move the ejector all the way to the back. Unless otherwise directed, leave these three components in the position listed here. Preparations for some repair activities may also require that one or more of these components be left in a specific position.
- Engage the parking brake.
- Shut down the engine.

Unless some aspect of the inspection requires power, also switch the electrical system and fuel disconnects to Off. Some companies may also require that a Do Not Operate tag be placed in the cab or another prominent location any time that work is being conducted on the unit.

The general walk-around inspection of the equipment includes looking for the following:

- Leaks (oil, fuel, hydraulic, or coolant)
- Worn or cut hoses
- Damaged tires or low air pressure.
- Loose or missing bolts
- Trash or dirt buildup around any component
- Broken or missing parts
- Damage to gauges or indicators
- Circuit breakers tripped
- Chafed or damaged fluid lines of any kind
- Dull, worn, or damaged blades
- General wear and tear

WARNING!

Do not check for or locate hydraulic leaks with bare hands. Use cardboard or a similar material passed over an area. Scraper hydraulic systems operating at high pressures can cause severe injuries to unprotected flesh.

Some manufacturers require that daily maintenance be performed on specific parts. These are generally parts that are exposed to dirt or dust and may malfunction if not cleaned or serviced. For example, the service manual may recommend lubricating specific bearings every 10 hours of operation, or always cleaning the air pre-filter before starting the engine.

To reduce the possibility of a breakdown or malfunction of a component, pay particular attention to the following areas during a prestart inspection. The operator manual for the specific scraper in use provides the locations of important maintenance aids, such as sight glasses and dipsticks.

- *Air cleaner* – See *Figure 21.* If the machine is equipped with an air cleaner service indicator, check the indicator. If the indicator shows red, the air filter and intake chamber need to be cleaned. If the machine does not have a service indicator attachment, then the air cleaner cover must be removed, the filter visually inspected, and any dirt at the bottom of the bowl cleaned out. Note that there may be both a pre-filter and a main filter. Check the operator manual for the specific procedures to follow. The prefilter element may be washable. The air cleaner needs more consistent attention when working in dry or dusty conditions.
- *Air reservoirs* – Check the reservoirs for signs of damage. Most reservoirs have a drain valve to drain off any water that condensed out of the compressed air supply. Allowing water to build up in the reservoir eventually causes the entire air system to flood.
- *Battery* – Check the battery cable connections. Make sure that the terminals are free of corrosion and the clamps are tight. Distilled water is required in most engine batteries. Also ensure that any battery supports or hold-down straps are secure.

22204-12_F21.EPS

Figure 21 Air cleaner housing.

WARNING!

Batteries can produce flammable or explosive fumes. The electrolyte is acid-based and can also cause physical injury. When working near a battery, extinguish all open flames and smoking materials, and wear safety glasses. Failure to take the proper precautions can result in serious personal injury or death.

- *Circuit breakers* – Circuit breakers for various electrical circuits may be located behind a panel outside of the cab, near the engine compartment (*Figure 22*). Check to ensure they are properly set. A tripped breaker typically protrudes farther and shows a band of different color.
- *Cutting edge* – The condition and type of cutting edge used significantly affects the performance of the scraper. A blunt and damaged cutting edge requires a lot more horsepower and fuel consumption to move through the soil. The cutting edge should certainly be replaced before it has deteriorated so far back that the base of the bowl it mounts to can be damaged.
- *Engine cooling system* – Check the coolant level. Make sure it is at the level specified in the operator manual. In many cases, it is unnecessary to remove the radiator cap. A surge tank is often provided; it should be used to determine the coolant level and add coolant when needed. Also check the radiator and other heat transfer surfaces to ensure they are clean (*Figure 23*).
- *Drive belts* – Check the condition and adjustment of drive belts on the engine.

22204-12_F22.EPS

Figure 22 Circuit breakers.

22204-12_F23.EPS

Figure 23 Radiator, condenser coil, and fluid coolers.

- *Engine oil* – Check the engine oil level using the dipstick to make sure it is within the safe operating range. Most dipsticks have markings indicating the proper level for the engine when it is stopped. The level reads differently in an operating engine.
- *Cab environmental controls* – The scraper is likely equipped with lights and windshield wipers. Make sure they work. Also check the wiper blades to ensure their condition is satisfactory.
- *Fuel level* – Check the fuel level in the fuel tank(s). Do this manually with the aid of a fuel dipstick or similar device. Do not rely on the fuel gauge during each walk-around. Check the fuel pump sediment bowl if one is available. A drain valve at the base allows the operator to drain any collected sediment or water.
- *Grease* – There are numerous points on the scraper that require greasing. Although the interval may vary considerably, some grease points may require attention daily. Check the equipment carefully for the location and interval of all grease points.
- *Hydraulic fluid level* – Check the hydraulic fluid level in the reservoir (*Figure 24*).
- *Hydraulic lines and couplings* – Check the lines that run between the tractor and the scraper unit (*Figure 25*). They are subjected to continuous flexing and can easily be damaged.
- *Transmission fluid* – Check the level of the transmission fluid with a dipstick to make sure it is in the operating range. Look on the ground for signs of leakage. Note that most transmissions require the engine to be running and at a normal operating temperature while the fluid level is checked. In this case, the operator may be required to check the fluid at the end of the warm-up period, but before beginning work.
- *Tires* – Check tires for cuts, excessive wear, or improper pressure. Uneven pressure in tires

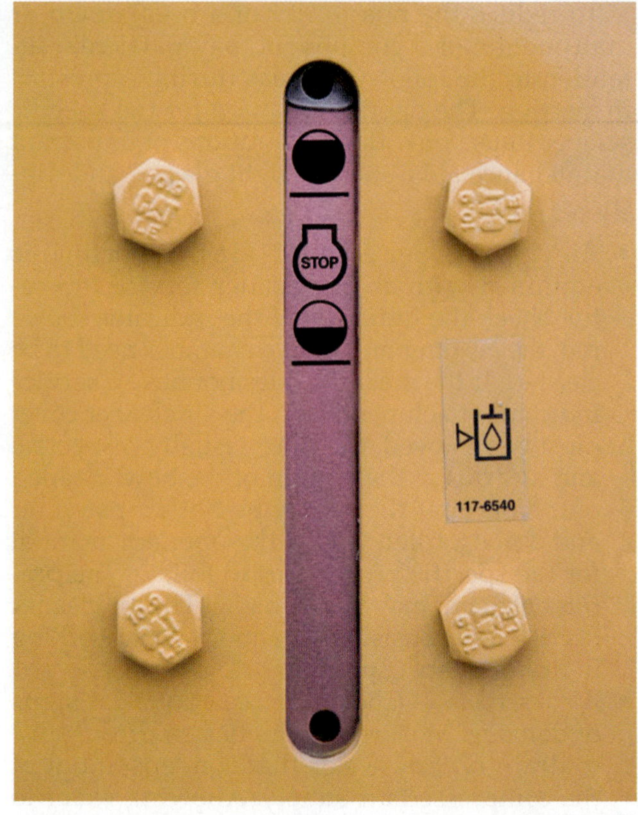

22204-12_F24.EPS

Figure 24 Hydraulic fluid sight glass.

can cause poor scraper performance and handling. The proper tire pressure can be found in the manual. Some manufacturers recommend dry nitrogen for tire inflation. Also check the wheels for any missing lug nuts or flange damage.

The pressure of a fully-charged nitrogen cylinder is approximately 2,500 psig. Improper use and handling of the nitrogen regulator or tank can result in serious injury or death.

- *Wires, insulation, and connections* – Check wires and their terminations in the engine compartment and in the cab.

If any problems with the scraper are found that cannot be taken care of immediately, inform the site supervisor or the equipment mechanic. Have the problem corrected before beginning operation.

2.2.0 Maintenance and Service

Maintenance time intervals for most scrapers are established by the Society of Automotive Engineers (SAE) and adopted by most equipment manufacturers. Instructions for preventive

22204-12_F25.EPS

Figure 25 Hydraulic lines.

maintenance are usually in the operator manual of each piece of equipment. Most maintenance is based on a number of operating hours. Common service intervals range from 10 hours (daily) to 12,000 hours. It is generally accepted that a 2,000-hour service interval represents one normal year of service. Therefore, an interval of 1,000 hours is also considered a semi-annual requirement. It is essential that operators and equipment maintenance personnel refer to the specific service manual for the scraper in use. Any special requirements for that particular piece of equipment are highlighted in the manual.

Normally, the service chart recommends specific intervals based on hours of run time. However, operators may find conditions during an inspection that must be addressed, regardless of the service interval. For example, hydraulic fluids should be changed whenever they become dirty or break down due to overheating. Continuous and hard operation of the hydraulic system can heat the hydraulic fluid to the boiling point and cause it to break down rapidly. Operating in extreme environments may also cause filters to become fouled very quickly.

Some scrapers have computer diagnostics. When the machine needs service, a code flashes on the instrument panel. These codes are cross-referenced in the manual for the scraper. On some machines, a computer can be plugged into the system for more advanced diagnostics and resetting of any alerts (*Figure 26*). The diagnostic system may keep track of long-interval activities and provide a reminder when the service is due.

22204-12_F26.EPS

Figure 26 Retrieving and analyzing diagnostic codes.

2.2.1 Fluid Sampling

Although many long-interval maintenance activities may be done in the service shop, fluid samples (*Figure 27*) may need to be drawn in the field by the operator. The samples are then sent to a lab for an analysis of the wear metals and other fluid characteristics. Testing fluid samples can identify contamination and even the potential sources for it.

Although engine oils and hydraulic fluids are the focus of most sampling and testing, fuel testing is increasing along with the use of new fuels. The sulfur content of diesel fuel has dropped substantially due to new EPA requirements. In addition, biodiesel fuel has entered the market. These fuel changes may result in a loss of lubricating performance that previously came from the fuel itself. Fuel sampling and records help managers of heavy equipment fleets maintain a proactive position on the condition of very expensive and mission-critical heavy equipment such as scrapers.

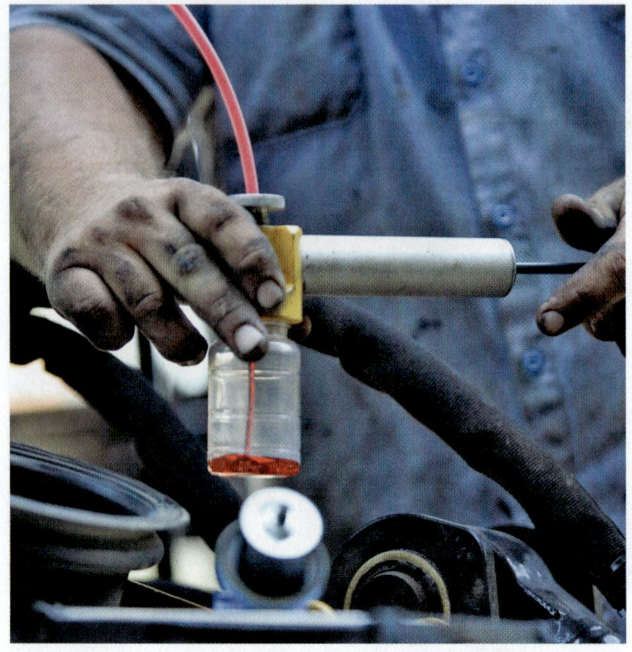

22204-12_F27.EPS

Figure 27 Obtaining a hydraulic fluid sample.

Sample kits are often provided by the lab. A handheld pump (*Figure 28*) is a very convenient method. As a general rule, taking the fluid from a drain valve or similar location results in a sample that doesn't represent the batch. Any debris allowed into the sample can produce misleading results. Sampling labs can provide excellent guidance on collecting samples to avoid contamination and inconsistent results. Proper sampling techniques are very important.

In most cases, taking a fluid sample requires that the system be started and brought to a normal operating temperature. Once the sample is collected and sealed in the appropriate container, a label is attached to identify the equipment it was drawn from, the date, and any extraordinary conditions.

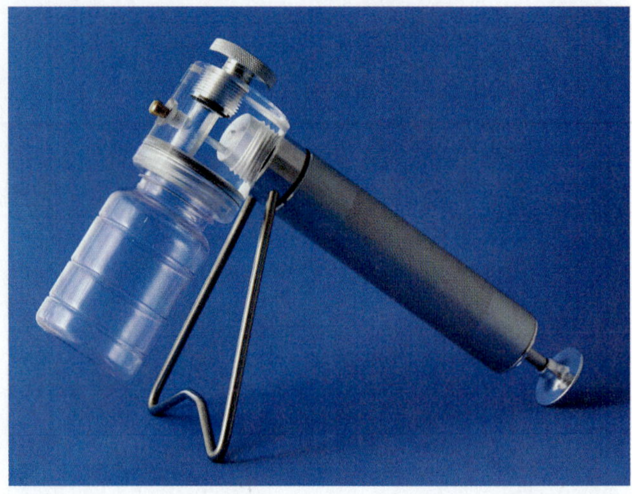

22204-12_F28.EPS

Figure 28 Handheld fluid sample pump.

2.2.2 Preventive Maintenance Records

Accurate, up-to-date maintenance records are essential for knowing the history of the equipment. Each machine should have a record that describes inspection and service that is to be done, along with the corresponding time intervals. Typically, an operator manual and some type of inspection sheet are kept with the equipment at all times. As maintenance activities are completed, operators make the necessary dated entries.

2.0.0 Section Review

1. The pressure of a fully charged nitrogen cylinder is approximately _____.
   a. 1,500 psig
   b. 2,500 psig
   c. 3,500 psig
   d. 5,000 psig

2. Changes in diesel fuels as required by the EPA had a negative effect on this fuel's _____.
   a. viscosity
   b. flashpoint
   c. lubricating qualities
   d. ability to repel water

SECTION THREE

3.0.0 SCRAPER OPERATION

Objective 3

Describe the startup, shutdown, and operating procedures for a scraper.

a. State scraper-related safety guidelines.
b. Describe startup, warm-up, and shutdown procedures.
c. Describe basic maneuvers and operations.
d. Describe related work activities.

Performance Tasks 2 and 3

Perform proper startup, warm-up, and shutdown procedures on a scraper.

Execute basic maneuvers with a scraper, including forward/backward movement, turning, loading, and unloading.

Trade Terms

End shoes: Flat pieces of steel on each side of the scraper bowl that help keep the material confined to the front of the cutting edge until the paddles can scoop it up; sometimes referred to as slobber bits.

Grade checker: Person who checks elevations and grades, and then gives signals to the equipment operators.

Lug down: A slowdown in engine speed (rpm) due to increasing the load. This usually occurs when heavy machinery is crossing soft or unstable soil or is pushing or pulling very hard. When a diesel engine lugs down, there is generally an increase in visible smoke and an audible change in the sound of the engine.

Pay material: Soil materials valuable enough to stockpile for future use, on or off site. Good topsoil is an example of a pay material, so named because it may otherwise have to be purchased.

Ripping: Loosening hard soil, concrete, asphalt, or soft rock with a ripping attachment mounted to a dozer or similar equipment.

Stripping: : Removing thin layers of pay material, which are then typically stockpiled or hauled elsewhere for use.

The operation of a scraper requires constant attention to the controls and the surrounding environment. Scrapers have a tremendous amount of power and force. Once that force is put in motion, it is not an easy task to stop or redirect the scraper quickly. Do not take risks. If there is doubt about the capability of the machine to do some work, stop the equipment and investigate the situation by discussing it with the foreman or engineer in charge.

Whether it is a slope that may be too steep or an area that looks like the material is too unstable to work, operators must know the limitations of the equipment and decide how to do the job ahead of time. Once the operation has begun, it may be too late to stop or change direction. This could cause extra work or result in an unsafe condition.

In addition to considering the safety aspect of a scraper project, operators must have a thorough understanding of how the vehicle is designed to operate. However, true skill at guiding scrapers and efficiently conducting land-moving work comes only with experience behind the wheel in a variety of conditions.

3.1.0 Safety Guidelines

The ultimate responsibility for safely controlling a scraper lies with the operator. Operators must develop safe working habits and recognize hazardous conditions to protect themselves and others from injury or death. Always watch for unsafe conditions to protect both the scraper and other workers. Before working in a new area, walk around to locate any cliffs, steep banks, obstacles, or other unusual conditions. Become familiar with the operation and function of all controls and instruments before operating the equipment. Read and fully understand the operator manual for the specific scraper in use.

3.1.1 Operator Safety

No worker wants to be responsible for an accident or be hurt. There are a number of things a scraper operator can do to avoid getting hurt on the job.

Know and follow the employer's safety rules. The employer or supervisor should provide a list of the requirements for proper dress, PPE, and job-safety equipment. It is part of a worker's job to become thoroughly familiar with the employer's policies and safety guidelines. The following are recommended safety procedures for all occasions:

- Only operate the machine from the operator's cab. There are few controls that can be accessed outside of the cab on a scraper, and for good reason. Operating the unit from the cab provides the operator access to all the other controls that may be needed when things go wrong.
- Mount and dismount the equipment carefully using three points of contact. At least two

hands and one foot, or one hand and both feet, should be on the personnel ladder or scraper body at any given moment. Across a number of trades, this rule is known as the three-point rule.
- Wear the PPE required by the employer when operating the equipment.
- Do not wear loose clothing or jewelry that could catch on controls or moving parts.
- Keep the windshield, windows, and mirrors clean at all times. A complete view of the work area is essential to avoid serious accidents.
- Never operate equipment under the influence of alcohol or drugs. Indeed, never enter a job site at all when under the influence of any substance that alters the mind or impedes physical performance.
- Never smoke while working with any fuels or other fluids, or while in the immediate area of the battery.
- Do not use a cell phone and avoid other potential sources of static electricity while refueling to avoid igniting the vapors.
- Never operate without protective guards or panels designed to protect personnel from moving parts.
- Always lower the bowl to the ground before performing any service or when leaving the scraper unattended.

Regardless of how much a worker studies and understands the safety guidelines of an employer, the guidelines must be coupled with a safe attitude and a conscious effort to be effective. To avoid accidents, be alert at all times.

3.1.2 Safety of Co-Workers and the Public

An operator is not only responsible for his or her own personal safety, but also for the safety of other workers and observers. The work area is often very congested. In these areas, take the time to be aware of all the other movement in the vicinity. It is often difficult to hear exterior sounds when operating a scraper. Use a spotter and a radio in crowded conditions.

The main safety points to remember when working around other people and equipment include the following:

- Walk around the equipment to make sure that everyone is clear before starting and moving it.
- Never let anyone in or near the hinge or pivot area of an articulated machine while the unit is running.
- Always look in the direction of travel, and rapidly scan the mirrors and gauge panels often. This helps to form a mental three-dimensional image of the surroundings while still being aware of the scraper's operating status.
- Know and understand the traffic rules for the job site. The travel path and work area of specific equipment is often planned in advance. Planning the movement of heavy equipment ahead of time serves the purposes of both safety and efficiency.
- Exercise particular care at blind spots, crossings, and other locations where traffic or pedestrians may step into the travel path.
- Scrapers and other heavy equipment that are loaded always have the right-of-way on job site haul roads and inside excavation areas.
- Maintain a safe distance between other machines and vehicles.
- Pass cautiously and only when necessary.
- Stay in gear when driving downhill. Do not allow the scraper to coast in neutral. Many scrapers do not allow a shift to neutral while the vehicle is significantly moving.
- Avoid traveling horizontally across or around the face of a slope. The scraper may roll over. Drive up or down the slope.
- When descending a steep slope, do not allow the engine to overspeed. Select the proper gear before starting down the slope, and use the transmission lock to prevent automatic shift-

Scraper Rollover

In 2005, a scraper operator was killed while unloading at a stockpile. The scraper rolled over the side of the stockpile, which was only about 6 feet high. In spite of the rollover protective system (ROPS), which is a roll cage around the cab, the operator was killed. The operator was not wearing a seat belt, and was thrown out of the cab door as the unit rolled over. It was also found that the brakes were seriously deficient. One air hose had been disconnected from a front brake. The air hose feeding both rear brakes was closed off with a pair of locking pliers, due to an air leak in one brake assembly. It was determined that the lack of brakes probably did contribute to the accident. However, there was no protective berm or build-up of soil along the edge of the stockpile, as is commonly required.

Take the time required for safety. The scraper cab is built to protect the operator. Buckle up and allow it to do its job. Further, in case it has not been made clear enough by this example, a functional brake on only one wheel of a scraper is not enough.

ing. Use the hydraulic retard instead of the brakes when possible. A scraper that is out of control is a significant hazard to other workers.

3.1.3 Equipment Safety

The scraper has been designed with certain safety features to protect the operator as well as the equipment. For example, there are guards, canopies, shields, roll-over protection, and seat belts. Know the equipment's safety devices and be sure they are in working order. Then make use of them.

Use to the following guidelines to keep the equipment in safe working order:

- Perform prestart inspections and lubrication activities completely and consistently.
- Look and listen to make sure the equipment is functioning normally. Most operators quickly learn the normal sound of their equipment and can hear when something abnormal occurs. Stop if the scraper is malfunctioning. Correct or report trouble immediately.
- Use caution when backing up to a hitch. Use a spotter if necessary.
- Keep the machine under control. Do not try to work the machine beyond its capacity.
- Keep the work areas smooth and level, allowing for easier maneuvering and greater stability.

One basic safety rule is to know the equipment in use. Learn the purpose and use of all gauges, controls, and safety devices. Commit the equipment limitations to memory so that they are instantly recalled as the work progresses. Never operate the scraper if it is not in good working order.

3.2.0 Startup, Warm Up, and Shutdown

Before starting the scraper, the operator must climb aboard and get organized in the cab. The prestart inspection should now be complete and any necessary documentation done. No other workers should be too close to the scraper or hidden from the operator's view. Mount the equipment using the grab rails and personnel steps. Adjust the seat to a comfortable operating position (*Figure 29*). The seat should be adjusted to allow full brake pedal travel with the operator's back comfortably against the seat back. This permits application of maximum force on the brake pedals if necessary. Adjust the seat's lumbar support if it has one. Adjust the steering column and all mirrors. Fasten the seat belt. The scraper's roll-over protection is useless unless the driver is secured in the seat.

22204-12_F29.EPS

Figure 29 Operator in position.

Operator stations vary, depending on the manufacturer, size, and age of the equipment. However, all stations have gauges, indicators, switches, levers, and pedals. Gauges tell operators the specific status of critical items, such as water temperature, oil pressure, battery voltage, and fuel. Typically, there is a tachometer, voltmeter, temperature gauge, oil pressure gauge, and an hour meter at a minimum. Read the operator manual and learn the normal operating range for each gauge on the machine.

3.2.1 Startup

There are specific startup procedures for the piece of equipment in use. The following sequence represents a common startup procedure.

Some scrapers have special features for starting the engine in cold temperatures. These features include glow plugs and ether starting aids. Glow plugs heat the engine to build up heat for ignition. Ether is far more flammable than diesel fuel, and helps to initiate combustion on a cold start. However, it is important to check the operating instructions and ensure that the use of ether is acceptable. Newer models may have automated starting aids. Remember to switch any disconnect switches located outside of the cab to On before climbing aboard. If the machine has dual engines, the tractor engine is started before the scraper engine.

Step 1 Turn the battery and/or fuel disconnect switches for the tractor engine and the scraper engine to the On position.

Step 2 Move the transmission control to the neutral position.

Step 3 Ensure the parking brake is still engaged.

Step 4 Make sure the joystick or bowl-operating levers are in a position that do not cause movement of the controlled components; all devices should hold their present position during startup.

Step 5 Turn the key switch to the On position. Observe the initial indicator and warning light test that may occur, looking for lights that are not working.

Step 6 Wait until the glow plug indicator light goes out (if equipped).

Step 7 Turn the key switch to the Start position. Release the key when the engine starts, allowing it to return to the On position.

> **CAUTION**
>
> Never operate the starter for more than 30 seconds at a time. If the engine fails to start, wait two to five minutes before cranking again. Overuse of the starter can cause it to overheat and fail.

Step 8 Check the oil pressure gauge immediately after the engine starts. If oil pressure does not develop, shut off the engine and look for the cause.

Step 9 Allow the engine to warm up. Begin the activities required of the operator during the warm-up period.

For dual-engine scrapers, start the scraper engine after starting the tractor engine. Repeat the procedure above. However, use the controls to switch the instrument display over to the scraper engine first so that the oil pressure and other values can be seen immediately after startup.

All engines must be allowed to warm up before they are operated. Until a normal operating temperature is reached, applying a load to the engine may cause damage. The critical clearances of moving parts, such as bearings and piston rings, are not normal while the engine is cold.

3.2.2 Warm Up

When the outside temperature is above 60°F (15.5°C), allow at least five minutes for warm up before moving the scraper. In colder temperatures, allow at least 10 to 15 minutes. Extremely cold conditions may require a warm-up period of 20 minutes or more.

While the machine is warming up, check all the gauges and indicators. Keep the engine speed low until the oil pressure registers. If oil pressure has not registered within 10 seconds, stop the engine and investigate.

Once the warm-up period has elapsed, check that all systems are working properly before operating the machine under a load. Follow these steps to check that all of the systems are working properly:

Step 1 Check the air pressure for the brake system.

Step 2 Determine that all gauges are functioning properly.

Step 3 Cycle all the scraper controls to circulate warm oil through the system components. Manipulate each control to be sure all components are working properly.

Step 4 Unlock the shifter and shift the gears to low range.

Step 5 Lightly depress the accelerator.

Step 6 Manipulate the controls to move the scraper forward and backward, and make a short turn in each direction. When backing up, make sure the audible back-up alarm can be heard as the scraper shifts into Reverse. This announces to nearby workers that the scraper is backing up.

Step 7 Check the service brake.

Step 8 Shift the transmission to neutral and lock.

Step 9 Make a final visual check for leaks and listen for any abnormal knocks or noises.

Check the other gauges and indicators to see that the engine is operating normally. Check that all of the gauges are in the normal range. If there are any problems, shut the engine down and investigate, or get a mechanic to look into the problem.

One fluid level check is commonly done while the engine is warm and idling. The transmission fluid in many transmissions must be checked in these conditions, rather than when the engine is off and/or cold. Check the operator manual to be sure of the correct procedure and safety guidelines for checking the transmission fluid.

3.2.3 Shutdown

Shutdown should also follow a specific procedure. Proper shutdown reduces engine wear and prevents possible damage to the machine.

Step 1 Find a dry, level spot to park the scraper.

Step 2 Bring the scraper to a stop by decreasing the engine speed and gently applying the brakes.

Step 3 Place the transmission lever in neutral and engage the parking brake.

Step 4 Allow the engine to run at a low idle for approximately five minutes.

Step 5 Turn the engine start switch off and remove the key.

Step 6 Lower the scraper bowl and apron to the ground or on wooden blocks. Release any remaining hydraulic pressure by moving the control levers until all movement stops.

Step 7 Turn the battery and fuel disconnect switches to Off.

Don't forget to report any malfunctions to the supervisor or the field mechanic.

Refueling at the end of the workday is generally a better choice than waiting until the next morning. Maintaining a full tank helps prevent water from condensing inside the tank as temperatures change overnight. When temperatures are below freezing, it is also a good idea to drain any collected water from the air system when shutting down for the day.

3.3.0 Basic Maneuvering

Basic maneuvering of the scraper consists of moving forward, moving backward, turning from side-to-side, and operating the apron, bowl, and ejector.

3.3.1 Moving Forward

Before moving the scraper, move the bowl and apron to the desired position. For hauling or returning to the site of soil removal, the bowl should be up. The apron can be closed, open, or set to float which allows it to maintain its position relative to the bowl. When the bowl is full and hauling begins, the apron is normally closed to avoid losing the load. The ejector would normally be retracted to its normal position for scraping and loading. The ejector remains in this position except when the load is being pushed out.

Apply the service brake and release the parking brake. Move the scraper forward by setting the gear selection lever to the desired gear and depressing the engine throttle. Change the transmission gear to move faster or slower at a given engine speed. Remember that many scrapers shift automatically; the gear set by the operator is the highest gear the transmission is allowed to choose.

To load the scraper bowl, apply power to the engine(s) while keeping the front wheels straight. Sometimes it is difficult to hold the scraper in a straight line because of the force of the blade digging into the earth. It may try to pull the scraper to one side or the other. This is especially true for uneven terrain. Adjust the gears and engine speed so that the machine can load without strain. Remember that only the two lower gears are designed for scraping.

Sometimes a scraper is pushed by a dozer or other equipment (*Figure 30*). The operator must make sure everything is lined up properly so the dozer pushes in a straight line. If the scraper and dozer are not aligned properly, the cutting edge

22204-12_F30.EPS

Figure 30 Dozer-assisted scraper.

may cut into the ground on one side only and create an unbalanced load in the bowl. This affects the operator's ability to steer in a straight line and results in an uneven cut.

3.3.2 Moving Backward

Moving backward in a straight line is not easy. Because the scraper is articulated, it has a tendency to move from side to side, depending on which way the tractor wheels are pointed. The process is the same as trying to back up a trailer attached to a car or truck. Once the scraper is moving backwards in the wrong direction, it is hard to straighten the rig out without stopping and pulling forward to get everything in line again.

Before beginning to back the scraper, always look around to make sure there is nothing behind the equipment. Also, allow enough room on either side of the scraper so there is enough lateral clearance to correct the direction should it start moving to the right or left. If the scraper starts to move at a sharp angle in the wrong direction, stop and pull forward enough to straighten out the scraper. Then start to back again in a straight line.

3.3.3 Turning

Because scrapers are such large pieces of equipment, turning can be tricky. Like all articulated equipment, they have a pivot point, or hinge, that allows them to turn more sharply than if they were on a fixed, rigid frame. Motorized scrapers have a highly-arched gooseneck hitch. The reason for this hitch design is so the tractor wheels can move under it during a sharp turn (*Figure 31*). The tractor can turn up to a 90-degree angle to the scraper unit. This significantly increases the range of motion and maneuverability of the scraper. At the same time, it also makes it more challenging to handle. Turning too tightly can jam the neck and stress the hydraulic lines. This may cause damage to the lines and couplings.

Remember that the differential should not be locked during turns. When the scraper turns, one drive wheel must rotate faster than the other. The differential is designed to allow this to happen. When the differential lock is engaged, both drive wheels are forced to turn the same number of revolutions.

Steering can be used to walk the tractor out of soft spots that may be encountered, with or without blocking the wheels. This technique is done by turning the wheels of the tractor first one way and then the other while applying reasonable power. This puts each drive wheel on a new footing and may provide enough traction to get out of the soft spot. This technique does put a tremendous strain on the motor and gears, and may cause the engine to run hotter if it must be done repeatedly. If so, allow frequent rests, operating at idle speed, for the engine to cool down.

3.3.4 Operating the Bowl, Apron, and Ejector

When loading, the operator must be able to position the bowl and apron while driving forward. This takes a lot of coordination of the hands and feet. When making a cut, the operator must lower

22204-12_F31.EPS

Figure 31 Scraper in a sharp turn.

the bowl to the correct depth using the joystick or other control. The operator must also raise the apron while moving forward at a speed sufficient to keep the operation moving. At the same time, the operator tries to keep the wheels from spinning by not applying too much power.

When unloading or spreading material, the operator must position the bowl, apron, and ejector. Using the ejector to unload the scraper while moving is less challenging than loading to coordinate. However, the operator still has to control the apron and ejector simultaneously so the material is dumped in a smooth, even manner. Once the proper bowl position is set, determining the maximum depth of the unloaded material, it is likely to remain in this position until the bowl has emptied completely. The bowl position determines the depth by pushing the material aside or down as it passes over the freshly dumped material.

3.4.0 Work Activities

The activities that a scraper can perform are rather limited. The machine was designed basically to do one activity well. It can collect and move large quantities of material from one place to another. This involves loading the material, hauling it to another location, and then dumping the material or spreading it over an area to build the area up. The variations of this activity and the details of the process are discussed below. Remember that a cycle refers to a complete trip, typically starting at the load point with an empty bowl and ending at the same point after the load has been collected, transported, and dumped or spread.

The scraper is an expensive machine to buy, operate, and maintain. Before beginning an operation, the use of scrapers to move material must be carefully planned. The plan will be based on the type of material, haul distances, volume of material, and expected results.

Increasing the load size decreases the hauling cost per yard of material, unless the speed and acceleration of the scraper are greatly reduced. In general, a light load can be hauled rapidly, but the operating cost is divided among fewer yards, possibly making the cost per yard high. Larger loads move more slowly, but costs are spread over more yards, so the cost per yard is probably lower.

If a push dozer is not required for the operation, the most efficient loading time would also give the greatest production. The hauling time should be the same in this situation, load after load. However, when push dozers must be used, their added operating cost demands that the loading time be decreased enough to compensate for their cost. For example, a planner may need to make calculations to determine if taking lighter cuts with scrapers alone is better than heavier cuts with the assistance of a dozer. Scrapers, though, cannot afford to sit and wait for an available dozer if one is not readily available in the cycle. In such a case, it is best to begin a shallower cut right away and consider a lighter load in order to keep the operation moving.

It should be noted that the scraper can be mechanically loaded, rather than load itself by scraping. For example, large piles of soil may have to be spread. An appropriate loader simply places the soil into the bowl, and the scraper hauls it to the proper location and spreads it.

3.4.1 Cutting, Loading, and Spreading

Making cuts and fills with a scraper is a repetitive job that can be organized into a pattern for the scraper operator to follow. The pattern is established based on the number of scrapers available, the terrain, and the required results. Generally, ma-

Grading Made Easy

Technology has had a tremendous impact on the management and use of earthmoving equipment. With today's GPS and laser guidance systems, operators can see exactly where they are on the site at any time. Not only can operators see their location and movement in real time, they can see the exact elevation of the scraper blade in relation to the desired finish grade. Although not inexpensive, such systems can save money by avoiding mistakes such as overfilling or undercutting an area. Even 3D displays are readily available.

For a fleet of earthmoving equipment working a site together, these systems provide the needed technology for all the operators to function as a single unit.

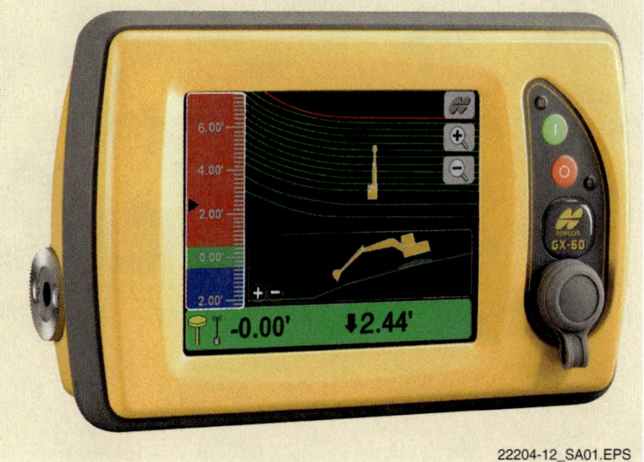

22204-12_SA01.EPS

terial scraped from high spots is used to fill in lower spots to create a large area of consistent elevation.

The first cut starts at the top of the first hill. Remember that it is hazardous to operate the scraper horizontally on a slope; uphill or downhill travel is the safest approach. The operator cuts as deeply as possible across the top. As the scraper starts down the hill, the operator raises the bowl slightly. Continuing this sequence on each cut down the slope eventually flattens the top of the hill, giving the operator a good flat surface from which to work. Then smooth, light cuts across a level plain make loading easier.

Start placing material at the lowest point of the first fill area. Bring the fill up from the lowest point in smooth level lifts until the desired grade is reached (*Figure 32*). The grade may be marked with stakes to provide a visual depth target for scraper operators. Laser-guidance or GPS systems may be in place, eliminating the need for elevation stakes. The operator is responsible for keeping the fills and cuts level.

The remaining hills would be worked the same way in sequence. When everything has been roughly leveled, some general smoothing from one end to the other may be required.

Try to avoid scraping uphill if at all possible. It is more difficult for the scraper engines, and it increases the loading cycle time. Any time scraper engines are working harder than necessary, more fuel than necessary is also being consumed. Level the top of the hill with several initial across the top and downhill passes. Then the scraper can be loaded quickly and the remaining high and low areas can be smoothed out.

One popular approach to common scraper work is called straddle loading (*Figure 33*). After one cutting pass is complete, a second pass is made about 8 feet away. On the third pass, the scraper collects the material between the first two passes. The width of material left for the third pass should be one-half to two-thirds the width of the scraper cutting edge. This should leave a clean, scraped surface with a minimum of windrow.

22204-12_F32.EPS

Figure 32 Spreading the load.

22204-12_F33.EPS

Figure 33 Straddle loading.

3.4.2 Strip and Stockpile Topsoil

For large construction projects where the topsoil needs to be saved for later use, the soil can be stripped away with a scraper and stockpiled until it is needed. Scraping cannot be done until the clearing and grubbing of stumps and undergrowth is first completed. Typically, topsoil is considered a **pay material**, so keeping it intact is also an economic consideration.

Before the **stripping** work begins, the stockpile area is identified and marked. The size and location of the stockpile area depend on how much material has to be moved and what the haul distance will be. To begin removing the topsoil, use the following procedures:

Step 1 Move the machine to one end of the work area.

Step 2 Downshift to a loading gear.

Step 3 Open the apron to the desired height.

Step 4 Lower the cutting edge to a 3- to 6-inch cut depth. The scraper should be able to maintain a level cut with little tire spinning.

Step 5 Close the apron when loaded and raise the bowl gradually while still in motion to avoid a shelving, or drop-off, effect at the end of the cut.

Step 6 Shift to an appropriate higher gear. Move to the stockpile area with the bowl just high enough to avoid obstructions.

Step 7 Raise or lower the bowl to the desired depth of soil lift.

Step 8 Raise the apron to the full open position.

Step 9 Use the joystick ejector control or ejector lever to push material toward the apron opening, maintaining a smooth, constant flow of material while remaining in motion.

Step 10 Lower the apron when the load is spread and return the ejector to full back position.

Step 11 Return to the cut area and repeat the cycle.

Note that stockpiles created by scrapers are often required to have an earth berm around the sides to help prevent the scraper from rolling over near the edges.

3.4.3 Scraper Operation Using a Bulldozer to Push

Assistance is sometimes required to load self-propelled rubber-tired scrapers. Tracked models are rare and much slower, but they are not designed to be pushed. If tractors with tracks are the desired approach, then a towed scraper is likely to be used with a tracked vehicle. While the travel speed of a scraper is increased by the use of wheeled tractors, this is at the expense of traction. For the loading operation where more power is needed, bulldozers are often used to push. The bulldozer blade is reinforced at the center with a heavy steel plate (*Figure 34*). All scrapers that can be pushed have a pusher block, projecting from the rear, which is engaged by the dozer. As a general rule, scrapers with paddle-wheel elevators self-load only. The elevators can be damaged by forcing too much soil into them from the push of the dozer.

Because loading time is of prime importance, it is necessary to organize operations at the loading site so time is not lost in engaging scrapers as they pull into loading position. There are three ways to organize the work. The space available determines which method is used. Refer to *Figure 35*.

- *Back-track loading* – The back-track loading method is used if the loading area is short and wide. After each push, the dozer swings through an arc of 180 degrees and returns to the point adjacent to its original starting point. From there, it turns 180 degrees again to line up behind the next scraper entering the cut. Each scraper and dozer combination makes a cut together from one end to the other.
- *Chain loading* – Use the chain-loading method where the loading area is longer and may be worked from end to end. In this case, several scraper runs are required to complete a single cut from one end to the other. The dozer completes a run with a partner, then immediately changes lanes, moving behind another scraper that is beginning where a previous scraper ended a cut. A number of scrapers can be loaded before the dozer needs to return to the starting point. All scrapers are traveling in the same direction.

22204-12_F34.EPS

Figure 34 Steel reinforcement on a dozer blade.

- *Shuttle loading* – The shuttle-loading method would also be used in limited areas or in areas that may be short but wide. Scrapers are moving in both directions while shuttle loading. This means that the scrapers should be divided into two groups, with each group moving in opposite directions. A dozer pushes a scraper, and then makes a 180-degree turn to engage another scraper traveling in the opposite direction.

Regardless of the loading arrangement, the push dozer and scraper must work together in a smooth and uniform manner. Both operators must understand what is expected of them. It is important that the dozer lines up centered on the scraper push block, or the scraper will be pushed sideways instead of straight ahead. The basic procedure for pushing and loading is as follows:

Step 1 Move the scraper in position for the next cut close to the dozer. Using hand signals, a good dozer operator can direct the scrapers to the area to be cut next.

Step 2 If the scraper has a cushioned hitch, lock the hitch to prevent an uneven cut, unless the area is very smooth.

Step 3 If at all possible, always move in from the right side of the dozer so that the front of the dozer blade can be seen, avoiding damage to tires or scraper body parts.

Step 4 Continue moving forward while lowering the cutting edge to the desired depth of cut.

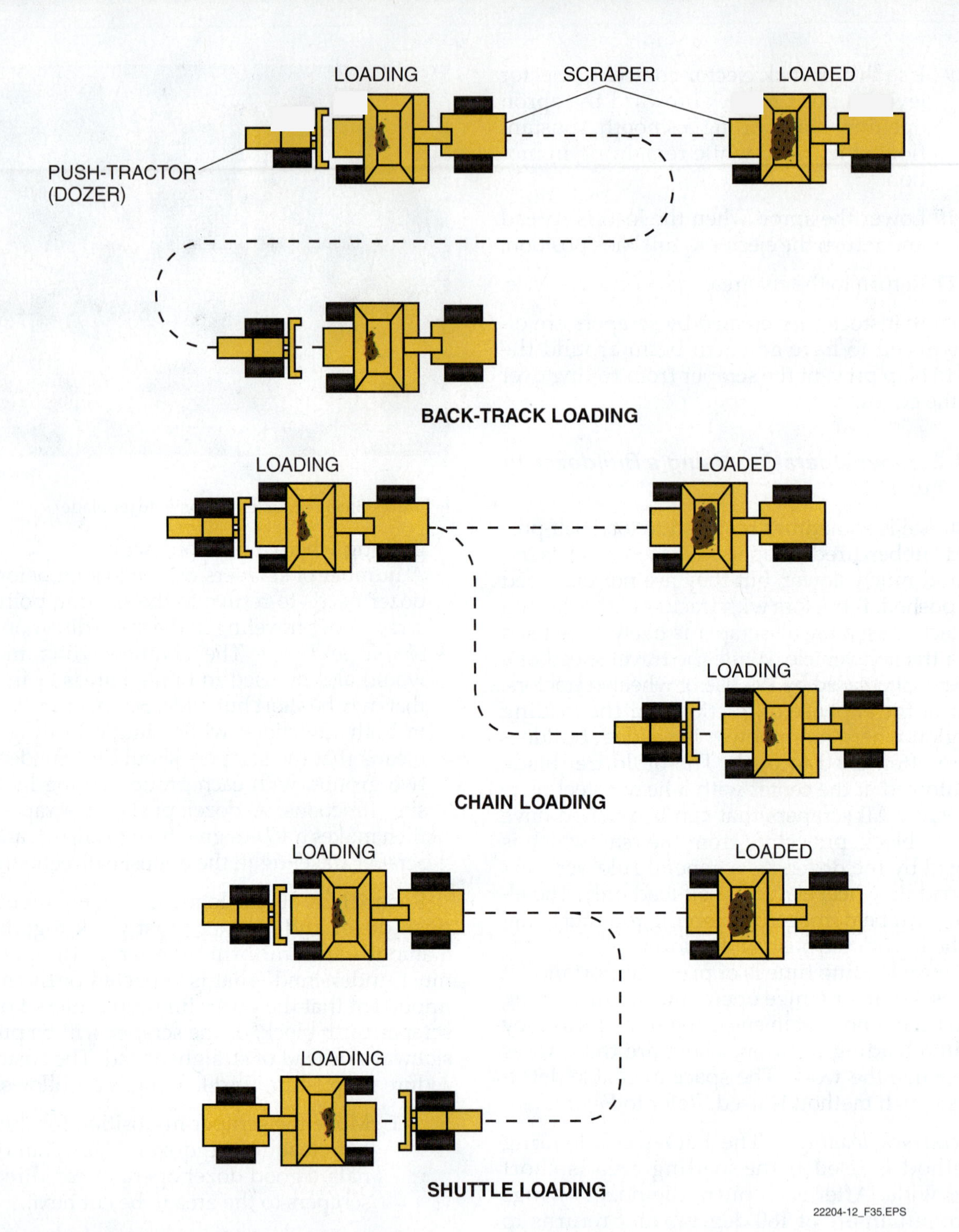

Figure 35 Approaches to scraper loading with a pusher.

Step 5 Raise the apron to allow material to begin flowing into the bowl.

Step 6 Maintain the cut at the desired depth, using only the throttle to aid the machine and to provide hydraulic power for the attachments.

Step 7 Lower the apron when the bowl is full; gradually raise the cutting edge from the cut to avoid a shelving effect at the end of the cut.

Step 8 Select a travel gear and move to the fill area as quickly and safely as possible. Once the travel gear is selected and the scraper begins to move away, the dozer stops pushing and returns to the starting point for the next push.

3.4.4 Using Two Scrapers in Tandem

Scrapers often work together. Two scrapers equipped with a bail and push plate can be linked together so that the engines can provide supplemental power to each other. Two scrapers working in tandem are shown in *Figure 36*. They can be used instead of, or in addition to, a dozer to supply additional power to load the bowl. This can be more cost effective than using a dozer, since two bowls are filled for each hauling cycle.

The second scraper moves directly behind the first. The bail of the second scraper is lowered onto the push plate of the scraper in front (*Figure 37*). The two scrapers select a drive gear and move off together. The bowl of the front scraper is always filled first. Assuming the cut area is long enough, the second scraper begins lowering its bowl for a cut as the front scraper raises its full bowl.

3.4.5 Collecting Windrows

Self-loading scrapers are sometimes used for fine trimming and smoothing. The operator must be able to pick up any excess dirt without cutting into the trimmed surface. A self-loading scraper can pick up a small windrow of dirt that has been left after the grader has trimmed the subgrade. The operator must keep the windrow centered on the scraper bowl so one side of the bowl does not fill faster and cause the scraper to lean and dig deeper on one edge (*Figure 38*).

3.4.6 Finish Grading and Trimming Subgrade

Check the air pressure in the scraper tires before any finish grading or trimming begins. The air pressure in all tires must be the same. If one tire has as little as 10 pounds less pressure than the others, the scraper will lean and dig deeper on the soft tire side as the scraper bowl fills.

22204-12_F36.EPS

Figure 36 Tandem scrapers in action.

22204-12_F37.EPS

Figure 37 Connecting two scrapers.

22204-12_F38.EPS

Figure 38 Collecting windrow with a paddle scraper.

Always keep a close watch on the cutting edge of the equipment. It is costly to repair a worn or damaged scraper pan, and a worn cutting edge will reduce efficiency. For finish work, a good cutting edge is essential. Watch the cutting edge on any paddle wheel scraper that is picking up soil in a trim operation. A scraper with a worn cutting edge or with worn **end shoes** will not make a clean pass, and will leave more work for the grader operator. A technician is replacing the end shoes in *Figure 39*.

Be sure that the end shoes are extended so they are close to the bottom of the cutting edge, but never lower. When they are in good condition, the cutting edge and end shoes should all touch the ground evenly.

Working with a **grade checker** is common when using a scraper to finish grades. The grade checker stands on the ground and gives directions regarding the depth of cut required or other trimming requirements. Here is how the operation typically works:

22204-12_F39.EPS

Figure 39 Replacing the end shoes.

Step 1 Approach the work area and downshift to a proper cutting gear.

Step 2 Observe the grade checker's hand signals and lower the bowl to begin cutting finish grade.

> **NOTE**
> If a fill is to be made instead of a cut, the bowl would first be loaded with material from a stockpile or a cut from another area.

Step 3 Continue cutting or filling, observing the grade checker's signals and making necessary corrections, until the bowl is full or empty.

Step 4 Shut the apron when the bowl is loaded and raise the bowl slowly out of the cut.

Step 5 Raise the bowl to a safe travel height and proceed to the spread area or stockpile.

Step 6 Lower the bowl to the amount of lift to be added following the grade checker's signals. Select the correct speed and raise the apron to the open position. Begin ejecting material from the bowl.

Step 7 Return to the work area once the load is spread, and again observe the grade checker.

Step 8 Line up on the next cut area by putting one wheel on the edge of the previous cut.

Step 9 Select the proper working speed and lower the bowl to begin the cut.

Step 10 Observe the grade checker's signals and proceed with the next cut.

3.4.7 Working in Unstable Material

Most of the time a scraper cannot load in a muddy area. A dozer is better suited for this type of work. If a scraper must be used to move mud with a dozer to push, there is a technique that works safely. Follow these basic steps:

Step 1 Move into the soft area with the scraper bowl down.

Step 2 Move ahead slowly until the tires start to slip and then stop.

Step 3 Move the dozer into position to push the scraper.

Step 4 Once contact is made, start to move ahead slowly. Avoid spinning the scraper wheels and don't try to cut too deeply because this causes the dozer to spin its tracks and get stuck. Use the differential lock, but remember not to execute a turn with the lock engaged.

Step 5 Once the scraper is full, apply a little more power. Again, be careful not to spin the wheels.

Step 6 The dozer operator should keep pushing until the scraper operator can get enough traction to pull away.

In very soft areas, the operator may have to take only half-loads until traction improves.

Always watch for soft spots during dumping or loading. If there is a soft area that must eventually be worked through, do not drive right into the center. Drive along the edge first. Move closer to the soft area with each pass. It can then be decided if it is becoming too soft to support the scraper on each pass.

CAUTION

When working over an area where the mud is at least 3 feet deep and the scraper loses traction, stop and wait to be pulled or pushed out. If the wheels continue to spin, the scraper will dig deeper into the mud. It will be a major job to winch it out, especially if there is a load in the bowl.

3.4.8 Loading a Scraper in Rock

Loading a scraper in rock is a delicate procedure. The operator must be very alert to keep from spinning the tires or damaging the cutting edge. Spinning the tires on rocks may ruin them. Apply little or no power while being pushed. Letting the dozer do the work helps to avoid any tire spin and saves the tires.

If a hard spot is hit and the scraper stops or nearly stops, pull the bowl up a little until the rock that was causing the trouble clears the cutting edge. Then lower the bowl slowly until the push dozer has to strain slightly to keep going. The operator can tell when the push dozer's engine is starting to **lug down** by watching the exhaust. If the exhaust smoke is visibly increasing, lift the scraper bowl to reduce the resistance on the dozer.

In loading rock, the scraper operator must raise and lower the scraper bowl constantly. The cutting edge frequently catches on rock that does not budge. Do not try to cut through it. Let a dozer perform a **ripping** operation. A dozer equipped with a ripper (*Figure 40*) loosens the rock to allow the scraper to load much more effectively. Pick up the ripped material on the next pass. The ripper of the dozer should cut a little deeper than the scraper's depth of cut.

Occasionally the scraper will load a rock so large that it won't pass under the bowl when the load is dumped. If the bowl can't be raised high enough to pass over it, dump the boulder on the ground and stop. Then back up slightly to give the scraper a little turning room, and then turn sharply so the bowl passes beside the boulder instead of over it.

Always work slowly and carefully in rock. Don't rush. An operator cannot work fast enough to make up for the premature loss of a tire or damaged equipment.

22204-12_F40.EPS

Figure 40 Dozer ripper.

Additional Resources

The Occupational Safety and Health Administration (OSHA) publishes safety requirements for scrapers and related material handling equipment in OSHA *Standard 1926.602*, found at **www.osha.gov**.

3.0.0 Section Review

1. When descending a hill, the scraper transmission should be shifted to neutral.
   a. True
   b. False

2. How much time should be allowed for the engine to warm up when the outdoor temperature is 50°F?
   a. At least 3 minutes
   b. At least 5 minutes
   c. At least 10 to 15 minutes
   d. At least 20 minutes

3. Why is the hitch of a wheeled scraper made with a high arch?
   a. It is stronger than a straight, low hitch design.
   b. It allows the operator to see the apron more clearly.
   c. It allows room for the tractor wheels during sharp turns.
   d. It provides a means for workers to move from side-to-side without having to walk around.

4. How does an operator know if a dozer is beginning to lug down?
   a. The smoke from the exhaust visibly increases.
   b. The grade checker signals the scraper driver.
   c. The dozer operator gives a specific hand signal to the scraper.
   d. The scraper engine controls are synchronized with the engine control of the dozer.

SUMMARY

Scrapers are large machines designed to move large quantities of soil or rock from one place to another. All models of scrapers perform only this one basic function well. In doing its work, a scraper is sometimes assisted by a bulldozer that pushes the scraper, allowing it to make deeper cuts into the material.

There are several different scraper designs. These include standard wheel tractor scrapers, elevating scrapers, towed scrapers, and tandem scrapers. The elevating scraper uses a rotating paddle-wheel arrangement or an auger in the bowl to move the soil to the back. The tandem scraper has two engines, one for the tractor in the front, and another at the back which powers the rear wheels on the bowl. Tandem scrapers can be linked together to push and pull each other. Towed scrapers are pulled by a tractor or dozer unit and can be economical on smaller jobs. These designs also come in many different sizes.

To be an effective scraper operator, the driver must know how to load; haul; spread material; create a stockpile; and do finish grade work. Scrapers are used to move material from one place to another for the purpose of trimming or leveling uneven terrain. Because of their large size and carrying capacity, they are a good choice for the job, as long as they don't have to work horizontally across a slope.

Finish grading and trimming is one activity where the scraper operator may work with another person, called a grade checker, who controls the cutting or trimming operation from the ground. Scraper operators must learn how to follow the guidance of a grade checker to produce smooth and accurate grades.

Review Questions

1. Another common name for a scraper is _____.
   a. pan
   b. scoop
   c. grader
   d. load-all

2. When the tractor engine projects forward of the drive axle, it is known as a(n) _____.
   a. tipped configuration
   b. overhung configuration
   c. weight-over configuration
   d. free balance configuration

3. What is the capacity of the largest scrapers?
   a. About 10 cubic yards
   b. About 20 cubic yards
   c. About 40 cubic yards
   d. About 80 cubic yards

4. The rear engine of a tandem scraper is the one with the most horsepower.
   a. True
   b. False

5. When the scraper is not equipped with a joystick, devices such as the ejector are generally controlled by _____.
   a. levers
   b. foot pedals
   c. slide switches
   d. ceiling-suspended pull handles

6. Which of the following, when illuminated, would be considered only an indicator light, rather than a warning?
   a. Charging system light
   b. Transmission hold light
   c. Engine oil pressure light
   d. Engine coolant temperature light

7. Other than air, the gas commonly preferred by manufacturers for tire inflation is _____.
   a. argon
   b. helium
   c. nitrogen
   d. nitrous oxide

8. The number of operating hours generally associated with one normal year of operation is _____.
   a. 500 hours
   b. 1,000 hours
   c. 2,000 hours
   d. 3,000 hours

9. The vehicle that would typically have the right-of-way at construction sites is a(n) _____.
   a. excavator
   b. loaded scraper
   c. empty scraper
   d. tracked bulldozer

10. At what point might a short test of the indicator and warning lights automatically occur?
    a. When the key is turned to On
    b. When the key is turned to Start
    c. When the Master switch is first turned on
    d. When the key is placed into the ignition switch

11. Even when the outdoor temperature is above 60°F, a scraper engine should be allowed to warm up for at least 10 to 15 minutes.
    a. True
    b. False

12. When starting the engine, what is the maximum amount of time an operator can allow for the oil pressure to register on the gauge?
    a. 3 seconds
    b. 5 seconds
    c. 10 seconds
    d. 15 seconds

13. When scrapers are working with push dozers, but a dozer is not available when the scraper returns to the soil removal area, the operator should _____.
    a. begin a lighter, shallower cut right away
    b. stop and wait until another scraper returns to push
    c. stop and wait for the next available dozer to ensure the largest possible load
    d. begin at the same cut depth as before, and hope the dozer shows up to push the scraper through the cut before it stalls

14. When straddle-loading a scraper, how wide an area should the operator leave for the third pass?
    a. 12 feet
    b. Just the width of the windrow
    c. One-fourth to one-third the width of the cutting edge
    d. One-half to two-thirds the width of the scraper cutting edge

15. In scraper operations, stripping refers to _____.
    a. the removal of thin layers of pay material for stockpiling
    b. the removal of undergrowth and trees before scraping can begin
    c. separating and taking away only the rocks, leaving the soil behind
    d. breaking up very hard and/or rocky soil so it can be scraped more easily

Trade Terms Introduced in This Module

Apron: A movable metal plate in front of the scraper bowl that can be raised or lowered to control the flow of material into or out of the bowl.

Bowl: The main component at the center or back of a scraper where the material is collected and hauled.

Cycle: One complete trip made by a scraper starting from the load point and returning to the same or adjacent point. A cycle typically includes scraping, hauling, spreading the contents, and returning to the origin to begin the next load.

Ejector: A large metal plate inside the bowl of a scraper that can be activated to push the material forward, causing it to fall from the front of the bowl.

End shoes: Flat pieces of steel on each side of the scraper bowl that help keep the material confined to the front of the cutting edge until the paddles can scoop it up; sometimes referred to as slobber bits.

Governor: A control that limits the maximum speed of an engine or vehicle.

Grade checker: Person who checks elevations and grades, and then gives signals to the equipment operators.

Lug down: A slowdown in engine speed (rpm) due to increasing the load. This usually occurs when heavy machinery is crossing soft or unstable soil or is pushing or pulling very hard. When a diesel engine lugs down, there is generally an increase in visible smoke and an audible change in the sound of the engine.

Pay material: Deposit of soil valuable enough to stockpile for future use, on or off site.

Rimpull: The torque applied to the axle from the engine that is then converted into pulling force at the face of the tire; more commonly known as gross traction in mathematical calculations.

Stockpile: Material put into a pile and saved for future use.

Ripping: Loosening hard soil, concrete, asphalt, or soft rock with a ripping attachment.

Stripping: Removal of overburden or thin layers of pay material.

Additional Resources

This module presents thorough resources for task training. The following resource material is suggested for further study.

The Occupational Safety and Health Administration (OSHA) publishes safety requirements for scrapers and related material handling equipment in OSHA *Standard 1926.602*, found at **www.osha.gov**.

Figure Credits

Reprinted courtesy of Caterpillar Inc., Module opener, Figures 1–5, 7–10, 12, 13, 15–27, 29, 32, 36–39

Courtesy of Deere & Company, Figures 6 and 11

Courtesy of ABC Parts Online, Figure 14

Courtesy of Analysts, Inc., Figure 28

Courtesy of David Fasules, Figures 30 and 34

Robert I. Carr, Ph.D., P.E., Figure 31

Courtesy of Topcon Positioning Systems, Inc., SA01

Courtesy of Robert P. VanNatta, Figure 40

Section Review Answers

Answer	Section Reference	Objective
Section One		
1 b	1.0.0; 1.1.0; Figure 7	1a
2 a	1.2.6	1b
3 a	1.3.0	1c
Section Two		
1 b	2.1.0	2a
2 c	2.2.1	2b
Section Three		
1 b	3.1.2	3a
2 c	3.2.2	3b
3 c	3.3.3	3c
4 a	3.4.8	3d

NCCER CURRICULA — USER UPDATE

NCCER makes every effort to keep its textbooks up-to-date and free of technical errors. We appreciate your help in this process. If you find an error, a typographical mistake, or an inaccuracy in NCCER's curricula, please fill out this form (or a photocopy), or complete the online form at **www.nccer.org/olf**. Be sure to include the exact module ID number, page number, a detailed description, and your recommended correction. Your input will be brought to the attention of the Authoring Team. Thank you for your assistance.

Instructors – If you have an idea for improving this textbook, or have found that additional materials were necessary to teach this module effectively, please let us know so that we may present your suggestions to the Authoring Team.

NCCER Product Development and Revision

13614 Progress Blvd., Alachua, FL 32615

Email: curriculum@nccer.org

Online: www.nccer.org/olf

❏ Trainee Guide ❏ AIG ❏ Exam ❏ PowerPoints Other ______________________________

Craft / Level: Copyright Date:

Module ID Number / Title:

Section Number(s):

Description:

Recommended Correction:

Your Name:

Address:

Email: Phone:

Glossary

American Association of State Highway and Transportation Officials (AASHTO): An organization representing the interest of all state government highway and transportation agencies throughout the United States. This organization establishes design standards, materials-testing requirements, and other technical specifications concerning highway planning, design, construction, and maintenance.

American Society of Testing Materials (ASTM): A national organization that establishes standards for testing and evaluation of manufactured and raw materials.

Apron: A movable metal plate in front of the scraper bowl that can be raised or lowered to control the flow of material into or out of the bowl.

Aquifer: An underground layer of water-bearing permeable rock or unconsolidated materials through which water can easily move.

Auxiliary axle: An additional axle that is mounted behind or in front of the truck's drive axles and is used to increase the safe weight capacity of the truck.

Average: The middle point between two numbers or the mean of two or more numbers. It is calculated by adding all numbers together, and then dividing the sum by the quantity of numbers added. For example, the average (or mean) of 3, 7, 11 is 7 ($3 + 7 + 11 = 21$; $21 \div 3 = 7$).

Balance point: The location on the ground that marks the change from a cut to a fill. On large excavation projects there may be several balance points.

Banked: Any soil mass that is to be excavated from its natural position.

Bedding material: Select material that is used on the floor of a trench to support the weight of pipe. Bedding material serves as a base for the pipe.

Bedrock: The solid layer of rock under Earth's surface. Its solid-rock state distinguishes it from boulders.

Berm: A raised bank of earth.

Blade: An attachment on the front end of a loader for scraping and pushing material.

Boot: A special name for laths that are placed by a grade setter to help control the grading operation. The boot can also be the mark on the lath, usually 3, 4, or 5 feet above the finish grade elevation, which can be easily sighted. This allows the grade setter to check the grade alone instead of having to use another person to hold a level rod on the top of the grade stake.

Bowl: The main component at the center or back of a scraper where the material is collected and hauled.

Breakout force: The maximum vertical upward force created by the curling action of a bucket attached to an excavator, backhoe, or loader, measured 4 inches behind the tip of the bucket's cutting edge.

Bucket: A U-shaped closed-end scoop that is attached to the front of the loader.

Cab guard: Protects the truck cab from falling rocks and load shift.

Capillary action: The tendency of water to move into free space or between soil particles, regardless of gravity.

Change order: A formal instruction describing and authorizing a project change.

Clutch: Device used to disengage the transmission from the engine.

Cohesive: The ability to bond together in a permanent or semipermanent state. To stick together.

Competent person: A person who is capable of identifying existing and predictable hazards in the area or working conditions that are unsanitary, hazardous, or dangerous to employees, and who has the authority to take prompt corrective measures to fix the problem.

Consolidation: To become firm by compacting the particles so they will be closer together.

Constant: A value in an equation that is always the same; for example pi is always 3.14.

Contour lines: Imaginary lines on a site/plot plan that connect points of the same elevation. Contour lines never cross each other.

Crab steering: A steering mode where all wheels may move in the same direction, allowing the machine to move sideways on a diagonal, also known as oblique steering.

Cross braces: The horizontal members of a shoring system installed perpendicular to the sides of the excavation, the ends of which bear against either uprights or walers.

Cycle: One complete trip made by a scraper starting from the load point and returning to the same or adjacent point. A cycle typically includes scraping, hauling, spreading the contents, and returning to the origin to begin the next load.

Debris: Rough broken bits of material such as stone, wood, glass, rubbish, and litter after demolition.

Density: Ratio of the weight of material to its volume.

Dewater: To remove water from a site.

Dozing: Using a blade to scrape or excavate material and move it to another place.

Easement: A legal right-of-way provision on another person's property (for example, the right of a neighbor to build a driveway or a public utility to install water and gas lines on the property). A property owner cannot build on an area where an easement has been identified.

Ejector: A large metal plate inside the bowl of a scraper that can be activated to push the material forward, causing it to fall from the front of the bowl.

Elasticity: The property of a soil that allows it to return to its original shape after a force is removed.

Elevation view: A drawing giving a view from the front or side of a structure.

End shoes: Flat pieces of steel on each side of the scraper bowl that help keep the material confined to the front of the cutting edge until the paddles can scoop it up; sometimes referred to as slobber bits.

Engine retarder: An alternate braking system activated from the cab that slows down the vehicle by reducing engine power.

Erosion: The removal of soil from an area by water or wind.

Expansive soil: A soil that expands and shrinks with moisture. Clay is an expansive soil.

Fines: Very small particles of soil. Usually particles that pass the No. 200 sieve.

Fixed displacement pump: A pump that cannot be adjusted to deliver more or less fluid volume in a pumping cycle.

Float mode: Placing the bucket or other attachment into the float mode allows the attachment to automatically follow the contour of the ground, remaining at the same elevation above it.

Four-wheel steering: A steering mode where the front and rear wheels may move in opposite directions, allowing for very tight turns, also known as independent steering or circle steering.

Friable: Crumbles easily.

Fulcrum: A point or structure on which a lever sits and pivots.

Governor: A control that limits the maximum speed of an engine or vehicle; device for automatic control of speed, pressure, or temperature.

Grade checker: Person who checks elevations and grades, and then gives signals to the equipment operators.

Groundwater: Water beneath the surface of the ground.

Grouser: A ridge or cleat across a track that improves the track's grip on the ground.

Grubbing: Digging out roots and other buried material.

Hoist: Mechanism used to raise and lower the dump bed.

Horizon: Layers of soil that develop over time.

Humus: Dark swamp soil or decaying organic matter. Also called peat.

Hydrostatic drive: A system that relies on the flow and pressure of a fluid to transfer energy. In skid steers, the transfer of energy is from the engine, driven by diesel fuel, to the four drive wheels through hydraulic motors.

Hypotenuse: The long dimension of a right triangle and always the side opposite the right angle.

In situ: In the natural or original place on site.

Inorganic: Derived from other than living organisms, such as rock.

Invert: The lowest portion of the interior of a pipe, also called the flow line.

Liquid limit: The amount of moisture that causes a soil to become a fluid.

Loadbearing: A base designed to support the weight of an object of structure.

Loading: Applying a force to soil. A building can be a permanent load at a site, and a truck can be a passing load on a roadway.

Lug: Effect produced when engine is operating in too high a transmission gear. Engine rotation is jerky, and the engine sounds heavy and labored.

Lug down: A slowdown in engine speed (rpm) due to increasing the load. This usually occurs when heavy machinery is crossing soft or unstable soil or is pushing or pulling very hard. When a diesel engine lugs down, there is generally an increase in visible smoke and an audible change in the sound of the engine.

Monuments: Physical structures that mark the locations of survey points.

Oblique steering: A steering mode where all wheels may move in the same direction, al-lowing the machine to move sideways on a diagonal. Also known as crab steering.

Optimum moisture: The percent of moisture at which the greatest density of a particular soil can be obtained through compaction.

Organic: Derived from living organisms, such as plants and animals.

Parallel: Two lines that are always the same distance apart even if they go on into infinity (forever is called infinity in mathematics).

Parallelogram: A two-dimensional shape that has two sets of parallel lines.

Pay material: Deposit of soil valuable enough to stockpile for future use, on or off site.

Peat: Dark swamp soil or decaying organic matter. Also called humus.

Plan view: A drawing that represents a view looking down on an object.

Plastic limit: The amount of water that causes a soil to become plastic (easily shaped without crumbling).

Plasticity: The range of water content in which a soil remains plastic or is easily shaped without crumbling.

Power hop: Action in heavy equipment that uses pneumatic tires to create a bouncing motion between the fore and aft axles. Once started, the oscillation back and forth usually continues until the operator either stops or slows down significantly to change the dynamics.

Powered industrial trucks: An OSHA term for several types of light equipment that include forklifts.

Property lines: The recorded legal boundaries of a piece of property.

Quadrilateral: A four-sided, closed shape with four angles whose sum is 360 degrees.

Rated operating capacity: The amount of weight that a skid steer is projected to handle safely through common maneuvers. The rated operating capacity is typically derived from the tipping load test, and is equal to 50 percent of the tipping load value.

Request for information (RFI): A form used to question discrepancies on the drawings or to ask for clarification.

Rimpull: The torque applied to the axle from the engine that is then converted into pulling force at the face of the tire; more commonly known as gross traction in mathematical calculations.

Ripping: Loosening hard soil, concrete, asphalt, or soft rock with a ripping attachment.

Roading: Driving a piece of construction equipment, on a public road, from one job site to another.

Rubble: Fragments of stone, brick, or rock that have broken apart from larger pieces.

Sedimentation: Soil particles that are removed from their original location by water, wind, or mechanical means.

Setback: The distance from a property line in which no structures are permitted.

Settlement: To become firm by compacting the particles so they will be closer together.

Shielding: A structure that is able to withstand the forces imposed on it by a cave-in and thereby protect employees within the structure.

Shrinkage: Decrease in volume when soil is compacted.

Spot: To line up the haul unit so that it is in the proper position.

Squared: Multiplied by itself.

Stations: Designated points along a line or a network of points used to survey and lay out construction work. The distance between two stations is normally 100 feet or 100 meters, depending on the measurement system used.

Stockpile: Material put into a pile and saved for future use.

Stormwater: Water from rain or melting snow.

String line: A tough cord or small diameter wire stretched between posts or pins to designate the line and elevation of a grade. String lines take the place of hubs and stakes for some operations.

Stripping: Removal of overburden or thin layers of pay material.

Subsidence: Pressure created by the weight of the soil pushing on the walls of the excavation. It stresses the excavation walls and can cause them to bulge.

Sump: A small excavation dug below grade for the purpose of draining or retaining subsurface water. The water is then usually pumped out of the sump by mechanical means.

Swale: A shallow trench used to direct the flow of water.

Swell: Increase in volume when soil is excavated.

Swell factor: The ratio of the banked weight of a soil to the loose weight of a soil.

Tag axle: Auxiliary axle that is mounted behind the truck's drive wheels. It may be called a pusher axle if it is placed in front of the drive axle.

Tandem-axle: Usually a double-axle drive unit.

Telehandler: A type of powered industrial truck characterized by a boom with several extendable sections known as a telescoping boom. Another name for a shooting boom forklift.

Tines: A prong of an implement such as a fork. For forklifts, tines are often called forks.

Tipping load: The loaded weight that will lift the rear wheels off the ground with the machine in a static (not moving) condition; the maximum load that a skid steer can hold in the air, with the coupler out-stretched, before the rear wheels begin to leave the ground.

Topographic survey: The process of surveying a geographic area to collect data indicating the shape of the terrain and the location of natural and man-made objects.

Uniform Construction Index: The construction specification format adopted by the Construction Specification Institute (CSI). Known as the CSI format.

Uprights: The vertical members of a trench shoring system placed in contact with the earth and usually positioned so that individual members do not contact each other. Uprights placed so that individual members are closely spaced, in contact with, or interconnected to each other, are often called sheeting.

Variable: A value in an equation that depends on the factors being considered; for example, the lengths of the sides of a triangle may vary from one triangle to another.

Variable displacement pump: A pump that can deliver a variable fluid volume in a pumping cycle to support the changing needs of a hydraulically operated device.

Voids: Open space between soil or aggregate particles. A reference to voids usually means that there are air pockets or open spaces between particles.

Walers: Horizontal members of a shoring system or coffer dam placed parallel to the excavation face whose sides bear against the vertical members of the shoring system or the earth. Also, supports for piles in a coffer dam.

Water table: The depth below the ground's surface at which the soil is saturated with water.

Well-graded: Soil that contains enough small particles to fill the voids between larger ones. Accessories: Attachments used to expand the use of a loader.

Index

C

D

E

F

M

N

O

S

T